SIMULATION
MODELING
HANDBOOK
A Practical Approach

INDUSTRIAL AND MANUFACTURING ENGINEERING SERIES
SERIES EDITOR Hamid R. Parsaei

SIMULATION MODELING HANDBOOK

A Practical Approach

Christopher A. Chung

CRC Press
Taylor & Francis Group
Boca Raton London New York

CRC Press is an imprint of the
Taylor & Francis Group, an **informa** business

CRC Press
Taylor & Francis Group
6000 Broken Sound Parkway NW, Suite 300
Boca Raton, FL 33487-2742

First issued in paperback 2019

© 2004 by Taylor & Francis Group, LLC
CRC Press is an imprint of Taylor & Francis Group, an Informa business

No claim to original U.S. Government works

ISBN-13: 978-0-8493-1241-0 (hbk)
ISBN-13: 978-0-367-39511-7 (pbk)
Library of Congress Card Number 2003046280

Library of Congress Cataloging-in-Publication Data
Chung, Chris.
Simulation modeling handbook : a practical approach / Christopher A. Chung.
p. cm.
Includes bibliographical references and index.
ISBN 0-8493-1241-8 (alk. paper)
1. Digital computer simulation. I. Title.
QA76.9.C65C49 2003
003'.3--dc21   2003046280

Visit the Taylor & Francis Web site at
http://www.taylorandfrancis.com

and the CRC Press Web site at
http://www.crcpress.com

Preface

Why a Practitioner's Handbook?

Simulation modeling and analysis is becoming increasingly popular as a technique for improving or investigating process performance. It is a cost-effective method for evaluating the performance of resource allocation and alternative operating policies. In addition, it may also be used to evaluate the performance of capital equipment before investment. These benefits have resulted in simulation modeling and analysis projects in virtually every service and manufacturing sector.

As the popularity of simulation modeling and analysis grows, the planning and execution of a simulation project will no longer be restricted to specially trained and educated simulation analysts. In the future, a much larger segment of engineering and management professionals will be called on to perform these tasks.

Although any professional faced with conducting a simulation project is encouraged to receive formal simulation training, other demands are likely to exist. Practitioners with limited or dated training may need resources other than theory-based academically oriented texts or software-specific manuals. This handbook is intended to provide the practitioner with an easy-to-follow reference for each step in conducting a simulation modeling and analysis project.

How This Handbook Differs from Other Simulation Texts

This handbook differs from other simulation texts in several major ways. First, the handbook was written to insulate practitioners from unnecessary simulation theory. Many currently available simulation publications are long on theory and short on application. For example, one competing simulation handbook is entirely based on conference proceedings. Another well-respected simulation text used in graduate-level simulation courses is primarily based on refereed journal publications. These types of simulation books have value to some very experienced analysts, but they are not focused on the needs of the average practitioner.

Many other simulation texts are actually expanded software manuals. These types of texts present most if not all of a specific simulation software's functionality. A problem with this approach is that a text with useful simulation concepts may be written for a software package other than the one that the practitioner has access to. Trying to detach different simulation concepts from a specific software implementation can be a daunting task. This is a particularly sensitive issue because some full software packages can represent a significant investment for a small organization.

This handbook has been specifically designed so that it may be utilized independently by practitioners, regardless of the simulation software package that may be used for modeling. One motivation behind this feature is the relentless upgrading of the major simulation software packages. It sometimes seems that simulation packages come out with new features and capabilities nearly every year. Insulating the handbook from these continuous upgrades assures the future utility of the handbook. This is not to say that the handbook does not include any simulation software-related content. There are tutorials on each of the major software packages in the handbook appendix.

Another motivation for insulating the handbook from specific software packages is the availability of on-line help. Most simulation software packages possess extensive modeling examples and function references. Explanations of the various software functions usually constitute a large component of most conventional simulation texts. By directing practitioners to utilize on-line help capabilities of their preferred software, this handbook has reduced volume, and users have the most current documentation.

Another distinguishing feature of this handbook is the inclusion of sample simulation project support material. This includes checklists, data collection forms, and sample simulation project reports and publications. This material will greatly facilitate the practitioner's efforts to conduct a simulation modeling and analysis project.

Last, this handbook also includes a set of course notes that can be used for delivering a short course on simulation to aspiring practitioners. The course notes are formatted so that they can also be photocopied onto transparencies. This material follows the basic organization of the handbook and allows the practitioner to use the handbook as a textbook in conjunction with the course notes and transparencies.

Handbook Organization

The handbook is organized into an introduction, a set of practical sections, and the appendix. The introduction includes basic information about simulation modeling. The practical sections directly correspond to the process steps for conducting a simulation modeling and analysis project. The appendix contains software manuals and statistical tables. Practitioners are encouraged to bypass unnecessary sections and focus on those of immediate relevance. Most of the simulation process step chapters are independent of the others. The practitioner may even find that some of the chapters are applicable to efforts outside of simulation projects. In all of these chapters, a simple no-nonsense approach is used as the guiding philosophy for the manner in which the material is presented. The general approach is to discuss the concepts at a high level to provide a basic understanding. Only after a basic understanding is achieved are the necessary theory and mathematics presented.

The "Introduction" (Chapter 1) begins with information on the application, advantages, and disadvantages of simulation modeling. The basic components of a simulation model are presented. These include:

- Entities
- Resources
- Queues
- Statistical measures of performance

In the practical sections, the steps included in the handbook are:

- Problem formulation
- Project planning
- System definition
- Input data collection
- Model translation
- Verification, validation
- Experimental design
- Analysis
- Presenting conclusions and results

Chapter 2, "Problem Formulation," includes material on project orientation issues and establishing the project objectives. Project objective selection techniques are presented to insure that the most important problems are addressed in the study.

Chapter 3, "Project Planning," includes project management techniques that will assist the user in planning a successful simulation project. This includes organizing the simulation project tasks with a work breakdown structure, assigning responsibility for the tasks with linear responsibility charts, and sequencing the tasks with Gantt charts.

Chapter 4, "System Definition," includes identification of the system components to be modeled in the simulation. These include identifying the important system processes, input data requirements, and output measures of performance.

Chapter 5, "Input Data Collection and Analysis," discusses collection of original data, use of existing data, and input data analysis techniques. Input data analysis techniques include the use of the chi-square goodness-of-fit test and currently available data-fitting software.

Chapter 6, "Model Translation," presents information on how to make simulation software selection decisions. Users will be able to understand the advantages and disadvantages of using general purpose programming languages versus simulation-specific software. This section also includes a brief summary of the capabilities of a few of the more established simulation-specific software packages that are available to practitioners. The section closes with guidance on programming the actual simulation model.

Chapter 7, "Verification," discusses a variety of techniques available for the user to help insure that the simulation model operates as intended. These include the use of entity animation and variable displays for debugging purposes.

Chapter 8, "Validation," presents a variety of techniques to determine whether or not the model represents reality. This section includes both qualitative and quantitative techniques available to the user. The primary qualitative technique discussed is face validation. The quantitative techniques include F tests, t-tests, and nonparametric tests.

Chapter 9, "Experimental Design," covers different techniques for determining which model alternatives will be beneficial to investigate. The section includes both simple one-to-one comparisons and multiple comparisons.

Chapter 10, "Analysis," includes techniques for making statistically robust comparisons between alternatives. This includes determining the number of simulation model replication runs that are necessary to conduct valid comparisons. It also includes confidence interval, analysis of variance, and Duncan multiple-range test statistical analysis techniques for comparing the alternatives identified in Chapter 9. This chapter section also includes information on performing economic comparisons of alternatives.

Chapter 11, "Project Reports and Presentations" includes information on conducting appropriate presentations and how to report the results of the simulation study. This includes what content to include and how to prepare the presentation or report.

Following the simulation process step sections, additional material is presented that represents new developments in the field of interactive multimedia computerized training simulators (Chapter 12). This includes both management process-based and equipment operation-based training simulators.

Chapter 13 includes a number of technical reports to assist the practitioner in reporting the results of a simulation study. Included in this section is an actual master's thesis that was conducted in the area of simulation modeling and analysis. This model can be used by practitioners aspiring to work on a graduate degree in this area.

The remaining chapters of the book (Chapters 14 through 16) consist of a variety of minimanuals for popular simulation software packages. These can be used to develop models to familiarize the practitioner with the capabilities of each of these different software packages.

The Appendix includes course notes and statistical tables. The course notes can be used to prepare lectures on the chapters in this handbook. The statistical tables include the t, chi-square, normal, and Duncan Multiple-Range tables.

Editor

Dr. Christopher Chung is currently an associate professor in the Department of Industrial Engineering at the University of Houston, Houston, TX. At the University of Houston, Dr. Chung instructs both undergraduate and graduate courses in computer simulation and management and training simulator software engineering. In addition to his courses in simulation and simulators, Dr. Chung has also performed a variety of projects and research for both major corporations and the U.S. Government. Dr. Chung's publications can be found in *Simulation*, the *International Journal of Simulation and Modeling*, the *ASCE Journal of Transportation Engineering*, and the *Security Journal*. Prior to becoming a university professor, Dr. Chung was a manufacturing quality engineer for the Michelin Tire Corporation, and prior to that, a U.S. Army bomb disposal officer. Dr. Chung has M.S. and Ph.D. degrees from the University of Pittsburgh and a B.E.S. from Johns Hopkins University.

Editor

Christopher A. Chung
University of Houston
Department of Industrial Engineering
Houston, Texas

Contributors

Charles E. Donaghey
University of Houston
Department of Industrial Engineering
Houston, Texas

Somasundaram Gopalakrishnan
University of Houston
Department of Industrial Engineering
Houston, Texas

Abu M. Huda
Continental Airlines
Houston, Texas

Erick C. Jones
University of Houston
Department of Industrial Engineering
Houston, Texas

Matt Rohrer
Brooks-PRI Automation
Salt Lake City, Utah

Randal W. Sitton
University of Houston
Department of Industrial Engineering
Houston, Texas

Contents

15 Simulation Using AutoMod and AutoStat

Matt Rohrer

16 Simpak User's Manual

Charles E. Donaghey and Randal W. Sitton

1

Introduction

"Where do we begin?"

1.1 Introduction

The objective of this chapter is to provide the simulation practitioner with some basic information about simulation modeling and analysis. Experienced practitioners using the handbook for reference purposes are advised to bypass this chapter and proceed to the appropriate chapter. Practitioners who have never received training in simulation or whose training is dated are strongly recommended to work through not only the examples but also the sample problems at the end of the chapter.

The chapter includes:

- An introduction to simulation modeling and analysis
- Other types of simulation

- Purposes of simulation
- Advantages and disadvantages of simulation
- Famous simulation quotes
- Basic simulation concepts
- A comprehensive example of a manual simulation

1.2   Simulation Modeling and Analysis

Simulation modeling and analysis is the process of creating and experimenting with a computerized mathematical model of a physical system. For the purposes of this handbook, a system is defined as a collection of interacting components that receives input and provides output for some purpose. Included within this field are traditional simulation and training simulators. In general, the distinction is as follows. Traditional simulation is used for analyzing systems and making operating or resource policy decisions. Training simulators are used for training users to make better decisions or improving individual process performance. A short section is included at the end of the handbook that discusses the basic nature of simulators.

The vast majority of this handbook concentrates on the field of simulation versus simulators. Although many different types of systems can be simulated, the majority of the systems that we discuss in this handbook are manufacturing, service, or transportation related.

Examples of manufacturing systems include:

- Machining operations
- Assembly operations
- Materials-handling equipment
- Warehousing

Machining operation simulations can include processes involving either manually or computer numerically controlled factory equipment for machining, turning, bending, cutting, welding, and fabricating. Assembly operations can cover any type of assembly line or manufacturing operation that requires the assembly of multiple components into a single piece of work. Material-handling simulations have included analysis of cranes, forklifts, and automatically guided vehicles. Warehousing simulations have involved the manual or automated storage and retrieval of raw materials or finished goods.

Examples of service systems include:

- Hospitals and medical clinics
- Retail stores
- Food or entertainment facilities
- Information technology
- Customer order systems

Hospital and medical clinic models can be simulated to determine the number of rooms, nurses, and physicians for a particular location. Retail stores may need to know how many checkout locations to utilize. Entertainment facilities such as multitheater movie complexes may be interested in how many ticket sellers, ticket checkers, or concession stand clerks to employ. Information technology models typically involve how many and what type of network or support resources to have available. Customer order systems may need to know how many customer order representatives are needed to be on duty.

Examples of transportation systems include:

- Airport operations
- Port shipping operations
- Train and bus transportation
- Distribution and logistics

Airport operations simulations have been performed on airport security checkpoints, check-in counters, and gate assignments. Port shipping operations can include how many cranes and trucks are needed to offload transportation ships. Train and bus transportation can include analysis involving routes. Distribution and logistics studies have included the analysis of shipping center design and location.

1.3 Other Types of Simulation Models

The types of simulation models previously discussed are not the only types of simulation model that the practitioner may encounter or have a need for. Another type of computer simulation model is the computer simulator. Though the distinction between simulation models and computer simulators may differ somewhat among practitioners, the following discussion may help differentiate these two types of simulation.

So far, the types of simulation models that we have discussed have been models of actual or proposed systems. Models of the systems are normally created with different resource or operating policies that have been previously determined to be of interest. After the simulation runs, the output measures of performance are compared between or among the models. Thus, the ultimate use of the models is to make resource or operating policy decisions concerning the system.

Simulators are also models of existing or proposed systems. In contrast to simulation models, resource and operating policy decisions are not made beforehand. These types of decisions are actually made during the simulation run. Thus, the output measures are observed not only at the end of the run but, more importantly, during the simulation run. The practitioner or user can see the effects of executing different resource and operating policy decisions in real time. Thus, the purpose of the simulator is not to make a decision but to expose the users to the system and to train them on how to make decisions. These types of simulators are often referred to as training simulators.

Although the principal focus of this book is on developing and analyzing simulation models, a short section has been included on simulators in Chapter 12, "Training Simulators."

1.4 Purposes of Simulation

The simulation modeling and analysis of different types of systems are conducted for the purposes of (Pedgen et al., 1995):

- Gaining insight into the operation of a system
- Developing operating or resource policies to improve system performance
- Testing new concepts and/or systems before implementation
- Gaining information without disturbing the actual system

1.4.1 Gaining Insight into the Operation of a System

Some systems are so complex that it is difficult to understand the operation of and interactions within the system without a dynamic model. In other words, it may be impossible to study the system by stopping it or by examining individual components in isolation. A typical example of this would be to try to understand how manufacturing process bottlenecks occur.

1.4.2 Developing Operating and Resource Policies

You may also have an existing system that you understand but wish to improve. Two fundamental ways of doing this are to change operating or resource policies. Changes in operating policies could include different scheduling priorities for work orders. Changes in resource policies could include staffing levels or break scheduling.

1.4.3 Testing New Concepts

If a system does not yet exist, or you are considering purchasing new systems, a simulation model can help give you an idea how well the proposed system will perform. The cost of modeling a new system can be very small in comparison to the capital investment involved in installing any significant manufacturing process. The effects of different levels and expenses of equipment can be evaluated. In addition, the use of a simulation model before implementation can help refine the configuration of the chosen equipment.

Currently, a number of companies require vendors of material-handling equipment to develop a simulation of their proposed systems before purchase. The simulation model is used to evaluate the various vendors' claims. Even after the installation, the simulation model can be useful. The company can use the simulation model to help identify problems should the installed system not operate as promised.

1.4.4 Gaining Information without Disturbing the Actual System

Simulation models are possibly the only method available for experimentation with systems that cannot be disturbed. Some systems are so critical or sensitive that it is not possible to make any types of operating or resource policy changes to analyze the system. The classical example of this type of system would be the security checkpoint at a commercial airport. Conducting operating policy or resource level experimentation would have serious impact on the operational capability or security effectiveness of the system.

1.5 Advantages to Simulation

In addition to the capabilities previously described, simulation modeling has specific benefits. These include:

- Experimentation in compressed time
- Reduced analytic requirements
- Easily demonstrated models

1.5.1 Experimentation in Compressed Time

Because the model is simulated on a computer, experimental simulation runs may be made in compressed time. This is a major advantage because some processes may take months or even years to complete. Lengthy system processing times may make robust analysis difficult or even impossible to perform. With a computer model, the operation and interaction of lengthy processes can be simulated in seconds. This also means that multiple replications of each simulation can easily be run to increase the statistical reliability of the analysis. Thus, systems that were previously impossible to analyze robustly can now be studied.

1.5.2 Reduced Analytic Requirements

Before the existence of computer simulation, practitioners were forced to use other, more analytically demanding tools. Even then, only simple systems that involved probabilistic elements could be analyzed by the average practitioner. More complex systems were strictly the domain of the mathematician or operations research analyst. In addition, systems could be analyzed only with a static approach at a given point in time. In contrast, the advent of simulation methodologies has allowed practitioners to study systems dynamically in real time during simulation runs. Furthermore, the development of simulation-specific software packages has helped insulate practitioners from many of the complicated background calculations and programming requirements that might otherwise be needed. These reduced analytic requirements have provided more practitioners, with a wider variety of backgrounds, with the opportunity to analyze many more different types of systems than was previously possible.

1.5.3 Easily Demonstrated Models

Most simulation-specific software packages possess the capability of dynamically animating the model operation. Animation is useful both for debugging the model and also for demonstrating how the model works. Animation-based debugging allows the practitioner to observe flaws in the model logic easily. The use of an animation during a presentation can help establish model credibility. Animation can also be used to describe the operation and interaction of the system processes simultaneously. This includes dynamically demonstrating how the system model handles different situations. Without the capability of animation, practitioners would be limited to less effective textually and numerically based presentations.

1.6 Disadvantages to Simulation

Although simulation has many advantages, there are also some disadvantages of which the simulation practitioner should be aware. These disadvantages are not really directly associated with the modeling and analysis of a system but rather with the expectations associated with simulation projects. These disadvantages include the following:

- Simulation cannot give accurate results when the input data are inaccurate.
- Simulation cannot provide easy answers to complex problems.
- Simulation cannot solve problems by itself.

1.6.1 Simulation Cannot Give Accurate Results When the Input Data Are Inaccurate

The first statement can be paraphrased as "garbage in, garbage out." No matter how good a model is developed, if the model does not have accurate input data, the practitioner cannot reasonably expect to obtain accurate output data. Unfortunately, data collection is considered the most difficult part of the simulation process. Despite this common knowledge, it is typical for too little time to be allocated for this process. This problem is aggravated by the fact that many practitioners probably prefer to develop a simulation model rather than collect mundane data.

Many simulation practitioners are lured into accepting historical data of dubious quality in order to save input data collection time. All too often the exact nature or the conditions under which these data were collected is unknown. In more than one case, the use of externally collected historical data has been the foundation of an unsuccessful simulation project.

1.6.2 Simulation Cannot Provide Easy Answers to Complex Problems

Some analysts may believe that a simulation analysis will provide simple answers to complex problems. In fact, it is more likely that complex answers are required for complex problems. If the system analyzed has many components and interactions, the best alternative operating or resource policy is likely to consider each element of the system. It is possible to make simplifying assumptions for the purpose of developing a reasonable model in a reasonable amount of time. However, if critical elements of the system are ignored, then any operating or resource policy is likely to be less effective.

1.6.3 Simulation Alone Cannot Solve Problems

Some managers, on the other hand, may believe that conducting a simulation model and analysis project will solve the problem. Simulation by itself does not actually solve the problem. It provides the management with potential solutions to solve the problem. It is up to the responsible management individuals to actually implement the proposed changes. For this reason, it is to the advantage of the practitioner to keep the manager or customer stakeholders as involved in the project as much as possible. The practitioner should strive to have the stakeholders develop a sense of ownership in the simulation process. All too often, potential solutions are developed but are never or only poorly implemented because of organizational inertia or political considerations.

1.7   Other Considerations

In addition to the advantages and disadvantages to simulation modeling and analysis previously discussed, the practitioner should be aware of some other serious considerations when embarking on a project. These considerations may influence the practitioner's decision whether or not to undertake the project alone or even to undertake the project at all. These include the following:

- Simulation model building can require specialized training.
- Simulation modeling and analysis can be costly.
- Simulation results involve many statistics.

1.7.1   Simulation Model Building Can Require Specialized Training

In the past, simulation modeling used to be extremely difficult to perform. In the days before graphic displays, all modeling was performed with arcane simulation languages with a text editor. It was necessary to create the source code, compile, link, and run the program. If any comma, colon, or period was out of place, the practitioner received a slew of compiling errors. Thus, without strong computer programming skills it was difficult to build successfully anything other than the simplest model. Fortunately for us, the advent of the powerful multimedia personal computer has brought simulation modeling more into the realm of the practitioner. The arcane simulation programming languages have given way to reasonably easy-to-use graphic interfaces. However, the overall simulation modeling and analysis process can still be complex. Many accomplished simulation analysts do have engineering, computer science, mathematics, or operations research degrees with specific coursework in simulation modeling and analysis.

1.7.2   Simulation Modeling and Analysis Can Be Very Costly

There is no question that the development of a complex simulation model can be very time consuming and hence costly. Even if the practitioner is proficient with a given simulation software package, a complex system still will require a proportionally larger amount of time for data collection, modeling building, and analysis. Some models have a way of initially appearing to be relatively simple. However, once the practitioner actually begins modeling, he or she may realize that the system is far more complex than it originally appeared. Although many simplifying modeling assumptions may be made to reduce the amount of development time, the assumptions may also be so extreme as to render the model invalid. So, just as with the collection of input data, there is a limit to how much time may be saved by cutting corners with the model development process.

1.7.3   Simulation Results Involve Many Statistics

Finally, simulation results are usually in the form of summary statistics. For this reason, simulation results may be difficult for individuals without any statistical knowledge to interpret. It is assumed that any practitioner utilizing this handbook have at least a limited knowledge of statistics. Some of the simulation-specific types of statistics presented in this handbook may not be familiar at all, even to statistically proficient practitioners. Although a few of the statistical techniques presented in this handbook may be new to many practitioners, a step-by-step format has been incorporated to minimize potential difficulties.

1.8 Famous Simulation Quotes

Now that we have some familiarity with simulation, it is time to introduce a few famous simulation quotations that have been recorded over the years:

- "You cannot study a system by stopping it."
- "Run it again."
- "Is that some kind of game you're playing?"

1.8.1 "You Cannot Study a System by Stopping It"

The first quotation is believed to have originated from the science fiction series *Dune* by Frank Herbert. It was in reference to a particular problem that the futuristic scientists were attempting to analyze. If the system was stopped long enough to analyze it, it would alter the nature of the system. Thus, the only way to study it properly was while it was in motion. Computer simulation allows the model of the system to be in motion. Any output measures of performance are acquired while the system is in operation. This quotation also focuses on computer simulation's ability to study an actual system while it is still in operation.

1.8.2 "Run It Again"

The second quotation is from an episode of *StarTrek: The Next Generation*. In the episode, the starship Enterprise was damaged and trapped inside a debris field. The chief engineering officer Geordi LaForge developed a plan to extricate the Enterprise from the debris field. The difficulty with the plan was that if it failed, it would deplete all of the Enterprise's energy resources and the fate of the Enterprise would be in question. The Enterprise's Captain Jean Luc Picard instructed Commander LaForge to run a simulation of his proposed plan. The simulation run indicated that Commander LaForge's plan would successfully remove the Enterprise from the debris field. However, Captain Picard, having some familiarity with simulation, ordered Commander LaForge to "run it [the simulation] again." The result of the second simulation run was that the Enterprise would not be able to remove itself successfully from the debris field. These conflicting simulation results convinced Captain Picard to seek an alternate plan for the Enterprise.

This example emphasizes that probabilistic systems cannot be analyzed with a single simulation run. The innate variability of probabilistic systems results in corresponding variability in any system output. Thus, in most of the systems that simulation is used to analyze, a single simple simulation run is not sufficient for making serious resource or operating policy decisions.

1.8.3 "Is That Some Kind of Game You Are Playing?"

The last quotation was directed toward one of us during work on a project that involved modeling and analyzing a multitheater movie complex. While in the process of animating this system another individual happened to pass by. The animation illustrated customers purchasing tickets, having their tickets collected, going to the concession area, and finally entering the theater. On observing the animation, the individual asked if the computer model was actually some kind of game. After recovering from his initial shock, the practitioner replied that it was actually a highly complex mathematical model based on probability distributions and statistics. Apparently unimpressed, the visitor shrugged his shoulders and left the practitioner to complete the model.

This example illustrates the danger that uneducated or uninformed individuals will fail to appreciate the complexity behind a sophisticated simulation model. Many of these types of individuals have the capacity to comment only on the animation aspect of the simulation model. Unfortunately, many of these individuals are the same ones whom you must convince that your model is useful for making significant resource and operating policy decisions. Not only must your model be mathematically and logically correct, it must also look convincing to the uninformed.

1.9 Basic Simulation Concepts

Virtually all simulation modeling that the practitioner is likely to be involved in will be implemented in a simulation specific-software package such as ARENA, AutoMod, Simscript, or SimPak. Although these simulation packages facilitate and speed the development time for a simulation model, they may insulate the practitioner from an understanding of the basic concepts of simulation modeling. The following section can introduce or refamiliarize the practitioner with these concepts before he or she proceeds with the remainder of the handbook. In this section we discuss:

- Basic simulation model components
- Simulation event lists
- Measures of performance statistics

1.9.1 Basic Simulation Model Components

For demonstration purposes, consider the simplest possible system that may be of interest to the practitioner. Examples of this simple type of system would include, but not be limited to:

- A customer service center with one representative
- A barber shop with one barber
- A mortgage loan officer in a bank
- A piece of computer-controlled machine in a factory
- An ATM machine

Each of these simple systems consists of three types of major components:

- Entities
- Queues
- Resources

The relationships among these components are illustrated in Figure 1.1.

1.9.1.1 Entities

The first type of component is an entity: something that changes the state of the system. In many cases, particularly those involving service systems, the entity may be a person. In the customer service center, the entities are the customers. Entities do not necessarily have to be people; they can also be objects. The entities that the mortgage loan officer deals with are loan applications. Similarly, in the factory example, the entities are components waiting to be machined.

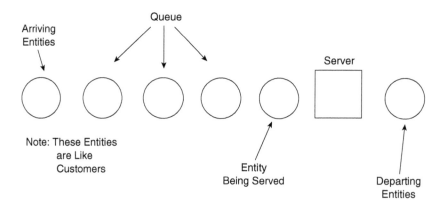

FIGURE 1.1 Basic simulation model components.

1.9.1.1.1   Entity Batches

The number of entities that arrive in the system at the same given time is known as the batch size. In some systems, the batch size is always one. In others, the entities may arrive in groups of different sizes. Examples of batch arrivals are families going to a movie theater. The batch sizes may be two, three, four, or more.

1.9.1.1.2   Entity Interarrival Times

The amount of time between batch arrivals is known as the interarrival time. It does not matter whether the normal batch size is one or more. We are interested only in the interval from when the last batch arrived to when the current batch arrives. The previous batch may have had only one entity, whereas the next batch has more than one. Interarrival time is also the reciprocal of the arrival rate. In collecting entity arrival data it is usually easier to collect the batch interarrival time. However, some historical data may be in arrival rate format. Interarrival time is considered input data that the practitioner would have to provide for the model.

1.9.1.1.3   Entity Attributes

Entities may also possess attributes. These are variables that have values unique to each entity in the system. Even though the entity attribute will have the same name, there could be as many different values as there are entities. An example of an attribute of this type involves the entity's arrival time. Each entity's attribute ARRTIME would store the simulation system time that the entity arrived in the system. So, unless a batch of entities arrived at the same time, each entity would have a unique value in its attribute ARRTIME. Some entity attributes may have the same value. In the case of airline passengers, the attribute PASSTYPE could hold a value corresponding to the type of passenger the entity represents. A value of 1 in PASSTYPE could represent a first-class passenger, and a value of 2 could represent a coach-class passenger. Thus, 20% of the entities in a simulation might have a value of 1 for PASSTYPE, and the remaining 80% would have a value of 2 for PASSTYPE. In the actual model, the attribute PASSTYPE would be used to prioritize the servicing and loading of passengers.

Simulation programs may also utilize global variables. Global variables are not to be confused with entity attributes. These variables differ from entity attributes in that each global variable can maintain only one value at a given time. A typical use of a global variable in a simulation program is the variable that keeps track of the simulation run time.

1.9.1.2   Queues

The second major type of components that simple systems possess is queues. Queues are the simulation term for lines. Entities generally wait in a queue until it is their turn to be processed. Simple systems generally use first-in-first-out (FIFO) queue priorities. Another characteristic of simple systems is that once customers enter the system, they must enter the queue. Furthermore, once entities enter the queue, they cannot depart before receiving service. We explore different variations of queue priorities and queue behavior in Chapter 4, "System Definition."

1.9.1.3   Resources

The third component that simple systems contain is resources. Resources process or serve the entities that are in the queue. Examples of resources are:

- Customer service representatives
- Barbers
- Loan officers
- Factory machines
- ATMs

In simple models, resources can be either idle or busy. Resources are idle when they are available for processing, but there are no more entities waiting in the queue. Resources are busy when they are processing entities. In more complex models, resources may also be temporarily inactive or failed. Inactive resources are unavailable because of:

- Scheduled work breaks
- Meals
- Vacations
- Preventive maintenance periods

Failed resources would correspond to:

- Broken machines
- Inoperative equipment

Resources take a certain amount of processing time to service the entities, for example, the time to total an order and receive payment, process a loan, or machine a part. The processing time is also frequently referred to as processing delay time or service time. Processing time is considered input data that the practitioner would normally have to collect by observation or otherwise acquire.

1.9.2   The Simulation Event List

The simulation event list is a means of keeping track of the different things that occur during a simulation run (Law and Kelton, 2000). Anything that occurs during the simulation run that can affect the state of the system is defined as an event. Typical events in a simple simulation include entity arrivals to the queue, the beginning of service times for entities, and the ending of service times for entities. These events change the state of the system because they can increase or decrease the number of entities in the system or queue or change the state of the resources between idle and busy.

The event list is controlled by advances in the simulation clock. In our basic simulation model, the simulation clock advances in discrete jumps to each event on the event list. This type of model is called a discrete event simulation. In more sophisticated models, the simulation clock may operate continuously. This type of model is usually associated with processes involving fluid or material that could be modeled as fluids. These types of models involve continuous event simulation. Systems that require continuous event simulation are usually significantly more difficult to model because they involve the use of differential equations. It is also possible to model a system that involves both discrete and continuous components. An example of this would be a refinery that fills tanker trucks. The refinery tanks that store liquid would require continuous simulation, while the individual tanker trucks would need to be modeled discretely.

Regardless of whether the model is discrete, continuous, or combined, the simulation event list is extremely important to the practitioner. In even our very simple simulation model, many different events can occur simultaneously. For example, entities may arrive at any give time, or a service period may end at any given time. This means that one moment an entity may arrive, and a second entity may arrive before the first entity receives service. Similarly, the first entity may arrive, receive processing, and depart before the second entity arrives. The arrival, service start, and service end processes can take on an infinite number of possible sequences. Without a formal means of keeping track of these events, the output measures of performance of the system would become hopelessly complicated. This is, in fact, the reason it is virtually essential to implement any simulation on a computer system.

1.9.3   Measures of Performance Statistics

We are almost always interested in how well the actual system and hence the system model performs. In order to ascertain this we will need to calculate some sort of output measure to eventually compare with other alternative forms of the model. Output measures of performance can be either observational or time dependent. Observational performance measures are based on the number of entities observed going through the process. Conversely, time-dependent measures are based on the length of time the statistics are collected. There are four commonly utilized measures of performance (Kelton et al., 2002):

1. System time
2. Queue time
3. Time average number in queue
4. Utilization

1.9.3.1 System Time

System time is an observational output measure. It is the total amount of time that the entity spends in the system. System time begins when the entity arrives in the system and enters the queue. It ends when the entity's service time is complete and it exits the system. The average system time for all of the entities is of most importance to the practitioner. The mathematical representation of the average system time is

$$\text{Average System Time} = \frac{\sum_{i=1}^{n} T_i}{n}$$

where T_i = the system time for an individual entity (arrival time – departure time) and n = the number of entities that are processed through the system.

1.9.3.2 Queue Time

Queue time is also an observational measure. It is similar to system time, except it accounts only for the time that an entity spends in the queue. Queue time is preferred by some practitioners because they suspect that the most objectionable time period, at least in customer-oriented service processes, is the waiting time in the queue. Many customers are at least partially satisfied when their service times begin, even though the service time itself may be lengthy. The formula for queue time is

$$\text{Average Queue Time} = \frac{\sum_{i=1}^{n} D_i}{n}$$

where D_i = the queue time for an individual entity (queue arrival time – service begin time) and N = *the* number of entities that are processed through the queue.

1.9.3.3 Time-Average Number in Queue

The time-average number in queue is a time-dependent statistic. As a time-dependent statistic, the time-average number in queue is not directly a function of the number of entities that have been processed through the queue. It is rather the average number of entities that you could expect to see in the queue at any given time during the period of interest. At any given time the queue will actually have a discrete number of entities. However, because the time-average number in queue is an average value, it will usually yield a number that also has a fractional value. For lightly loaded queues it is actually possible for the time average number in queue to be less than 1. The formula for calculating the time-average number in queue is

$$\text{Time Average Number in } Q = \frac{\int_{0}^{T} Q\,dt}{T}$$

where

Q = the number in the queue for a given length of time.

dt = the length of time that Q is observed.

T = the total length of time for the simulation.

Because the equation for the time-average number in queue is time dependent, further explanation is warranted. The equation essentially calculates the total entity-time in the queue that is observed during the simulation run divided by the total simulation run time. In an entity customer simulation, this would correspond to the customer waiting time by all of the customers that were waiting in line. Each period of time is calculated by multiplying the number of customers waiting in line for the amount of time that number of customers waited in line. A change in the number of customers waiting in line triggers the beginning of a new period of calculation. At the end, all of the periods with customer-minutes are totaled and divided by the length of the simulation.

Manual calculations of the time-average number in queue are best handled by drawing a two-axis graph of the system. The vertical axis records the number of entities in the queue. The horizontal axis records the simulation time. A line is drawn at the number of entities in the queue for the length of time that number of entities is in the queue. The entity time is calculated by calculating the area of each box, which is the number of entities waiting multiplied by the ending time for that number of entities in the queue minus the starting time for that number of entities in the queue. The total area is calculated by summing all of the individual areas. The time average number in queue is then calculated by dividing the total area by the length of the simulation run. This method is illustrated in Figure 1.2.

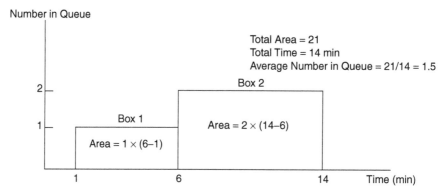

FIGURE 1.2   Resource utilization calculations.

1.9.3.4   Resource Utilization

Resource utilization is also a time-dependent statistic. At any given time a single resource can be either idle or busy. The idle state corresponds to a resource utilization level of 0. Naturally, the busy state corresponds to a resource utilization level of 1. The length of time that the resource is either at a 0 level or a 1 level is a function of the entities that come into the system. The formula for average resource utilization is

$$\text{Average Resource Utilization} = \frac{\int_0^T B\, dt}{T}$$

where

B = either 0 for idle or 1 for busy.

dt = the length of time that *B* is observed.

T = the total length of time for the simulation.

As with the time-average number in queue, we are summing the length of time that the resource is either busy or idle and then dividing by the total time of the simulation run. The average utilization rate can be calculated using a bar chart in the same manner as for the time-average number in queue. The only difference is that the vertical axis of the graph can take only a value of either 0 or 1.

1.9.3.5 Manual Simulation Example

The following interarrival and service times were observed in a single-server, single-queue system:

Interarrival times (min)	1, 4, 2, 1, 8, 2, 4, 3
Service times (min)	2, 5, 4, 1, 3, 2, 1, 3

Calculate summary statistics for the time-average number in queue, average system time, and average utilization based on 20 min.

1.9.3.6 Example Solution

It is best to begin by organizing our data in a chart with headings for the number arrival, arrival time, begin-service time, end-service time, and total system time. We can then populate the chart with our input data:

Arrival num.	Arrival time	Begin sv. time	End sv. time	System time
1	1	1	3	2

The first event is the first arrival, which occurs at 1 min after the system starts. Because there is no one else in the queue, and the server is idle, the customer can immediately seize the resource and begin the service time. This means that the service time also begins 1 min after the system starts. The service start is the second system event. The service time for the first customer was 2 min. This means that the service end occurs at 3 min. The service end is the third system event. Note also that the total time that the customer was in the system is the service end or departure time of 3 min minus the arrival time of 1 min.

The fourth event that occurs is the arrival of the second entity. Because there is no one in the queue, and the server is idle, this entity also goes directly to the server.

Arrival num.	Arrival time	Begin sv. time	End sv. time	System time
1	1	1	3	2
2	5	5		

The second entity is scheduled to have a service time of 5 min. This means that the second entity's service time would end at 10 min. However, the interarrival time between the second and third entities is 2 min. This results in the arrival of the third entity at 7 min. Thus, the next event is not the end of the service of the second entity but the arrival in the system of the third entity. Because the server is busy with the second entity, the third entity takes the first position in the queue.

Arrival num.	Arrival time	Begin sv. time	End sv. time	System time
1	1	1	3	2
2	5	5		
3	7			

At this point two different things could occur. The service time end for entity 2 could occur at 10 min, or another entity could arrive in the system before 10 min. As it turns out, the interarrival time between the third and fourth entities is 1 min. This means that the fourth entity arrives at 8 min. This is before the service time ends for the second entity. The fourth entity enters the queue behind the third entity. There are now two entities in the queue.

Arrival num.	Arrival time	Begin sv. time	End sv. time	System time
1	1	1	3	2
2	5	5		
3	7			
4	8			

Now two other events could occur. The service time for entity 2 could end at 10 min, or the fifth entity could arrive in the system. As it turns out, the interarrival time for the fifth entity is 8 min. This means that the service end for the second entity will be the next event.

Arrival num.	Arrival time	Begin sv. time	End sv. time	System time
1	1	1	3	2
2	5	5	10	5
3	7	10		
4	8			

We can calculate the system time for the second entity in the same manner as the first entity. Because the second entity did not wait in the queue before being serviced, its system time is the same as its service time. At 10 min a second event occurs. Because entity 3 was waiting in the queue, as soon as entity 2 was finished, entity 3 immediately seized the resource at 10 min. With entity 3 now being served, only entity 4 is now waiting in the queue.

Entity 3 has a service time of 4 min. This means that its service time will be completed at 14 min. Because the interarrival time for entity 5 was 8 min, the service time for entity 3 will be completed before entity 5 arrives at 16 min.

Arrival num.	Arrival time	Begin sv. time	End sv. time	System time
1	1	1	3	2
2	5	5	10	5
3	7	10	14	7
4	8	14		

The system time calculation for the third entity is slightly different. Entity 3 actually waited in the queue for 3 min. So in this case, the system time is 14 min minus 7 min, for a system time of 7 min. Because entity 4 was waiting in the queue when entity 3 finished, it immediately left the queue and seized the server. The queue now has no entities.

The next event turns out to be the end of the service time for entity 4. This is because entity four has a short service time of only 1 min. Entity 4 also waited in the queue. The system time for entity four is 15 min minus 8 or 7 min.

Arrival num.	Arrival time	Begin sv. time	End sv. time	System time
1	1	1	3	2
2	5	5	10	5
3	7	10	14	7
4	8	14	15	7

Because the server is now idle at 15 min, the only event that can occur next is an arrival. The next event is the arrival of entity 5 at 16 min. No one is in the queue, so entity 5 immediately seizes the resource at 16 min.

Arrival num.	Arrival time	Begin sv. time	End sv. time	System time
1	1	1	3	2
2	5	5	10	5
3	7	10	14	7
4	8	14	15	7
5	16	16		

At 16 min, two different events could occur. The service time for entity 5 could end, or entity 6 could arrive. The interarrival time for entity 6 is 2 min. The service time for entity 5 is 3 min. This means that the next event is the arrival of entity 6 at 18 min. The server is busy with entity 5, so entity 6 enters the queue.

Arrival num.	Arrival time	Begin sv. time	End sv. time	System time
1	1	1	3	2
2	5	5	10	5
3	7	10	14	7
4	8	14	15	7
5	16	16		
6	18			

The interarrival time for entity 7 is 4 min for an arrival at 22 min, whereas the service end for entity 6 is 19 min. Because no more entities can arrive before 22 min, the next event is the service end for entity 6. The system time can be calculated for entity 5 as 3 min. Because entity 6 is waiting in the queue, it can immediately seize the resource at 19 min.

Arrival num.	Arrival time	Begin sv. time	End sv. time	System time
1	1	1	3	2
2	5	5	10	5
3	7	10	14	7
4	8	14	15	7
5	16	16	19	3
6	18	19		

The next event can be either the arrival of entity 7 or the service end for entity 6. The interarrival time for entity 7 is 4 min for an arrival at 22 min. The service time of entity 6 is 2 min. This would make the next event be the end of service for entity 6 at 21 min.

We actually specified in the model that we wanted to calculate the summary statistics based on 20 min. This means that we do not need to worry about the arrival of entity 7 at 22 min. We also do not need to worry about the end of service for entity 6 at 21 min.

Arrival num.	Arrival time	Begin sv. time	End sv. time	System time
1	1	1	3	2
2	5	5	10	5
3	7	10	14	7
4	8	14	15	7
5	16	16	19	3
6	18	19	21	
7	22			

Our event list is now complete, and we can turn our attention to calculating the output measures of performance. The easiest measure of performance to calculate is the system time of the entities. At time

20, only five different customers have exited the system. We do not need to be concerned about any entities still in the system when we are calculating an observational type of measure such as system time. We can calculate the average system time by summing all of the individual system times and dividing by 5. Thus, the average system time for all of the entities that were processed through the system was 4.8 min.

$$\text{Average System Time} = \frac{2+5+7+7+3}{5} = 4.8$$

To calculate the time-dependent statistics, time-average number in queue and average resource utilization, it is best to resort to the use of a chart. However, we also include a textual description of the relevant events.

1.9.3.6.1   Time Average Number in Queue Calculations

For entity 1 at 1 min and entity 2 at 5 min, service is available immediately. When entity 3 arrives at 7 min, it must wait in line. Entity 3 waits in line until entity 4 arrives at 8 min. This means that for 1 min between 7 and 8 min there was one entity in line:

$$1 \text{ entity in the queue} \times (8 - 7) \text{ min} = 1 \text{ entity-min}$$

The next relevant event is the end of service of entity 2 at 10 min. This means that two entities, entities 3 and 4, waited in the queue for the period between 8 and 10 min:

$$2 \text{ entities in the queue} \times (10 - 8) \text{ min} = 4 \text{ entity-min}$$

At 10 min there is only one entity, entity 4, in the queue. The next relevant event is the service end of entity 3 at 14 min. This means that entity 4 waited in the queue alone between 10 and 14 min:

$$1 \text{ entity in the queue} \times (14 - 10) \text{ min} = 4 \text{ entity-min}$$

At 14 min there is no one waiting in the queue. The next event is the arrival of entity 5 at 16 min. Because there is no one in the queue, and the server is idle, entity 5 does not have any queue time. The next event that occurs is the arrival of entity 6 at 18 min. The server is busy, so entity 6 enters the queue. The next relevant event is the end of service time for entity 5 at 19 min. This event causes entity 6 to leave the queue. Entity 6 was the single entity in the queue for 1 min:

$$1 \text{ entity in the queue} \times (19 - 18) \text{ min} = 1 \text{ entity-min}$$

When entity 6 leaves the queue, there are no other entities in the queue. The next entity does not arrive until 22 min. This means that for the remainder of the simulation, up to 20 min, the queue is 0.

The average time in queue can now be calculated by summing the periods of time that entities were waiting in the queue and dividing by the simulation time:

$$\text{Time Average Number in } Q = \frac{1+4+4+1}{20} = 0.50$$

This means that the average number of entities in the queue at any given time would be one half of an entity. Obviously, this is a mathematical representation because it is impossible to see half of an entity. It does indicate that the system is not very stressed if there is only half a person waiting in the queue at any given moment. The graphic representation of the time average number in queue is illustrated in Figure 1.3.

1.9.3.6.2   Average Resource Utilization Calculations

The average utilization calculations are somewhat easier to compute because the single resource can be only idle or busy (see Figure 1.4). The relevant resource utilization events begin at 1 min. This is when

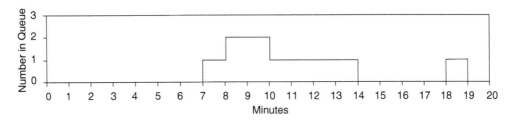

FIGURE 1.3 Average number in queue calculations.

entity 1 arrives and finds the queue empty and the server idle. So between 0 and 1 min, the utilization rate is 0:

$$0 \text{ resources busy} \times (1 - 0) \text{ min}$$

The resource remains busy until the end of the service time for entity 1. This occurs at 3 min. This area is calculated by:

$$1 \text{ resource busy} \times (3 - 1) \text{ min} = 2 \text{ resource-min}$$

At 3 min, there are no entities waiting in the queue. The next resource event occurs at 5 min, when entity 2 arrives to an empty queue and an idle server:

$$0 \text{ resources busy} \times (5 - 3) \text{ min} = 0 \text{ resource-min}$$

The server stays busy with entity 2 until the end of entity 2's service time at 10 min:

$$1 \text{ resource busy} \times (10 - 5) \text{ min} = 5 \text{ resource-min}$$

At the end of entity 2's service time, entity 3 is already waiting in the queue. Entity 3 immediately seizes the resource for its service time of 4 min, until 14 min into the simulation:

$$1 \text{ resource busy} \times (14 - 10) \text{ min} = 4 \text{ resource-min}$$

At the end of entity 3's service time, entity 4 is waiting in line. Entity 4 immediately seizes the resource for its service time of 1 min, until 15 min into the simulation:

$$1 \text{ resource busy} \times (15 - 14) \text{ min} = 1 \text{ resource-min}$$

At 15 min, there are no entities waiting in the queue. The next entity arrives at 16 min:

$$0 \text{ resources busy} \times (16 - 15) \text{ min} = 0 \text{ resource-min}$$

At 16 min, entity 5 arrives and immediately seizes the resource. Entity 5 uses the resource for 3 min, until 19 min into the simulation:

$$1 \text{ resource busy} \times (19 - 16) \text{ min} = 3 \text{ resource-min}$$

At 19 min, entity 6 is already waiting in the queue. It immediately seizes the resource at 19 min and keeps it past our cutoff time of 20 min. The average utilization of the single resource can be calculated to be an 80% utilization rate:

$$\text{Average Resource Utilization} = \frac{2+5+4+1+3+1}{20} = 0.80$$

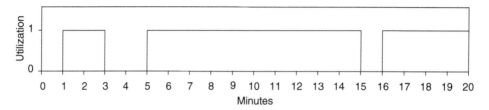

FIGURE 1.4   Average utilization.

It should be obvious to the practitioner that event lists involving anything more than a single queue and single server system run for more than a short period of time can become quite complex. The individual calculations themselves are not particularly cumbersome; however, there are a large number of events that could be happening nearly simultaneously. The consequences of a single sequencing or calculation error anywhere in the list could completely skew the output measures of performance. The advantages of utilizing a simulation-specific software package that automatically handles the event list should be readily apparent at this point.

1.10 Additional Basic Simulation Issues

Occasionally practitioners have difficulty understanding how even simple simulation models and event lists work. Sometimes this is because the practitioner has had significant experience in general-purpose programming languages. This is particularly noticeable when the practitioner's programming skills were developed using procedural or structured programming with languages such as FORTRAN, and the difficulty results from the experienced programmer's tendency to think about the program from an external viewpoint. Practitioners who become involved in simulation modeling need to look at the simulation model from the viewpoint of the entity. For example, the entity arrives in the system, the entity enters the queue, the entity seizes the resource, the entity releases the resource, and the entity is disposed of. As the practitioner develops more sophisticated and larger models, this issue becomes increasingly critical.

1.11 Summary

The objective of this chapter was to provide the novice and rusty practitioner with a beginning point from which to proceed with the simulation project. In this chapter we discussed the purposes of simulation, the advantages and disadvantages of simulation, and basic simulation concepts. In addition, we provided a few humorous quotations with what we hope was an educational message.

The section on basic simulation concepts included a step-by-step manual simulation model to provide experience with handling the event list and calculating summary statistics. Although the small example may have little actual operational value, it clearly demonstrates the complexity of handling the event list and calculating even a few summary statistics. Fortunately, most simulation projects will use a simulation software package that will completely insulate the practitioner from manually maintaining the event list or calculating any summary statistics. The practitioner can always refer to this chapter when questions arise with the reams of statistics that can be automatically generated by most simulation software packages.

Chapter Problems

1. What is the difference in use between traditional simulation models and training simulators?

2. Name a few typical manufacturing applications that can be modeled and analyzed using simulation.

3. Name a few typical service applications that can be modeled and analyzed using simulation.

4. Why might some systems be able to be analyzed only by using simulation?

5. What components does a simple system such as a one-chair barbershop contain?

6. What is the difference between observational and time-dependent data?

7. What does time-average number in queue mean?

8. What is the difference between an entity attribute and a global variable?

9. What is the difference between a discrete event simulation model and a continuous event simulation model?

10. Generate a manual event list for customers arriving at a single-queue, single-server system. Calculate system time, average number in queue, and resource utilization based on the system for 18 min.

    Interarrival times in minutes for 10 arrivals: 2, 1, 3, 1, 3, 2, 4, 2, 1, 1

    Service times in minutes for 10 arrivals: 2, 3, 1, 3, 2, 2, 1, 3, 2, 2

References

Kelton, D.W., Sadowski, R.P., and Sadowski, D.A. (2002), *Simulation with Arena,* 2nd ed., McGraw-Hill, New York.
Law, A.M. and Kelton, D.W. (2000), *Simulation Modeling and Analysis,* 3rd ed., McGraw-Hill, New York.

2

Problem Formulation

"We solved the wrong problem!"

2.1 Introduction

The first step in conducting a significant simulation project is to ensure that adequate attention has been directed toward understanding what is to be accomplished by performing the study. These activities are known as the problem formulation process and consist of:

- A formal problem statement
- Orientation of the system
- Establishment of specific project objectives

The importance of directing adequate attention toward the problem formulation process cannot be overemphasized. During this period, the simulation practitioner can firmly establish the practicality of using simulation to analyze the system. It may turn out that the system is deterministic or otherwise will not lend itself to simulation analysis-type approaches. If either of these situations exists, the practitioners can save themselves a significant amount of work and embarrassment early in the process. In the more favorable event that the system lends itself to simulation analysis, the practitioner should begin by developing a problem statement.

2.2 Formal Problem Statement

The problem statement component of the problem formulation step should provide both the practitioner and the potential audience with a clearly understandable high-level justification for the simulation.

Although each situation will be unique, problem statements commonly include text related to:

0-8493-1241-8/04/$0.00+$1.50
© 2004 by CRC Press LLC

Simulation Modeling Handbook: A Practical Approach

- Increasing customer satisfaction
- Increasing throughput
- Reducing waste
- Reducing work in progress

2.2.1   Increasing Customer Satisfaction

Increasing customer satisfaction is of fundamental interest in any system involving service operations. This type of system typically includes waiting or processing queues. Reductions in queue time usually result in increased customer satisfaction. Customer satisfaction may also involve delivering products when promised. Reductions in the number of tardy jobs will reduce operating costs associated with the loss of goodwill and will increase customer satisfaction.

2.2.2   Increasing Throughput

Increasing throughput involves the amount of products or number of jobs that can be processed over a given period of time. This can involve the elimination or improvement of different process operations. It can also include the identification and redesign of bottleneck processes.

2.2.3   Reducing Waste

Reducing waste results in reduced operating costs and increased net profits. Waste can be reduced through reductions in spoilage and obsolescence. Spoilage can involve processes that are time and temperature critical. Obsolescence waste can result from an organization's inability to bring its product to the market on time.

2.2.4   Reducing Work in Progress

Work in progress is work that requires further processing for completion. Work in progress is commonly found in processes that require multiple discrete operations. Work in progress typically requires storage before the next process can be carried out. Reducing work in progress reduces process costs associated with resource capacity and storage requirements. Large amounts of work in progress can result from insufficient resource capacity or poor operating policies. Reducing work in progress can decrease the space needed for manufacturing or distribution facilities.

2.2.5   Tools for Developing the Problem Statement

There are two common tools available to the practitioner for assisting with the problem statement. These are not actually problem-solving tools but rather are problem identification tools. Although these tools were originally developed for the manufacturing environment, they can easily be adapted to other sectors of industry. These tools are:

- Fishbone chart
- Pareto chart

2.2.5.1   Fishbone Chart

The Fishbone chart is also known as the cause-and-effect diagram, man–machine–material chart, and as the Isikawa chart (Suzaki, 1987). The purpose of this chart is to identify the cause of the problem or effect of interest. The Fishbone chart looks similar to the bones of a fish. The head of the fish is labeled with the problem or effect. Each major bone coming out of the spine is a possible source or cause of the problem. For example, in a manufacturing process there are major bones for man, machine, material, and methods. This is illustrated in Figure 2.1.

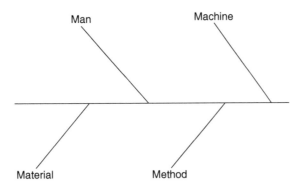

FIGURE 2.1 Fishbone chart.

Off each of the major bones, the practitioner is to add additional subbones. For the man bone, this might include supervisors, shift 1, shift 2, etc. It may also include other people such as maintenance or engineering. For the materials bone, the fish would include subbones for all of the raw materials that are present in the process. For the machine bone, the fish would have a subbone for each major piece of equipment involved in the manufacturing process. Finally, the method bone would have subbones for the work methods for the different manufacturing processes.

There is actually no limit to the number of subbones that can be created for a major bone or a subbone. However, if the process is complicated enough, it might be necessary to change the level of the fish from the complete product or process to a more focused component or subprocess.

When the Fishbone diagram is complete, the practitioner can concentrate on the most important sources or causes of the problem or problems. To assist the practitioner in this process, there is the Pareto chart.

2.2.5.2 Pareto Chart

The Pareto chart is a second technique to help the practitioner to develop the problem statement (Nahmias, 1987). It may turn out that there are several sources or causes of the problem or problems of interest. The concept of the Pareto chart was developed by the Italian economist Wilfredo Pareto, who postulated that only a few factors are the cause of many problems. This is frequently referred to as the 80–20 rule. The 80–20 rule states 80% of the problem is caused by 20% of the factors. The 80 and the 20 are not magic numbers but seem to be approximately observed in many instances.

Suppose you are interested in increasing the quality of a manufacturing operation. By performing the wishbone diagram analysis, you have narrowed down the quality problem to a particular segment of the manufacturing process. Defects can be a result of variations in temperature, humidity, product age, or worker error. Depending on the severity of the defect, the product may be reworked or have to be scrapped. If the product is reworked, additional cost is entailed to disassemble the product and repeat the defective part of the process. If the product is scrapped, loss is associated with both the raw materials and the work invested in the part up to the time that it is declared defective.

To implement a Pareto chart, you would need to know the number of defects from each source and the cost associated with the defect. If these are multiplied together, the true cost of each defect can be determined. It is possible to end up with four general combinations of true costs:

1. Large number of inexpensive, easily repairable defects
2. Large number of expensive defects or a nonrepairable product
3. Small number of inexpensive, easily repairable defects
4. Small number of expensive defects or a nonrepairable product

The Pareto chart would clearly indicate the amount of money lost to the large number of expensive defects or nonrepairable products, which should be the focus of the problem statement. The next question would be whether the small number of expensive defects or nonrepairable product cost is greater than the large number of inexpensive easily repairable defect cost. Last, small numbers of inexpensive easily repairable defects would be the subject for a later study. By understanding the true cost associated with the process, the simulation practitioner can focus his or her limited time on those few factors that would be expected to have the maximum benefit for the organization.

2.3   Orientation

Orientation is the second step in the problem formulation process. Orientation involves the practitioner's familiarizing himself or herself with the system. In most cases the practitioner will not have the depth of knowledge of the managers, technicians, or workers who are involved in the process on a daily basis. If the practitioner has had some contact with the system, the practitioner may feel as though he or she already has sufficient knowledge to bypass a formal orientation. Even if this is the case, abandoning the system orientation is likely to be a poor decision. The reason for this is that even though the practitioner may have general familiarity with the system, it is unlikely that the practitioner is familiar with the system from a simulation viewpoint. The practitioner must revisit the system keeping in mind how the system should be modeled and analyzed with respect to the nuances associated with simulation modeling.

2.3.1   Orientation Process

The orientation process should consist of a minimum of three different types of orientation visits. We recommend that the visits be conducted separately to increase the probability of understanding the operation of the system. However, in extreme cases, there is no operational reason why these visits cannot be performed during the same time period as long as the visit requirements can be met. The three different types of orientation visits are:

1. Initial orientation visit
2. Detailed flow orientation visit
3. Review orientation visit

2.3.1.1   Initial Orientation Visit

The purpose of the initial orientation visit is to obtain a high-level understanding of the basic inputs and outputs of the system. This type of orientation is typically conducted as a guided tour by a member of the commissioning organization. It is important not to attempt to understand too much detail of the system in the initial visit. If the guided tour lasts longer than a couple of hours, it is likely that too much information is being presented to be digested at this point in the orientation process.

After the initial orientation visit, the practitioner should immediately reflect on the major components of the process while recollection of the visit is still fresh. The practitioner should identify particularly complicated components of the system. In the first detailed flow orientation visit, the practitioner can request additional clarification on those components.

2.3.1.2   Detailed Flow Orientation Visit

After the practitioner has obtained a basic high-level understanding of the system, it is time to gather detailed information on the operation of the system. The detailed flow orientation visits may not necessarily be accompanied by a member of the commissioning organization. It is quite possible that the practitioner will be left on his or her own to conduct activities during this more time-consuming phase of the orientation process. If the system is complicated or contains a number of components, the practitioner may need to conduct multiple visits. The practitioner should make detailed notes on the system operation by recording:

- The types of entities that are processed by the system
- The number and type of queues
- The number and type of system resources

- The sequence of processes as experienced by the entities
- How the system performance can be measured

Note that it is not necessary to collect any sort of input data related to entity interarrival times or service times at this stage of the simulation study. Those activities will be conducted in subsequent visits to the system.

After the detailed flow orientation visits, the practitioner should begin developing an understanding of how the system operates. It may be beneficial to begin the development of the high-level flow chart ultimately required for the system definition phase of the project. Details on how to develop a high-level flow chart are included in Chapter 4, "System Definition." While the practitioner is attempting to understand all of the activities associated with the system, it is likely that a number of unanswered questions will arise. If this occurs, it may be necessary to conduct additional detailed flow orientation visits.

2.3.1.3 Review Orientation Visit

Once the practitioner has an understanding of the system operation, it is time to return to the system for the review orientation visit. The purpose of this visit is to ensure that the understanding of the system operation is consistent with the practitioners' understanding of the system and/or flow chart. It is also not unusual for the practitioner to observe new aspects of the system even after the review orientation visit has been conducted. During the review orientation visit, it would be helpful to reengage the services of the system representative to answer any last minute questions and to insure that the practitioner's understanding of the system is correct.

2.3.2 Orientation Example

To demonstrate the orientation process, consider the relatively simple process of airline passengers taking their seats in a commercial jetliner. We will assume that we have undertaken this process ourselves many times. Given our familiarity with the general process, it will not be necessary to conduct an initial orientation visit.

In our subsequent detailed flow visits, we will need to observe the process from the standpoint of the passengers. Close observation of this process will indicate that we will need to have knowledge of the following items:

- The number of passengers on the plane
- The type of passenger: regular, family, handicapped
- The rate at which passengers travel down the aisle
- How far back the passenger must travel down the aisle
- Whether the passenger has an aisle, center, or window seat
- The percentage of passengers with carry-on bags
- The time it takes passengers to load each piece of luggage in the overhead bins

It is obvious from this list of observations that the seemingly simple process of taking a seat in an aircraft suddenly becomes more complex when viewed from a simulation modeling standpoint. In addition, each of these processes may take place during a very short time frame. Thus, it may be possible to observe only a small part of each complete process at a given time. Note that we do not actually need to collect any of the data listed. However, we do need to be aware of the fact that different types of data are associated with each of these processes.

If we took our list of processes for a review orientation visit, we might discover that we had neglected to identify the following processes:

- The number of times a passenger must get out of his or her seat to let other passengers by
- How long it takes a passenger from an aisle or center seat to let the other passenger by
- Whether the passenger must search the luggage compartments for space
- How long it takes the passenger to search for the space and place the luggage

Only by carefully considering the original list and revisiting the system may we actually be able to identify a more complete set of processes that need to be modeled or data that need to be collected. Although it is not necessary to develop a flow chart of the process at this stage of the simulation study, there is little doubt that it would have been beneficial, even in this simple example.

2.3.3　Tools for the Practitioner's Orientation

Because the simulation study is most likely commissioned by the system's organization, the practitioner should have a high degree of access to the system. A high degree of access may enable the practitioner to employ electronic recording devices such as a digital camera or camcorder. The purpose of using these devices is not to record system input data but to allow the practitioner to review complex processes as necessary to gain a fundamental understanding of the system. System input data will be collected at a later point in time. The purpose of the orientation phase is primarily to understand the system to be modeled.

2.4　Project Objectives

Once the practitioner has completed his or her orientation of the system, thought may be given to specific project objectives. This is the final step in the problem statement process of the simulation study. When the project objectives are being established, it should be obvious that the practitioner develops the objectives in close cooperation with the individuals commissioning the simulation study. Though this may seem obvious, any breakdown in communication during this critical phase can have a devastating impact on the value of the simulation to the commissioning organization.

The project objectives should also be considered as dynamic. During the course of the study, the practitioner may discover other information that may prove to be more profitable to pursue. Sometimes the commissioning organization does not initially have as much interest as would be beneficial. This means that the practitioner, who may or may not have sufficient domain knowledge of the system, is forced to suggest the initial project objectives. Only after the credibility of the practitioner and simulation modeling process has been established does the necessary level of interest appear. If this does occur, one hopes it will not be too late to adjust the simulation project objectives.

2.4.1　Examples of Project Objectives

The following list of common project objectives can be used by the practitioner as a starting point to develop a project-specific set of objectives. Common project objectives may involve

- Performance-related operating policies
- Performance-related resources policies
- Cost-related resource policies
- Equipment capabilities evaluation

2.4.1.1　Performance-Related Operating Policies

If the system is already in place, the practitioner may be interested in making the best use of the existing resources. The analysis of different types of scheduling priorities falls in this category of project objectives. In this case, the practitioner would specifically be interested in which scheduling priority policies result in either the smallest average system times or the fewest jobs that are tardy with respect to their due dates.

The opposite objective approach could also be used for scheduling. In this case, the practitioner could use a simulation model to establish how to quote system times or delivery dates. Here, the practitioner examines the system with a particular scheduling policy. Given the different types of jobs that the system will receive, the practitioner can provide summary statistics on how long each type of job will take. Because the state of the actual system is already known, the practitioner can determine the due date for each different type of job with a certain level of confidence.

Another type of performance-related operating policy can involve preventive maintenance. In systems that require regular maintenance, the duration of the preventive maintenance is frequently related to the interval between preventive maintenance periods. This type of operating policy is particularly common in process industries, including the petrochemical, chemical, and food industries. For example, in a chicken parts processing plant, the cleanup operation is considered a preventive maintenance operation. The more chicken parts that are processed, the more contaminated the processing stations become. As the processing stations become more contaminated, the throughput of the stations is reduced. The preventive maintenance cleanup operation requires that the processing stations reduce or even stop production. In some processing plants, the first two 8-h shifts are dedicated to production. The level of contamination requires that the majority of the third shift be dedicated to cleanup preventive maintenance. The practitioner can use simulation to develop a more effective cleanup preventive maintenance policy. Alternative policies involving time or cycle number intervals before preventive maintenance is performed can be examined without interfering with the existing operations.

Yet another type of performance-related operating policy objective involves the layout of the process. One of the most common types of objective is to try to determine the best alternative among many different queue configurations. For example, banking facilities utilize both a number of different parallel queues that feed to a particular clerk and a single snake-type queue that feeds to a number of different clerks. In addition, some banking facilities may include express lines in addition to the regular transaction lines. So for a given level of resources, the objective is to determine the best queue configuration.

2.4.1.2 Cost-Related Resource Policies

Many simulation project objectives focus on reducing costs while still maintaining a given level of performance. Here, the project objective may be how many or how few resources are needed at any given point in time in order to keep the system performance at a given acceptable level. System performance can be defined as a specific level of system time or as a certain percentage of properly processed jobs.

This type of objective approach is typically used in the passenger airlines industry. Passengers must not miss flights frequently because of excess security, ticket, baggage, or gate processes. The consequence is reduced repeat business. On the other hand, excess resource capacity does not necessarily translate into additional business. Provided that the required level of service is provided, the objective becomes determining the most cost-effective manner to staff or resource the processing stations.

2.4.1.3 Performance-Related Resource Policies

A different type of alternative resource policy objective involves the level and distribution of resources with less consideration to cost. In this case, the focus is on determining the best performance among different resource level alternatives. This means that the objective of the project is to identify the point of diminishing performance return with respect to resource costs. The highest level of performance is required, but not at the cost of noncontributing resources.

This type of project objective is frequently found in service-type industries. These include retail stores, restaurants, and entertainment facilities. Here the best service process will impact on the performance of the system. For example, if the wait time for a restaurant becomes excessive, customers will go elsewhere. The loss of goodwill can impact the restaurant not only at that time but also in the future. Thus, it is to the benefit of the restaurant to provide the best possible service without having unnecessary staff.

A slightly different type of situation may exist with service systems where the resources are interchangeable. Here the objective would be to determine how to distribute different levels of resources to increase operating profits. A situation where this has been studied involved the staffing of a multitheater movie complex. The staff consists of ticket sellers, ticket checkers, and concession sellers. The objective was to balance the staffing levels at each of these different positions so that the theater could operate effectively while having high profits. In this example, the objectives could actually be in conflict with one another. The reason for this is that the most profit is made from the concession sales. Because of the limited amount of time between the sale of the ticket and the movie start, more concession clerks would translate into more concession sales. However, in order to get to the concession area, the customers will have had to purchase their tickets. This means that a sufficient number of ticket clerks must also exist to allow the customers to have an adequate amount of time to purchase concessions.

This type of project objective also includes manufacturing applications. Here the objective would be the type and number of pieces of manufacturing equipment that should be utilized by a particular system. The type of manufacturing equipment could specifically include different processing capabilities. The objective may be to determine whether it is better to use a larger number of less capable equipment or a smaller number of more capable equipment. In this type of project, the average utilization of the equipment would be of particular interest. The initial capital investment and the operating cost of the equipment would need to be taken into consideration.

2.4.1.4 Equipment Capabilities Evaluation

The objective of the simulation project could also be to test or determine the capabilities of a proposed type of new equipment. Without a simulation model of the new equipment, there is no way of determining if the equipment will actually function as claimed. Some organizations require equipment vendors to develop and demonstrate simulation models of their equipment. This helps prove that the equipment can meet the necessary requirements for the purchasing organization's particular application.

An example of this is a barcode-reading and sorting system that is used for different types of packages. The orientation and size of the packages can affect the performance of the barcode-reading system. Because the orientation and size of the packages are probabilistic, the read rate and the reliability of the reading will also vary. Simulation offers the opportunity to test the long-term performance of this type of system.

2.4.2 Decision-Making Tools for Determining Project Objectives

When there is no clearly defined set of project objectives, the practitioner must develop a set of objectives that are satisfactory for both internal and external project stakeholders. Internal stakeholders include the practitioner responsible for the project and any other individuals in the organization who will be directly involved in completing the project. The external stakeholders include the customer commissioning the project, which could include organizational management or a consulting client.

One common way of deciding on appropriate project objectives is to utilize engineering management decision-making tools. These tools can include (Gibson et al., 2003):

- Brainstorming
- Nominal group technique
- Delphi process

Brainstorming and the nominal group technique work most effectively when conducted in conjunction with each other. Most decision-making sessions will begin with a brainstorming session and then use the nominal group technique to come to a consensus as to what action should be taken.

2.4.2.1 Brainstorming

The idea behind brainstorming is to generate a storm of ideas that are created by being exposed to other ideas. In a brainstorming session, participants will often come up with ideas that otherwise they would not have. Discussion of other ideas gives birth to new ideas. A fundamental aspect of brainstorming is that no participant should criticize the validity of another participant's ideas. If any criticism exists, some participants may feel uncomfortable and be reluctant to share openly their ideas.

2.4.2.1.1 Brainstorming Preparation

To administer a brainstorming session, the project manager must arrange for all of the participants to meet in one physical location. The number of participants may be dependent on who is involved in the project. However, some studies have shown that the optimal number of participants in an activity of this sort is between 5 and 12. With fewer than five participants, there is not sufficient synergy. With more than 12 it becomes too difficult to maintain control of the brainstorming process.

The administrator must also have some sort of large-format recording medium that all of the participants can simultaneously see. This can be a whiteboard, an overhead, or an easel with a large pad of paper. The overhead is probably the best medium because the transparencies can easily be kept for future record. A technologically advanced organization may also have access to an electronic whiteboard. On this type of equipment, the contents of the whiteboard may be electronically recorded. In addition to the large-format medium, the administrator should also have small notecards. These will be used for the nominal group activities following the brainstorming sessions.

2.4.2.1.2 Brainstorming Process
In the brainstorming process, the administrator will have all of the participants gathered in one area such as a conference room. The administrator will provide all of the participants with some sort of orientation describing the rules and purpose of the session. It is most important that the administrator emphasize the rule that in the brainstorming stage of the session, all ideas are considered valid and should not be criticized. The purpose would be to identify a set of agreed-on simulation project objectives. At the end of the orientation, the administrator should explicitly solicit the participants for questions. When the participants are ready to start, the administrator begins the process by listing one idea for the project objectives on the large-format medium. The administrator would then go around to each member in the group to offer additional ideas for the project objectives. Here is when a member may offer a completely new idea or an idea that was created by seeing another group member's idea. If a member does not have any ideas at that time, the member passes. This cycle is repeated until all of the participants pass on offering new ideas. At the end of the brainstorming phase of the session, there should be a storm of ideas for further processing.

2.4.2.1.3 Electronic Brainstorming
It is also possible for the practitioner to use the concept of electronic brainstorming. There are a number of advantages to the use of electronic brainstorming over conventional brainstorming. These include the ability to include more individuals in the discussion and the participation of individuals who could not attend the session because of distance or schedule limitations. It also allows for an electronically recorded history of the discussions. The primary disadvantage is the unsynchronized nature of implementing electronic brainstorming. Unless everyone in the session is simultaneously participating, the discussion does not occur in real time. Thus, it may be necessary to allow for several days before the session can be concluded. Electronic brainstorming can be implemented in either of the following ways:

- E-mail discussion list
- Electronic bulletin board

An e-mail discussion list is the easier of the two electronic brainstorming methods to implement. With this method, the practitioner sets up an e-mail address list that includes the e-mail addresses of all of the individuals with potential interest in the project. These individuals are commonly known as stakeholders. The practitioner is directed to Chapter 3, "Project Planning," for an in-depth discussion of the different types of stakeholders. As each individual thinks of a new idea, the concept is entered in an e-mail reply to the list and is automatically redistributed to all of the members of the e-mail list.

A small irritation that has been identified with these types of e-mail list brainstorming sessions is when the participants set their e-mail package to notify them automatically of incoming e-mail. If there is high volume on the list, the constant notification of e-mail messages can become disturbing. Many otherwise useful individuals are more likely to request to be removed from the list rather than be continuously disturbed by the series of e-mail messages.

The electronic bulletin board approach to brainstorming eliminates the continuous e-mail message notification problem. Here, the practitioner sets up a Web-based bulletin board to control the brainstorming discussion. Individuals must take the initiative to regularly check the bulletin board for new developments. The potential success of this approach may at first seem questionable. However, many user group bulletin boards typically experience high levels of activity. This activity includes a number of topic professionals who are completely willing to give expert technical advice free to complete strangers. If the practitioner's brainstorming group is as interested in the project as the average individual is in interest group bulletin boards, the necessary level of activity is assured.

2.4.2.2 Nominal Group Technique

Once the storm of ideas for the project objectives has been generated, it is the administrator's job to determine the most important project objectives. These objectives can be identified by using the "nominal group technique." In this procedure all of the participants have the opportunity to vote on what they believe is the most important project objective. All of the participants are given an equal number of votes to cast. Each vote holds the same weight regardless of the status of the voter. This equality will encourage the less senior individuals in the group to vote without undue influence. Before the nominal group technique process begins, participants may ask for clarification on different project objectives. Should this occur, the parent of the idea should provide a brief response.

Once the process is ready to start, the voters are asked to cast one vote per cycle for an individual project objective idea. This cycle repeats until all the votes are cast. Voters may vote for the same idea until all of their votes are cast, or they may split their votes among different ideas. At the end of the nominal group technique process, the administrator should compile the votes and identify the most popular project objective ideas. These project objectives will then guide the rest of the project activities.

2.4.2.3 Delphi Process

In some organizations or situations undue political influence interferes with the selection of the project objectives. If it is desirable to minimize the effect of these influences, the administrator may elect to utilize the Delphi process technique. This process serves the same function as the nominal group technique; however, all voting is conducted anonymously after the brainstorming session has concluded. The process begins by having the administrator distribute a list of all of the brainstorming ideas. The participants then cast one vote for a particular idea. At the end of the first cycle, the administrator retains only a set of the most popular ideas. The new set of ideas is redistributed to the group. This process continues until the set of project objective ideas is a suitable number. The number of ideas to advance to the next round is at the discretion of the administrator. One method is to eliminate any ideas that receive no more than one vote. This means that that idea is of predominant interest to only one individual. As with the nominal group technique, the final list of project objectives guides the rest of the project activities.

Like brainstorming sessions, the Delphi process can also be implemented electronically. The difference is that e-mail messages are sent directly and individually to the individuals participating in the Delphi process. Members should not be encouraged to engage in direct e-mail communication among themselves. For this reason, e-mail lists should be avoided. The rest of the process is similar to the normal Delphi process. When the e-mail responses are received, the administrator compiles and individually rebroadcasts the results.

2.5 Summary

In this chapter we presented the steps involved in the problem formulation process of the simulation study. The problem formulation process is the first step in the simulation study process. During this phase the practitioner should determine if simulation is the appropriate tool to analyze the study under consideration. A principal concern is whether the system is deterministic or probabilistic. If the system is deterministic, then it would not necessarily be beneficial to conduct a simulation study. However, if the system is probabilistic, then it is possible to conduct the simulation study. These steps involve a formal problem statement, a system orientation, and project objectives.

The formal problem statement process includes identifying the nature of the problem. It is essential that both the practitioner and the commissioning organization properly understand what the simulation is trying to study. Failure to come to an agreement during the problem statement process can result in the practitioner modeling and analyzing the wrong system or an incomplete part of the system.

The formal problem statement process is followed by the orientation process. Here, the practitioner is attempting to gain an understanding of how the system operates. The orientation process is best broken down into three visits with distinct purposes. In the initial visit, the practitioner should attempt to obtain only the most basic understanding of the system. In the subsequent visits the practitioner should attempt to identify all of the system components and determine how the entities flow through the system. After the practitioner has obtained an understanding of the system, the system should be visited a final time. During this review visit, the practitioner makes a final comparison between his records of the system and the actual system. A high-level flow chart of the system is not required until the system definition phase. However, it would be helpful to begin work on this flow chart during the orientation process.

The final step of the problem formulation phase is to explicitly identify the objectives of the project. These should be developed in close conjunction with the previously established problem statement. The project objective phase specifies exactly what the simulation project is supposed to accomplish. Project objectives can be related to improving the system as a result of examining different operating policies, resource policies with respect to performance, resource policies with respect to cost, or evaluation of new system capabilities.

The proper execution of the problem formulation phase is critical to the overall success of the project. It could be extremely detrimental to both the practitioner's individual credibility and the credibility of simulation as an analysis tool if the practitioner did not have a proper understanding of the problem of interest. If there is some question as to the actual problem, it will be even more difficult to properly develop a set of project objectives. To a great extent, the problem formulation phase drives the rest of the project. It is better to spend a greater length of time during the problem formation phase if it will help to insure that the correct problem will be completely addressed.

Once the project formulation phase is complete, the practitioner can focus on the next phase of the simulation project process. This is the project planning phase. In this next phase, we introduce the practitioner to basic project management processes including the development of the project work breakdown structure, linear responsibility chart, and Gantt chart schedule.

Chapter Problems

1. What are the three major components to the problem formulation process?

2. Under what circumstances would it not be beneficial to conduct a simulation study of a system?

3. Why would it probably not be effective to attempt to conduct all orientation activities in one visit?

4. What tools are available to the practitioner to conduct the orientation visit?

5. Why do you not want to criticize someone's ideas during a brainstorming session?

6. Under what circumstances is it better to utilize the Delphi technique over the nominal group technique?

7. What is the advantage of using electronic resources to conduct either the nominal group technique or the Delphi technique?

References

Gibson, J.L., Konopaske, R., Donnelly, J.H., and Ivancevich, J.H. (2003), *Organizations: Behavior, Structure, and Processes,* 11th ed., McGraw-Hill, New York.
Nahmias, S. (1997), *Production and Operations Analysis,* 3rd ed., Richard D. Irwin, New York.
Suzaki, K. (1987), *The New Manufacturing Challenge,* Macmillan, New York.

3

Project Planning

"The plan is to have no plan."

3.1   Introduction

When the commissioning organization has decided to proceed with the simulation project, the practitioner may suddenly be thrust into the position of project manager. If you are using this handbook, there is a good probability that you are a novice simulation practitioner. Unfortunately, many novice simulation practitioners have little formal training in the way of project management skills. The skills that are acquired are obtained through trial and error or on-the-job training. This is unfortunate because the success or failure of a simulation project can depend heavily on the amount of time and skill devoted to the planning phase of the project.

To give the practitioner a jump start in the project planning and project management process, we provide a brief discussion of the basic issues necessary to manage successfully projects in general and then walk through, step by step, the development process for planning and organizing a simulation project, in particular. Practitioners who do not need to review Section 3.2, Project Management Concepts, should move directly to Section 3.4, Developing the Simulation Project Plan.

These topics are organized into the following sections:

- Project management concepts
- Project manager functions
- Work breakdown structures
- Linear responsibility charts
- Scheduling and the Gantt chart

Knowledge of these topics will allow the practitioner to develop a comprehensive project management plan that will help increase the probability of success of the project.

The chapter concludes with a summary and a set of blank templates for developing work breakdown structures, linear responsibility charts, and Gantt charts. At the very end of the chapter, you will also find a short list of project management-related Web sites for further investigation.

3.2 Project Management Concepts

In this section we discuss concepts fundamental to the practice of project management. These concepts include:

- Project parameters
- Project life cycles
- Project stakeholders

3.2.1 Project Parameters

The success or failure of a project is frequently measured by three standard project parameters. These are:

- Time
- Cost
- Technical performance

3.2.1.1 Time Parameter

The time parameter is associated with the project schedule, which is most frequently implemented as a Gantt chart. As we will later see, a Gantt chart is a horizontal bar chart on which each bar represents the sequence and duration of a different task in the project. The project schedule may also contain milestones. These are specific points in time at which some activities must be completed or conducted.

While the project is progressing, the project manager should be regularly comparing the actual project progress with the project schedule. If there are significant differences between the actual project progress and the project schedule, something may be amiss. This means that being continuously ahead of schedule is not necessarily advantageous. If the project is continuously ahead of schedule, the following situations may exist:

- The budget is being consumed at an excessive rate.
- Excess resources are assigned to the project.
- The technical performance standards are being compromised.
- The project is much less complex than originally estimated.

If the project does continuously run ahead of schedule, the project manager can investigate these possible causes. In most cases, the situation should be corrected. However, if the project is actually much less complex than originally estimated, the project manager can consider advancing the project completion date.

Conversely, if the project is continuously behind schedule, any of the following situations may be present:

- Expenditures are being delayed.
- Insufficient resources are assigned to the project.
- The technical performance standards are being exceeded.
- The project is much more complex than originally estimated.

If the project does run continuously behind schedule as a result of any of these causes, the project manager must take immediate corrective action. The only exception might be the situation in which the project was originally very tightly planned. It is not unusual for a tightly planned schedule to begin to slip. If the slip becomes too large, the project manager will need to consider revising the schedule for the remaining tasks in the project.

A delayed project schedule can also take on additional importance in some cases where the simulation project is just a small part of an overall effort. If the simulation component falls behind schedule, it may prevent other important decisions from being made. The project must be completed on time according to the schedule in order to be successful from a time standpoint.

3.2.1.2 Cost Parameter

The cost parameter means that there is a budget associated with the project. Simulation project costs may include computer hardware and software. It may also include the practitioner's time or the time of other resources needed to complete the project. A very definite expense can be associated with the cost of collecting original data. Additional personnel may need to be hired specifically for collecting data. If the project becomes extremely complex, the practitioner may need to resort to hiring simulation consultants to assist with the project.

The budget is perhaps the most easily tracked project parameter. Obviously, the project manager should regularly monitor the project budget. As with the time parameter, it is not necessarily advantageous to be continuously under budget. In the event that the project is continuously under budget, the following situations may be present:

- Expenditures are being delayed.
- Insufficient or inexperienced resources are assigned to the project.
- The technical performance standards are being compromised.
- The project is much smaller or less complex than originally estimated.

If the project is running under budget, the project manager should investigate the above possible causes and hope that the cause of being under budget is that the project is much less complex than originally estimated.

Similarly, problems may also exist if the project is continuously overbudget. Typical causes are:

- The budget is being consumed at an excessive rate.
- Excess resources are assigned to the project.
- The technical performance standards are being exceeded.
- The project is much larger or more complex than originally estimated.

If the project is running over budget, the project manager can reduce expenditures and resources at the risk of finishing the project behind schedule or compromising the technical performance standards. Alternatively, the project manager can attempt to increase the existing project budget.

3.2.1.3 Technical Performance Parameter

Finally, the project will have technical performance objectives. These would normally be specified in the problem statement phase of the project. These objectives may require that the simulation model be used to make operational or capital investment decisions. The model must both operate as intended and represent reality in order to make accurate and informed decisions. The project must achieve these objectives in order to be considered successful from a technical performance standpoint.

If the technical performance parameters are not achieved, then the project manager should first check whether:

- Expenditures are being delayed.
- Insufficient or inexperienced resources are assigned to the project.
- Technical performance standards are unnecessarily demanding.
- The project is much larger or more complex than originally estimated.

Inadequate technical performance can seriously endanger the success of the project. The project manager must immediately investigate and address the cause. Only in very rare cases does a situation exist where the technical performance standards are unnecessarily demanding and need to be relaxed.

On the other hand, if the technical performance parameters are constantly exceeded, any of the following situations could be occurring:

- The budget is being consumed at an excessive rate.
- Excess resources are assigned to the project.
- Technical performance standards were originally insufficient and must be revised.
- The project is much smaller or less complex than originally estimated.

The situation in which properly specified technical performance parameters are continuously exceeded is probably a rare event. A more likely situation is that limited information was available when the insufficient technical performance parameters were originally specified.

3.2.1.4   Project Parameter Model

As you may suspect by this point, the three project parameters are interrelated. Gross variation of any of the individual parameters may be an indicator of problems in the other parameters. The three project parameters are frequently illustrated in the form of a triangle. Each side of the triangle represents one project parameter. This concept is illustrated in Figure 3.1.

It is not enough to be successful in only one or two of these parameters. Each of these three project parameters must be satisfactorily fulfilled in order for the practitioner to consider the project a success. Sometimes the project manager may be tempted to sacrifice one or even two of the parameters to insure success in the third. For example, a project manager may be willing to accept cost overruns. By adding additional expensive resources to the project, the project manager may be able to finish the project on time and to the necessary technical performance standards. Conversely, to stay on budget, the project manager may be willing to adjust the project schedule or the project technical performance objectives. In a really serious case, both the schedule and the budget may have to be compromised in order to achieve important technical performance objectives.

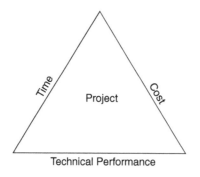

FIGURE 3.1 Project parameters.

3.2.2   Project Life Cycles

There is some thought that all projects proceed through certain life-cycle phases in some form or another. The following phases are commonly recognized:

- Conceptual
- Planning
- Execution
- Completion

3.2.2.1   Conceptual

The conceptual phase involves the initial decision by someone in the organization to consider conducting the simulation study. During this phase, the organization will formally assign the project to the project manager. The problem formulation process may be completed during this life cycle phase.

3.2.2.2   Planning

During the planning phase, the project manager will identify all of the project team members. The simulation-planning process activities involving the work breakdown structure, linear responsibility chart, and Gantt chart are conducted during this life-cycle phase.

3.2.2.3   Execution

Most of the simulation project activities for the project will be completed during the execution life-cycle phase. These include the system definition, input data collection and analysis, model translation, verification, validation, experimentation, and analysis.

3.2.2.4   Completion

The completion phase activities are normally conducted after most of the technical project activities have been concluded. These activities are associated with turning over the results of the project and primarily include the simulation project report and presentation activities.

3.2.3   Project Stakeholders

Every project has what are known as stakeholders (Cleland and Ireland, 2002). Stakeholders are individuals or organizations who have some sort of vested interest in the ongoing operations or completion of the project. The vested interest of the stakeholders may result in some action or lack of action on their behalf that may affect the project. Because their actions, or lack thereof, may affect the well-being of the project, the practitioner should be at a minimum aware of their existence.

Stakeholders can be categorized as either internal or external. The perspective is from the viewpoint of the simulation project team.

3.2.3.1   Internal Stakeholders

For our purposes, we define internal stakeholders as individuals who are directly associated with the simulation project team. This can also be interpreted as individuals over whom the practitioner/project manager exercises some degree of control. This means that the internal stakeholders will most likely be part of the parent organization. However, consultants brought in to assist the project team may also be considered internal stakeholders. Specific examples of internal stakeholders include:

- Practitioner/project manager
- Analysts
- Statisticians
- Data collectors
- Functional managers who supervise the project team members

3.2.3.2 External Stakeholders

We define external stakeholders as individuals or organizations who are not directly associated with the simulation project team. As a result, they can be within the parent organization but outside of the simulation project team, or they can be completely external to the parent organization. Specific examples of external stakeholders include:

- Practitioner/project manager's manager
- Managers of the process modeled
- Workers in the process modeled
- Simulation software vendor/developer
- Equipment vendors and manufacturers
- Regulatory agencies
- Competitors
- Public or community in general

3.2.3.3 Stakeholder Strategies

When there is a high potential for project disruption, the practitioner should consider developing a stakeholder strategy. Stakeholder strategies can include:

- Identification of the stakeholders
- Determining the strengths and weaknesses of the stakeholders
- Likely strategies of the stakeholders
- Counterstrategies to the stakeholder strategies

For the simulation practitioner/project manager, special consideration should be given toward developing stakeholder strategies for the manager and workers involved in the process that is simulated. In particular, the workers are not given adequate briefings as to the origin and purpose of the simulation study. A not uncommon assumption on the part of workers is that the simulation team's intention is to study the level of worker effectiveness or productivity. This can easily result in the workers unintentionally artificially skewing the processing or service times. This is known as the "Hawthorne effect," in which the act of observing an individual can skew the observations. If this occurs, the practitioner will not know exactly what was observed. In addition, uncooperative workers may also deliberately resist assisting the simulation team in the modeling process.

3.3   Simulation Project Manager Functions

The importance of the simulation practitioner as a project manager cannot be discounted. As the primary individual responsible for the project, the practitioner is largely responsible for the overall success or failure of the simulation project. There are five generally accepted project manager functions that will affect the success or failure of the simulation project:

- Planning
- Organizing
- Motivating
- Directing
- Controlling

3.3.1   Planning

The first responsibility of the project manager is to ensure that a proper and effective plan for the simulation project is in place. This process includes the development of a work breakdown structure and a Gantt chart. A work breakdown structure is the successive division of project tasks to the point that individual responsibility and accountability can be assigned. A Gantt chart illustrates the duration and relationships among the work breakdown structure tasks. These activities are discussed at length later in this chapter.

3.3.2 Organizing

The second management function involves identifying, acquiring, and aligning the simulation project team. This is an important issue because most of the simulation project team is likely to be available to the project manager on only a temporary and limited basis. Many project managers believe that this process is best accomplished with a traditional organizational chart. Although this type of chart does provide a basic framework for the project team, it does not provide insight into the relationship between the team members' host organizations and the different projects that the team members are assigned to. The traditional organizational chart also does not provide details on the relationships between project tasks and individuals.

The relationship between the host organizations and the different projects that the team members are assigned can be clarified through the use of a matrix organizational chart. This is a two-axis chart. On the horizontal axis, the chart contains the host organization or department for each team member. On the vertical axis, the chart contains the different projects that the team member or members are assigned to. The intersection of the horizontal and the vertical axis lists the team member's identity. With the matrix organizational chart, the project manager can quickly identify what other responsibilities may be competing for each team member's time. Figure 3.2 illustrates the concept of the matrix organizational chart.

The details of the relationships between project tasks and individuals are better served with the development of a linear responsibility chart. A linear responsibility chart indicates who is responsible and to what degree for each of the work breakdown structure tasks. This enables the project manager to hold individuals accountable for the performance or nonperformance of these tasks. Additional discussion on linear responsibility charts can be found later in this chapter.

The project manager should also be prepared for the fact that the simulation project team may continuously change in composition as the project proceeds through different phases.

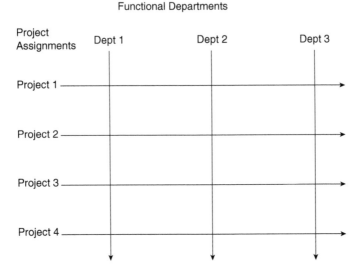

FIGURE 3.2 Matrix organizational chart.

3.3.3   Motivating

The third project manager function is motivating the simulation project team. By ensuring that the project team is highly motivated, the project manager has a greater probability of completing a successful simulation project. There are a number of well-regarded motivation theories. Three of the most popular ones include (Gibson et al., 2003):

1. Maslow's hierarchy of needs theory
2. Alderfer's ERG theory
3. Herzberg's two-factor theory

Since motivating the project team members is one of the simulation project manager's functions, we will briefly review some of the basic principles behind each of these three well-established motivation theories.

3.3.3.1   Maslow's Hierarchy of Needs

Maslow's hierarchy of needs theory is based on a five-level pyramid. According to the theory, the lowest level consists of physiologic needs such as food, water, and shelter. The second level involves safety and security. The third level includes friendships, affiliation, and interaction with others. The fourth level involves self-esteem and respect from others. The fifth and final need is to self-actualize through maximizing one's abilities, skills, and potentials. This pyramid is illustrated in Figure 3.3.

According to the theory, individuals are motivated to satisfy the needs at each level of the pyramid and progress upwards. Individuals must satisfy the needs at each level before proceeding to the next level. This means that the simulation project manager must identify on which level each of the team members currently resides. Motivation efforts should be directed toward satisfying those needs at that level. Efforts on the part of the manager to motivate the individual by satisfying previously satisfied lower-level needs will not necessarily be effective.

3.3.3.2   Alderfer's ERG Theory

Alderfer's motivation theory involves three levels in contrast to Maslow's five levels. These levels are:

1. Growth
2. Relatedness
3. Existence

FIGURE 3.3   Maslow's hierarchy.

The existence level corresponds to Maslow's physiologic, safety, and security levels. The relatedness level is similar to Maslow's friendship, affiliation, and interaction level and his esteem level. Last, the growth level corresponds to Maslow's self-esteem level. As with Maslow's theory, Alderfer believed that it is necessary to fulfill lower-level needs before being interested in fulfilling higher-level needs.

Where Alderfer's theory differs substantially from Maslow's theory is the additional concept of a frustration–regression process. What this means is that if a person cannot fulfill his or her needs at a given level, then the previous lower-level needs regain importance to the individual. For example, if a person is frustrated in his or her growth attempts, it becomes more important for him or her to fulfill relatedness needs. Similarly, if a person is frustrated in his or her relatedness needs, then the existence needs take on increased importance.

The frustration–regression process has particular value for the simulation project manager in motivating team members. If the project manager can recognize that an individual cannot fulfill needs at a particular level, the project manager can keep the person motivated by helping to fulfill his or her lower-level needs. For example, if the project manager cannot give the team member the responsibility or sense of achievement that he or she needs in the growth level, the project manager may still be able to motivate the individual by increasing the individual's opportunity to interact with other team members. This fulfills the frustrated team member's increased need for relatedness because he or she cannot satisfy growth needs.

3.3.3.3   Herzberg's Two-Factor Theory

Herzberg's two-factor theory is slightly more complex than Maslow's theory. This theory has two independently operating factors. The first factor involves what are known as satisfiers or motivators. The theory states that the absence of motivators does not mean that an individual will be highly dissatisfied. However, it does mean that the individual will have a low level of satisfaction. Conversely, the strong presence of motivators helps insure that the individual has the will to perform. Examples of motivators include:

- Feelings of achievement
- Recognition
- Responsibility
- Opportunity for advancement
- Opportunity for personal growth

The second factor involves what are known as hygiene factors. The theory states that an individual who does not have the presence of hygiene factors will be highly dissatisfied. However, the presence of hygiene factors does not mean that the individual will be motivated. There is instead a low level of dissatisfaction. Examples of hygiene factors include:

- Good working conditions
- Pay
- Fringe benefits
- Job security

The Herzberg two-factor theory is illustrated in Figure 3.4.

It is not enough for the project manager to ensure that just hygiene factors or motivators are present in the work environment. To ensure a high level of performance, the project manager must insure that both hygiene factors and motivators are present. Fortunately, most working professionals are likely to already have reasonable pay, benefits, and job security. It is up to the project manager to create a work environment where the team members can have the opportunity to take on responsibility, advance, grow, and feel a sense of accomplishment.

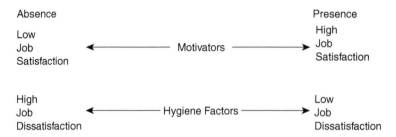

FIGURE 3.4 Herzberg's two-factor theory.

3.3.4 Directing

The fourth project manager function is directing or leading the actions of the simulation project team. This can be particularly difficult because of the nature of the practitioner's relationship with other individuals who may be involved with the project. Thus, the success or failure of the project may depend on the cooperation of a variety of resources over which the practitioner may or may not necessarily have authority.

Authority is defined as the power to command others to act or not to act. With respect to project management, there are two different types of authority. The first type is *de jure* authority. This type of authority is that awarded by some official organizational document. *De jure* authority gives the project manager the legal power to acquire and utilize organizational resources. This type of authority would typically be held by an engineering or production manager.

The second type of authority is *de facto* authority. This type of authority is based on an individual's personal knowledge or skills related to a particular task. *De facto* authority depends on other individuals complying out of respect for the individual's unique knowledge or skills. This type of authority is typically the kind of authority held by some sort of analyst or engineer.

The simulation practitioner may have either *de jure*, *de facto*, or both types of authority. However, the most likely scenario is that the simulation practitioner is not a manager but an analyst or engineer who must undertake the simulation project. As an analyst or engineer without *de jure* authority, the practitioner must take particular care in project management, as he will not have the power to command others to act or not to act. In other words, the practitioner is more likely to have to use his *de facto* authority and interpersonal skills to gain the cooperation of the different individuals involved at different levels and phases in the project.

There are a large number of leadership-motivation theories available to the practitioner. A few of the more popular theories include:

- Expectancy theory
- Equity theory
- Contingency leadership theory
- Hersey–Blanchard situational leadership theory

The practitioner is invited to become familiar with each of these theories. However, of all of these theories, the Hersey–Blanchard situational leadership theory (SLT) appears to be particularly applicable to motivating simulation project team members.

The Hersey-Blanchard SLT stipulates that the leadership technique should be dependent on the technical capability and level of willingness of the project team members (Gibson et al., 2003). Team members can be classified as either capable or not capable of performing the necessary tasks. The team members are also classified as either willing or unwilling to perform the necessary tasks. These two classifications result in a total of four different combinations of team members:

1. Able and willing
2. Able but unwilling
3. Unable but willing
4. Unable and unwilling

The classification of each of the team members will dictate how the project manager should interact with each individual to obtain the best performance.

With able and willing team members, the project manager should feel very fortunate. The team members have the necessary technical proficiency and are motivated to apply these skills to the project. In this situation, the project manager can turn over limited responsibility for different parts of the project. This is a follower-directed type of project management approach. This means that the project manager will delegate responsibility for decisions and implementation to the project team members.

With able but unwilling team members, the project manager is primarily faced with a means of motivating technically proficient team members. In this situation, the project manager must increase the sense of participation and ownership in the project among the project team members. This is considered more of a follower type of project management approach. By sharing ideas on how to accomplish different project tasks and involving the project team members in decisions, the project manager can increase the willingness of the project team.

With unable but willing team members, the project manager has technically unproficient but motivated team members. The major issue is helping to ensure that the team members acquire the necessary technical skills. The project manager must provide explicit instructions on how to accomplish specific tasks. This is considered more of a coaching approach to project management. The project manager must also be readily available for questions that will arise during the course of the project.

With unable and unwilling team members, the project is in the worst possible situation. Not only do the team members not have the technical skills for the project, but they are also not motivated to learn the necessary skills or apply the skills once acquired. In this situation, the project manager should first attempt to change the composition of the project team. However, if this is impossible, the project manager should be prepared to provide explicit instructions and provide close supervision to all of the project tasks. This is considered a leader-directed approach. This approach will increase the probability of taking corrective action before catastrophic damage occurs to the project.

The Hersey–Blanchard situational leadership theory is summed up in Figure 3.5.

3.3.5 Controlling

The final project manager function is the process of controlling the project. The control process consists of:

- Setting project standards
- Observing performance
- Comparing the observed performance with the project standards
- Taking corrective action

The project manager will have already set project standards as part of the time, cost, and technical performance parameters. However, as the project progresses, the project manager may provide additional detailed project standards or adjust the project standards.

Observing performance can take many forms. In the simplest, most informal form, the project manager simply asks the project members for a verbal status report. More formal forms of observing performance included written progress reports and visual presentations.

The project manager must compare the observed performance with the previously defined project standards. This may include comparing the schedule to the current progress, the planned budget with the actual expenditures, and expected and achieved technical performance standards. In each of these parameters, the project manager will need to assess any serious gaps.

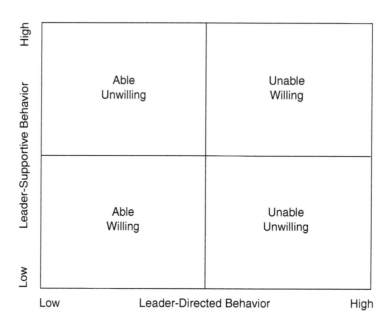

FIGURE 3.5 Situational leadership theory.

Finally, the project manager must take whatever ethical actions are deemed necessary to close the gap between the time, cost, or technical performance standards and the observed performance. The project manager may need to perform personnel, organizational, or technical interventions as necessary.

These actions actually constitute a cycle, as the entire process continuously repeats throughout the project. This means that the time, cost, and technical performance standards may need to be periodically adjusted to get or keep the project on track. The control cycle is illustrated in Figure 3.6.

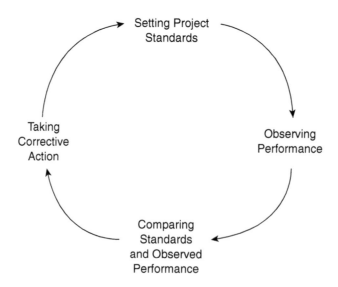

FIGURE 3.6 Control cycle.

3.4 Developing the Simulation Project Plan

The purpose of this section is to provide the practitioner with all of the necessary information to successfully plan the simulation project. Although successful planning does not guarantee that the simulation project will be successful, it does increase the probability that the project resources will be effectively utilized. The project-planning process as a minimum consists of developing:

- A work breakdown structure (WBS)
- A linear responsibility chart (LRC)
- A Gantt chart

In addition to these fundamental project-planning components, subsequent sections also introduce the more sophisticated project network techniques. These techniques are not absolutely necessary to help insure that the simulation project will be planned successfully. However, these techniques are included for reference purposes.

3.4.1 Work Breakdown Structure

A work breakdown structure (WBS) is a successively lower division of project tasks into subtasks. On the first or topmost level, the WBS might consist of only the nine simulation study steps outlined in this handbook. How many lower levels to use is at the discretion of the practitioner. However, for the WBS to be meaningful, it is likely to require at least two levels. The following outline illustrates a two-level WBS for a simulation project.

WBS Activity

1.0 Problem formulation
    1.1   Orientation
    1.2   Problem statement
    1.3   Objectives
2.0 Project planning
    2.1   Work breakdown structure
    2.2   Linear responsibility chart
    2.3   Gantt chart
3.0 System definition
    3.1   Identify system components to model
    3.2   Identify input and output variables
4.0 Input data collection
    4.1   Collect data
    4.2   Analyze data
5.0 Model translation
    5.1   Select modeling language
    5.2   High-level flow chart of model
    5.3   Develop model in selected language
6.0 Verification
    6.1   Debug
    6.2   Animate
7.0 Validation
    7.1   Face validity
    7.2   Statistical validity
8.0 Experimental design
    8.1   Select model alternative configurations
    8.2   Conduct production simulation runs

 9.0 Analysis
     9.1   Conduct *t*-test or analysis of variance
     9.2   Conduct Duncan multiple-ranges test
 10.0 Recommendations and conclusions
     10.1 Recommendations
     10.2 Conclusions

If a third level were utilized, it would be annotated with a second digit. For example, a particular type of input data collection might be represented by the number "4.1.1 Collect order interarrival times." Another type of input data would be represented by "4.1.2 Collect order entry service times."

The level of planning in some projects is so detailed that the division of tasks into subtasks continues until the point at which a single work package is reached. A work package is a discrete unit of work for which authority and responsibility can be assigned. An example of a three-level work breakdown structure for a simulation project is included for reference purposes. This file may be modified by the practitioner as a starting point for the simulation-planning process.

3.4.2 Linear Responsibility Charts

Most business entities utilize an organizational chart to some extent. Even though a trendy organization may deny the existence of such a chart, there will still be some sort of organizational framework in place. These charts depict a hierarchic relationship between different organizational divisions and personnel. Although these types of charts may be of some value, a simulation project is in need of a type of chart that illustrates the relationship between an activity on the WBS and the different individuals involved in the project.

This type of chart is commonly known as a linear responsibility chart (LRC). The LRC shows who participates in each activity and to what degree. For example, there may be a number of individuals involved in collecting data, but only one individual holds primary responsibility for completing the activity. The LRC can be superimposed on the existing WBS as the *x* axis going horizontally across the chart. Across the top each individual's name or title is listed. There is an intersection between the vertical WBS tasks and the horizontal project personnel designations. At each of these intersections, the project manager can place a code corresponding to the individuals' level of participation. Although the simulation project manager may choose any codes, the following codes are commonly employed:

P = primary responsibility
S = secondary responsibility
W = worker
A = approval
R = review

Primary responsibility means that the individual has the primary authority and responsibility for completing that work package. Secondary responsibility is assigned as a backup to the individual with primary responsibility. If utilized in the chart, "W" means that the individual is to assist the primary responsibility individual with completing the task. Approval involves going over the end results of the task and providing feedback to the primary responsibility individual. Review, if utilized, may involve that individual reviewing the results for possible effects on other ongoing activities. Not all tasks are so complex that both a primary and a secondary responsibility individual are required. In addition, if an individual is not involved in a particular task at all, that intersection is simply left blank.

The following table illustrates the use of a LRC with the previously existing WBS.

WBS	Activity	Eng. Mgr.	Analyst 1	Analyst 2	Analyst 3
1.0 Problem formulation					
1.1	Orientation	A	P		
1.2	Problem statement	A	P		
1.3	Objectives	A	P		
2.0 Project planning					
2.1	WBS	A	P	S	
2.2	Linear resp. chart	A	P	S	
2.3	Gantt chart	A	P	S	
3.0 System definition					
3.1	ID system comp.	A	P	S	
3.2	ID I/Output vars.	A	P	S	
4.0 Input data collection					
4.1	Collect data	A	P	S	W
4.2	Analyze data	A	S	P	W
5.0 Model translation					
5.1	Select language	A	P		
5.2	High-level flow chart	A	P		
5.3	Develop model	A	P	S	P
6.0 Verification					
6.1	Debug	A	W	S	P
6.2	Animate	A	W	S	P
7.0 Validation					
7.1	Face validity	A	P	S	
7.2	Statistical validity	A	P	S	
8.0 Experimental design					
8.1	Select configs.	A	P		
8.2	Conduct runs	A	P	S	W
9.0 Analysis					
9.1	*t*-test or ANOVA	A	S	P	
9.2	Duncan MR test	A	S	P	
10.0 Recs. and conclusions					
10.1	Recommendations	A	P	S	
10.2	Conclusions	A	P	S	

Note how, in this case, the engineering manager wants to approve everything. Analyst 1, the practitioner, is mostly responsible for planning and control issues. Analyst 2 is good at statistical analysis. Analyst 3 must be junior to analysts 1 and 2 as well as a skilled programmer.

3.4.3   Scheduling and the Gantt Chart

Up to this point, we have not discussed the project schedule. As we know, the project consists of a number of different activities. Each of these activities is expected to take a certain amount of time. In order to determine how long the project will take and when each project task needs to be started and completed, we can use two different techniques in conjunction with each other. These techniques are:

1. Back planning
2. Gantt chart

3.4.3.1   Back Planning

When the decision is made to proceed with the project, the time parameter of the project can be defined by either a fixed duration or scheduled completion date. In the event of a fixed duration period, there will also be a specified work start date. Either way, a scheduled completion date can be calculated for the project. The concept of back planning involves setting the completion date and working backward according to how much time is required for each of the tasks in the project. If it turns out that the project requires a start date earlier than the current date, the estimate for the length of the individual project task lengths must be modified. On the other hand, if it turns out that the start date is after the current date, two options exist. The first option is to delay the project start until the scheduled start date. This will allow the practitioner to complete other responsibilities between the current date and the scheduled start date for the project. The second alternative is to start the project on the current date and anticipate a completion date before the needed completion date. This option allows the project to slip somewhat in the event that difficulties arise. To assist in the process of scheduling each of the tasks, we can use a Gantt chart.

3.4.3.2   Gantt Chart

A Gantt chart illustrates the durations and relationships among different project activities. The durations of individual tasks are represented by horizontal bars. The relationships between activities are illustrated by connecting lines with arrows between the dependent activities. With a simulation project, the Gantt chart can be appended to the right of the LRC or interchanged with the LRC if presentation space is at a premium. This enables the simulation practitioner to identify the task, individuals involved, duration, and task relationships at a glance.

Project tasks may have what are known as relationships. These are connections between the tasks that dictate what sequences must be observed between a preceding task and a succeeding task. Common relationships include:

- Finish to start
- Start to start
- Finish to finish

3.4.3.2.1   *Finish-to-Start Relationship*

The most common relationship is the finish-to-start relationship. In the finish-to-start relationship, it is necessary for the preceding task to finish before the succeeding task may start. This type of relationship occurs in a simulation project between verification and validation. Verification is the process of ensuring that the model operates as intended. This includes having all the intended components in the model and debugging the model. Validation is the process of ensuring that the model represents reality. In order to test the model to see if it represents reality, if must first be complete and be able to run.

Under most circumstances, the only way to reduce the overall time for a finish-to-start predecessor and a successor is to reduce the individual time of either the predecessor or the successor or both. If the tasks could be performed in parallel, then the relationship would not be finish to start in the first place. A finish-to-start relationship is represented by an arrow that leaves the rightmost part of the predecessor and enters the leftmost part of the successor.

The first-level finish-to-start relationship between the verification and validation phases is illustrated in Figure 3.7.

3.4.3.2.2   *Start-to-Start Relationship*

Less common than the finish-to-start relationship is the start-to-start relationship. The start-to-start relationship means that the predecessor and the successor must start at the same time. This sort of situation may occur when a single previous process splits into two different tasks that must be worked on simultaneously. In the simulation project process there are actually no two tasks that must be started at the same time. It will often be desirable to begin working on two tasks as soon as possible, but it is not absolutely necessary to have any start-to-start relationships. The start-to-start relationship is represented by two arrows feeding into the leading edge of the Gantt chart task bars. The start-to-start relationship between two tasks is illustrated in Figure 3.8. Note that the project manager may specify that the data collection phase and the model translation phase start at the same time. This means that the model can be developed even though not all of the input data have been collected. However, before the model translation phase can be finished, the input data phase has to be completed.

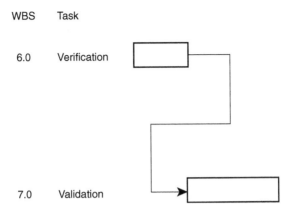

FIGURE 3.7 Finish-to-start relationship.

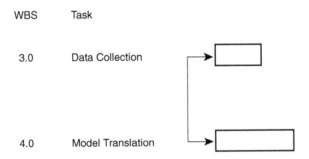

FIGURE 3.8 Start-to-start relationship.

3.4.3.2.3 *Finish-to-Finish Relationship*

A finish-to-finish relationship between a predecessor and a successor is found when both tasks are desired to be completed at the same time. This means that if the tasks are of different duration, then the longer-duration task must be started before the shorter-duration task. The finish-to-finish relationship is most often found in situations where the pot lives of the two processes are limited and must be combined in a following task at the same time.

Like the start-to-start relationship, there is not necessarily an absolute requirement to finish two different tasks at the same time in a simulation project. However, the project manager may determine that simultaneous finish of two tasks is desirable and designate a finish-to-finish relationship. The finish-to-finish relationship is represented by two arrows linking the end of the Gantt chart task bars. Figure 3.9 illustrates a finish-to-finish relationship between the conclusions and recommendation tasks. The project manager can force both the conclusions and recommendations to be complete at the same time in order to be included in the project report.

3.4.3.3 Lags

A possible variation to the above relationships is the concept of a lag. A lag is a required time period between the predecessor and the successor. For example, it may be desirable to begin the input data analysis as soon as possible after the start of the input data collection. However, in order to begin the analysis, it is necessary to have input data to analyze. Thus, for project-planning purposes there could be a small lag between these two tasks. Similarly, it would be expected that the data analysis would finish somewhat behind the data collection process (Figure 3.10).

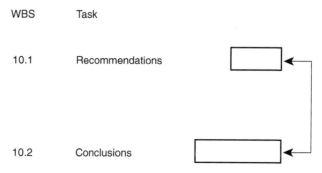

FIGURE 3.9 Finish-to-finish relationship.

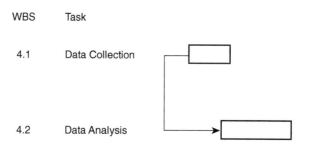

FIGURE 3.10 Lag.

3.5 Compressing Projects

A natural question both during the planning phase and also during the project execution is "How can the project be compressed?"

This question can arise as a result of a requirement to move on to another phase in the overall project or when the project begins to slip behind schedule. An extreme case of compression is known as crashing the projects. This is an analysis of how quickly the project can be complete without respect to cost. In other words, crashing the project optimizes the time parameter of the project at the expense of resource utilization.

Whether the practitioner is simply interested in compressing the project time or crashing the project, the approach is similar. The general objective is to attempt to sequence as many tasks as possible at the same time or in parallel with each other. Because a number of tasks in the simulation project process are in whole or part independent of each other, it is possible to perform significant project compression without affecting the quality of the project. On the other hand, some simulation project tasks simply cannot be significantly compressed because it is necessary to have the predecessor complete before the successor can begin.

Simulation project tasks that can be compressed are problem statement and project planning. The problem statement phase and the project-planning phase can be performed in parallel since the project planning activities are fairly consistent across projects. Regardless of the project, the same tasks will have to be performed, it is probably known who will work on the project, and the relative length of the tasks can be estimated regardless of the overall amount of time allocated to the project.

Simulation project tasks that can be partially compressed are:

- System definition and input data collection/analysis
- Input data collection/analysis and model translation

- Model translation and verification
- Experimental design and analysis
- The rest of the project and the project report/presentation

3.5.1 Partially Compressing the System Definition and Input Data Collection Phases

Once the system definition process has determined what kind of input data is necessary, the practitioner can begin the input data collection process. The practitioner does not need to know every component of the system that is going to be modeled. Similarly, the practitioner does not need to know what output measures of performance will be used when collecting input data.

3.5.2 Partially Compressing the Input Data Collection and Model Translation Phases

Even though the practitioner has not completed the input data collection and analysis phase, the practitioner can still develop a high-level model flow chart and begin the model translation. When it is necessary to provide an input data parameter, the practitioner can use a deterministic value or a dummy distribution so that the modeling does not become delayed. When the input data parameters are available, they can be inserted. As long as the actual input data parameters are inserted before the validation phase, the practitioner can continue to develop the model in parallel with the input data collection and analysis process.

3.5.3 Partially Compressing the Model Translation and Verification Phases

Practitioners are recommended to utilize the "divide and conquer" approach to model development. This means that the practitioner is going to build the model small pieces at a time. The small pieces are easier to control and debug. A practitioner following this approach is actually performing some of the model translation phase activities at the same time the model is being verified as operating as intended.

3.5.4 Partially Compressing the Experimental Design and Analysis Phases

In the experimental design phase, the practitioner determines which alternative configurations should be examined. In the analysis phase, the practitioner performs statistical comparative analysis on the alternative configurations. On the surface, it might seem impossible to begin the analysis phase if the experimental design phase is not complete. However, the analysis phase actually consists of several steps. As the practitioner decides which alternative configurations will be utilized, it is possible to begin immediately performing the preliminary steps in the analysis phase. As the remaining alternative configurations are identified, the practitioner can completely move into the analysis phase.

3.5.5 The Rest of the Project and the Project Report/Presentation

The simulation project report and presentation contain a significant amount of information that is available early in the project. The sections such as the problem statement, project planning, system definition, input data collection and analysis, model translation, verification, validation, and experimental design can all be incorporated into the project report and presentation while the practitioner is still analyzing the project experimental data. By working on the project report and presentation as they occur, the practitioner can save a significant amount of time over sequentially preparing the report and presentation after the project is nearly complete. The experimental results and the conclusions can be inserted into blank spaces left in the report and the presentation.

Simulation project tasks that cannot be compressed significantly are verification and validation.

As previously discussed, verification is the process of insuring that the model operates as intended, and validation is the process of insuring that the model represents reality. Because the model cannot be determined to represent reality until it operates properly, it is not possible to validate the model before it has been completely verified.

3.6   Example Gantt Chart

The following simplified Gantt chart illustrates a project that has been compressed as previously described. The relative durations of the tasks are indicated by shading. The time periods are listed as 1 through 12. Tasks that can be performed in parallel are shaded in the same time periods. Tasks that can be compressed to some extent have overlapping shaded time periods. Tasks that have finish-to-start relationships and cannot be performed simultaneously have separately shaded time periods. In reality, the length of each time period will have to be estimated by the practitioner.

WBS	Activity	1	2	3	4	5	6	7	8	9	10	11	12
1.0	Problem formulation												
1.1	Orientation												
1.2	Prob. stat.												
1.3	Objectives												
2.0	Project planning												
2.1	WBS												
2.2	LRC												
2.3	Gantt chart												
3.0	System definition												
3.1	Id. comp. to model												
3.2	Id. I/O variables												
4.0	Input data collection												
4.1	Collect data												
4.2	Analyze data												
5.0	Model translation												
5.1	Select language												
5.2	Flowchart of model												
5.3	Develop model												
6.0	Verification												
6.1	Debug												
6.2	Animate												
7.0	Validation												
7.1	Face validity												
7.2	Statistical validity												
8.0	Experimentation												
8.1	Configurations												
8.2	Simulation runs												
9.0	Analysis												
9.1	*t*-Test or ANOVA												
9.2	Duncan test												
10.0	Recs. and conclusions												
10.1	Recommendations												
10.2	Conclusions												

3.7   Advanced Project Management Concepts

Many of the more advanced project management concepts are outside the scope of this handbook. For more in-depth descriptions of some of these techniques, practitioners are directed to any one of a number of project management or operations management references such as Meredith and Mantel (2000) or

Krajewski and Ritzman (2001). We briefly review the critical path method (CPM) and program evaluation and review technique (PERT).

PERT and CPM were both developed in the 1950s as project management tools. The primary difference between PERT and CPM lies in their abilities to handle randomness in managing project tasks. CPM focuses on project task durations that are deterministic or, in other words, do not vary in time. Conversely, PERT was specifically designed to handle variations in project task durations. Fifty years later, the distinction between PERT and CPM has blurred, and sometimes these techniques are referred to as PERT/CPM. For the remainder of this section, we refer to these techniques as networks. To describe the project in network terminology, we have:

- Activities
- Events
- Nodes
- Arcs
- Paths
- Critical path

An activity is a task that requires time to complete. An event is defined as the completion of one or more activities at a particular time. A project is represented by a network consisting of nodes and arcs. The nodes in the network are connected by one or more arcs. Each arc has a directional arrow. The arc arrow represents the required sequence of the activities. A path is a sequence of nodes and arcs. The critical path is the path that, if delayed, results in an increase in the overall project time.

The project generally starts with a node on the left and proceeds through the network to the right. The rightmost node signals the completion of the project. This concept is illustrated in Figure 3.11.

The actual representation of the project with respect to the nodes and arcs can take two different forms:

1. Activity on node (AON)
2. Activity on arc (AOA)

3.7.1 Activity on Node

In the activity-on-node or AON network approach, the project activities are represented by the nodes on the network. The AON-type network is illustrated in Figure 3.12.

In this AON representation, we start at the leftmost node. From the start node, two arcs are possible. There is an arc to node 1 and there is an arc to node 2. This means that there is no precedence relationship between the activities represented by nodes 1 and 2. There is also a single arc leaving node 1 and a single arc leaving node 2. Both of these arcs lead to node 3. This representation means that in order to begin the activity represented by node 3, both the activities represented by node 1 and node 2 must first be completed. Once the activity at node 3 is complete, there is one arc that leads to node 4. The activity on node 4 can begin only when the activity represented by node 3 is completed. When the activity on node 4 is completed, there are two different arcs. One arc leads directly to the finish, and the other arc leads

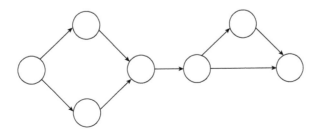

FIGURE 3.11 Project network representation.

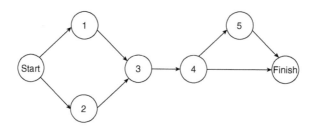

FIGURE 3.12 Activity-on-node representation.

to node 5. Before either of these activities can be conducted, the activity on node 4 must be completed. There is a single arc from node 5 that also leads to the finish node. This means that in order to finish the network, both the activity of node 4 and the activity of node 5 must first be completed.

3.7.2 Activity on Arc

The activity-on-arc (AOA) project representation utilizes nodes to represent finished events and arcs to represent activities. This means that both the nodes and arcs are identified. The AOA approach is illustrated in Figure 3.13.

   With this representation, we start at event node 1. At node 1 two different activity arcs lead to event nodes 2 and 3. Arc A leads between node 1 and node 2. Arc B leads between node 1 and node 3. This means that both of these activities can be performed at the same time. The next node is node 4. A single arc, C, leads from node 2 to node 4. Another single arc, D, leads from node 3 to node 4. This means that both activity arc C and activity arc D must be completed in order to reach event node 4. Once at event node 4, there is activity arc E to event node 5. At event node 5 there are activity arcs F to node 6 and G to the finish node. Arc F and arc G can be performed at the same time. At event node 6, there is a final activity arc, H, to the finish event. Before the finish event can be reached, both activity arcs G and H must be complete.

   With the AOA approach it is sometimes necessary to insert a dummy activity. This is an activity that actually has no duration and uses no resources. Its sole purpose is to provide a logical connection between two nodes. Dummy activities are typically represented by a dashed-line arc. The orientation of the arc still has significance in terms of precedence. For example, two different activities may be needed for the same event node. Suppose, in the previous example, both activites A and B are needed for event 3. In this case, the original activity arc B still exists, but there is a new dummy activity arc with no label between event node 2 and 3. This situation is illustrated in Figure 3.14.

   Of the two types of approaches there is some indication that practitioners prefer the AOA method over the AON method. The thought behind this preference is that the AOA approach is somewhat more intuitive than the AON approach. We utilize the AOA approach to illustrate our simulation project network example.

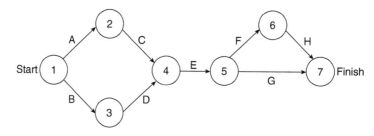

FIGURE 3.13 Activity-on-arc representation.

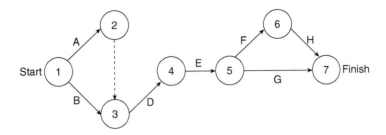

FIGURE 3.14 Dummy arc.

3.7.3 Simulation Project Network Example

We use the AOA approach to illustrate one possible network for part of a simulation project (Figure 3.15). In this network, the arcs correspond to the following first-level project tasks:

A = Problem statement
B = Project planning
C = System definition
D = Input data
E = Model translation
F = Verification
G = Validation
H = Experimental design
I = Analysis
J = Conclusions

In this particular simulation project network model, we begin by simultaneously working on activity A, the problem formulation phase, and activity B, the project plan. We can begin the project planning before completing the problem formulation phase because we are already familiar with most of the simulation project tasks. However, before we can reach event node 3 to begin activity C, the system definition, we must complete both activity A, the problem formulation, and activity B, the project plan.

Once both of these are completed at node 3, we can proceed with activity C, the system definition. When we complete activity C, the system definition, we know what we need to model, and we know what data need to be collected. We can begin these tasks at the same time as represented by activity D, input data, and activity E, model translation. While we are building the model, we can use dummy or estimate input data until the actual data are ready. When both activity D, input data, and activity E, model translation, are compete, we reach event node 6.

With the initially complete model, we are in the verification phase represented by activity F. On the completion of activity F, verification, we are at event node 7. With the verified model, we can begin the validation process represented by activity arc G. With the validated model at event node 8, we are ready to develop our experimental model alternatives with activity arc H. When we have identified our

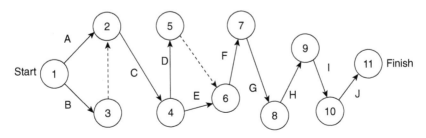

FIGURE 3.15 Simulation project network.

experimental models, we reach node 9. At node 9, we can begin the model analysis process including replication analysis, production runs, and statistical comparison of alternatives represented by activity arc I. When this is complete at node 10, we can complete our conclusions, write the report, and present the project results in activity arc J. At node 11, the project is complete.

This network is only one of many ways that the simulation project might have been represented. Depending on practitioner's willingness to compress some tasks, it would be possible to develop a more complicated network.

3.7.4   Calculating the Critical Path

Once we have developed our AOA network and have estimated the duration of each activity arc in the network, we can consider calculating the critical path through the network. As we have previously defined, the critical path is the list of activities that, if delayed, will extend the overall length of time to complete the project.

To compute the critical path, we will need to calculate the following values for each activity in the network:

ES = Earliest start time
EF = Earliest finish time
LS = Latest starting time
LF = Latest finish time

For each of the activities in the network, we can easily calculate the earliest finish time as:

$$EF = ES + \text{task duration}$$

Similarly, the latest start time for each of the activities in the network can be calculated with the following equation:

$$LS = LF - \text{task duration}$$

In general, we make a forward pass and a backward pass to calculate all of theses values and eventually obtain the critical path. In the simplest implementation of calculating the critical path, we can use deterministic estimates of each of the activity durations. For our activities we can use:

Activity	Task	Duration
A	Problem formulation	5
B	Project planning	2
C	System definition	5
D	Input data collection and analysis	25
E	Model translation	20
F	Verification	10
G	Validation	5
H	Experimental design	2
I	Analysis	10
J	Conclusion, report, presentation	10

We now begin our calculations with a chart with the following headings:

- Activity
- Duration
- Predecessor
- ES
- EF

We begin by listing the data for activities A and B. As the network chart indicated, both of these activities do not have any predecessors. Because neither of these activities has a predecessor, the earliest start date is time 0. The earliest finish times for each activity are the same as the duration of the activities.

Activity	Name	Duration	Predecessor	ES	EF
A	Prob. Formulation	5	None	0	5
B	Proj. Planning	2	None	0	2

The next activity is C, System Definition. This activity has both activity A, Problem Formulation, and activity B, Project Planning, as predecessors. This means that the earliest start for activity C, System Definition, is the later of activities A and B. Because activity A has the earliest finish of 5, the earliest start for activity C is 5. The earliest finish for the System Definition is the earliest start plus the duration of the activity. Thus, the earliest finish is 10 days.

Activity	Name	Duration	Predecessor	ES	EF
A	Prob. formulation	5	None	0	5
B	Proj. planning	2	None	0	2
C	System definition	5	A, B	5	10

Activities D, Input Data Collection and Analysis, and E, Model Translation, both have activity C, System Definition, as a predecessor. Because the earliest finish date of activity C, System Definition, is 10, the earliest start date of activity D, Input Data Collection, and activity E, Model Translation, is 10. The duration of the Input Data Collection and Analysis activity is 25, so the earliest finish date is 35. Similarly, the duration of the Model Translation phase is 20 days, and the earliest finish date is 30.

Activity	Name	Duration	Predecessor	ES	EF
A	Prob. formulation	5	None	0	5
B	Proj. planning	2	None	0	2
C	System definition	5	A, B	5	10
D	Input data	25	C	10	35
E	Model translation	20	C	10	30

Activity F, Verification, requires that both activity D, Input Data Collection and Analysis, and activity E, Model Translation, be complete. Because the later of these two activities is 35 days, activity F, Verification, cannot begin until the end of activity D, Input Data Collection and Analysis.

Activity	Name	Duration	Predecessor	ES	EF
A	Prob. formulation	5	None	0	5
B	Proj. planning	2	None	0	2
C	System definition	5	A, B	5	10
D	Input data	25	C	10	35
E	Model translation	20	C	10	30
F	Verification	10	D, E	35	45

Activity G, Validation, has activity F, Verification, as a predecessor. Since activity F, Verification, does not end until 45, the earliest start date for activity G, Validation, is 45. The duration of activity G is 5 days, so the earliest finish date is 50 days.

Activity	Name	Duration	Predecessor	ES	EF
A	Prob. formulation	5	None	0	5
B	Proj. planning	2	None	0	2
C	System definition	5	A, B	5	10
D	Input data	25	C	10	35
E	Model translation	20	C	10	30
F	Verification	10	D, E	35	45
G	Validation	5	F	45	50

Activity H, Experimental Design, has activity G, Validation, as a predecessor. Because activity G, Validation, does not end until 50, the earliest start date for activity H, Experimental Design, is 50. The duration of activity H is 2 days, so the earliest finish date is 52 days.

Activity	Name	Duration	Predecessor	ES	EF
A	Prob. formulation	5	None	0	5
B	Proj. planning	2	None	0	2
C	System definition	5	A, B	5	10
D	Input data	25	C	10	35
E	Model translation	20	C	10	30
F	Verification	10	D, E	35	45
G	Validation	5	F	45	50
H	Exper. design	2	C, F	50	52

Activity I, Analysis, has activity H, Experimental Design, as a predecessor. Because the earliest finish time for activity H, Experimental Design, is 52, the earliest start date for activity I, Analysis, is 52. The duration of activity I is 10 days, so the earliest finish date is 62 days.

Activity	Name	Duration	Predecessor	ES	EF
A	Prob. formulation	5	None	0	5
B	Proj. planning	2	None	0	2
C	System definition	5	A, B	5	10
D	Input data	25	C	10	35
E	Model translation	20	C	10	30
F	Verification	10	D, E	35	45
G	Validation	5	F	45	50
H	Exper. design	2	C, G	50	52
I	Analysis	10	H	52	62

Last, Activity J, Report and Presentation, has activity I, Analysis, as a predecessor. Because the earliest finish time for activity I, Analysis, is 62, the earliest start date for activity J, Report and Presentation, is 62. The duration of activity J is 10 days, so the earliest finish date for activity J, Report and Presentation, is 72 days.

Activity	Name	Duration	Predecessor	ES	EF
A	Prob. formulation	5	None	0	5
B	Proj. planning	2	None	0	2
C	System definition	5	A, B	5	10
D	Input data	25	C	10	35
E	Model translation	20	C	10	30
F	Verification	10	D, E	35	45
G	Validation	5	F	45	50
H	Exper. design	2	C, G	50	52
I	Analysis	10	H	52	62
J	Report and pres.	10	I	62	72

This completes the calculation of the critical path. The minimum amount of time that this particular project can be completed in is a total of 72 days. Once we have determined the length of the critical path, we can perform our backward pass. We can now add the latest start and the latest finish columns to our table. We begin the backward pass by fixing the latest finish value of activity J at 72 days. Because activity J, Report and Presentation, is 10 days in duration, the latest start for activity J is 62 days. From this point on, the latest finish date for each remaining task is generally going to be the earliest of all of the latest start dates of its successors.

Activity	Name	Duration	Predecessor	ES	EF	LS	LF
A	Prob. formulation	5	None	0	5		
B	Proj. planning	2	None	0	2		
C	System definition	5	A, B	5	10		
D	Input data	25	C	10	35		
E	Model translation	20	C	10	30		
F	Verification	10	D, E	35	45		
G	Validation	5	F	45	50		
H	Exper. design	2	C, G	50	52		
I	Analysis	10	H	52	62		
J	Report and pres.	10	I	62	72	62	72

The only successor to activity I, Analysis, is activity J, Report and Presentation. Because the latest start date of activity J is 62, the latest finish date of activity I, Analysis, is 62. The latest start date of activity I, Analysis, is 52 days.

Activity	Name	Duration	Predecessor	ES	EF	LS	LF
A	Prob. formulation	5	None	0	5		
B	Proj. planning	2	None	0	2		
C	System definition	5	A, B	5	10		
D	Input data	25	C	10	35		
E	Model translation	20	C	10	30		
F	Verification	10	D, E	35	45		
G	Validation	5	F	45	50		
H	Exper. design	2	C, G	50	52		
I	Analysis	10	H	52	62	52	62
J	Report and pres.	10	I	62	72	62	72

Activity H, Experimental Design, has only one successor. This is activity I, Analysis. The latest start time for activity I, Analysis, is 52 days. This makes the latest finish date for activity H, Experimental Design, 52 days. The latest start date for activity H becomes 52 − 2 days or 50 days.

Activity	Name	Duration	Predecessor	ES	EF	LS	LF
A	Prob. formulation	5	None	0	5		
B	Proj. planning	2	None	0	2		
C	System definition	5	A, B	5	10		
D	Input data	25	C	10	35		
E	Model translation	20	C	10	30		
F	Verification	10	D, E	35	45		
G	Validation	5	F	45	50		
H	Exper. design	2	G	50	52	50	52
I	Analysis	10	H	52	62	52	62
J	Report and pres.	10	I	62	72	62	72

Activity G, Validation, has one successor, activity H, Experimental Design. The latest start time for activity H is 50 days. This means that the latest finish date for activity G, Validation, is 50 days. The latest start date for activity G, Validation, becomes 45 days.

Activity	Name	Duration	Predecessor	ES	EF	LS	LF
A	Prob. formulation	5	None	0	5		
B	Proj. planning	2	None	0	2		
C	System definition	5	A, B	5	10		
D	Input data	25	C	10	35		
E	Model translation	20	C	10	30		
F	Verification	10	D, E	35	45		
G	Validation	5	F	45	50	45	50
H	Exper. design	2	C, G	50	52	50	52
I	Analysis	10	H	52	62	52	62
J	Report and pres.	10	I	62	72	62	72

The only successor to activity F, Verification, is activity G, Validation. Because the latest start date for activity G is 45 days, the latest finish date for activity F, Verification, is also 45 days. The latest start date for activity F, Verification, is 45 − 10 days or 35 days.

Activity	Name	Duration	Predecessor	ES	EF	LS	LF
A	Prob. formulation	5	None	0	5		
B	Proj. planning	2	None	0	2		
C	System definition	5	A, B	5	10		
D	Input data	25	C	10	35		
E	Model translation	20	C	10	30		
F	Verification	10	D, E	35	45	35	45
G	Validation	5	F	45	50	45	50
H	Exper. design	2	C, G	50	52	50	52
I	Analysis	10	H	52	62	52	62
J	Report and pres.	10	I	62	72	62	72

Activity E, Model Translation, has only activity F, Verification, as a successor. The latest start time for activity F, Verification, is 35 days. This means that the latest finish time for activity E, Model Translation, is 35 days. The duration of activity E is 20 days. The latest start date for activity E, Model Translation, becomes 15 days.

Activity	Name	Duration	Predecessor	ES	EF	LS	LF
A	Prob. formulation	5	None	0	5		
B	Proj. planning	2	None	0	2		
C	System definition	5	A, B	5	10		
D	Input data	25	C	10	35		
E	Model translation	20	C	10	30	15	35
F	Verification	10	D, E	35	45	35	45
G	Validation	5	F	45	50	45	50
H	Exper. design	2	C, G	50	52	50	52
I	Analysis	10	H	52	62	52	62
J	Report and pres.	10	I	62	72	62	72

Activity D, Input Data Collection and Analysis, has only one successor. This is activity F, Verification. The latest start time for activity F, Verification, is 35 days. The latest finish time for activity D, Input Data Collection and Analysis, becomes 35 days. The duration of activity D is 25 days. This means that the latest start date for activity D, Input Data Collection and Analysis, becomes 10 days.

Activity	Name	Duration	Predecessor	ES	EF	LS	LF
A	Prob. formulation	5	None	0	5		
B	Proj. planning	2	None	0	2		
C	System definition	5	A, B	5	10		
D	Input data	25	C	10	35	10	35
E	Model translation	20	C	10	30	15	35
F	Verification	10	D, E	35	45	35	45
G	Validation	5	F	45	50	45	50
H	Exper. design	2	C, G	50	52	50	52
I	Analysis	10	H	52	62	52	62
J	Report and pres.	10	I	62	72	62	72

The latest start and latest finish times for activity C, System Definition, are somewhat more complicated to determine than the previous activities. Activity C actually has two successors. The first successor is activity D, Data Collection and Analysis. The second successor is activity E, Model Translation. The latest start time for activity D is 10, and the latest start time for activity E is 15. Because we need the earliest of the latest start times, the latest finish time for activity C, System Definition, becomes 10 days. Because the duration of activity C is 5 days, the latest start date is 5 days.

Activity	Name	Duration	Predecessor	ES	EF	LS	LF
A	Prob. formulation	5	None	0	5		
B	Proj. planning	2	None	0	2		
C	System definition	5	A, B	5	10	5	10
D	Input data	25	C	10	35	10	35
E	Model translation	20	C	10	30	15	35
F	Verification	10	D, E	35	45	35	45
G	Validation	5	F	45	50	45	50
H	Exper. design	2	C, G	50	52	50	52
I	Analysis	10	H	52	62	52	62
J	Report and pres.	10	I	62	72	62	72

Activity B, Project Planning, has only one successor, activity C, System Definition. The latest start time for activity C is 5, so the latest finish time for activity B, Project Planning, becomes 5 days. The duration for activity B is 2 days, so the latest start date becomes 3 days.

Activity	Name	Duration	Predecessor	ES	EF	LS	LF
A	Prob. formulation	5	None	0	5		
B	Proj. planning	2	None	0	2	3	5
C	System definition	5	A, B	5	10	5	10
D	Input data	25	C	10	35	10	35
E	Model translation	20	C	10	30	15	35
F	Verification	10	D, E	35	45	35	45
G	Validation	5	F	45	50	45	50
H	Exper. design	2	C, G	50	52	50	52
I	Analysis	10	H	52	62	52	62
J	Report and pres.	10	I	62	72	62	72

The final activity, activity A, Problem Formulation, has only one successor. This is activity C, System Definition. The latest start date of activity C, System Definition, is 5 days. This means that the latest finish date for activity A is also 5 days. Because activity A is 5 days in duration, the latest start date is 0 days.

Activity	Name	Duration	Predecessor	ES	EF	LS	LF
A	Prob. formulation	5	None	0	5	0	5
B	Proj. planning	2	None	0	2	3	5
C	System definition	5	A, B	5	10	5	10
D	Input data	25	C	10	35	10	35
E	Model translation	20	C	10	30	15	35
F	Verification	10	D, E	35	45	35	45
G	Validation	5	F	45	50	45	50
H	Exper. design	2	C, G	50	52	50	52
I	Analysis	10	H	52	62	52	62
J	Report and pres.	10	I	62	72	62	72

The backward pass is complete. We can now calculate what is known as the slack for each activity. This is the difference between the latest start and the earliest start. It is also the difference between the latest finish and the latest start. The slack is listed in an additional column in our table.

Activity	Name	Duration	Predecessor	ES	EF	LS	LF	Slack
A	Prob. formulation	5	None	0	5	0	5	0
B	Proj. planning	2	None	0	2	3	5	3
C	System definition	5	A, B	5	10	5	10	0
D	Input data	25	C	10	35	10	35	0
E	Model translation	20	C	10	30	15	35	5
F	Verification	10	D, E	35	45	35	45	0
G	Validation	5	F	45	50	45	50	0
H	Exper. design	2	C, G	50	52	50	52	0
I	Analysis	10	H	52	62	52	62	0
J	Report and pres.	10	I	62	72	62	72	0

The critical path can be identified by the activities with 0-day values in the slack column. This means that the critical path is:

- Activity A – Problem formulation
- Activity C – System definition
- Activity D – Input data collection and analysis
- Activity F – Verification
- Activity G – Validation
- Activity H – Experimental design
- Activity I – Analysis
- Activity J – Report and presentation

If any of the above critical path tasks exceeds the estimated activity duration, the duration of the whole project will be extended. Conversely, any of the noncritical task durations can be extended by the slack time without extending the overall length of the project. If any of the noncritical tasks do exceed slack time, then these tasks become critical. These previously noncritical tasks that have become critical can now extend the project if their activity times are extended.

The critical path calculated for this particular simulation project example will differ from that for any other simulation project if the precedence of activities or the duration of the activities is changed. If the practitioner decides to utilize this sort of network technique, a different network and activity table will have to be generated. The critical path for this new network will also have to be resolved.

3.8   Project Management Software Packages

For demanding simulation projects the practitioner/project manager may want to consider utilizing a project management-specific software package. These packages provide great assistance in developing the work breakdown structure, the linear responsibility chart, and the Gantt chart. In addition, a project management software package is the only type of automated tool that can assist the practitioner/project manager with the more advanced project management techniques.

3.8.1   Advantages to Using Project Management Software

The advantages to using a project management software package are many. For the simulation practitioner, a few of the more important advantages are:

- Integrated project documents
- Easy updates or changes in the project
- Actual performance tracking
- Presentation and report graphics
- Project progress reports

As we have determined, effectively managing a project can become a complex process. Because the software package offers a single integrated solution to the practitioner/project manager, it will be easier to keep track of different aspects of the project. Instead of having to consult between individual word processor, spreadsheet, and presentation graphics software packages, the project manager has to contend with only the project management software package.

It is relatively easy to provide updates or changes in the project work breakdown structure, linear responsibility chart, and Gantt chart. These changes can include the addition or deletion of work packages, reassignment of responsibilities, or schedule changes. Any changes that are made to the task sequencing will be automatically updated with respect to the relationships to the other tasks.

Another significant advantage is that it is possible to use the software package to compare actual performance with planned performance. Many project managers create a Gantt chart and schedule at the beginning of the project but never consistently track the progress of the project. Because the progress on individual tasks can be easily tracked with a project management software package, it is more likely that the project manager will track the project.

Project management software packages can also create very refined graphics associated with the project. It is simple to cut and paste the different types of individual graphics produced by these software packages into electronic presentations and project reports. It is even possible to compress a high-level work breakdown structure, linear responsibility chart, and Gantt chart into a single screen image for use in other documents.

Finally, project management software can simplify the project-reporting process. It is relatively easy to generate a wide variety of reports related to the project parameters of time, cost, and technical performance. Once a particular type of project report is set up, updated copies of the reports can be easily generated. This can substantially reduce the administrative burden of reporting the project status to the project stakeholders.

The use of project management software packages has enabled some fairly remarkable projects to be completed, particularly in the construction field. In one legendary example, a fast food franchise outlet was destroyed during the devastating Los Angeles riots. In an effort to restore a market presence quickly, the franchise chain used project management software to crash the construction project. As you may recall, crashing means executing the project in the least possible amount of time without respect to cost. By carefully sequencing and preparing successor tasks, it is rumored that the franchise chain was able to rebuild the restaurant in 24 hours. It is unknown how much additional cost the franchise chain incurred in order to execute this feat. However, the restaurant did not have any fast food competition for a significant period of time.

3.8.2   Disadvantages to Using Project Management Software

There are at least two possible disadvantages to using project management software for typical simulation projects. These disadvantages are:

1. A steep learning curve
2. The temptation to overmanage the simulation project

The first disadvantage is related to the usually steep project management software package learning curve. Part of this is because project management software is significantly different from any other type of software the practitioner is likely to have used. Even relatively simple functions can present some difficulties for someone who has never used this type of software. For some of the more complex project management functions, using the software can actually be a project in itself.

The second potential disadvantage is that project management software can promote the overmanagement of small simulation projects. This is particularly dangerous for simulation practitioners who operate either by themselves or with a relatively small project team. In these situations, the practitioner/ project manager is likely to be more of a practitioner than a project manager. This category of practitioner/ project manager simply cannot afford to spend the same amount of time or gain the level of familiarity with the software as a dedicated or professional project manager. If an excess amount of time is spent overmanaging the project, the technical performance aspects of the project may be compromised. Even experienced project managers can be observed infinitely adjusting the project schedule and continuously updating the actual progress when their time could have been more effectively expended.

3.8.3   Major Software Packages

Currently, a large number of software management packages are available for the personal computer. Perhaps the two best known project management software packages are:

- Microsoft Project
- Primavera systems

Microsoft Project is a lower-end but fully capable project management software package that is readily available in any software store. It has been through many upgrades. The current version at the time the handbook went to press was Microsoft Project 2002. A 60-day trial period use of the standard edition of the software package was available for $8.00 through the Microsoft Project homepage at: *www.Microsoft.com/office/project/default.asp*. A typical MS Project screen is illustrated in Figure 3.16.

Primavera actually offers a number of different versions. The high-end software package Primavera Project Planner is designed for multiple complex projects with multiple simultaneous users. The commercial version retails for approximately $4000. The lower-end version of the Primavera software is SureTrack Project Manager. It was available at the time of publication for $499.00. A screen capture of Primavera's Project Planner is illustrated in Figure 3.17.

Many of the less known but still capable project management software packages are available over the Internet with cost-free trial periods. A particularly inexpensive piece of project management software that has received favorable reviews by a number of different organizations is the MinuteMan Systems Project Management software. At the time of publication, this software was available for $49.00. Additional details on this software package can be obtained from www.minute-man-systems.com. A screen capture of MinuteMan Systems Project Management software is illustrated in Figure 3.18.

3.9   Summary

We began this chapter by discussing the need for the practitioner to acquire basic project manager skills. The ability to apply these skills in a simulation project can have a significant effect on the success or

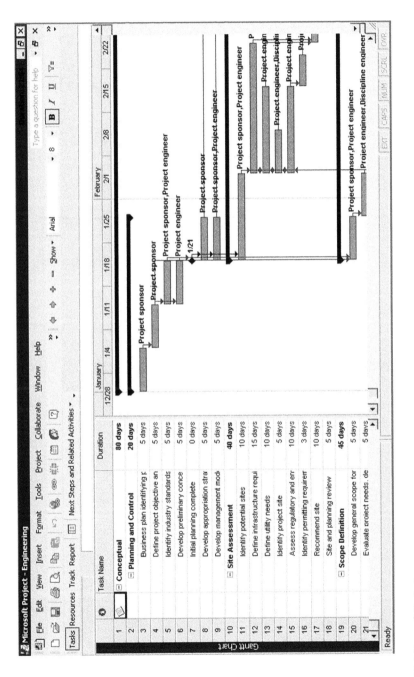

FIGURE 3.16 Microsoft Project.

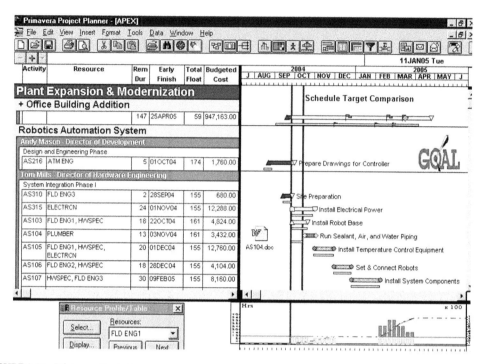

FIGURE 3.17   Primavera Project.

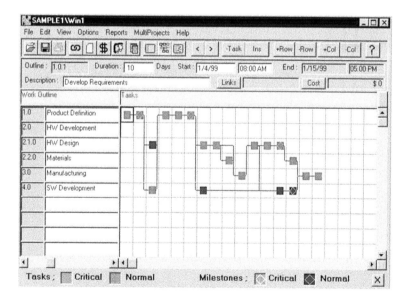

FIGURE 3.18   MinuteMan Systems Project Management.

failure of the project. To provide the practitioner with the necessary skills, we covered basic project management concepts, project stakeholders, project manager functions, work breakdown structures, linear responsibility charts, and Gantt charts.

The basic project management concepts included a discussion of the time, cost, and technical performance project parameters. The interdependent relationship among the three parameters was illustrated. The project stakeholder discussion included the identification of internal stakeholders, external stakeholders, and stakeholder strategies. The project manager function discussion focused on planning, organizing, motivating, directing, and controlling the project team. Planning includes the development of a work breakdown structure and Gantt chart. Organizing includes acquiring resources and creating a linear responsibility chart. The motivation function was presented through the use of a number of accepted motivation theories. Leadership styles were discussed as part of the directing function. The final management function discussion included material on controlling the performance of project resources.

Additional step-by-step detail was provided in Section 3.4.1 on work breakdown structure to illustrate how to divide the project into individually assignable tasks or work packages. The linear responsibility section (Section 3.4.2) provided information on how to assign responsibility to the individual work packages. The chapter concluded with a discussion of Gantt chart sequencing and relationships.

The practitioner can utilize these basic concepts to help plan and manage the simulation project. With proper project planning, the practitioner can increase the probability of achieving the desired level of project performance. Blank work breakdown structures, linear responsibility charts, and Gantt charts are available at the end of the chapter to assist the practitioner in developing a proper project management plan.

Chapter Problems

1. Why may a practitioner need to know something about project management?

2. What are the five management functions?

3. What is the difference between *de jure* and *de facto* authority?

4. What is the minimum number of levels necessary for a WBS to be useful?

5. Why is a linear responsibility chart superior to the traditional organizational chart?

6. What two things does a Gantt chart tell you about each specific task?

7. Why might there be a lag between two tasks?

8. What is a work package?

9. What is meant by crashing a project?

10. What does critical path mean?

Sample LRC

Work breakdown structure	Project activity	Team member 1	Team member 2	Team member 3	Team member 4
1.0	Problem formulation				
1.1	Orientation				
1.2	Problem statement				
1.3	Objectives				
2.0	Project planning				
2.1	Work breakdown structure				
2.2	Linear responsibility chart				
2.3	Gantt chart				
3.0	System definition				
3.1	Identify system components to model				
3.2	Identify input and output variables				
4.0	Input data collection				
4.1	Collect data				
4.2	Analyze data				
5.0	Model formulation				
5.1	Select modeling language				
5.2	High-level flowchart of model				
5.3	Develop model in selected language				
6.0	Verification				
6.1	Debug				
6.2	Animate				
7.0	Validation				
7.1	Face validity				
7.2	Statistical validity				
8.0	Experimentation				
8.1	Select model alternative configurations				
8.2	Conduct production simulation runs				
9.0	Analysis				
9.1	Conduct *t*-test or analysis of variance				
9.2	Conduct Duncan multiple-ranges test				
10.0	Recommendations and conclusions				
10.1	Recommendations				
10.2	Conclusions				

Code:

A = Approval; P = Primary Responsibility; S = Secondary Responsibility; R = Review.

Sample Gantt Chart

WBS	Activity	1	2	3	4	5	6	7	8	9	10	11	12
1.0	Problem formulation												
1.1	Orientation												
1.2	Prob. stat.												
1.3	Objectives												
2.0	Project planning												
2.1	WBS												
2.2	LRC												
2.3	Gantt chart												
3.0	System definition												
3.1	Id. comp. to model												
3.2	Id. I/O variables												
4.0	Input data collection												
4.1	Collect data												
4.2	Analyze data												
5.0	Model translation												
5.1	Select language												
5.2	Flowchart of model												
5.3	Develop model												
6.0	Verification												
6.1	Debug												
6.2	Animate												
7.0	Validation												
7.1	Face validity												
7.2	Statistical validity												
8.0	Experimentation												
8.1	Configurations												
8.2	Simulation runs												
9.0	Analysis												
9.1	t-Test or ANOVA												
9.2	Duncan test												
10.0	Recs. and conclusions												
10.1	Recommendations												
10.2	Conclusions												

References

Cleland, D.I. and Ireland, L.R. (2002), *Project Management: Strategic Design and Implementation,* 4th ed., McGraw-Hill, New York.

Gibson, J.L. et al. (2003), *Organizations: Behavior, Structure, and Processes,* 11th ed., McGraw-Hill, New York.

Krajewski, L.J. and Ritzman, L.P. (2001), *Operations Management,* Pearson Education, New York.

Meredith, J.R. and Mantel, S. (2000), *Project Management,* John Wiley & Sons, New York.

Project Management-Related Web Sites

Microsoft Project *www.Microsoft.com/office/project/default.asp*

Minute Man Systems Project *www.minuteman-systems.com*

Primavera Project *www.primavera.com*

Project Management Institute *www.pmi.org*

Project Management Software *www.project-management-software.org*

Windows-Based Project Management Software List *www.infogoal.com/pmc/pmcswr.htm*

4

System Definition

"How do you define that?"

4.1 Introduction

Once the problem has been properly formulated and the project plan is complete, the simulation practitioner should direct his or her attention toward defining the system that should be modeled. For our purposes, a system is a group of components that receives inputs and delivers outputs. The components determine how the system converts inputs to outputs. The system may exist in an environment that could also potentially affect the inputs and outputs. The system definition and model formulation process consists of determining:

- The system classification
- How much of the system to model
- What components and events to model
- What input data to collect
- What output data to generate with the model

Note that we are going to define our system and formulate the model at the same time. In many other simulation texts, these two processes are treated completely independently. By defining our system and formulating the model nearly simultaneously, we are in a better position to understand how the system components will be modeled. In this chapter we focus on what we are going to model. In the following model translation chapter, we discuss the actual process of how we are going to model these components.

4.2   System Classifications

One of the first steps that the practitioner must perform in the system definition phase is to classify the system. Systems can be classified with respect to two different dimensions. First, a system may be discrete, continuous, or combined. Second, a system is either terminating or nonterminating. These classifications are a significant issue; they affect how the practitioner will conduct the modeling and analysis for the project (Law and Kelton, 2000).

4.2.1   Discrete versus Continuous versus Combined

The classification of the system as being discrete, continuous, or combined is a function of how the simulation clock will function. As you may recall from Chapter 1, the simulation clock manages system events. System events are occurrences that change the status of some component in the system.

4.2.1.1   Discrete Event Systems

In Chapter 1, "Introduction," the example that was used was a discrete system. The example system events occurred according to discrete jumps in the time clock. In that example, the system events were arrivals, service starts, and service ends. These events occurred at particular points in time on the event list. Between events on the event list, the system did not change with respect to the number of entities in the system. So, systems that jump between events are considered as discrete event systems. It is important to note that the type of entities in the system can cause the way that the system jumps between events. Entities that are also individual or discrete in nature promote the advancement of the clock in discrete jumps. Generally speaking, any system that uses people as an entity will be a discrete event system.

Examples of discrete event systems include:

- Stores
- Service centers
- Manufacturing facilities
- Transportation centers

4.2.1.2   Continuous Event Systems

In contrast to discrete event systems, in continuous event systems, some event is always occurring. This means that the status of some component in the system is continuously changing with respect to time. Systems that are continuous usually involve some sort of fluid or fluid-like substance. Fluid-like substances could include any type of small-particle, high-volume material that flows like a fluid. These types of materials are usually measured by weight rather than count. An example of a fluid-like substance is coffee. Coffee starts out as a bean and then is continuously processed, eventually resulting in a grain.

Continuous event systems must be modeled with differential equations. Because of the additional complexity presented by the use of differential equations, these models are typically more difficult to model accurately. The fluid or fluid-like materials are not modeled as entities but as either volume or weight. The volume or weight flows through the model.

Examples of continuous event systems include:

- Water treatment plants
- Chemical industries not including distribution points

4.2.1.3   Combined Event Models

Combined event models contain both discrete and continuous components. This means that the entities in the model may exist in an individual countable form in one part of the model and a fluid-like form in another part of the model. This type of situation occurs in many processing plants where the fluid or fluid-like substance is canned or packaged. Examples can be found in:

- Food industries
- Chemical distribution points

A specific example of a combined model in the food industry is the coffee plant that was previously discussed. The system begins with fluid-like beans in storage tanks. The beans go through a series of roasting and grinding processes. The fluid-like coffee particles are eventually packaged into either cans or bags. As soon as the coffee is canned or packaged, that part of the model becomes discrete. The individual can then flows though the system as an entity until it departs.

In the chemical industry, a petroleum product simulation can begin with oil in a tank. Obviously, the oil is a fluid. It must therefore be modeled using a continuous event model. The oil will eventually be transported by tanker trucks. The process of filling and emptying changes the continuous model into a discrete model. In this case, the discrete component is the tanker truck. When the tanker truck is filled, it can individually be modeled for further processing.

Combined event systems are typically the most difficult type of system to model. The practitioner has to handle not only the increased complexity of the continuous event modeling but also the transformation of the fluid or fluid-like material to discrete entities. The practitioner must also decide how to handle the output measures of performance. Will they be based on the continuous portion of the model, the discrete portion of the model, or the interface between the continuous and discrete portions of the model?

4.2.2 Terminating versus Nonterminating

The second means of classifying the system is whether it is terminating or nonterminating in nature. Terminating and nonterminating systems are distinguished by:

- Initial starting conditions
- Existence of a natural terminating event

4.2.2.1 Initial Starting Conditions

The status of the system at the beginning of the period of interest is one means of distinguishing between a terminating and a nonterminating system. Terminating systems generally start each time period without any influence from the previous time period. This means that the system cleans itself of entities between time periods. Many systems that utilize customer-type entities are considered terminating-type systems. These systems do not have any customers left in the system from the previous time period. A bank, for example, does not let customers stay in the bank overnight. This means that each new day or time period, the system starts empty.

In contrast to the terminating system, the nonterminating system may begin with entities already in the system from the previous time period. In this case, the system may close with entities still in the system. At the beginning of the next period, the system starts with whatever was left from the previous period. Even though the system actually closes between time periods, it starts again as though it had never closed. Another type of nonterminating system actually has no beginning time period and no closing time. These types of systems just continuously run. In this case the system never stops, so it may never clean itself of entities.

4.2.2.2 Existence of a Natural Terminating Event

A second means of identifying whether a system is terminating or nonterminating is the existence of a natural terminating event. This may be the shutting down of the system at a particular point in time or the end of a busy period that is of specific interest to the practitioner. Examples of systems shutting down are service systems that close at the end of the day. Examples of systems that have busy periods are those that exhibit a high degree of activity during specific meal times. The existence of naturally occurring terminating events means that these systems may be classified as terminating systems, subject to other terminating system requirements.

4.2.2.3   Types of Terminating Systems

There are many different types of terminating systems. However, in general, terminating systems must:

- Have a natural terminating event
- Not keep entities in the system from one time period to the next

Examples of terminating-type systems include:

- Stores
- Restaurants
- Banks
- Airline ticket counters

Stores and restaurants generally close at the end of the day. This constitutes a terminating event. All of the customers are forced out of the store. So, each new time period, the system starts empty.

Banks are similar to stores; however, the period of interest would be shorter, perhaps the period between 10 a.m. and 2 p.m. The terminating event may be considered as the end of the busy period at 2 p.m. In many cases, this would be the end of the lunch period, at which time the customers would have to return to work.

Airline ticket counters can also be modeled as terminating systems. Here, the period of interest might be the time between one batch of planes arriving and the next batch of planes departing. Thus, the terminating event could be the departure of the batch of planes. Several time period cycles may exist in a single day as a result of the use of hub-type air transportation systems.

4.2.2.4   Types of Nonterminating Systems

There are also many different types of nonterminating systems. In general, nonterminating systems can either:

- Have a terminating event but keep entities in the system between time periods
- Not have a terminating event and run continuously

This means that a number of systems that may actually appear to be terminating systems are actually nonterminating systems because they keep entities in the system between time periods. It is also possible for a nonterminating system to appear as a terminating system because the system may be temporarily emptied of entities during nonbusy periods. Examples of nonterminating-type systems include

- Most manufacturing facilities
- Repair facilities
- Hospitals

Many manufacturing systems run continuously 24 h/day, 7 days/week. The only time they shut down is for annual maintenance. Other manufacturing systems do shut down after either one or two shifts per day. However, these systems probably keep work in progress between shutdowns.

Repair facilities that keep the end item in the facility between shutdown periods are nonterminating systems. Even though they terminate at the end of each day, the work not repaired each day remains in the system. An example of this type of system is a car dealership service center.

Hospitals do not normally close. Although the activity may be reduced during the early hours of the morning, because the system does not close, it can be considered as a nonterminating-type simulation.

4.2.2.5   Statistical Analysis Approach Based on Type of Simulation

As we point out in Chapter 5, "Input Data Collection and Analysis," there is a great difference in the manner in which terminating and nonterminating systems are analyzed. Terminating system analysis is significantly easier to perform than nonterminating analysis. For this reason, many practitioners incorrectly model and analyze nonterminating systems as terminating systems. Practitioners who are not confident with the nonterminating system analysis approach may attempt to modify the system in order to use the less demanding terminating system approach. The usual technique is to look at only a small period during the long nonterminating system run and to use a terminating system analysis approach.

4.3 High-Level Flow Chart Basics

An essential tool for defining the system is a high-level flow chart. A high-level flow chart helps the practitioner obtain a fundamental understanding of the system logic. It graphically depicts how the major components and events interact. It also illustrates how the input and output data play a role in the model. To help insure that your flow chart is properly developed, we begin this section with basic information on the design of high-level flow charts.

The components and logic of the system should have been previously observed in the orientation activities associated with the problem statement phase. Because the simulation flow chart is a high-level flow chart, the practitioner should endeavor not to include so much data that it is difficult to follow. The chart is intended only to provide general information as to how the process flows. It is understood that the actual programming of the simulation model will require additional detail.

Unfortunately, the development of even relatively simple high-level flow charts requires a certain level of discipline. All too often, practitioners attempt to sit down and begin modeling the system without a functional high-level flow chart. This approach usually has an unhappy ending. Without the flow chart it is doubtful that the practitioner possesses a fundamental understanding of the model. The writing of the simulation program is difficult enough. This process will become significantly more difficult if the practitioner attempts simultaneously to address the logic and programming issues.

4.3.1 Standard Flow Chart Symbols

There are four basic flow chart process symbols. These are the:

1. Oval
2. Rectangle
3. Tilted parallelogram
4. Diamond

There are specific ANSI standards for the use of these flow chart symbols. For practical use by simulation practitioners, these standards can be reduced to a couple of common-sense guidelines. The first guideline is that the symbols should be arranged so that the sequence of processes flows downward and to the right as much as possible. The second guideline is that any given symbol should have only one connecting path into the symbol and only one connecting path out of the symbol. The only exception to this rule is the decision icon, which has two different output paths. However, even with the decision icon, only one output path can be taken at a given time. Last, it is helpful to try to keep the flow chart on as few sheets of paper as possible. If additional detail is required, it should involve a lower-level flow chart of the particular process, not additional detail in the high-level flow chart.

4.3.1.1 Start and Stop Oval

The oval symbol is used to designate both the start and stop processes. The start symbol is the first symbol that should be present on the flow chart. The start symbol generally has only a single output connector. The stop or end symbol is normally the last symbol present on the flow chart. Even if the process has several different possible paths, there should still only be one stop or end symbol. All of the processes should terminate in this one symbol. The start oval is illustrated in Figure 4.1. The stop oval is identical except for labeling.

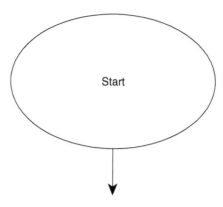

FIGURE 4.1   Start flow chart symbol.

4.3.1.2   Process Rectangle

The rectangle is used to represent general-purpose processes that are not specifically covered by any of the other flow chart symbols. The process rectangle is normally entered from the top or the left side. It is exited from either the bottom or the right side. In a high-level flow chart, a service time delay would be a common example of a process (Figure 4.2).

4.3.1.3   Input/Output Tilted Parallelogram

The tilted parallelogram is used for processes that involve some sort of input or output. The tilted parallelogram is normally entered from the top or the left side (Figure 4.3). It is exited from either the bottom or right side. An example of an input process symbol in a simulation program flow chart would be the creation or arrival of entities into a system.

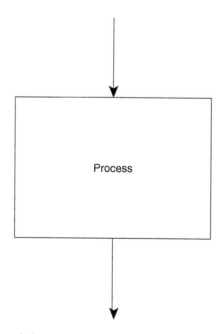

FIGURE 4.2   Process flow chart symbol.

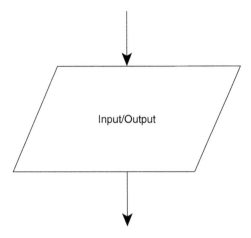

FIGURE 4.3 Input/output process flow chart symbol.

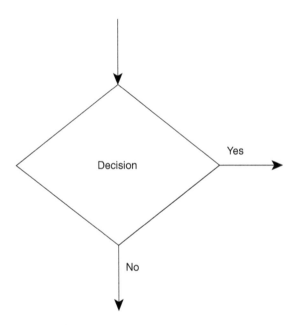

FIGURE 4.4 Decision flow chart symbol.

4.3.1.4 Decision Diamond

The diamond is used to represent a decision in the flow chart logic. The diamond has one input connector but two output connectors (Figure 4.4). The single input connector should come into the top vertex of the diamond symbol. The output connectors can leave through either of the side vertices or the bottom vertex. The output connectors must specifically be labeled as either true or false or yes or no. Because these responses are mutually exclusive, only one output path may be taken at a given time.

4.3.2 Sample Flow Chart

The sample flow chart in Figure 4.5 represents a simple system example such as the one used in Chapter 1. This is a single-queue, single-server system.

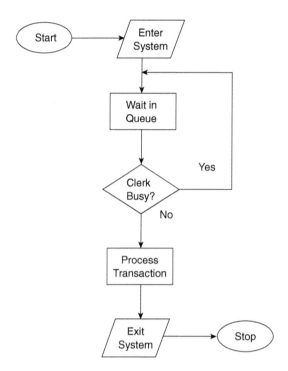

FIGURE 4.5 Sample flow chart.

The system flow chart begins with the start oval. The next symbol is an input/output tilted parallel-ogram, which represents the customer entering the system. After the customers enter the system, they enter into the queue. Waiting in the queue is a process represented by a rectangle. The customer next determines whether or not the clerk is busy. This operation is represented by a diamond. If the clerk is not busy, then the customer undergoes the transaction process represented by another rectangle. However, if the clerk is busy, the customer stays in the queue. When the customer finally finishes the transaction, the customer exits the system. This operation is represented by another tilted parallelogram. The flow chart finishes with an end oval symbol.

Note how the flow of the process is to the right and down as much as possible. The only place where the flow is not down is when the customer remains in the queue. Also note how most of the symbols have only one entry and one exit. The only exceptions are the start with a single exit, the stop with a single entrance, and the decision. Even though the decision has two exits, only one exit can be taken at a time. Either the clerk is busy or the clerk is not busy. Note also how when the clerk is busy that the flow line does not go directly back to the queue. Instead, the flow line intersects with the flow line that goes into the queue. This preserves the principle that the queue symbol can be entered in only one place.

In this particular example, the flow chart utilized six symbols. The flow chart for a more complicated system could easily require tens if not hundreds of flow chart symbols. As long as the practitioner strives to keep the flow chart organized, a large number of symbols can be managed. Keeping the flow chart organized means flowing to the right and down and only entering and exiting each symbol in one place at a given time.

4.4 Components and Events to Model

Now that we have reviewed the basic concepts of creating a flow chart, we can turn our attention toward what components and events should be modeled and represented in our high-level flow chart.

In a small independently operating system, there is little question that the practitioner should attempt to model the entire system. However, in a large interrelated system, it may be very difficult to identify which parts of the system to model. On one hand, the model must contain enough content so that the behavior of the system is properly understood. On the other hand, only a limited amount of time is available for most projects. Thus, in the beginning it is natural that some question may exist as to the appropriateness of the initial model. The suggested approach is to begin with a coarse model and add refinements as you proceed. It is better to have a higher-level model that can be used for more limited analysis than it is to have a lower-level model that cannot be used for analysis. Many practitioners have also found that the increased modeling demands of the lower-level model do not allow much time for analysis or reporting activities.

4.4.1 Components

Service models and manufacturing system models will naturally contain different system components. Some more common system components that may be modeled include (Kelton et al., 2002):

- Personnel
- Machines
- Transporters
- Conveyors

4.4.1.1 Personnel

In service-related systems, personnel may include:

- Sales clerks
- Customer service representatives

In manufacturing systems, personnel would probably include:

- Machine operators
- Material handlers

4.4.1.2 Machines

Machines in service-related systems may include:

- Computer or network systems
- Automatic teller
- Ticket machines
- Scanners
- X-ray machines

Machines in manufacturing-related systems could include:

- Computer numerically controlled mills
- Machining centers
- Lathes
- Turning centers
- Robots

4.4.1.3 Transporters

A transporter is any kind of vehicle that is used to move entities from one location to another. In transportation service models, transporters may include vehicles such as

- Airplanes
- Buses
- Trains

In manufacturing-related systems, there may be

- Forklifts
- Hand trucks
- Dollies
- Automatically guided vehicles

Note that there is also a distinction between free-path and fixed-path transporters. Free-path transporters can go between different locations without respect to following a specific path or the presence of other vehicles. A typical example of a free path transporter is a fork lift. Fork lifts typically can move between points without a specific track. Forklifts can also go around other transporters because they are not restricted to movement on rails. On the other hand, fixed-path or guided transporters must follow a specific rail or track and can be affected by the presence of other vehicles. A typical example of a fixed-path transporter would be a rail train. The forward movement of any rail train is dependent on clear track ahead of the rail train.

4.4.1.4 Conveyors

Conveyors are devices that can move entities between locations without the conveyor device itself physically moving. The conveyor actually physically stays in the same place, but the belt or rack does the moving. Examples of conveyors in service systems include:

- Moving sidewalks
- Escalators
- Chair lifts

Conveyors in manufacturing systems include:

- Overhead crane systems
- Fixed production assembly lines

Conveyors can also be classified as:

- Nonaccumulating
- Accumulating

Nonaccumulating conveyors maintain the spacing on the conveyor belt or track between entities. New entities can be placed on the conveyor only if there is sufficient space between other entities. Nonaccumulating conveyors are commonly circular in shape. This means that if the entities cannot be removed, they can ride around the conveyor an unlimited number of times. One typical example of a nonaccumulating conveyor is the suspension system that is used to retrieve clothes in a dry cleaning store.

Accumulating conveyors, on the other hand, allow the entities to become compressed on the conveyor belt. This usually means that there is some mechanism on the conveyor that causes the relative position between the entities on the belt to be altered. Accumulating conveyors are also usually linear in shape and transport entities in only one direction. An example of an accumulating conveyor is the cash register point in many grocery stores. The clerk operating the belt can run the system until the goods are compressed at the end of the belt.

Some conveyors appear to have a combination of characteristics. The most immediate one that comes to mind is the baggage conveyor in the baggage claim area of a airport. Here, bags can come down a chute onto the main circular conveyor. The impact of the bags can cause the density of the bags in a section of conveyor belt to increase. On the other hand, the belt is circular, so there is no other specific mechanism that would cause the bags on the belt to become compressed.

4.4.2 Processes and Events

The practitioner must also decide what system events should be included in the model. One way of determining which processes to model is to include any process that is capable of being performed differently over time.

4.4.2.1 Service System Processes and Events

In a service system, customers may wait in line for different amounts of time, have different services or amounts of goods to purchase, and the customers may pay differently. Thus, the model must at least include:

- Arrival of customers at a processing area
- Customer queue behavior
- Service processing
- Payment for the goods

4.4.2.1.1 Arrival of Customers at a Processing Area
The event of customers initially arriving into the system is perhaps one of the most important events in service-type models. In simple models, some practitioners will not properly model the arrival process. As we will see in the input data collection section of this chapter, it is not only important to model the arrival of the arrival batches, but also the size of the arrival batches.

4.4.2.1.2 Customer Queue Behavior
Most service-type systems include waiting queues for customers. Customer queues can either be

- Parallel queues
- Single snake queues

Parallel queues are found in systems that have multiple server resources. Each server has a dedicated queue. Customers enter the tail end of one of several queues and proceed to the head of the queue as previously queued customers leave the queue to receive service. Once at the head of the parallel queue, the customer will wait until the dedicated server is available for processing.

In contrast to the multiple parallel queues, a single snake queue is often used to model complex systems. Snake queues earn their name for their back and forth layout. This layout is used in order to make the best use of available space. With a snake queue, a single queue holds customers for a number of server resources. Once the customer reaches the head of the queue, the customer will wait until the first server resource is available for processing. An issue with snake queues with multiple servers is what occurs when one or more resources are available. In this situation, the system can operate in different ways. These are:

- Preferred order
- Random

In the preferred order system, when two or more service resources are available, the customer entity will always pick specific resources over the others. This typically occurs when the customer entity picks the available resource physically closest to the head of the snake queue. This means that in a lightly loaded system, the one or two closest resources will have much higher utilization levels than the furthest resources.

When the head of the queue is centered with respect to the multiple resources, the customer entities may be less prone to pick particular service resources. In this case, the customer entity process for selecting resources may be more random in nature.

Both parallel and snake queues may also exhibit different types of queue behavior. These include:

- Queue priorities
- Queue entity behavior

Queue priority means that the order of the entities in the queue may change according to the priority scheme. These are also sometimes called ranking criterion orders. In any event, there are many different types of queue priorities:

- First In–First Out (FIFO)
- Last In–First Out (LIFO)
- Shortest Processing Time (SPT)
- Longest Processing Time (LPT)
- Lowest Value First (LVF)
- Highest Value First (HVF)
- User-Defined Rules

For most simple systems entities are served on a first-come first-served basis (FIFO). That is, whoever is first in line will be served before later arriving entities. This type of queue priority system is the most commonly encountered in service-type systems. LIFO is another type of queue priority. It is the direct opposite of the FIFO priority. This means that whoever entered the queue last will be the first entity to be processed. The use of LIFO is not nearly as common as FIFO. Most LIFO applications involve some sort of penalty where the less senior members of the queue are faced with some undesirable processing.

Two other queue priority rules that the practitioner may encounter are SPT and LPT. The SPT algorithm can be effectively utilized where there is a service cutoff for the system. Here customers who have the shortest processing time are sent to the head of the queue and are processed first. This type of situation exists in the commercial airline industry. Passengers will arrive at the ticket counter on different flights. It is most important to processes the customers who have flights with imminent departures. If these customers are not given priority, then it is possible that they may miss their flights. Waiting a little extra time in the line will not affect the other passengers with later flights. The LPT algorithm means that customers with the most complicated transactions are handled first. This algorithm is much less common in service systems. It may be encountered only when the longer processing time is of much greater economic benefit to the system. The use of the LPT algorithm will endanger the processing of the customers with shorter processing times. These customers will be prone to reneging.

There is also the LVF priority scheme. LVF priorities are often used to model commercial passenger transportation systems. Passengers can be categorized as first, second, and third class. The value of the class can correspond to 1, 2, and 3. The LVF queue priority would result in first class being at the head of the queue, second class in the middle, and third class at the tail. Second-class passengers can receive service only if there are no first-class customers in the queue. Similarly, third-class passengers can receive service only if there are no first- or second-class passengers in the queue. Any time a first-class passenger arrives in the system, he or she will automatically go to the head of the queue. Similarly, any second-class passenger will automatically enter the queue behind any first-class passenger but before any third-class passenger. The LVF queue priority may also be used to model systems that use a service ticket. Here, customers are served in the order of the lowest ticket. This is similar to the FIFO priority, except there may be customers who temporarily leave the area and then come back for service.

The last normal queue priority method is the HVF priority scheme. Here, each customer has a value assigned that will reorder the position of the customers in the queue in descending order of the value. This type of queue priority might be used when certain individuals are repeat customers. In this situation, the service system may want to provide priority handling to those customers who have done the most business in the past.

It is possible that none of the normal queue priority or criterion-ranking rules adequately models the type of queue priority that the actual system utilizes. In the event that the actual system does use a complicated queue priority rule, most simulation languages provide the opportunity to program in any calculations that are required. These are typically known as user-defined rules.

Queue entity behavior involves the actions of the entities with respect to entering and remaining in the system queues. There are three different types of queue entity behaviors that the practitioner should be familiar with. These are:

1. Balking
2. Reneging
3. Jockeying

Balking occurs when a customer enters the system but leaves before entering a queue. Balking is the result of facing a long queue wait or limited queue capacity. If the customer believes that there are so many individuals in line that an unacceptably long wait will ensue, the customer may depart before getting in line. The assessment of when to balk when facing a long queue wait is entirely individually determined. No two customers can be expected to balk the same. This means that this type of balking must be modeled with a probabilistic distribution. On the other hand, limited queue capacity is usually associated with space constraints. If the length of the queue exceeds the physical space available for the system, customers may not be able to wait in the queue and must leave the system. In this case, the decision to balk will be a specific determination based on physical capacity.

A second type of entity behavior in both parallel queues and snake queues is called reneging. Reneging is when an entity enters the line but leaves before being processed. This would correspond to a customer who is tired of waiting in line and leaves. The decision to renege is also an individual decision. No two individuals will wait the same time before reneging. As a result, the length of time before reneging must be modeled with a probabilistic distribution. Modeling reneging can be slightly more complicated than just waiting a period of time before departing. For instance, a customer who is considering reneging will probably delay this decision if he or she is next in line. Thus, a really sophisticated model will monitor both the current queue time and queue position of each entity.

A final type of queue behavior is jockeying. Jockeying is associated only with parallel queues. This is when an entity switches between two different queues. The decision to jockey is usually triggered by the end of a service period with the resource related to the other queue. The end of the service period results in the entity in the queue leaving the queue and engaging the resource. If the other queue still has more entities or the same number of entities, the entity will not jockey. On the other hand, if the other queue has one fewer entity than the current queue, the entity will jockey. This type of jockeying is complicated but can be modeled in a number of different manners. Jockeying may also occur if the entity perceives that the other queue is moving faster. Jockeying based on this type of perception may not necessarily be able to be modeled.

4.4.2.2 Service Processing

Another event that should be modeled in a service-type system is associated with the type of service that the entity is to receive. To illustrate the different components associated with service processing, we examine a few of the most commonly simulated service system processes. The types of service systems are only representative. The different types of service processes within each type of system should be viewed as providing a starting point for similar simulation system models. Particular service systems may be more or less complex than the examples provided here. We examine the following types of service processes:

- Retail service checkout
- Banking service
- Restaurant service
- Airline ticket counter service

4.4.2.2.1 *Retail Service Checkout Processes*

Retail service checkout processes are the simplest type of service process that the practitioner is likely to encounter. This type of process is widely observable in department stores, discount stores, specialty stores, and grocery stores. These types of processes simply consist of:

- Calculating the cost of goods
- Payment

Note how the single overall process of buying goods is broken down into two separate service processes. The reason for this is that these two different activities can follow completely independent servicing times. The calculation time for the goods purchased would be expected to be a function of:

- Number of goods purchased
- Type of goods purchased

The greater the number of goods purchased, the longer the process would be expected to take. However, if on the average the goods are large heavy items, the calculation times for the goods may be much greater than those for smaller, more easily handled goods.

For the payment process, it may take just as long to pay for one item as it does for many items. As we see in a following section of this chapter, the service time for payment processes follows its own type of distribution according to whether the customer pays by:

- Cash
- Check
- Credit
- Debit
- Account

As a result, the most realistic type of model will break down these two processes into distinct components. This will increase the probability that our model represents reality.

4.4.2.2.2 Banking Service Processes

In a service-type system such as a bank, there are actually many different types of transactions that the customer may require. For example, for customers seeing the tellers, we might expect:

- Deposits
- Withdrawals
- Money orders
- Cashier's checks

There is likely to be a separate service time distribution for each of these different types of transactions. This means that the practitioner must first determine which types of transactions are to be modeled. The service time delay must be modeled so that it corresponds to the specific type of transaction that is taking place. Here we are not including other types of customers who will also be using other bank services for:

- Opening new accounts
- Closing current accounts
- Safety deposit box transactions
- Mortgages

Obviously, each of these different types of transactions will also require additional modeling and data collection and analysis. The major point in identifying these different types of customers and processing events is that even the relatively simple operation of modeling a bank can become complicated.

4.4.2.2.3 Restaurant Service Processes

Consider a second type of common service system such a restaurant or deli. In this type of service processing we have:

- Placing an order
- Waiting for the order
- Consuming the order

- Augmenting the order
- Paying for the order

In the first part of the process we will be placing an order. The complexity of the order-placing process can depend on whether or not the group orders drinks, appetizers, or main meals. The service processing time may also be dependent on the number of people that are in the group.

Once the order is placed, there will be a probabilistic waiting time for the main order to arrive. During this period, if the restaurant is a sit-down type restaurant, the group may consume their drinks, appetizers, or salads. The length of time that the waiting period takes can be a function of the level of staffing and the quantity of customers already in the restaurant.

Time for consuming the order can vary widely. Some customers will just consume the food while others will include other activities. These include business transactions or discussions or social conversations. The length of time to consume the order may also be a function of the size of the group and the composition of the group. For example, a large family with young children might be expected to take more time than an individual person.

The group may decide to augment the order with after-dinner drinks or dessert. If the group augments the order, the overall service processing time can increase dramatically. Time will be needed for processing the augmented order and consuming the additional food or drink.

Last, the service process will have to include payment for the food and drink. This part of the service process must also include getting the server's attention to calculate the bill and the server's time to generate the bill. As we will discuss later, the overall time for the actual payment can also vary according to the type of payment that the group elects to use.

Note that in some restaurant systems, more than one type of activity may be occurring at the same time. For example in a deli model or a fast-food drive-through, the following activities may be happening simultaneously:

- Waiting for the order
- Paying for the order

In these types of situations, the order is placed and is immediately processed. During the order waiting time, the customer is also either waiting to pay or is actually paying. This has the effect of compressing the overall time that the entire service processing time would have taken if the individual processes were performed sequentially.

4.4.2.2.4 *Airline Ticket Counter Service Processing*
Airline ticket counter service processing is another representative type of service process. In an airport ticketing system you might expect the following different types of transactions:

- Purchasing tickets
- Checking in
- Changing tickets
- Changing seats

The process of purchasing the tickets would include:

- Determining a suitable flight itinerary
- Payment
- Issuing the tickets
- Checking in luggage

The service delay for determining a suitable flight itinerary would obviously be a function of the complexity of the travel connections with respect to airports and air carriers. The payment would be subject to the same variations as described in the payment service process section of this chapter. Issuing tickets could be expected to be a relatively short and consistent service period. Lastly, the checking in of luggage would be dependent on the type and number of pieces of check-in luggage.

The process of checking in would be similar to the purchasing ticket process, but include only:

- Issuing tickets
- Checking in luggage

The process of changing tickets could be the result of missing a flight connection. This process would be expected to include:

- Determining the new flight itinerary
- Canceling the old ticket
- Issuing the tickets
- Rerouting baggage

Since this type of transaction is performed in real time at the last minute, the service time for determining the new flight itinerary might take a different service distribution than that of purchasing a new ticket. Canceling the old ticket should require minimal processing time as will issuing the new tickets. Rerouting the baggage, however, may require a different service time than any other process previously described.

The service process of changing seats would be expected to be relatively simple. Here the ticket agent would simply be checking for the availability of other seats. However, two different outcomes could occur. These are:

1. Successful seat change
2. Unsuccessful seat change

The whole process will take a minimum amount of time to check for the seat availability. If the seat change is unsuccessful, then no more processing time is required. However, if the seat change is successful, then the ticket agent will have to reissue a ticket and or boarding pass. This will require additional processing time.

Again, these types of transactions would be expected to have completely different distributions for each type of transaction. Some customers might actually want to perform different types of transactions at the same time.

In the bank, airline, and restaurant examples, it would be generally wrong to assume that there is no difference between service types and to attempt to model the service transaction as a single type with the same service time distribution. While there are cases where it may be necessary to make this assumption, it can lead to other problems, the most serious of which is the inability to be able to state that the model is valid or represents reality.

4.4.2.3 Payment for the Goods or Services

Another system event that the practitioner must observe closely and model carefully is the process of paying for goods or services. This process can be considered as independent of the service processing time because it takes just as long to pay for one item as it does many items. Here the customer's payment options are likely to be:

- Cash
- Check
- Credit
- Debit
- Account

The difference in the time required for these different types of payments can be significant. We would expect that the cash transaction be the fastest, while the check transaction would be the slowest. Credit, debit, and account-type transactions would normally fall somewhere in the middle. However, we must also account for extraordinarily long credit and debit transactions if there is some sort of communications error. The use of probability distributions will assist us in this process. There are distributions with very few observations with long values.

Note that if the payment time is large in comparison to the service-processing time, the payment time can have an overwhelming influence on the overall service time. As with the service-processing event, the practitioner will first have to determine what type of payment transaction will be utilized. A separate payment transaction time distribution will have to be modeled for each type of payment transaction. Again, the consequences of assuming that the payment transaction time distribution is the same for all types of payment can prevent the model from being validated.

4.4.3 Manufacturing System Processes and Events

Let us now turn our attention to manufacturing-type systems. In a manufacturing system the entities are likely to be considered as work in progress or products. The product entities need to be processed as work orders, enter machine queues, and undergo processing times and movement. In addition, machine reliability and product inspection processes must also be modeled. Thus, the practitioner would be interested in including the following components:

- Types of job orders
- Machine queue behavior
- Machine processing
- Machine buffers
- Material transportation
- Machine failures
- Preventive maintenance
- Product inspection failures

4.4.3.1 Types of Job Orders

In a manufacturing system, there are likely to be different types of products that are going to be produced. The products can vary widely in terms of:

- Raw materials
- Components
- Manufacturing processes

It would be expected that these differences would result in different processing times and paths through the manufacturing system. The practitioner will have to determine what parameters exist that differentiate each different type of job order that flows through the system. The differences between the types of jobs can be recorded as a set of entity attributes or a sequence associated with each particular type of job. Sometimes it will also be necessary to include an attribute that uniquely identifies each particular end product. This may be useful in the event that a number of different components for the product will have to be assembled in a later section of the model.

4.4.3.2 Machine Queue Behavior

Just as customers must wait in queues for service, products may have to wait in queues for machine processing. The difference is that a product may ultimately need to enter a large number of queues in a manufacturing system because the manufacturing process for the product may consist of several steps before the product is complete. Each time the product completes one step, it may have to wait for processing for the next step.

In general, most of the types of queue priority or ranking criterion schemes that were discussed in the service queue section are also applicable to manufacturing queues. There are two notable exceptions. In some job order systems, priority algorithms may be based on total processing time or remaining process time. If shortest processing time (SPT) is used as a priority scheme, rapid turnarounds are not held in the queue for unreasonable amounts of time. This can keep up a steady flow of completed jobs.

Another way that the shortest processing time may be utilized in a manufacturing system is with respect to products that are subject to spoilage. The manufacturing system will want to process the products that have the least shelf time remaining before spoiling. This type of queue priority is consistent with the total quality management concept of just-in-time. This concept attempts to minimize the work in progress so storage issues do not become critical.

The second queue exception is associated with manufacturing systems that require a minimum amount of time between processes. This might include a system in which a housing is finished with paint or a resin needs to cure. The longer the time allowed for the paint to dry or the resin to cure, the fewer subsequent finishing or structural defects will result. In this case, the queue priority system would be highest-value-first (HVF). The value would be the amount of time since the last process was completed. If an insufficient interval has passed, the next process may experience what is known as starvation. This is where the process could be performed but cannot because of lack of suitable incoming work in progress.

4.4.3.3 Machine Processing

Perhaps the most important processes to model in manufacturing systems are the machines. Machine processing involves the sequence of machines needed by the product entity. Just like service resources, machines can be laid out in:

- Parallel
- Serial

4.4.3.3.1 Parallel Manufacturing Systems

When laid out in parallel, the product entity can choose from one of the parallel machine sequences. Many large mass production systems are set up in this manner. Sometimes mass production chains are set up in discrete sequences. This means that a particular work in progress goes though one parallel sequence. When it goes through the next parallel sequence, it may shift lines according to the sequence loading and availability. It is also possible that the manufacturing process has some portions laid out in parallel while others are sequential. When a manufacturing system is laid out in a parallel fashion, the resource selection process can become a significant modeling issue. When the entity may select from more than one resource, the selection process must be observed and modeled accordingly. Some of the most common resource selection methods include:

- Preferred order
- Cyclic
- Smallest remaining capacity
- Largest remaining capacity
- Random

When one of the parallel machines has a greater capacity or processing speed, the manufacturing system manager may want to keep the utilization of that machine at a higher level than that of the other machines. Although this will normally increase system throughput, it will cause extra wear on the primary machine. If this is acceptable and actually practiced by the manufacturing system manager, the practitioner will have to use a preferred order approach to modeling the set of parallel machines. This means that the system will always attempt to use the preferred machine unless it is already busy. In that case, the next most preferred machine will be used, and so on.

Sometimes, the system manager does not want any one particular machine among several to receive an inordinate amount of wear. By equally rotating the use of each machine, the system manager can ensure that each machine receives approximately the same utilization. When this is the case, the practitioner should model the parallel resources using a cyclic resource selection rule.

Some manufacturing system components strive to collect a certain sized batch before the process can begin. This is usually present in a manufacturing process with a large per-cycle setup cost. Here, the objective is to reduce the number of total cycles by making each batch as close to capacity as possible before beginning the cycle. Manufacturing processes that operate in this manner utilize a smallest remaining capacity resource selection rule. As with the actual system, a model that utilizes this rule will try to send the work in progress to the machine that is loaded closest to the specified operating batch size.

The opposite of the smallest remaining capacity resource selection rule is the largest remaining capacity resource selection rule. The use of this rule appears to be much less prevalent than the smallest remaining capacity rule. This rule can be found when the parallel resources each have a large excess capacity and it is better not to have to switch continuously between resources. This situation can be found in process-type industries where it is important to maintain the integrity of individual batches in separate tanks or hoppers.

Occasionally, there does not seem to be any rational method to determining how the work in progress is assigned to a particular parallel machine. In this case, the practitioner can utilize a random resource selection rule to model the system accurately.

4.4.3.3.2 *Serial Manufacturing Systems*
When laid out in a serial manner, the product entity has no choice but to proceed through the sequence of machines. In the event that the serial layout experiences machine failures, the entire process upstream from the failed machine can come to a complete halt.

The machines in the manufacturing sequence may differ in terms of:

- Capacity
- Processing speed

Machine capacity refers to the number of different work-in-progress entities that the machine can process at a given time. Many machining operations that use computer numerically controlled (CNC) machines can only process one part at a time. However, the single machine can perform many different types of sequential machining operations. On the other hand, it is possible for a machine like a paint sprayer or a curing oven to process many different parts at the same time. One of the common types of simulation studies is to determine if the capacity of a particular manufacturing operation can have an effect on either reducing process bottlenecks or increasing the process throughput.

Manufacturing machines can also vary in processing speed or rate. This is how fast a particular machine can perform its operation. For example, in the tire industry, many types of machines are rated by the number of tires the equipment can produce per hour. With processing speeds or rates, faster is generally better than slower. Sometimes in a simulation model, it is easier to describe the capability of manufacturing machines in terms of processing time or cycle time. This is how long it takes an individual machine to perform one cycle. If you know the processing speed or rate, the processing time or cycle time can be easily calculated. Here smaller values are generally better than larger values. The smaller the processing time or cycle time, the larger is the number of products that can be produced in a given amount of time. This means that if there are a number of different parallel machines in the manufacturing system, the practitioner can model the machines the same way only if the capability of the machines is identical. If the performance of the machines is different because of age or features, the machines must be modeled independently even though they may be positioned in parallel.

The accurate modeling of machines may require not only an evaluation of the machine's capability. In any type of manufacturing system where there is a human operator, the practitioner must also account for the operator's effect on the machine's performance. A less-capable machine with an expert operator could easily equal the performance of a much higher-capacity machine with a less-experienced operator.

The simulation study could easily indicate that a more capable machine is needed when, in reality, the operator just needs to be trained better.

4.4.3.4 Machine Buffers

As we previously discussed, a manufacturing system may consist of a number of individual manufacturing processes. The product entity will flow between the processes. In many systems of this type, there is a limited amount of space between the individual processing machines. In this case, the limited queue capacity would block the forward movement or further processing of the work-in-progress parts. This process is known as blocking. Blocking can occur both upstream and downstream of a particular process.

4.4.3.5 Material Movement

In a large manufacturing facility, some transportation time may be needed to be modeled. This may be in the form of conveyor or transporter movement.

4.4.3.5.1 *Material Movement by Transporter*

The actual process of movement by a transporter in a manufacturing system may include the following processes:

- Waiting until the transporter arrives at the work-in-progess's location
- Loading the work in progress onto the transporter
- Actual transportation process
- Unloading from the transporter

In the component section of this chapter, we identified transporters as forklifts, dollies, and vehicles. When these types of transporters are used to move a work in progress between locations, the work in progress must first make a request to be transported. In the actual system, this might consist of a batch of components completed in one part of the manufacturing plant. The completion of the batch means that the work in progress must be transported to either another process or a storage area. Either way, the batch will remain in its current position until the appropriate transporter is available.

When the transporter is available, there will ordinarily be some sort of delay while the work in progress is loaded onto the transporter. During this period of time, the transporter is not available for any other tasks and must also normally remain stationary. In addition to the transporter, there may also be other types of material-handling equipment such as a crane and also possibly operators involved in the process. The larger, heavier, more numerous, and more difficult to handle the work in progress, the longer the loading time would be expected to be. At the completion of the loading process, the other material-handling equipment and the operators if any will be released for other activities.

When the transporter is ready to move, the practitioner must determine all of the possible end locations for the work in progress. The work in progress may go to one of any number of other machine locations according to the work process or to a storage location. The distances and or travel times between these different possible locations must be determined as well.

When the transporter ultimately arrives at the proper location, there will be another service delay to model the unloading process. As with the loading process, the unloading process may also require the use of additional material-handling equipment as well as human operators. Again, the larger, heavier, more numerous, and more difficult to handle the work in progress, the longer the unloading time would be expected to be. Once the unloading process is complete, the other material-handling equipment and the human operators may be released for other transportation requests.

4.4.3.5.2 *Material Movement by Conveyor*

The actual process of movement by a conveyor in a manufacturing system may include the following processes:

- Waiting to access the conveyor
- Loading the conveyor

- Actual transportation process
- Unloading from the conveyor

In a heavily loaded conveyor system, there may be some delay before the work in progress is able to access the conveyor. It could turn out that the conveyor is temporarily not moving because an unloading process is taking place, or it could be that there is no space on the conveyor. Either way, until the work in progress is able to access the conveyor, there may be a delay with the work in progress remaining in its current position.

When the conveyor is operating and there is space for the work in progress, there may be an additional delay for the work in progress to be loaded onto the conveyor. As with the transporters, the loading process for a conveyor may also require some sort of material-handling equipment as well as human operators. A typical type of material-handling equipment for a conveyor is a small robot arm. The arm may remove work in progress from a pallet or other temporary storage onto the conveyor. In a similar manner as with transporters, the heavier, bulkier, and more difficult to handle the work in progress is, the longer the conveyor-loading time may be.

Once on the conveyor, the work in progress will be moved to its final destination. In many cases, the conveyor will move the work in progress to a specific location rather than to one of several locations as with transporters. When modeling transporters, it is necessary to measure the distance between the conveyor loading and unloading areas. It is also necessary to record the movement velocity of the conveyor during operation.

When the work in progress arrives at the unloading area, there may also be some delay while the work in progress is unloaded. As with the loading process, the work in progress may require material-handling equipment and a human operator to be available. Once all of the necessary unloading resources are available, the work in progress can actually be unloaded. Our heavier, bulkier, more difficult to handle issues must also be taken into consideration when modeling the unloading process.

Failure to model the overall transportation time will make the model appear as though the product is moving instantaneously between areas. This can have a negative effect on the overall system time and possibly prevent model validation.

A final conveyor issue involves the size of the work in progress with respect to the actual conveyor, not the material-handling equipment. In instances where differently sized objects are placed on the conveyor, it is possible that an insufficient amount of space will be available on the conveyor belt when the work in progress is to be loaded on the conveyor. The situation might actually occur that a subsequent piece of work in progress could be loaded onto the conveyor instead. Thus, subsequent smaller work-in-progress entities are blocked by the large first-in-line work-in-progress entities. If, for example, a human being is involved in the system, it is possible to circumvent these types of blockages. If the actual system has this sort of capability, the practitioner will also have to model this logic. The resulting model can become complex. These types of situations can occur in shipping-type industries that handle different sizes and shapes of packages. It can also occur in airport baggage-handling systems.

4.4.3.6 Machine Failures

Anytime there is a machine involved in a manufacturing process, there is the possibility of some sort of machine failure. The machine failure will prevent any additional processing to the product currently being produced. In addition, no other product entities will be allowed to be processed until the machine failure is resolved. These machine failures can include:

- Broken components
- Jammed machines

The practitioner will have to identify a list of common machine failures resulting from broken components. Each of these different types of machine failures can have different time distributions to repair.

Some of these broken components will be relatively quick to repair, such as a broken machine tool cutting bit. On the other hand, other machine tool failures may result in the complete shutdown of that machine for an extended period of time. An example of this is a burnt-out circuit board.

A jammed machine will also prevent the machine from processing the product entity. However, in this case, the machine failure could be limited to the removal of the product entity from the machine. The machine may then be able to continue on as before. The product entity that was being processed will be an inspection failure as discussed in the next section.

4.4.3.7 Preventive Maintenance

Preventive maintenance is performed in many manufacturing systems to help reduce the probability of machine failures. Preventive maintenance processes may be required after so many manufacturing cycles or after so many hours of machine times. Other preventive maintenance processes are required on a calendar basis. Generally speaking, preventive maintenance processes will require that the manufacturing operation be temporarily suspended. Preventive maintenance can come in the form of:

- Calibration
- Tooling replacement
- Lubrication
- Cleaning

High-accuracy manufacturing operations will require that the manufacturing equipment be periodically calibrated. This will ensure that the machine settings are set correctly and have not accidentally drifted, or the tooling has not become distorted.

When manufacturing processes include any type of machining operations, tooling replacement processes are regularly performed as preventive maintenance. Manufacturing tooling will wear according to the raw material, cutting feed rates, and the amount of machining time.

Manufacturing equipment generally requires some sort of lubrication in order to function correctly. Proper lubrication will maintain cycle rates and reduce wear on the manufacturing equipment. Proper lubrication can also be required to reduce corrosion.

In many manufacturing operations, it is essential that the manufacturing equipment be cleaned. In the food-processing industry, cleaning may involve the removal of food byproducts. In machining operations, cleaning processes can include the removal of metal chips, ribbons, or slag.

When attempting to model these types of preventive maintenance operations, the practitioner can make use of special preventive maintenance entities. These entities are not physical in the same sense as normal job, order, or product entities. Preventive maintenance entities are more analogous to events that occur during the course of the simulation run. The preventive maintenance entity is created by being either scheduled or generated as some other event occurs, such as the completion of the required number of cycles. When the preventive maintenance entity is created, it will take control over the manufacturing resource. The entity retains control over the manufacturing resource until the end of the preventive maintenance operation. At the end of the preventive maintenance operation, the preventive maintenance entity releases control over the resource. At this point, if the entity is no longer needed, it is discarded.

4.4.3.8 Inspection Failures

In a manufacturing process, the product entity will go through a series of processes that add value to the product. In many processes, it is possible that the product becomes defective in some manner or another. For this reason, the practitioner must determine what types of product inspection failures may exist. These product failures may be either

- Reworked
- Scrapped

In the event that the product can be reworked, the practitioner must determine what additional processing is necessary to be able to insert the product back into the overall manufacturing process. This additional processing could include additional testing to determine the exact nature of the failure or the disassembly and removal of the defective component.

In the event that the product is to be scrapped, it may be necessary to update a variety of statistical counters or system variables. The use of counters and variables will assist the practitioner in keeping track of manufacturing costs that result from scrapped defects.

4.4.4 Events Not to Model

The practitioner may deliberately elect not to model some events. These would typically include events with very limited impact on the system outputs. Limited impact may be the result of the small importance or infrequent occurrence of the event. Some events that would not normally be included in a system include:

- Discovery of an explosive device in a security checkpoint system
- Power outage in a manufacturing facility
- Bus involvment in an accident
- Workers' strike

4.5 Data to Be Included in the Model

As you are probably already aware from either the introduction chapter or your previous experience, there are input data and output data. Input data are what drive the system, and output data are what result from the system. In this section, we will take a high-level view of the model data requirements. In Chapter 5, "Input Data Collection and Analysis," we cover this subject in greater detail. The subjects covered in this section include:

- Input data
- Input data considerations

4.5.1 Input Data

In the system definition process, the practitioner is interested in identifying a preliminary list of the types of input data that affect the system's output data performance. As the practitioner begins to collect data or model the system, the preliminary list may be modified. In this section, we discuss some data collection principles and identify some of the more common types of input data.

4.5.1.1 Input Data Collection Principles

A fundamental concept in this process is to break down the types of data into as many different independent types as possible. Consider the processing of commercial air passengers through a security checkpoint system. One practitioner created a model that took into account only the total processing time for each customer. Subsequent analysis of this security process indicated that the same processing time actually consisted of several different types of input data. These included, among others:

- Number of carry-on bags
- Time to load each bag on the x-ray machine
- Time to convey and x-ray a bag
- Processing time for the metal detector including emptying pockets
- Pass/fail rate for the metal detector
- Time to remove each bag from the x-ray conveyor

Consider another example with the staffing requirements for a ticket booth at a movie theater. We would not necessarily be interested in the number of tickets sold per hour. We would really be interested in the number of individuals who buy tickets and how many tickets each individual buys. By breaking the data down in this manner, we can always find the total number of tickets sold. However, if we had only counted the number of tickets, we would not be able to determine the number of individuals who had purchased tickets. It is the number of individuals purchasing tickets that would most affect the staffing requirements.

4.5.1.2   Types of Input Data

There are two general categories of input data. The first category is related to the concept of system entities. Entities can be thought of as the elements that are processed by the system. Common examples of entities include customers, passengers, and job orders. Typical types of input data associated with entities are the time between arrivals to the system, the number of entities in an arrival, and entity processing times for various operations.

The second category involves system resources. System resources are the parts of the system that process the system entities. Common examples of system resources are personnel, manufacturing equipment, and vehicles. System resource input data include break times, breakdown or failure rates, operating capacities, and movement speeds.

In the interest of facilitating the input data collection process, some of the most commonly encountered types of input data are identified in this section. The selection of these common types of input data does not mean that other types of input data will not be encountered. The analyst must be prepared to identify types of input data particular to the specific project.

4.5.1.3   Interarrival Times

Interarrival times are the amount of time that passes between the arrivals of batches or groups of entities into the system. Entities can include customers, orders, jobs, or breakdowns. Data of this type are also commonly found in the form of arrival rates. The interarrival times are the inverse of the arrival rate. For example, if there are five arrivals per minute, the interarrival time is 0.2 min. This is the same as an interarrival time of 12 s.

Interarrival times are normally considered to be probabilistic in nature. Many events follow a random arrival process. In this event, mathematical proofs can demonstrate that the interarrival times follow an exponential distribution. Because this handbook focuses on practitioner needs, readers will be spared this demonstration. Interarrival times are not always probabilistic. In the case of production scheduling, the release of a production order may be decided in advance according to needs.

4.5.1.4   Batch Sizes

Batch sizes involve the number of individual entities that arrive at the same time to be processed by a system. An example of a batch size is the number of airline passengers that are traveling together and arrive at a security checkpoint system. Although the batch arrives at the same time for processing, each individual in the batch must be processed individually. Batch sizes are generally probabilistic and discrete in nature.

4.5.1.5   Balking, Reneging, and Jockeying

As discussed earlier in this chapter, balking, reneging, and jockeying are queue-related behaviors exhibited by system entities. Recall that balking involves the entity's observing the operation of the queue and leaving the system before entering the queue. The decision to balk can be either probabilistic or deterministic. If a customer decides that the line is too long and leaves, a subjective probabilistic decision was involved. Conversely, if the balking is caused by limited space in the line, then the decision to balk would be deterministic. As previously introduced, reneging involves the entity's actually entering the queue but leaving the queue before receiving service. Reneging would normally be considered to be a subjective probabilistic decision. Last, we defined jockeying as the movement of entities between similar queues in the expectation of receiving earlier service. Jockeying would also normally be considered to be a subjective probabilistic decision. However, in actual practice, many individuals will choose to seek the queue that holds the promise of earlier processing.

4.5.1.6 Classifications

Classifications include types or priorities of entities arriving in the system. Classifications are commonly used to determine the process scheduling of jobs or customers. Classifications would normally be expected to be probabilistic in nature.

4.5.1.7 Service Times

Service times include processing times that a job or customer undergoes. This includes time during which the entity occupies the attention of a resource such as a machine, operator, or clerk. It specifically does not include the time during which an entity waits in a queue for processing. Service times may be either probabilistic or deterministic. Probabilistic service times are likely to include the presence of humans. A deterministic service time is more likely to involve some sort of automatic processing such as running a computer numerically controlled machining program.

4.5.1.8 Failure Rates

These rates involve the frequency of process failure or resource unavailability. Failure rates are normally probabilistic in nature. Examples of failure rates involve inspection processes and machine breakdowns. Both of these rates would be expected to follow some sort of probability distribution. In the case of a machine breakdown, there would also likely be a repair time. This type of input data can be considered to be similar to a service time.

4.5.1.9 Scheduled Maintenance

This involves reducing the availability of resources such as machines to perform preventive maintenance to reduce the probability of equipment failures. Scheduled maintenance is normally deterministic in nature. However, the implementation of the scheduled maintenance may be according to calendar time, operating time, or the number of jobs that have been processed.

4.5.1.10 Break Times

Break times primarily pertain to system resources such as operators and clerks. In most cases, these individuals will be unavailable because of breaks and meal periods. It is likely that the duration of the break times will be deterministic. One issue that the practitioner will need to model is exactly how the resources operate when the break occurs. There are three basic methods of modeling breaks.

1. Wait until the end of service
2. Ignore the start of the break
3. Preempt the current service

In the first type of break model, if the resource is busy, the resource will wait until any entity being served has completed its time before going on the break. When the entity does go on break, the duration of the break is as long as it was originally scheduled for. In the event that the resource is not busy, the resource will go on its break immediately for the full duration.

In the second type of break model, if the resource is busy, the resource will continue to service the entity until the end of the service period. At the end of the current service, the resource then goes on break. However, in this type of model, the entity is only entitled to the remaining break time after the beginning of the scheduled break. In the event that the service time exceeds the break, the resource is not allowed to go on break. If the resource is not busy, the resource will go on break immediately for the full duration as with the first type of break model.

In the final type of break model, if the resource is busy, the resource will immediately go on the break. The entity that was being served will be put aside and will wait until the resource comes off the break. When the resource comes off the break, the remaining service time will be continued. If the resource is not busy, the resource will immediately go on break for the full duration.

4.5.1.11 Movement Times

Movement times can include the duration for an entity to travel between areas of a system. This can include movement on foot, by vehicle, or conveyor. The movement times may be deterministic in the case of a conveyor or probabilistic in the case of an individual walking.

4.5.2 Other Input Data Considerations

The importance of collecting accurate and complete input data cannot be overemphasized. However, the practitioner may be interested in a high-level model and may have only limited project time. In this situation, collecting an exhaustive amount of data may not be in the best interest of the practitioner. When this occurs, the practitioner should make note of what input data collection compromises were necessary. These compromises should be clearly documented under the assumptions and limitations section of the project report.

4.6 Output Data

Output data are generally used to measure performance levels. These measures of performance are used to validate the model and conduct model experimentation. In the model validation process, output measures of performance are collected and compared from both the actual system and the simulation model. In model experimentation, we create models with different operating and resource policies. We then compare the output measures of performance between the different models.

4.6.1 Primary Measure of Performance

Practitioners frequently collect multiple types of output measures of performance for a single simulation study. This is not normally a difficult task because of the output data collection abilities of most modern simulation software packages. Although many different types of output measures of performance may be easily collected, the practitioner will still have to decide which measure to use as the principal means of comparison. The decision as to the most appropriate output measure of performance to use can be made only by the practitioner. In some cases, the practitioner can return to the original project objectives. If, for example, the driving force was to reduce operating costs, the practitioner could focus on resource utilization. If resource utilization was low, fewer resources may be needed. Conversely, if customer satisfaction needed to be improved, the practitioner might focus on system time as the primary output measure of performance.

Some of the commonly used output measures of performance and their mathematical definitions were introduced for simple systems in the Introduction (Chapter 1). The following section provides additional information on these measures:

- Average time in the system
- Average time in a queue
- Time average number in queue
- Average utilization rates
- Counters

4.6.1.1 Average System Time

System time is defined as the complete time each entity is in the entire system or a component of the system. It typically includes queue time, service time, and transportation time within the system. Queue time is the length of time the entity waits in a specific queue. Service time is the period during which the entity is utilizing a resource. Transportation time is the length of time that an entity is transported between components within the system. In some cases, the practitioner may choose also to include transportation time to enter the system and transportation time to exit the system. These would typically be included in facilities design-type simulations. The following formula is used to calculate average system time:

$$\text{Average System Time} = \frac{\sum_{i=1}^{n} T_i}{n}$$

where T_i is the system time for an individual entity (arrival time – departure time), and n is the number of entities that are processed through the system.

4.6.1.2 Average Queue Time

Queue time is defined as the length of time that each entity waits in a particular queue. It begins when the entity enters the queue, and it ends when the entity leaves the queue for processing or transporting by a resource. Average queue time is simply the total time that all entities wait in a particular queue divided by the total number of entities that were processed through the queue. If multiple queues exist in a model, it makes more sense to calculate the averages of each individual queue. This method makes more sense because unless the system is extremely well balanced, particular queues will have longer average queue times than others. If the average queue times are lumped together, any differences between the queues will be lost.

$$\text{Average Queue Time} = \frac{\sum_{i=1}^{n} D_i}{n}$$

where D_i = queue time for an individual entity (queue arrival time – service begin time), and n = number of entities that are processed through the queue.

4.6.1.3 Time-Average Number in Queue

The time-average number in queue is the number of entities that can be expected to be observed in the queue at any given time. Time-average number in queue is frequently confused with the average queue time by novice practitioners. As Chapter 1 illustrated, the time-average number in queue is time dependent, whereas average time in queue is observational:

$$\text{Time Average Number in Q} = \frac{\int_{0}^{T} Q\,dt}{T}$$

where

Q = number in the queue for a given length of time

dt = length of time that Q is observed

T = total length of time for the simulation

The time-average number in queue calculations rarely result in discrete numbers. Novice practitioners are also frequently disturbed by the seeming impossibility of observing a fractional entity in the queue at any given time. Only when the practitioner understands what this measure represents do fractional numbers make sense.

When multiple queues within the model for processes are involved, separate time-average number in queue calculations should be made. The only exception to this would be if there were a number of identical queues serving the same process within the model. In this case, the average of the multiple queue values could be calculated.

4.6.1.4   Average Utilization Rates

The utilization rate for a resource is the percentage of time that the resource was being utilized by an entity. If there is more than one resource for a given process, the practitioner may proceed in one of two manners. If the resources are identical with respect to processing capability, the practitioner can add up all of the average utilization rates then divide by the number of resources. In the other situation where resources are not identical with respect to processing capability it would be better to collect average utilization rate data for each individual resource:

$$\text{Average Resource Utilization} = \frac{\int_0^T B\, dt}{T}$$

where

$B = 0$ for idle or 1 for busy

dt = length of time that B is observed

T = total length of time for the simulation

4.6.1.5   Average Utilization Rates for Multiple Resources

Sometimes the practitioner will choose to model several parallel identical system resources with one model resource with a capacity larger than 1. The advantage of this is a far more simplistic model. As long as the capabilities of the different parallel system resources are actually identical, this simplifying modeling technique can be utilized. However, when using this approach, the practitioner must realize that the model statistics associated with the average utilization rate must also be properly translated. This means that the overall utilization values must be divided by the capacity of the single multiple-capacity resource.

For example, if there are two identical resources modeled as a single resource with a capacity of 2, the instantaneous utilization value will be between 0 and 2. This means that any one of the following situations may be occurring:

- Both resources are idle.
- One resource is busy and the other is not.
- Both resources are busy.

In order to translate properly the resource utilization statistics, the practitioner must insure that the values are divided by 2. This means that the instantaneous value will now fluctuate as either 0, 0.5, or 1. Similarly, the average utilization value will now take a value between 0 and 1. These numbers can actually be misleading because one of the two resources may be doing the bulk of the work. If the values are consistently around 0.5, it could mean that one resource is almost always busy while the other resource is almost always idle.

4.6.2   Counters

Counters can be used to keep track of the number of entities that have been processed through either the entire system or a particular component of the system. Counters can be either incremented or decremented. An example of an incrementing counter would be the number of customers entering a store. An example of a decrementing counter would be the number of seats available on a bus. Counter increments and decrements are normally one unit. However, in some cases multiple-step values can be used when entities represent more than one unit. Multiple step values would be appropriate to calculate the number of individual components in a case of components.

In addition to performance purposes, counters are also frequently used for verification purposes. In this case, the practitioner would use a counter to ensure that entities are processed through particular parts of the system. If entities should be coming through a component of the system, but no entities are counted, the practitioner will need to investigate the model for logic errors.

4.7 Summary

In this chapter we discussed the procedure to define the system that we are interested in modeling. This process began with classifying the type of system. Systems can be discrete, continuous, or combined with respect to the system events. One way of determining the type of system is by examining the type of entities that flow through the system. Entities like people automatically make the system discrete. Entities that are fluids or act as fluids are generally continuous event simulations. Entities that start out as fluids and end up as discrete units are combined systems.

Another way that systems are classified is as terminating or nonterminating. Terminating systems have an event that ends the time period of interest. In general, terminating systems are also cleared of entities at the end of the time period. In contrast, nonterminating systems do not close, or, if they do close, the entities remain in the system until the system reopens. The difference between terminating and nonterminating systems is significant because each type of system requires a different type of analysis approach.

Another important component of the system definition process is to decide what components and events to model. The practitioner must make decisions on how wide a scope the model should cover and how great the model detail should be. The practitioner also needs to determine what type of input data needs to be collected and analyzed. Finally, the practitioner must decide what sort of output data needs to be generated. These output data will eventually be used to make statistical comparisons between model alternatives. At the end of the system definition process, the practitioner can begin collecting input data and translating the model into a simulation programming language.

Chapter Problems

1. What differentiates a terminating system from a nonterminating system?

2. What is the difference between a discrete and a continuous event system?

3. Identify five different types of input data for an airport security checkpoint system?

4. What other types of processing times are included in an entity's system time?

5. What is meant by an arrival batch size?

6. What kinds of events would you not need to model in an airport security checkpoint system?

7. What are the differences between the preferred order rule and the cyclic resource selection rules?

8. What process can cause a continuous event model to need to be modeled as a combined model?

9. In what different ways can a resource's scheduled break be handled with respect to customers?

10. Why would it be important to record how a customer pays in a retail store simulation model?

References

Kelton, D.W., Sadowski, R.P., and Sadowski, D.A. (2002), *Simulation with Arena*, 2nd ed., McGraw-Hill, New York.

Law, A.M. and Kelton, D.W. (2000), *Simulation Modeling and Analysis*, 3rd ed., McGraw-Hill, New York.

5

Input Data Collection and Analysis

"Garbage in, garbage out"

5.1 Introduction

In a simulation project, the ultimate use of input data is to drive the simulation. This process involves the collection of input data, analysis of the input data, and use of the analysis of the input data in the simulation model. The input data may be either obtained from historical records or collected in real time

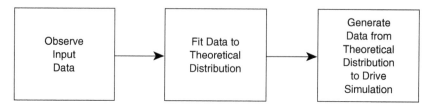

FIGURE 5.1   Role of theoretical probability distributions in simulation.

as a task in the simulation project. The analysis involves the identification of the theoretical distribution that represents the input data. The use of the input data in the model involves specifying the theoretical distributions in the simulation program code. This process is represented in Figure 5.1.

A common question is why we need even to consider fitting the data to a theoretical distribution if we have already collected data values. The fundamental reason is that when the practitioner collects data, only a sample of the actual data distribution is collected. Even though the practitioner did not observe particular values, it does not mean that the unobserved data values do not actually exist in the system. If we successfully fit the observed data to a theoretical distribution, then any data value from the theoretical distribution may drive the simulation model. This is a much more realistic situation than driving the simulation with only data values that were observed.

One possible weakness of this approach is that the theoretical distribution may periodically generate an unusual value that might not actually ever be present in the real system. An example is an interarrival time with a very long value. Though this does not happen very often, the practitioner should be aware of this possibility. If the model does exhibit unusual behavior that could have been a result of a theoretical distribution, then the practitioner can take action. It is important to stress that this is a very unlikely event and that it is possible to make many simulation models over a period of years without encountering this situation.

The collection of input data is often considered the most difficult process involved in conducting a simulation modeling and analysis project. Part of this is attributable to the fact that the simulation analyst may be dependent on individuals and operations outside the control of the analyst. This is true regardless of whether the input data are historical in nature or will be collected in real time as part of the project. In other situations, the required input data may not even exist. This situation is frequently encountered when the project involves the analysis of new capital equipment.

To assist the practitioner in the data collection and analysis process, this section includes discussion on:

- Sources for input data
- Collecting input data
- Deterministic vs. probabilistic input data
- Discrete vs. continuous input data
- Common input data distributions
- Analyzing input data

5.2   Sources for Input Data

There are many sources that the practitioner can tap to acquire input data. This data can be historical, anecdotal, or observational. Even if an actual system model does not exist, it is possible for the practitioner to acquire the needed input data from other sources. Sources that are available to the practitioner include, but are not limited to, the following:

- Historical records
- Manufacturer specifications
- Vendor claims
- Operator estimates
- Management estimates
- Automatic data capture
- Direct observation

5.2.1 Historical Records

If the base system or a similar base system has been in existence for some time, it is likely that some form of historical records is available to the practitioner. Because the records are already in existence and will not require real-time data collection, this approach may appear to be a very attractive option to the practitioner. It is also frequently attractive to the commissioning organization, which believes that money is to be saved by using this approach. However, the use of historical records is not without great risk.

The first risk is that the original historical system has changed somehow or is not exactly identical to the current or proposed system. Basing any kind of comparison on data with questionable reliability is certainly asking for future difficulties. Predictably, because the system is not the same, any attempts to validate the model with current system data is doomed to fail.

Another problem is that the historical data may not have been collected with a simulation model specifically in mind. This means that the data that are really needed are not actually available. All too frequently, the practitioner does not realize this until well into the model translation or even the validation phase of the simulation project. At that time, the project may be so compressed that it is impossible to go back to the actual system and collect the correct model input data.

Because of the potentially substantial problems associated with using historical records, the practitioner should think long and hard at any suggestions to this effect. It is the practitioner who will be the recipient of the extra work if this approach goes sour.

5.2.2 Manufacturer Specifications

Manufacturer specifications also offer the practitioner the opportunity to use data that someone else has compiled. Obviously, most manufacturers will be providing a theoretically based specification for their equipment. Whether or not these claims can actually be achieved in a real environment has to be proven.

5.2.3 Vendor Claims

The vendor or distributor claims will probably fall between the manufacturer's specifications and reality. The vendor or distributor should already have some experience with the type of system that is being considered. As we will discuss in this handbook, some manufacturing organizations require vendors and distributors to create a simulation model to prove their equipment's capabilities.

5.2.4 Operator Estimates

Operators of existing equipment can be a valuable data resource when the practitioner does not have the time or data collection resources to collect actual data. If the operator is knowledgeable about the system, it may be possible to obtain some performance estimates that can be used as input data. For example, the operator could be questioned on the shortest, most common, and longest processing times for a particular operation. The practitioner could then use these data as a first estimate with the triangular distribution.

5.2.5   Management Estimates

The practitioner may also consider soliciting managers or engineers associated with the system. Though these individuals probably do not have the same proximity relationship to the process, their input may be helpful when an experienced operator is not available for input.

5.2.6   Automatic Data Capture

It may be possible to set up some sort of automatic data capture system. This is analogous to the traffic volume monitors that are frequently encountered on the road. These monitors count the frequency of cars passing by a specific point over a given time period. Electronic access or other information-based systems may also be able to capture the type of data that the practitioner needs for the simulation model.

5.2.7   Direct Observation

The most physically and mentally demanding form of data collection is direct observation. This is where the practitioner or another individual actually goes to the location of the system and visually collects data. The data are collected by pen and pad or perhaps with some technological assistance. If the low-tech pen-and-pad approach is used, the practitioner may want to develop some sort of data collection form to keep the process as organized as possible. An example of the headings on this sort of form for a customer-based system is provided below:

Arrival #   Batch size   Arr. Time.   Svc. Start   Svc. End   Comments

Note that there is a specific heading for comments. Sometimes it is necessary to record an identifying feature about the customer to keep track of each significant event in the customer's process. Sometimes the comment section can be used to include the customer's gender, age, ethnic origin, or clothing color to help in the identification process.

The use of the direct observation approach can be particularly grueling and costly when a large amount of data on infrequently occurring events must be captured. This is certainly how the data collection phase can earn its reputation as being the most difficult part of the simulation process.

5.3   Collecting Input Data

From the previous discussion, we should be firmly convinced that the best policy is to collect original observable input data. This is, of course, provided that time and resources permit such an approach. As we have come to this conclusion, it is now time to concentrate on the mechanics of collecting original input data.

In this section we discuss:

• Data collection devices
• Time collection mode and units
• Other data collection considerations

5.3.1   Data Collection Devices

If input data are collected in real time, it may be collected either manually or with the assistance of electronic devices. If data are collected manually with a timing device, it may be helpful to create a form to help keep the data collection process as organized as possible. The practitioner can keep things organized by easily securing the mechanical or electronic stopwatch to a clipboard for unobtrusive data collection.

A wide selection of electronic time-studies devices may be utilized. Alternatively, the analyst may choose to develop a simple program on a notebook computer to assist with the collection of data. By pressing programmed keys, the practitioner can keep track of interarrival rates without having to write down clock times. The beauty of this approach is that the data can be automatically recorded in a file for analysis. This spares the practitioner the exceedingly tedious process of data transcription from paper to computer.

The use of video recording devices offers yet another possibility for collecting data when privacy issues do not need to be considered. Most video recording devices contain movable viewing screens. So, it is possible to record different processes without appearing too obvious. Because the playback can be reviewed an infinite number of times, it is frequently possible to observe activities that would normally have been missed in real time.

5.3.2 Time Collection Mode and Units

An important issue for simulation input data concerning time intervals is the time unit that should be used. Novice practitioners can be frequently observed recording the absolute time or clock time of when different entities arrive into the system. If this approach is taken, extra work will be necessary to convert the absolute time to a relative time so that interarrival times may be calculated. It is usually less labor intensive to collect the data correctly in the first place using a relative, interarrival time approach.

A second time collection issue is what types of units to use. For calculation purposes, it may be difficult to use as small a time unit as seconds. It is more easily understood when a simulation run is performed for either 8 h or 480 min, not 28,800 s. If the middle ground were taken, and a simulation time unit of minutes were chosen, all of the data values would either have to be collected in the same time unit or converted at some point. For example, if service times were taken in minutes and seconds, the seconds would eventually have to be converted into fractions of minutes. Although the conversion of data in this manner can be performed in an electronic spreadsheet, it is also possible to acquire what is known as a decimal minute stopwatch. This type of special stopwatch records minutes as minutes and seconds as hundredths of minutes. When a decimal stop watch is used, the data are collected in an immediately usable format. Keep this in mind when the time unit conversion process becomes overly burdensome.

5.3.3 Other Data Collection Considerations

The simulation practitioner should also strive to be as open but as unobtrusive as possible when collecting data. There are at least two reasons for this:

1. You want unbiased data.
2. You do not want to disrupt the process.

5.3.3.1 Unbiased Data

First, if it is obvious that the practitioner is collecting certain types of performance data, some workers may attempt to bias data collection results. Some workers may temporarily speed up their work rate to appear to be productive. Others may deliberately slow their work rate to prevent high work standards from being set. If the data are biased in either of these manners, it can lead to a model that may yield inaccurate results. So the best course of action is to provide a short briefing to the workers. In this briefing, the practitioner should explain the purposes of the data collection.

5.3.3.2 Avoid Process Disruption

Another goal that the practitioner should strive for is not disrupting the ongoing operation. If the practitioner maintains a high profile in the center of the process, then customers or operators who have not received a briefing may become curious. These individuals may not only behave abnormally but may also distract the practitioner with questions. If the practitioner is distracted, then important input data may be lost or corrupted. In extreme cases, other problems may occur. Uninformed operators or customers may even contact security or law enforcement personnel and report your suspicious data collection activity.

5.4   Deterministic versus Probabilistic Data

While collecting the input data, the practitioner should realize that there are different classifications of data. One method of classifying data is whether it is deterministic or probabilistic. Each individual project will call for a unique set or type of input data. Some of the types of input data may be deterministic, and other types are probabilistic.

5.4.1   Deterministic Data

Deterministic data mean that the event involving the data occurs in the same manner or in a predictable manner each time. This means that this type of data needs to be collected only once because it never varies in value. Examples of deterministic input processes include:

- Computer numerically controlled machining program processing times
- Preventive maintenance intervals
- Conveyor velocities

Computer numerically controlled machining program processing times are deterministic because unless there is some sort of problem, the program always takes the same amount of time to run. These programs follow a set number of processing steps with predetermined feed rates as the machine tool is controlled by the computer. Unless there is some sort of machining problem, there is no possibility of deviation from the computer program.

Preventive maintenance schedules can also be predetermined with a specific interval. The interval could be after so many components have been processed or after a particular time interval has passed. In either case, the number or time interval is decided in advance and can be the same for each preventive maintenance cycle.

Similarly, most conveyors operate at a particular velocity. The velocity is determined by the motor speed and the gearing that drives the conveyor belt. This means that the velocity of the conveyor may be set to different values, but, while the conveyor is running, it will run at the set velocity. While the conveyor is running the velocity will change only if there is some sort of malfunction or breakdown.

5.4.2   Probabilistic Input Data

In contrast to deterministic processes, a probabilistic process does not occur with the same type of regularity. In this case, the process will follow some probabilistic distribution. Thus, it is not known with the same type of confidence that the process will follow an exactly known behavior. Examples of probabilistic data include:

- Interarrival times
- Customer service processes
- Repair times

Interarrival times for entities coming into a system are almost always probabilistic. The interval between the last entity and the next entity may be short, or it may be long. It is not known exactly when the next entity will arrive. However, by collecting a mass of interarrival data, it is possible to see if the data follow some sort of distribution. If the number of entity arrivals in a given time period is completely random, it turns out that the interarrival time between entities actually follows an exponential distribution.

Customer service times can also be expected to be probabilistic. In other words, the amount of time that it takes to process an individual customer at a service center will vary depending on what the customer needs. In addition, any time a human being is involved in a service process, there is likely to be some variance in the service times. Generally speaking, in a service process, a small number of customers will be processed more rapidly than others. Similarly, a small number of customers may take a much longer time to be processed. Most of the customers may take some time between the two extremes. The pattern describing the number of observations with respect to the processing time creates a probabilistic distribution. Again, although some information is available about the probabilistic pattern that the processing time follows, the processing time of an individual customer in the future cannot be predicted with exact certainty. However, it frequently turns out that many customer service times follow a normal distribution because the process is actually a sum of a number of smaller subprocess times.

As with service times, repair times are likely to be probabilistic in nature. This is because it is impossible to predict what sort of problem needs to be repaired. If the problem is not serious, it would be expected that the repair time would be shorter. Conversely, if the problem is severe or requires other parts, the repair time will be longer.

5.5 Discrete vs. Continuous Data

Another classification of input data is whether the data are discrete or continuous. Discrete-type data can take only certain values. Usually this means a whole number. Examples of discrete data in simulation applications would be:

- The number of people who arrive in a system as a group or batch
- The number of jobs processed before a machine experiences a breakdown

Batch arrivals typically occur in service and entertainment-type systems. The number of customers in a batch can assume only whole number values. It is not possible to have a fractional customer. Similarly, the number of jobs processed through a machine before it experiences a breakdown is a whole number. Jobs are either completed or not completed.

Continuous distributions, on the other hand, can take any value in the observed range. This means that fractional numbers are a definite possibility. Examples of continuous type data include:

- Time between arrivals
- Service times
- Route times

Obviously, the time between entity arrivals in a system can take on any value between 0 and infinity regardless of the time unit used. This definitely includes fractional time values. Although some service times may be more often observed than others, service times can also take on any value within the observed time range. Service times, however, may have some specific lower limit below which it is not possible to observe. Similarly, travel times for entities on foot may also take any value within a reasonable range.

5.6 Common Input Data Distributions

The purpose of this section is not to turn the practitioner into a statistician but to provide some level of familiarization with a few of the most common input data distributions. There are many more different types of probabilistic distributions that the practitioner may actually encounter. Sometimes the practitioner may encounter these distributions only as a result of a computerized data fitting program. These types of programs are geared toward returning the best mathematical fit among many possible theoretical distributions. In these types of cases, a particular result does not necessarily mean that there is a rational reason why the data best fit a specific distribution. Sometimes a theoretical distribution that does make sense will be almost as good a fit. In these cases, the practitioner will have to decide for himself or herself whether it makes more sense to use the best mathematical fit or a very close fit that makes sense.

Because the purpose of this handbook is not to make the user a statistician but rather a simulation practitioner, the discussion of probability distributions is restricted to the most common types. These include the common distributions:

- Bernoulli
- Uniform
- Exponential
- Normal
- Triangular

There is also limited discussion of the following less common distributions:

- Beta
- Gamma
- Weibull

For a more mathematically or statistically focused treatment of this subject, practitioners are directed to any one of the multitude of respected references on probability and statistics.

5.6.1 Bernoulli Distribution

The Bernoulli distribution is used to model a random occurrence with one of two possible outcomes. These are frequently referred to as a success or failure. The mean and variance of the Bernoulli distribution are:

$$mean = p$$

$$var. = p(1-p)$$

where p = the fraction of successes and $(1 - p)$ = the fraction of failures.

The Bernoulli distribution is illustrated in Figure 5.2.

Examples of the Bernoulli distribution in simulation can be found in:

- Pass/fail inspection processes
- First-class vs. coach passengers
- Rush vs. regular priority orders

In the case of pass vs. fail inspection processes, the part can be only one of these two states. If 95% of the parts successfully pass the inspection, then the p value would be 0.95. The 5% failure rate would be represented by 0.05.

The first-class vs. coach passenger process does not really have a success vs. failure categorization. In this case it is simply the percentage of first-class passengers and coach-class passengers that must add up to 1.0.

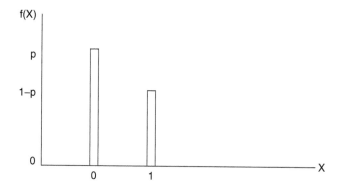

FIGURE 5.2 Bernoulli distribution.

A similar situation exists with the rush vs. regular priority orders. There is no success or failure associated with rush or regular orders. There are just two different types of orders with different percentages of likelihood.

5.6.2 Uniform Distribution

A uniform distribution means that over the range of possible values, each individual value is equally likely to be observed. The common example of a uniform distribution is the behavior of a single six-sided die. The minimum possible value is 1 and the maximum possible value is 6. Because all of the sides are the same, there is an equal probability of receiving any of the values between 1 and 6.

Although there is little possibility of encountering a simulation model that uses a single six-sided die, the uniform distribution does have some application in the world of simulation. Uniform distributions can be used as a first cut for modeling the input data of a process if there is little knowledge of the process. All that is needed to use the uniform distribution is the minimum time that a process has taken and the maximum time that the process has taken. Although it will not necessarily be a valid assumption that the data are uniformly distributed, the uniform distribution does allow the practitioner to begin building the simulation model. As the model becomes more sophisticated, the practitioner can give thought to using a more accurate or more appropriate distribution.

The uniform distribution may be either discrete or continuous. In the case of the six-sided die, the distribution is considered discrete. This means that the data values can take only whole numbers within the valid range. In reality, most simulation processes that are actually uniformly distributed can take any value including fractional numbers between the minimum and maximum values. This type of data is a continuous uniform distribution.

For the statistically inclined, the formulas for the mean and variance of a uniform distribution are:

$$mean = \frac{(a+b)}{2}$$

$$var. = \frac{(b-a)^2}{12}$$

where a is the minimum value and b is the maximum value.

The uniform distribution is illustrated in Figure 5.3.

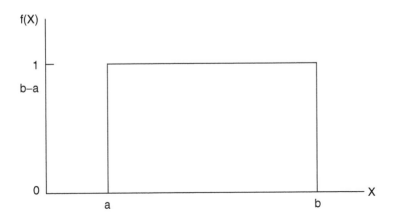

FIGURE 5.3 Uniform distribution.

5.6.3 Exponential Distribution

The exponential distribution is commonly utilized in conjunction with interarrival processes in simulation models because the arrival of entities in many systems has been either proven or assumed to be a random or Poisson process. This means that a random number of entities will arrive within a specific unit of time. Even though the process is random, there will still be some average number of arrivals within the unit of time. The number of arrivals that can be expected to arrive during the unit of time is randomly distributed around the average value.

When a system exhibits a random or Poisson distribution for the number of arrivals for a given period of time, the time between the arrivals turns out to be exponentially distributed. As we know, the time between the arrivals is known as the interarrival time. The exponential distribution has only a single parameter. This value is the mean of the data set. If the interarrival time does exhibit an exponential distribution, there will be a greater number of observations with interarrival times less than the mean value. There will also be fewer observations with interarrival times greater than the mean value. Thus, the number of observations continuously decreases as the interarrival time increases. The statistical equations for the mean and variance of the exponential distribution are:

$$mean = B$$

$$var. = B^2$$

The probability is represented by the following formula:

$$f(x) = \frac{1}{B} e^{-x/y}$$

where B is the average of the data sample and x is the data value.

It is also possible to manipulate the exponential distribution equation for other purposes. Specifically, if we integrate this equation and perform an inverse transformation, we can find out what cumulative percentage of observations is below a certain value. Conversely, if we specify a cumulative percentage, then we can also find the critical value. This means that for an exponentially distributed set of data with a mean of B, $F(x)$ percentage of the data will exist with a value less than or equal to x.

$$x = -B * \ln[1 - F(x)]$$

For example, if a data set is exponentially distributed with a mean of 5, then we can calculate the critical value for a cumulative percentage of 0.75:

$$6.93 = -5 * \ln(1 - 0.75)$$

This means that 75% of the observations in an exponentially distributed set of data with a mean of 5 will have a value of 6.93 or less.

The ability to transform a cumulative percentage into a critical value is particularly useful when one is attempting to perform a goodness of fit test to determine if a set of data could have come from an exponential distribution (Figure 5.4).

Examples of possible processes that would follow the exponential distribution include:

- Interarrival of customer
- Interarrival of orders
- Interarrival of machine breakdowns or failures

Note that each of these examples involves interarrival times. As previously discussed, the exponential distribution is applicable when the number of arrivals, whatever they may be, is randomly distributed

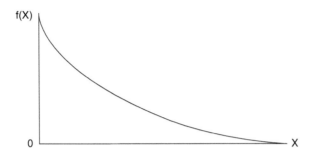

FIGURE 5.4 Exponential distribution.

with some mean over a given period of time. This does not mean that the exponential distribution cannot be used successfully to model other types of processes. However, if there is no underlying reason for a theoretical distribution to be used, the observed data will have to fit specifically the theoretical distribution in order for its use to be legitimate.

5.6.4 Triangular Distribution

The triangular distribution may be used in situations where the practitioner does not have complete knowledge of the system but suspects that the data are not uniformly distributed. In particular, if the practitioner suspects that the data are normally distributed, the triangular distribution may be a good first approximation. The triangular distribution has only three parameters: the minimum possible value, the most common value, and the maximum possible value (Figure 5.5). Because the most common value does not have to be equally between the minimum and the maximum value, the triangular distribution does not necessarily have to be symmetric.

If the practitioner is willing to assume that the distribution is triangularly distributed, it is not necessary to collect much data. In fact, the practitioner really needs to know only the three parameter values. In practice, these values may be acquired by asking a manager or machine operator to provide estimates.

The mean and variance of the triangular distribution are:

$$mean = \frac{a+m+b}{3}$$

$$Var. = \frac{(a^2+m^2+b^2-ma-ab-mb)}{18}$$

where

a = minimum value

m = most common value (the mode)

b = maximum value

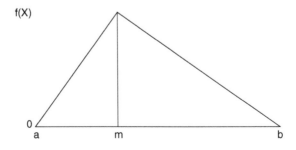

FIGURE 5.5 Triangular distribution.

Examples of uses of the triangular distribution include:

- Manufacturing processing times
- Customer service times
- Travel times

Manufacturing processing time can include probabilistic processes. This means that the manufacturing process would most likely require the presence of human beings somewhere in the process. It could also include a more deterministic process where there was still a variety of different jobs.

Customer service time can be successfully modeled with the triangular distribution. The minimum end of the distribution might represent the amount of time to record demographic data or access the customer's record in a data base. The most common time would reflect all of the processes of a typical transaction. The maximum time would represent a particularly complex service.

Travel time could also be modeled with a triangular distribution. In the case of modeling the travel time in an airport, the minimum time would correspond to someone running between points without luggage. The most common time would be someone with only one or two easily manageable bags. The maximum time would be someone walking with a large load of bulky luggage.

5.6.5 Normal Distribution

The time duration for many service processes follows the normal distribution. The reason for this is that many processes actually consist of a number of subprocesses. Regardless of the probability distribution of each individual subprocess, when the subprocess times are added together, the resulting time durations frequently become normally distributed.

The normal distribution has two parameters: the mean and the standard deviation. The normal distribution is also symmetric. This means that there are an equal number of observations less than and greater than the data mean. The pattern or distribution of the observations on each side is also similar.

The somewhat formidable mathematical formula for the normal distribution probability is:

$$f(x) = \frac{1}{\sigma\sqrt{2\pi}} e^{-(x-\mu)^2/2\sigma^2}$$

where μ is the mean and σ is the standard deviation. Figure 5.6 illustrates the major features of the normal distribution.

The normal distribution is frequently discussed in terms of the standard normal or Z distribution. This is a special form of the normal distribution where the mean is 0 and the standard deviation is 1. With the standard normal distribution 68% of the data lies between plus or minus one standard deviation of the mean; 95% of the data can be observed between plus or minus two standard deviations of the

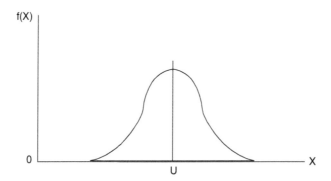

FIGURE 5.6 Normal distribution.

mean and 99.7% of the data is between plus or minus three standard deviations of the mean. For normally distributed data, the standard normal distribution can be converted to an actual data value with the following equation:

$$x = \mu \pm \sigma Z$$

where μ is the true population mean and σ is the true population standard deviation.

In practice, the true population mean is estimated with the sample mean, and the true population standard deviation is estimated with the sample standard deviation. This means that, for example, if we had a sample mean of 10 and a sample standard deviation of 2, we could calculate the 68% boundaries of a normal distribution with:

$$8 = 10 - 2\,(1)$$

$$12 = 10 + 2\,(1)$$

So approximately 68% of the observations of a normally distributed set of data with a mean of 10 and a standard deviation of 2 would be present between the values of 8 and 12. Similarly, we could calculate the boundaries for approximately 95% of the observations with:

$$6 = 10 - 2\,(2)$$

$$14 = 10 + 2\,(2)$$

This means that if the data were normally distributed with a mean of 10 and a standard deviation of 2, 95% of the observations would be between 6 and 14.

5.6.5.1 Using Normal Distribution Tables

We can also manipulate this equation to find out the values for specific cumulative percentages. This is performed with the use of the standard normal or Z table. A copy of this table is available in the appendix of the handbook. We illustrate the use of this table with a small example.

The following table is a common standard normal or Z table. These tables are readily available in most statistics books. The small table below was generated using Excel. It is an abbreviated version to illustrate how to use the table. A copy of the full table is easily generated with the function NORMDIST. This table is known as a right-hand tail table. The top row and left column contain the Z values. The interior of the table contains the cumulative percentage of observations for the standard normal distribution. The table is symmetric, so to obtain values on the left side of the distribution below the mean, the cumulative percentages are subtracted from 1.

Z value	0.0	0.01	0.02	0.03	0.04	0.05	0.06	0.07	0.08	0.09
0.0	0.500	0.504	0.508	0.512	0.516	0.520	0.524	0.528	0.532	0.536
0.1	0.540	0.544	0.548	0.552	0.556	0.560	0.564	0.567	0.571	0.575
0.2	0.579	0.583	0.587	0.591	0.595	0.599	0.603	0.606	0.610	0.614
0.3	0.618	0.622	0.626	0.629	0.633	0.637	0.641	0.644	0.648	0.652
0.4	0.655	0.659	0.663	0.666	0.670	0.674	0.677	0.681	0.684	0.688
0.5	0.691	0.695	0.698	0.702	0.705	0.709	0.712	0.716	0.719	0.722
0.6	0.726	0.729	0.732	0.736	0.739	0.742	0.745	0.749	0.752	0.755
0.7	0.758	0.761	0.764	0.767	0.770	0.773	0.776	0.779	0.782	0.785
0.8	0.788	0.791	0.794	0.797	0.800	0.802	0.805	0.808	0.811	0.813
0.9	0.816	0.819	0.821	0.824	0.826	0.829	0.831	0.834	0.836	0.839

To find out the standard normal *Z* value for a particular percentage on the right-hand side, we find the desired percentage on the interior of the table. We then move to the border to find the corresponding *Z* value. For example, if we wanted to know the value for having 80% of the observations for a normal distribution with the mean of 10 and a standard deviation of 2, we would follow these steps:

1. Look up the closest value in the interior of the table for 0.80. This value is in the ninth row, the fifth column. The *Z* value corresponding to this row is 0.8 from the ninth row and 0.04 from the fifth column. Our *Z* value is 0.84.
2. Obtain the actual *x* value by calculating: $x = 10+2*0.84 = 11.68$.

So, 80% of the observations of a normally distributed set of data with a mean of 10 and a standard deviation of 2 can be found below 11.68.

The procedure for finding values that are less than 50% is similar. For example, we would use the following procedure to find the value for a cumulative percentage of 0.39:

1. Because we have only a right-hand side table, we must convert our cumulative percentage value. A cumulative percentage of 0.39 is 0.11 to the left of the mean. To convert this to the right side, we add 0.11 to 0.50. We can now look up the cumulative value for 0.61. This value is found in the third row, ninth column. The corresponding Z value is 0.28.
2. Obtain the actual *x* value by calculating: $x = 10 - 2 * 0.28 = 9.44$.

This means that for a normally distributed set of data with a mean of 10 and a standard deviation of 2 that 39% of the observations will have a value of 9.44 or less.

The use of normal distribution tables is of specific value to the practitioner. Several subsequent procedures call for determining if a distribution is normal or not. In order to make these determinations, it is necessary to be able to identify the critical values for different normal distribution cumulative percentages.

5.6.5.2 Simulation Use of the Normal Distribution

When using the normal distribution to model service delays, the practitioner must display care when the mean service time is small. In this situation, the variance of the data may cause values less than the mean to become negative. Because service times cannot be less than zero, the practitioner must ensure that the service times generated during the simulation do not go negative. Fortunately, most simulation packages allow the practitioner to specify that any values generated by a normal distribution be discarded if they are negative.

Examples of processes frequently modeled with the normal distribution include:

- Manufacturing processing times
- Customer service times
- Travel times

Note that these are the same processes that the practitioner might initially estimate in the absence of data as the triangular distribution. In these cases, the use of the normal distribution would most likely occur as a result of the normal distribution fitting the empiric data rather than as an estimate. Without actually performing some sort of goodness of fit test, it would be questionable to automatically assume that a process was normally distributed and that the mean and standard deviation could be accurately estimated.

5.6.6 Poisson Distribution

The Poisson distribution (Figure 5.7) is used to model a random number of events that will occur in an interval of time. The Poisson distribution has only one parameter, λ. This distribution is unique in that the mean and variance are both equal to λ. The probability of observing a particular value is:

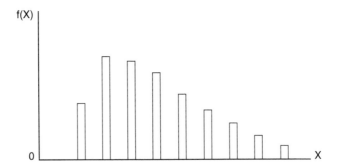

FIGURE 5.7 Poisson distribution.

$$p(x) = \frac{e^{-\lambda}\lambda^x}{x!}$$

where λ is both mean and variance and x is the value of the random variable.

Simulation uses of the Poisson distribution include:

- Number of arrivals in a given interval of time
- Number of entities in a batch

5.7 Less Common Distributions

Less commonly used distributions include the Weibull, gamma, beta, and geometric distributions. The practitioner is forewarned that the mathematics associated with these distributions is limited in actual application to the practitioner. In most instances, these distributions will be encountered only as a best fit from a computerized data-fitting software package. When a software package indicates that a particular distribution is the best fit, the package will also provide the distribution parameters associated with the fit. The normal practitioner use of this information is to plug these parameters into the simulation program. Although some will argue otherwise, the mathematics behind some of these less common distributions has limited meaning to the practitioner.

5.7.1 Weibull Distribution

The Weibull distribution is often used to represent distributions that cannot have values less than zero. This situation frequently exists with symmetric distributions like the normal distribution that represent service or process times. If the mean is small and the standard deviation is sufficiently large, many observations will accumulate on the left-hand side of the distribution near 0. This results in a non-symmetric distribution. The right-hand side of the distribution may still display a classic normal distribution tail.

The Weibull distribution possesses two parameters. These are an α shape parameter and a β scale parameter. Depending on the values of these two parameters, the Weibull can take on shapes ranging from that of the exponential distribution to the normal distribution. The mathematics associated with the Weibull distribution should not concern the practitioner. The equations are included here in the interest of completeness. In practice, the only information that the practitioner really need be concerned with are the α and β parameters. The lengthy probability function for the Weibull is:

$$f(x) = \alpha\beta^{-\alpha}x^{\alpha-1}e^{-(x/\beta)^{\alpha}}, \textit{ for } x > 0, 0 \textit{ otherwise}$$

where α is a shape parameter and β is a scale parameter.

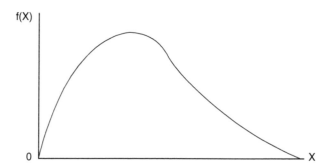

FIGURE 5.8 Weibull distribution.

The mean and variance for the Weibull distribution (Figure 5.8) are represented mathematically by:

$$mean = \frac{B}{\alpha}\Gamma(\frac{1}{\alpha})$$

$$variance = \frac{\beta^2}{\alpha}\left\{2\Gamma\left(\frac{2}{\alpha}\right) - \frac{1}{\alpha}\left[\Gamma\left(\frac{1}{\alpha}\right)\right]^2\right\}$$

where α is a shape parameter, β is a scale parameter, and Γ is given by

$$\Gamma = \int_0^\infty x^{\alpha-1}e^{-x}dx$$

5.7.2 Gamma Distribution

The gamma distribution is another distribution that may be less common to the practitioner. The gamma distribution (Figure 5.9) can also be somewhat intimidating to the practitioner. The same comments from the Weibull distribution apply to the gamma distribution. As with the Weibull distribution, the gamma distribution utilizes an α shape and a β scale parameter. The probability density equation for the gamma distribution is:

$$f(x) = \frac{1}{\beta^\alpha \Gamma(\alpha)} x^{\alpha-1} e^{-x/\beta}, \text{ for } x > 0, 0 \text{ otherwise}$$

where α, β, and Γ are defined as in the Weibull distribution.

Fortunately, the mean and variance of the gamma distribution are easily represented by $\alpha\beta$ and $\alpha\beta^2$, respectively.

The gamma distribution does possess one interesting characteristic of potential interest to the practitioner: under certain circumstances, the gamma distribution can degenerate to the same mathematical representation as the exponential distribution. These circumstances occur when the α shape parameter happens to be equal to 1. If the α shape parameter is close to 1, then exponential-like distributions may also be a close fit.

As with the Weibull distribution, the gamma distribution cannot go below 0. When the α shape parameter exceeds a value of 2, the gamma distribution can take on shapes similar to the Weibull distribution. This means that the gamma distribution may also be applicable when otherwise symmetric service or process distributions have small means and large variances.

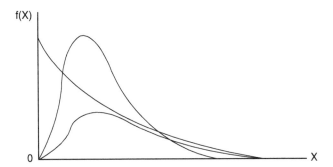

FIGURE 5.9 Gamma distribution.

5.7.3 Beta Distribution

A final equation that the practitioner may encounter is the beta distribution. This distribution is a bit different from most of the previously presented distributions. The beta distribution (Figure 5.10) holds the distinction of being able to cover only the range between 0 and 1. While the distribution itself can go only between 0 and 1, it is possible to offset it and or scale it with a multiplier value.

The beta distribution has two different parameters. These are the shape parameter α and the shape parameter β. The mathematical formula for the beta distribution probability density is:

$$f(x) = \frac{\Gamma(\alpha+\beta)}{\Gamma(a)\Gamma(\beta)} x^{\alpha-1}(1-x)^{\beta-1}, \text{ for } 0 < x < 1, 0 \text{ elsewhere}$$

where α and β are shape parameters 1 and 2, respectively, and Γ is defined as for the Weibull and gamma distributions.

The mean and variance of the beta distribution are:

$$mean = \frac{\alpha}{\alpha+\beta}$$

$$variance = \frac{\alpha\beta}{(\alpha+\beta)^2(\alpha+\beta+1)}$$

Aside from coincidentally having data that fit the beta distribution, the practitioner may encounter this distribution where it is necessary it fit data that is skewed to the right rather than the left as with the Weibull and gamma distributions. This represents some maximum time that can be taken for a particular process.

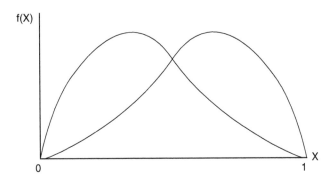

FIGURE 5.10 Beta distribution.

5.7.4   Geometric Distribution

In contrast to the previous less common distributions, the geometric distribution is discrete. As you may recall, this means that the values taken on by the geometric distribution must be whole numbers. The geometric distribution has one parameter p, which is considered the probability of success on any given attempt; $(1 - p)$ is the probability of failure on any given attempt. The probability of $x - 1$ failures before a success on the xth attempt is represented by:

$$p(x) = p(1 - p)^{x-1}, x = 1, 2...$$

The mean and variance of the geometric distribution are represented by:

$$mean = \frac{1-p}{p}$$

$$var. = \frac{1-p}{p^2}$$

The geometric distribution (Figure 5.11) can be used in simulation programs for:

- Arrival batch sizes
- Number of items inspected before a failure is encountered

The geometric distribution can model the size of batch arrivals. In particular, it has been found to model accurately the number of people in a batch arriving at airport security checkpoint systems. The batches represented single travelers, husbands and wives, families with children, and business travelers in groups. For arrival batch sizes, we are interested in the probability associated with a batch of x size.

Another use of the geometric distribution is to model the number of failures that are present before a success. To make more sense in a simulation program, the percentages between success and failure can be switched. Thus, we would actually be modeling the number of successes before the occurrence of the first failure.

5.8   Offset Combination Distributions

Some types of input data are actually a combination of a deterministic component and a probabilistic component. These types of processes generally have a minimum time, which constitutes the deterministic component. The remaining component of the time follows some sort of distribution. Examples of offset combination distributions are:

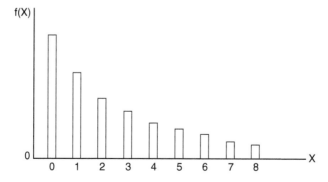

FIGURE 5.11   Geometric distribution.

- Technical telephone support
- Oil changes for cars
- Flexible manufacturing system cycles

In a typical telephone technical support system, a minimum amount of time is always required for each call. During the initial few minutes, the telephone technician must establish who the customer is and whether or not he or she is entitled to telephone technical support. If the customer is not eligible for support, that would be the minimum amount of possible time for the service process. The stochastic process would include the time solving the actual problem. Note that in a case like this, the practitioner may want to model the contact process as two separate processes: one for establishing qualifications and one for the service.

In the case of car oil changes, a minimum deterministic amount of time is required to raise and lower the car. The probabilistic component of the time would involve the removal of the old oil, the replacement with the new oil, and the time to change the oil filters.

Flexible manufacturing systems (FMSs) consist of two or more computer numerically controlled (CNC) machines and a common automatic materials-handling system. The processing times for the machining of the components and the transfer of the components is likely to be be deterministic because they are controlled by computer. However, the entire cycle will still require a human operator to initially load the components at the start of the cycle. Similarly, at the end of the cycle, a human operator will also be required to unload the components.

5.9 Analyzing Input Data

5.9.1 General Concepts

The process of determining the underlying theoretical distribution for a set of data usually involves what is known as a goodness of fit test (Johnson et al., 1999; Hildebrand and Ott, 1991). These tests are based on some sort of comparison between the observed data distribution and a corresponding theoretical distribution. If the difference between the observed data distribution and the corresponding theoretical distribution is small, then it may be stated with some level of certainty that the input data could have come from a set of data with the same parameters as the theoretical distribution. There are four different methods for conducting this comparison:

1. Graphic approach
2. Chi-square test
3. Kolmogorov–Smirnov test
4. Square error

Each of these approaches is examined in depth in the remainder of this chapter. Although the actual mechanics of each of the approaches is included, the practitioner may elect to go directly to the last section, which describes the software implementations of these techniques.

5.9.2 Graphic Approach

The most fundamental approach to attempting to fit input data is the graphic approach. This approach consists of a visual qualitative comparison between the actual data distribution and a theoretical distribution from which the observed data may have come. The steps for utilizing the graphic approach include:

- Create a histogram of observed data
- Create a histogram for the theoretical distribution
- Visually compare the two histograms for similarity
- Make a qualitative decision as to the similarity of the two data sets

In this technique, the practitioner must first decide how wide a data range each bar in the histogram covers and how many bars to graph. The lower and upper values in each data range form what is know as a data cell. The number of observations in each data cell is used to represent the height of the histogram bars. There are two common approaches for determining how to handle the cell issue:

1. Equal-interval approach
2. Equal-probability approach

5.9.2.1 Equal-Interval Approach

In the equal-interval approach, the practitioner sets the width of each data cell range to be the same value. This means that the practitioner then need decide only how many cells to utilize. The complete range of all the data values is then divided by the number of cells that are to be used. The resulting value is the width of each data range. Consider the following example of service times. The input data have a minimum value of 4 min and the maximum value is 24 min. The complete range is 20 min. If the practitioner decides to have five cells, the range for each of the five data cells is 4 min. This means that the first cell covers between 4 and 9 min, the second cell between 9 and 13 min, and so on.

The equal-interval approach requires that the practitioner decide how many cells to utilize. Although it would be possible to decide arbitrarily on a number, it would be better to derive this number in some sort of rational manner. One approach that has been taken in the past is to take the square root of the number of data points. The resulting value is the number of cells to use. With the number of cells, the cell ranges can be calculated as previously illustrated.

5.9.2.2 Equal-Probability Approach

A more statistically robust method for determining the number of cells is the equal-probability approach. With this method, the number of cells is dictated by the following algorithm:

- Use a maximum number of cells not to exceed 100.
- The expected number of observations in each cell must be at least 5.

To utilize this approach, the practitioner first needs to divide the total number of data points by 5. This will yield the number of cells to use. If the number of cells is greater than 100, then 100 is used for the number of cells. With this method it is likely that the number of data points is not going to be divided equally by 5. In this case, the practitioner can either discard the extra data or reduce the number of data cells to the next lower number. If the practitioner decides to discard the extra data, it is important to insure that the number of cells is not too small. On the other hand, if the practitioner rounds the number of cells down, the number of expected observations in each cell will be some number that is a fraction between 5 and 6.

Some statisticians would argue against discarding data, but if there is a large enough data set, the simplest approach would be to discard the few data points in order to have an even number of cells. So, for example, if there are 52 data points, there will need to be 52 divided by 5 data cells. This is 10 data cells with 2 data points left over. The discarding data approach would mean that the last 2 data points would not be used in the distribution-fitting process.

The practitioner can always round the number of cells down; however, this may require having to deal with a nonwhole number for the expected number of observations. The use of a nonwhole number for the expected number of observations is more statistically robust. However, it can be quite disconcerting to many practitioners. If this approach is used, then the 52 data points would result in a total of 52 divided by five cells. The resulting number of cells is 10.4. The number of cells is rounded down to 10. The number of expected observations is calculated by dividing 52 by 10. This means that each cell would be expected to have 5.2 observations. Naturally it will be impossible to observe 5.2 pieces of data in any data range because you cannot have a fractional observation.

The equal-probability approach possesses one other characteristic that can be disturbing to the novice practitioner. This characteristic is a result of the fact that each cell is equally probable in nature. This means that each cell is expected to have the same number of observations regardless of the distribution that is being examined. As a result, the width of different data cell intervals may vary.

5.9.2.3 Graphic Comparison of the Cells

Once the number of cells has been determined, the practitioner can decide what the cell boundaries are for each cell if the data came from a particular theoretical distribution. The practitioner then sorts the observed data and counts the frequency of observations corresponding to each cell. These frequencies are used to create the histogram. The final step involves a subjective visual comparison of the difference between the observed histogram and what might be expected from a theoretical distribution for each cell.

If this approach to data fitting appears to be questionable, you are correct. The subjective nature of the comparison between the observed and theoretical data does not result in a robust comparison. For this reason alone, the use of the graphic method to make conclusions about a particular data set is not recommended.

5.9.3 Chi-Square Goodness of Fit Test

The chi-square test is commonly accepted as the preferred goodness of fit technique. Like the graphic comparison test, the chi-square test is based on the comparison of the actual number of observations versus the expected number of observations. This means that the chi-square test also uses the equal-probability approach for determining the number of cells and the cell boundaries. The steps in executing the chi-square test are as follow:

- Establish null and alternative hypotheses
- Determine a level of test significance
- Calculate the critical value from the chi-square distribution
- Calculate the chi-square test statistic from the data
- Compare the test statistic with the critical value
- Accept or reject the null hypothesis

5.9.3.1 Hypotheses Statements

The null hypothesis will generally be a statement to the effect that the input data could have come from a particular theoretical distribution. Similarly, the alternate hypothesis is a statement that the input data could not have come from a particular theoretical distribution. In practice, the null and alternative hypotheses statements are condensed into the following format:

H_o: Distribution (parameter 1, parameter 2, …)
H_a: Not distribution (parameter 1, parameter 2, …)

Distribution would be the name of the actual theoretical distribution, and the parameters would correspond to the specific parameters associated with that particular theoretical distribution. If we were testing service data being normally distributed, we would need parameter data for the mean and standard deviation. For example, if we thought that the theoretical distribution was normal with a mean of 5 and a standard deviation of 2, the hypotheses statement would appear as:

H_o: Normal (5, 2)
H_a: Not normal (5, 2)

5.9.3.2 Determine a Level of Test Significance

It is necessary to establish a confidence level for the test. For example, if you want to be 95% confident of your test results, the level of significance is 0.05. The level of significance is often referred to as the α level. Other common α levels are 0.01 and 0.10.

5.9.3.3   Determine Critical Value for Chi-Square Distribution

This process consists of determining the critical value for the chi-square distribution. The critical value is the boundary between nonsignificant and significant halves of the chi-square distribution. In other words, with a 0.05 level of significance, the critical value would be the value where 95% of the distribution is to the left of the critical value, while 5% of the distribution is to the right of the critical value.

5.9.3.4   Calculate the Chi-Square Test Statistic from the Data

The test statistic is calculated by summing the squared differences between the observed number of data points and the expected number of data points divided by the expected number of data points for each individual data cell. This process is more easily explained with the following formula:

$$\chi^2 = \sum_{i=1}^{n} \frac{(O_i - E_i)^2}{E_i}$$

where

χ^2 = test statistic to be calculated and compared to the critical value.

O_i = number of observations of data points in the ith data cell.

E_i = number of expected data points in the ith data cell.

n = number of data cells.

For each cell in the above formula, the practitioner will take the number of observed values in the cell range and subtract the number of expected values. This term is then squared. The resulting positive value is divided by the same number of expected values just previously used. These calculations are made for each cell in the test and summed. The resulting value follows the chi-square distribution.

5.9.3.5   Compare the Test Statistic with the Critical Value

In this step, we compare the test statistic that was just calculated with the critical value that was previously determined. The test statistic will either be less than the critical value or greater than the critical value.

5.9.3.6   Accept or Reject the Null Hypotheses

If it turns out that the test statistic is less than the critical value, then the null hypothesis that the data could have come from that theoretical distribution cannot be rejected at the previously determined level of significance. Conversely, if the test statistic value is greater than the critical value, then the null hypothesis is rejected. That would mean that there is evidence that the data did not come from that theoretical distribution.

5.9.3.7   Minimum Number of Data Points for the Chi-Square Test

A possible weakness of the chi-square test is that it can be executed only if a sufficient quantity of data exists to apply the test. Generally speaking, it is necessary to have at least 20 data points in order for the test to mathematically function. If only 20 data points are available, a total of 4 data cells will be utilized. This means that a more realistic minimum would be at least 30 data points. With at least 30 data points, the practitioner can have reasonable confidence in the chi-square test results. In the event that an insufficient quantity of data exists to perform the chi-square test, the practitioner may consider utilizing the Kolmogorov–Smirnov, or KS, test.

5.9.3.8   Chi-Square Example

Consider the following example to demonstrate the use of the chi-square goodness of fit test. The data in the table below were obtained from interarrival times in minutes of customers at a service center. We would like to verify that the interarrival distribution is exponentially distributed.

0.87	2.57	3.23	3.94	0.06	0.95
2.48	1.43	1.63	15.80	1.50	1.36
3.43	0.25	1.04	5.53	0.54	1.41
2.68	0.80	3.86	2.23	2.00	1.88
2.73	0.17	0.01	0.55	3.48	0.77

To begin, we will need to calculate summary statistics for the data. Even though we can calculate both the mean and standard deviation of the data, the exponential distribution has only a single parameter, the mean.

Mean $= 2.31$

Std. Dev. $= 2.88$

Count $= 30$

We have 30 data points, and we are going to use the recommended equiprobable approach. This means that we will need to use a total of 30/5 or 6 cells in our test.

With the summary statistics, it is possible to set up the null and alternative hypotheses. These are:

Step 1. H_o: Expo (2.31). H_a: Not Expo (2.31).

Step 2. The level of significance is chosen as 0.05.

Step 3. The chi-square critical value is $6 - 1 - 1 = 4$. There are six cells, one parameter for the mean and one additional degree of freedom for the test. Looking at a chi-square table or using the Excel chiinv(0.05,4) function formula, the critical value is 9.49.

Step 4. We calculate the lower and upper percent boundaries for each cell, the upper and lower x values for each cell, and the observed and expected observations for each cell.

Cell	Lower Percent	Upper Percent	Lower x	Upper x	Observed	Expected	$\dfrac{(O-E)^2}{E}$
1	0.000	0.167	0.000	0.421	4	5	0.2
2	0.167	0.333	0.421	0.935	5	5	0
3	0.333	0.500	0.935	1.599	6	5	0.2
4	0.500	0.667	1.599	2.534	5	5	0
5	0.667	0.833	2.534	4.133	8	5	1.8
6	0.833	1.000	4.133	84.736	2	5	1.8

The values in the lower x and upper x columns are calculated using the formula:

$$x = -0.97 * \ln[1 - F(x)]$$

where $F(x)$ is the cumulative percentage from the lower and upper percentage columns.

The test statistic is the sum of the last column:

$$0.2 + 0.0 + 0.2 + 0.0 + 1.8 + 1.8 = 4.0$$

Step 5. The test statistic is less than the critical value at alpha 0.05, $4.0 < 9.4$.

Step 6. Cannot reject null hypothesis of data being exponentially distributed with a mean of 2.31.

5.9.3.9 Normal Distribution Chi-Square Example

The chi-square goodness of fit test is also easy to implement for the normal distribution. Here we may want to test the following service time data to see if it could have come from a normal distribution.

6.44	5.92	6.88	6.49	9.14	6.98
9.01	4.80	1.77	0.52	3.59	5.10
1.28	7.75	4.90	7.91	6.33	6.44
3.10	7.09	3.59	6.26	4.88	3.68
5.35	3.82	9.37	4.98	7.50	7.39

The summary statistics for this data set are:

Mean   = 5.61

St. Dev. = 2.25

Count = 30

5.9.4   Kolmogorov–Smirnov

The KS test should be utilized only when the number of data points is extremely limited and the chi-square test cannot be properly applied. The reason for this is that it is generally accepted that the KS test has less ability to properly fit data than other techniques such as the chi-square test. A final limitation of the KS test is that some references recommend against using the KS with discrete distributions.

There are actually many versions of this test with varying degrees of complexity. For a complete discussion of the KS test, practitioners are directed to the Law and Kelton simulation text. The version of the KS test presented in this handbook is the simplest to implement. Hard-core statisticians sometimes criticize this version of the test as statistically weak. However, for the practitioner, the difference is likely to be insignificant.

The concept behind the KS test is a comparison between the cumulative theoretical distribution and the cumulative observed distribution. If the maximum difference between the cumulative theoretical and observed distribution exceeds a critical KS value, then the observed distribution could not have come from the theoretical distribution. The steps for the KS are:

- Establish null and alternative hypotheses
- Determine a level of test significance
- Determine the critical ks value from the *d* table
- Determine the greatest absolute difference between the two cumulative distributions
- Compare the difference with the critical ks value
- Accept or reject the null hypothesis

5.9.4.1   Establish Null and Alternative Hypotheses

As with the chi-square test, the KS test begins with the establishment of the null and alternative hypotheses. For the KS test, we can also condense the null and alternative hypotheses to:

H_o: Distribution (parameter 1, parameter 2, ...)
H_a: Not distribution (parameter 1, parameter 2, ...)

5.9.4.2   Determine a Level of Test Significance

In the same fashion as the chi-square test, it is also necessary to establish a confidence level for the KS test. For example, if you want to be 95% confident of your test results, the level of significance is 0.05. The level of significance is often referred to as the α level. Other common α levels are 0.01 and 0.10.

5.9.4.3   Determine the Critical KS Value from the *D* Table

The critical value for the KS test is obtained from a *D* table. The *D* table has two parameters, the size of the sample and the level of significance. A copy of a *D* table is available in the appendix of this handbook.

5.9.4.4   Determine the Greatest Absolute Difference between the Two Cumulative Distributions

In this step, the cumulative probabilities of both the theoretical and observed distributions are determined. A simple plot may be used to assist in this process. The cumulative probability is plotted on the vertical axis, and the data value ranges are plotted on the horizontal axis. For the observed data distribution, the cumulative probability is the number of observations that are less than or equal to the data value divided by the total number of observations. For the theoretical distribution, the cumulative probability can be mathematically calculated.

Once the plot is complete, the objective is to determine the maximum absolute difference in the cumulative probability between the theoretical and observed distributions. This is simply performed by subtracting the cumulative distribution values.

5.9.4.5 Compare the Difference with the Critical KS Value

The maximum absolute difference in the cumulative probability between the theoretical and observed distributions is then compared.

5.9.4.6 Accept or Reject the Null Hypotheses

If the maximum absolute difference is less than the critical KS value, then the null hypotheses cannot be rejected. Conversely, if the maximum absolute difference is greater than the critical KS value, then the null hypothesis is rejected. If the null hypotheses cannot be rejected, then the sample could have come from the theoretical distribution with the specified parameters. Otherwise, if the null hypothesis is rejected, then the sample did not come from the theoretical distribution with the specified parameters.

5.9.5 Square Error

The square error approach utilizes the same previously describe equal-interval or equal-probability approach to determine the number of cells and cell boundaries. As the name implies, the square error approach uses a summed total of the square of the error between the observed and the theoretical distributions. The error is defined as the difference between the two distributions for each individual data cell. The square error approach is commonly used as a means of assessing the relative suitability of a variety of different theoretical distributions to represent the observed distribution. Thus, the best fitting theoretical distribution would be considered to be the one with the least summed square error.

5.10 How Much Data Needs to Be Collected

A very common question among novice practitioners concerns how much data needs to be collected. The likelihood of this question seems to be related to the degree of difficulty involved with the collection of the data. The harder or more unpleasant it is to collect the data, the greater the probability is that the question will arise. This is a difficult question to answer in isolation; however, the following observations may help:

- The right data
- The different values that are likely to occur
- The need to have enough data to perform a goodness of fit test

5.10.1 Observe the Right Data

An important data collection principle is that the data must not be biased in any manner. For example, if the practitioner attempts to collect interarrival data for a restaurant during the week, the data cannot be used to drive a model used to examine staffing procedures on the weekends. If the intention is to examine the performance of the system on the weekends, then the input data must also be collected during the same time. Furthermore, if the model is to be examined for both the weekday and weekend effects, then both types of data must be collected.

5.10.2 Observe Representative Data

Another important issue is that the practitioner collects data that is reasonably likely to contain the complete range of input data that might be present for a particular process. If extreme values are not observed but do occur and are important, then any resulting theoretical distribution may not be valid. This issue is not as easily resolved as the weekday/weekend issue.

5.10.3 Have Enough Data for Goodness of Fit Tests

In order to fit observed data to a theoretical distribution, a minimum amount of data is usually necessary. Although there are no magic numbers, having fewer than 25 to 30 data points may prevent the practitioner from properly executing a goodness of fit test. When a marginal number of data points is collected, the practitioner may be forced to use less desirable or less robust forms of analysis. It is always in the best interest of the practitioner to take a conservative approach and collect as much reliable data as is reasonably possible. It is usually much more difficult to go back and collect additional data than it is to collect a conservative amount of data in the first place.

5.11 What Happens If I Cannot Fit the Input Data?

Periodically the practitioner will encounter a situation where it is not possible to fit the observed data to a theoretical distribution. Assuming that the data were collected accurately, possible causes for this difficulty include:

- Not enough data were collected.
- Data are a combination of a number of different distributions.

Even though the practitioner may have been able to collect at least 30 data points, there is no guarantee that this data is enough to accurately fit a given distribution. So unless it is overly burdensome to collect additional data, the practitioner should be prepared to return to the system and collect more data. If it is not possible to collect the additional data, the practitioner may attempt to drive the simulation with observed data instead of theoretical distribution data. Most simulation packages have a provision for this eventuality. Although the implementation of the observed or empirical data distribution may vary, a common approach is to use a cumulative distribution approach to generate the data.

Another possibility for the failure to fit a theoretical distribution is that the observed data time may actually be a combination of several different processes. These processes could be either mutually exclusive or sequential. A typical example of where this might occur is when customer-type entities pay for goods at a checkout counter. Here, practitioners may lump several different sequential processing times into one set of data. The actual process of the checkout system consists of:

- Loading of goods on the conveyor
- Scanning the goods
- Payment for the goods

If all of these processes are lumped together, we will ignore the effect that more items are likely to take longer to load and scan. On the other hand, the payment of goods would not be expected to be a function of the number of goods that are being purchased. Thus, the practitioner may have to break up this process into more than one set of data in order to have reasonable hopes of fitting the data sets into theoretical distributions.

The other type of situation where it could be very difficult to fit the observed data is if different mutually exclusive processes are combined. This situation might also be present in the same checkout example. The methods of payment for the goods are mutually exclusive processes. That is, only one type of payment would normally be performed:

- Cash
- Check
- Credit
- Debit

The practitioner should expect that a cash transaction would take on the average less time than a more complicated credit or debit transaction. A check transaction might even take longer than either a credit or debit transactions. If each of these types of payments possesses a different theoretical distribution, then disregarding the type of payment could easily result in a combined distribution that cannot be fit to any single theoretical distribution. In this situation, the practitioner must collect individual data times for each of the different types of payment. Note that this can automatically result in a significantly greater amount of data collection than was originally performed. The data collection situation is actually a little bit worse. The practitioner must also have an additional distribution that determines the percentage of each different type of payment. If the practitioner is diligent in the data collection, it is possible that the percentage of type distribution may also be collected at the same time as the payment times.

5.12 Software Implementations for Data Fitting

Fitting a significant number of observed data sets to theoretical distributions can become a very time-consuming task. In some cases, it may be mathematically very difficult to fit the observed data to some of the more exotic probability distributions. For this reason, most practitioners utilize some sort of data-fitting software. Two commonly available programs among others to perform this function are:

1. Arena input analyzer
2. Expert fit

5.12.1 ARENA Input Analyzer

The Input Analyzer is part of the ARENA simulation software package available from Rockwell software. The Input Analyzer has the capability to:

- Determine the quality of fit of probability distribution functions to input data
- Examine a total of 15 distributions for data fitting
- Calculate chi-square, Kolmogorov–Smirnov, and square error tests
- Generate high-quality data plots

The Input Analyzer is available through *www.arenasimulation.com*. A screen capture from the Input Analyzer is illustrated in Figure 5.12.

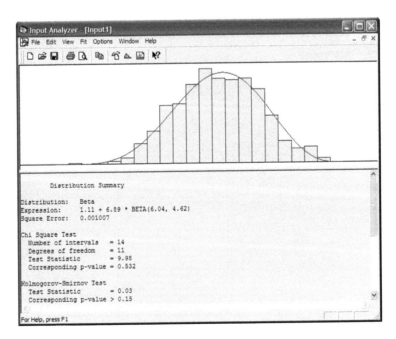

FIGURE 5.12 ARENA Input Analyzer.

A short tutorial on using the ARENA Input Analyzer is available in the User's Manuals section of this handbook. Practitioners can use the guide along with the online help system to analyze successfully input data.

5.12.2 Expert Fit

The Expert Fit data-fitting software is available through Averill M. Law & Associates. This software has the capability to:

- Automatically determine the best probability distribution for data sets
- Fit 40 probability distributions
- Conduct chi-square, Kolmogorov–Smirnov, and Anderson–Darling goodness of fit tests
- Provide high-quality plots
- Analyze a large number of data sets in batch mode

For additional information, practitioners are directed to the Expert Fit Web site at *www.averill-law.com*. Figure 5.13 is a screen capture from Expert Fit.

5.13 Summary

In this chapter we covered the collection and analysis of input data. This phase of the simulation project is often considered the most difficult. Data may be collected from historical records or can be collected in real time. In order to use observed data in a simulation model, it must preferably first be fit to a theoretical probability distribution. The theoretical probability distribution is then used to generate values to drive the simulation model.

There are a large number of theoretical distributions that the practitioner may encounter. The most common distributions are the Bernoulli, Uniform, Poisson, Exponential, Triangular, and Normal. The less common distributions are the Weibull, Gamma, Beta, and Geometric.

In order to determine the best theoretical distribution fit for the observed data, the practitioner can use one of many comparison methods. These include the graphic, chi-square, KS, and square error

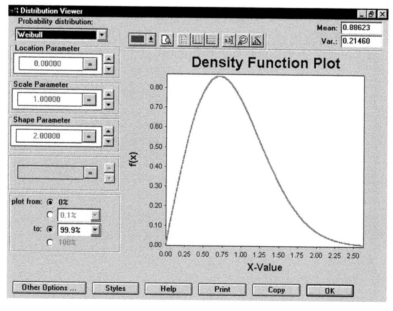

FIGURE 5.13   Expert Fit.

methods. In some cases, it may not be possible to obtain a satisfactory theoretical fit. If possible, additional data should be collected, and a new attempt to fit the data should be made. If this is not possible, it may be necessary to use observed data to drive the simulation model.

Chapter Questions

1. Why is it necessary to determine the underlying theoretical distribution for an input data set?

2. What is the difference between deterministic and probabilistic data?

3. What is the difference between discrete and continuous distributions?

4. Why do many service processes seem to be normally distributed?

5. What is a common cause of observed data's not being able to be fitted to a theoretical distribution.

6. The following interarrival times were observed. Determine the mean of the sample. Determine if the sample could have come from an exponentially distributed population with the same mean at an α level of 0.05.

   3.9, 4.1, 4.4, 6.0, 4.4, 4.5, 3.2, 4.0, 4.7, 5.0,

   3.7, 5.6, 3.3, 7.3, 5.1, 4.3, 3.3, 1.2, 5.3, 3.5,

   3.8, 3.0, 2.7, 3.6, 4.4, 4.5, 2.5, 1.8, 1.9, 2.1,

   1.8, 2.8, 2.9

7. The following service times were observed. Determine the mean and standard deviation of the sample. Determine if the sample could have come from a normally distributed population with the same mean and standard deviation at an α level of 0.05.

4.2, 3.1, 3.6, 6.2, 4.4, 4.2, 3.6, 4.0, 4.9, 5.1,

3.4, 6.1, 4.4, 4.3, 6.1, 5.3, 4.1, 0.2, 4.3, 3.5,

3.8, 3.3, 2.9, 3.8, 6.2

References

Hildebrand, D.K. and Ott, L. (1991), *Statistical Thinking for Managers,* PWS-Kent, Boston.

Johnson, R.A., Freund, J.E., and Miller, I. (1999), *Miller and Freund's Probability and Statistics for Engineers,* Pearson Education, New York.

6

Model Translation

"Is that some kind of game you are playing?"

6.1 Introduction

The objective of the preceding system definition and model formulation phase was to determine which components of the system should be included in the model and how the model should flow. The objective of the model translation phase is to translate the system into a computer model that can be used to generate experimental data. This is a two-step process. The first part of this process requires that the practitioner decide what type of computer program to utilize to model the system. The second part of the model translation phase is to actually perform the programming of the simulation model.

6.2 Simulation Program Selection

Today, the simulation practitioner has the opportunity to choose among a wide variety of different computer simulation programs (Banks et al., 2000). The practitioner's choice of a particular simulation software package may be based on:

- Advances in computer hardware
- Software cost
- Practitioner preference

6.2.1 Advances in Computer Hardware

Many advances have taken place in the microcomputer industry in the last two decades. Advances, particularly in the areas of computing power and graphics displays, have had a tremendous impact on the acceptance of simulation modeling. When the graphics capabilities of computer models were non-existent, the practitioner was forced to relate the significance of numerical data to his audience. With the advent of high-powered microcomputers and photorealistic graphics, the practitioner can now illustrate his or her point. This does not mean that nongraphical simulation models do not still have value to the practitioner. Even advanced graphical simulation packages may rely on or even force the practitioner to use general purpose programming languages for complicated heuristics that are not available as internal functions. Another point in favor of general purpose programming languages is calculation speed. Graphical simulation packages must carry user interface overhead calculations. This means that calculation-intensive simulation applications may actually favor the use of general purpose programming languages.

6.2.2 Software Acquisition and Costs

Another issue that may influence the decision to use either a simulation-specific program or a general purpose programming language is software cost. The typical commercial simulation package price is measured in thousands of dollars. Because this can amount to a sizable percentage of a small organization's budget, most simulation software companies offer both academic and training/evaluation versions.

The academic versions of the software are frequently available to registered students in educational institutions at no cost. Another way to acquire an academic version may be to purchase a textbook specifically designed for a university simulation course. Simulation textbooks frequently include a copy of simulation software as part of the textbook. The training and evaluation versions of simulation software may also be available online from the vendor at little or no cost.

The downside to acquiring either an academic or training/evaluation version of the software is that it is undoubtedly restricted in some manner. Some software companies restrict the program by limiting the number of entities that may be present in the model at any given time. This can have the effect of limiting the size of the model that the practitioner can develop. For example, the academic version of ARENA is limited to 150 entities at any given time.

Another approach commonly utilized by software companies is to keep full functionality of the program but to limit the amount of time that the software may be utilized. At the end of the evaluation period, the practitioner has the opportunity to purchase the full version of the software.

On the other end of the cost spectrum are general purpose programming languages. It is possible that copies of this type of software may already be found in-house or can be purchased for a few hundred dollars. In the event that the practitioner intends to perform all programming tasks himself or herself, there are no further software costs. However, if the practitioner plans to utilize previously developed simulation-specific subroutines, some additional costs may be incurred. These costs may come in the form of purchasing either hardcopy or electronic programming code.

6.2.3 Practitioner Preference

Sometimes the selection of the computer language or program to use comes down to the preference of the practitioner. If, for example, the practitioner is a very accomplished general purpose programmer and the project is calculation intensive and does not require animation, it may be a wise decision to use a general purpose programming language. On the other hand, if the practitioner is not familiar with programming and needs to be able to demonstrate the model to others, then it may be more appropriate to use a simulation-specific software package. To help decide which approach is more appropriate, we have included a short discussion and a number of user minimanuals.

6.2.4 Simulation-Specific Programming Languages

There a number of simulation-specific programming languages currently on the market. The number of these types of programs is probably increasing on a yearly basis. The selection of the software packages discussed in this handbook cover only a few of the most commonly used packages and operate on the IBM PC platform:

- Arena
- Automod
- Simscript

Copies of basic software manuals for each of these software packages are included at the end of this handbook. The software manuals are not intended to cover all of the capabilities of the software package. The software manuals are intended to provide enough instruction to provide the practitioner with a basic knowledge to develop a simple simulation model. If the practitioner decides that a particular software package is most appropriate, then the practitioner should consider acquiring the full version of that software package.

6.2.5 General Purpose Programming Languages

Virtually any general purpose high-level programming language can be used to create simulation models (Banks et al., 2000; Schriber, 1990). This includes programming languages such as Visual Basic, C++, FORTRAN, and Pascal. However, one aspect of using general purpose programming languages is that the practitioner must usually accede to the exclusion of highly detailed animation graphics. The return on programming time investment for animation graphics of this type would not ordinarily be justified.

The level of sophistication of the mathematical model along with moderate animation depends primarily on the practitioner's individual skill level. As illustrated in Chapter 1, "Introduction," all simulation programs must include some code to manage the event list. Even a seemingly simple task such as this may actually require a significant amount of programming and debugging.

The practitioner can leverage previously developed simulation-specific subroutines by other practitioners. This can significantly reduce the amount of programming effort required. However, the practitioner will still be responsible for integrating the subroutines into their main program. Most notable of these are the FORTRAN subroutines developed by Schriber and the BASIC subroutines developed by Donaghey of the University of Houston. An extensive description of the SIMPAK BASIC subroutines developed by Donaghey is included in the software manual section of this handbook.

6.3 Model Translation Section Content

The second part of the model translation process includes the actual programming of the system. Although it would be impossible to provide programming examples of every type of simulation model in every type of programming language, the following general simulation programming philosophy is provided to help insure that the practitioner develops robust and functional simulation models. This philosophy includes approaches for:

- Getting started
- Version management
- Programming commenting
- Program organization
- Mnemonic naming conventions
- Advanced program planning
- Multiple practitioner program development

6.3.1 Getting Started

Sometimes, the most difficult part of programming the simulation is getting started. Fortunately for the practitioner, at this point in the simulation program, a high-level flow chart should already exist. As you will recall from the system definition phase, the high-level flow chart represents a conceptual view of the major process component flow and interactions. Because the practitioner should have initially developed the high-level flow chart from the perspective of the system entities, there is a direct correlation to the actual simulation program. The major difference between the two is the degree of detail. If the practitioner has a conceptual high-level understanding of how the system operates, it will be far easier to expand the high-level flow chart to develop the lower-level simulation program code.

Unfortunately, many novice practitioners find themselves at the model translation phase without a functional high-level flow chart. These individuals attempt to attack the program from their mental understanding of the system process. If a high-level flow chart is actually required, it is usually developed after the program code has been completed. Because the high-level flow chart is developed at the end of the project, it is not always representative of the overall level of quality of the project. This approach inevitably wastes a great deal of time and effort that could have easily been avoided if a functional high-level flow chart had been available.

An equally avoidable situation is one in which the practitioner allows the model development to outrun the high-level flow chart. One cause for this is the discovery of additional system information after the high-level flow chart was initially developed. Another cause is an increased understanding of the system while modeling the process. When either of these events occurs, the practitioner is once again working in the dark, possibly without the benefit of a conceptual understanding of the newly discovered process. The best course of action is to temporarily halt the model translation process until the high-level flow chart of the system can be updated.

6.3.2 Version Management

This section addresses administrative issues to help assist the practitioner in creating and maintaining an organized program file system. Topics in this section include the use of:

- Project subdirectories
- Saving simulation programs
- File version management techniques
- Backing up simulation project files

6.3.3 Project Subdirectories

Practitioners should begin all simulation projects by creating a separate computer subdirectory for each individual project. These subdirectories can reside in either the native simulation program subdirectory, which holds the simulation program executable files, or in a dedicated set of data subdirectories. Each method has its advantages and corresponding disadvantages.

Advantages of the simulation program subdirectory are:

- Project files are immediately accessible after starting the simulation program.
- Project files are in the same place regardless of which specific computer is used.
- Fewer navigation issues exist for accessing supporting simulation libraries.

Advantages to using a dedicated project subdirectory are:

- Simulation program subdirectory remains uncluttered.
- It is easier to back up separate simulation project file subdirectories.
- Program reinstallations or deinstallations will not affect the subdirectories.

Although the practitioner will have to make an individual choice as to which subdirectory location to use, it will generally be more advantageous to use a separate dedicated data subdirectory. A major factor in this decision is that many practitioners already regularly back up data subdirectories but do not back up the entire computer drive. So, when disaster strikes, it will be necessary to reinstall only the software on the new local drive and copy the project files to a new data subdirectory.

6.3.4 Saving Simulation Programs

Practitioners should be utilizing a "simplify and embellish" and "divide and conquer" approach to developing complex simulation programs. As discussed in Chapter 7, "Verification," the "simplify and embellish" and "divide and conquer" approach can greatly reduce the development effort. With this approach, the practitioner will start out with a relatively simple simulation program. Once the program is running correctly, the practitioner can add additional components and sophistication. This will result in a steady stream of incrementally different models. The practitioner should be saving the model on a continuous basis, even during a single development session. The computer could crash at any moment, causing the practitioner to lose any data that have not been recently saved. This is not to say that simulation programs are inherently unreliable. However, whenever a complex program is used, perhaps simultaneously with other programs, the practitioner must be prepared for this eventuality. So save your model and save your model frequently to avoid this potential problem. When the practitioner does decide to save, there are two choices in terms of file version management.

6.3.5 Single-File Version Management

The first choice is to keep overwriting the original file with the same name. This means that the single version of the file will always be under development. If the file becomes corrupted, all work on the project may be lost. Almost as bad is a situation where the practitioner's latest modifications become untenable. Significant time can be wasted attempting to recover the code that does work properly. As ridiculous as this approach may appear, many practitioners resolutely subscribe to this method (or perhaps non-method) of file management.

Sometimes practitioners who use the single-file method attempt to address these deficiencies by backing up the file to another subdirectory or to a removable disk. In this case, the practitioner then has to continuously deal with which version is actually the latest model. With the extraordinarily inexpensive hard drive space available today, there is absolutely no need to practice this approach. Unfortunately, many novice practitioners take this approach. Even some experienced practitioners who have lost model data in this manner continue to develop their models in this manner. Perhaps these individuals are practicing some sort of psychological data file denial. If the single-file method appeals to you, now is the time to develop a more disciplined approach to version file management.

6.3.6 Multiple-File Version Management

The second and most appropriate version file management technique is continuously to save each new version under a different file name. Thus, an original file named "model" would become "model01," "model02," etc. The model should be saved under a new number every time either significant enhancements have been made or the current model has been successfully debugged. Should the model be very complex, the practitioner might even give thought to using separate subdirectories within the project subdirectory. At the end of the model translation phase, the practitioner should have a substantial number of models representing an increasingly sophisticated program. Should disaster strike in some form or another, it is obvious that much less work will be lost than with the single-file method.

Once the initial model is complete, the practitioner will also have to make modifications to represent the other experimental configurations. Each experimental configuration should be saved as a different model. Once again, the practitioner should use some sort of rational method to describe the different experimental versions. For example, the following table illustrates how to name different experimental versions for a study with two different factors at three different levels.

X-ray \ Metal det.	2 Metal operators	3 Metal operators	4 Metal operators
2 X-ray operators	Config22	Config23	Config24
3 X-ray operators	Config32	Config33	Config34
4 X-ray operators	Config42	Config43	Config44

The time to save the new configuration is not after the modifications have been made to the model. It is immediately after identifying that a new model needs to be made. It is far too easy to accidentally hit the save button rather than the save as button after the practitioner is engrossed in the model modifications.

6.3.7 Using Multiple Computers for Project Development

Another file version situation in which the practitioner may find himself or herself involved is the use of more than one computer to develop a simulation program. This situation can occur when the practitioner performs some development on a work office computer, a notebook computer, and a home computer. The use of multiple computers can be necessitated by the work office computer having the best monitor or processor, a lengthy trip needing a notebook computer, or working at home for the day. Regardless of the reason for using multiple computers, the effort to keep track of different versions of the model can quickly get out of hand. It is extremely easy to overwrite files accidentally while copying the project between computers.

To help avoid this type of version management problem, the practitioner can take several measures. These measures cannot guarantee that no data will be lost; however, they will minimize the possibility that it will occur. These include:

- Ensuring that all the computers clocks are synchronized
- Starting a new file number each session
- Employing synchronizing software

By ensuring that all of the computer clocks are exactly synchronized with respect to date and time, the practitioner can reference the date and time when the file was saved. In the event that the practitioner ends up with two versions of the same file with the same name, there may be hope that the later version can be identified. The date and time index may also be used in conjunction with the file size to assist in the proper identification of the latest file. This of course assumes that the practitioner is incrementally developing the model and that the model size is constantly increasing.

A more fail-safe approach is to use a new version file number immediately on beginning to work on the file with a new computer. This will go a long way to help guarantee that no two versions of the most current file can exist at any given time on two different computers. If there is any doubt as to the sequencing of the numbers, the practitioner should use an increment that is large enough to avoid confusion. So instead of incrementing the new version by a unit of 1, the practitioner may play it safe by creating a new sequence by incrementing to the next tenth unit. In the event that there is actually a problem, it is easy for the practitioner to go back to the last version of the previous sequence.

The final tactic that the practitioner may use is to employ some sort of file synchronization software. This type of software is frequently available as part of a removable media drive software package. The software will synchronize the different file versions with varying levels of automation. If the practitioner chooses to go this route, the software should be checked periodically to insure that it is actually doing what the practitioner intends.

6.3.8 Backing up Simulation Files

A final file-saving issue involves backing up the program files. Even though a practitioner may have many archival versions of the program, if all of the files are located on one physical device, the practitioner is once again putting them at risk. This practitioner once observed a simulation project team back up files in each other's file server subdirectory. The team thought that these multiple backups prevented data loss. This approach worked until the entire file server drive was lost without a recent tape backup. This demoralized project team was not able to recover fully from this loss.

For most personal computer hard drives, the question is not if the drive will crash, it is a question of when the hard drive will crash. So, it is advisable to ensure that you as a practitioner give thought to this issue. It is not enough just to back up the files to a second drive or storage device in the same computer. What happens if the computer is stolen or damaged in a fire? The only safe way to insure against these eventualities it either to back up the files to another system that is in a different physical location or to keep backup copies of the files separate from the computer.

Many sophisticated computer simulation projects will be well beyond the capacity of floppy disks or even zip disks. Fortunately, in the last few years, inexpensive rewriteable CD-ROM drives have become available. These drives are now available even in notebook computers. This form of medium offers a readily convenient and inexpensive means of keeping multiple backup copies separate from the host computer. In the event that the host computer goes down, is stolen, or is damaged, it is a simple matter to load the latest CD-ROM version onto the new host computer. The practitioner is not dependent on the computer manufacturer's support technicians or on the company's internal information technology group. It is not likely that either of these groups will be as sensitive as you are to your project deadlines.

6.3.9 Commenting

As with any other computer program, a simulation program can benefit greatly from liberal commenting. Liberal commenting helps not only while you are developing the program but also when you attempt to look at and understand the program years from now. Some practitioners, including this one, have used the argument of job security to justify the lack of programming commenting. After all, if you can barely understand your own program, what chance does anyone else have? Looking back on this situation, the only guidance for a situation of this type is to consider looking for another job where you can spend more time working on your job and less time worrying about keeping it.

Liberal commenting with simulation programs is actually somewhat more beneficial than commenting with general purpose programming languages. The reason for this is that most accomplished general purpose programmers attempt to develop elegant code where a particular code segment appears only once in the entire program. Any time the particular function is needed, the program diverts to the commonly accessed subroutine.

In contrast, many successful simulation projects will have code segments that are almost, but not quite, identical. However, the code segments will perform functions associated with much different parts of the model. Thus, it is not unusual for some segments of simulation code to appear to be very similar on the surface. It is only when the practitioner examines the code parameters that the differences become apparent. This can result in the practitioner constantly hunting for specific segments of code in large complex models. Under these circumstances, effective commenting will not only help explain the code but can also help the practitioner more rapidly access the desired segment of code among many other similar segments. Commenting in simulation programs can be performed in two distinct manners.

6.3.10 Commenting at the Structure Level

The most common and conventional manner of commenting in simulation programs is to include text messages about the purpose of a structure at the structure level. Some simulation programs actually include a comment field in each structure dialogue box. The practitioner should strive to make these comments explanatory rather than descriptive. If the practitioner is practicing proper mnemonic structure naming, there should be no need for additional descriptive text. On the other hand, explanatory text may be of great value, particularly when branching-type structures are utilized.

6.3.11 Commenting at the Model Level

The second method of commenting involves placing text messages in the program at the model level. This provides the practitioner with the ability to understand quickly what part of the model performs what functions. This commenting approach is particularly effective when utilized with the divide-and-conquer programming approach. In some cases, the practitioner may even want to use graphic elements such as boxes to keep model-level commenting associated with a particular group of program segments.

6.4 Program Organization

The simulation practitioner must wage an endless battle to keep the simulation program as organized as possible. In this section, we discuss:

- Why simulation programs can easily become disorganized
- How to use subroutines and subroutine views
- Programming hot keys
- Mnemonic naming conventions
- Thinking in advance

6.4.1 Why Simulation Programs Can Easily Become Disorganized

The graphic interface of many simulation-specific programs actually sometimes seems to promote the disorganized development of different model components. In contrast to text-based high-level programming languages, there are no discrete lines on which to place different program structures. In fact, it is possible to place the iconic representations of the different model components virtually anywhere on the computer window. The downside of this increased flexibility is that more of the burden of keeping the program organized is forced on the shoulders of the practitioner. Thus, the practitioner must make more of a continuous disciplined effort to keep the program organized than would be necessary with other types of programming languages.

6.4.2 Graphic Spaghetti Code

Even if the practitioner attempts to keep the initial program organized, the "simplify and embellish" philosophy may actively work against the practitioner. Rather than sticking in the enhanced code anywhere there is available space, the practitioner must make a disciplined attempt to rearrange all of the existing code structures so that the program appears to flow naturally. Failure to adhere to this principle will result in structure connectors appearing like spaghetti all over the program window. The spaghetti connectors can be the graphic equivalent of textual programming spaghetti code. For those of you who are too young to remember, spaghetti code was the result of the excess use of "goto" statements in a program. Just as it is impossible to follow individual strands of spaghetti, it was impossible to follow spaghetti code that jumped in and out of this and that part of a program. Figure 6.1 illustrates what can happen if the practitioner allows this to get out of control.

6.4.3 Subroutine Views

In attempting to keep the program organized, the practitioner should keep in mind the different levels that can be used to view the program. The practitioner should attempt to keep all like components or components that are associated with each other in the same view. Although all components need to be in the view, the view level cannot be so low that it is impossible to identify the individual program

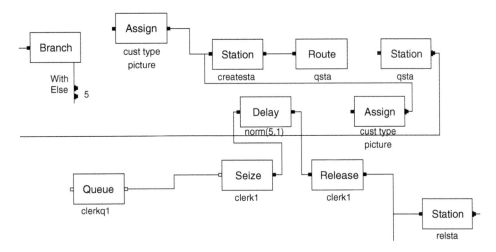

FIGURE 6.1 Goto approach to program organization.

components. The practitioner must also strive to make sure that no other nonassociated components for the other parts of the model are visible in the view. These other components would merely be a valueless distraction to the practitioner. In this sense, a certain level zoom view can be considered as the equivalent of a text-based program subroutine. Figure 6.2 illustrates a subroutine view.

Naturally, the practitioner should also make an attempt to make the spatial positioning of the subroutine view significant. In other words, even though subroutine views are used, related subroutines should be geographically adjacent. This will aid in the rapid movement between related subroutines. This subroutine view approach should be used throughout the model to keep the program as organized as possible. Figure 6.3 illustrates the geographically adjacent positioning of different subroutines at a high-level view.

6.4.4 Program Hot Keys

A very useful technique for navigating to the correct subroutine view of an organized simulation model is to program hot keys. Hot keys enable the practitioner to associate a particular view of the model with a key on the computer keyboard. So any time the practitioner wants to navigate to a particular area, all

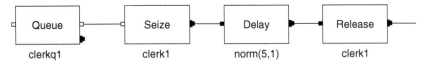

FIGURE 6.2 Subroutine view.

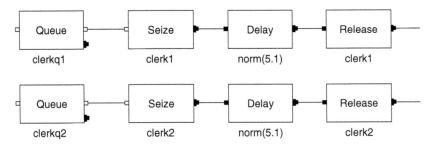

FIGURE 6.3 High level view of adjacent subroutines.

that is necessary is to press the key corresponding to that part of the model. For example, the key "c" could be programmed to navigate to the code that causes the system entities to be "created." Likewise, the key "a" could send the practitioner to the part of the program where the animation can be viewed. While the initial programming of hot keys may take a few minutes, the downstream time-saving benefits will be many times greater.

6.4.5   Mnemonic Naming Conventions

The practitioner should make every possible attempt to use consistent mnemonic naming conventions for the model queues, resources, transporters, and conveyors. The naming convention should describe the exact nature of the component as well as what type of component it is. The practitioner will find that these mnemonic techniques will greatly speed the rate at which the program can be developed. If the practitioner fails to use mnemonic techniques, he or she will be continuously forced to refer to other parts of the program. Examples of mnemonic naming include:

- Checkoutclrk
- Checkoutclerkq
- Golfcart
- Golfcartq
- Bagconv
- Bagconvq

Checkoutclrk would be the resource name for a checkout clerk. Checkoutclrkq could be used for the queue feeding the checkout clerk. Golfcart could be the name for a transporter. Golfcartq is the queue for the golf cart. Bagconv is the conveyor for baggage. Bagconvq is the name for the queue that precedes the baggage conveyor.

6.4.5.1   Naming Multiple Resources and Queues

If a consistent mnemonic method is used for naming the components in the model, then the same approach should be utilized for naming multiple resources and queues. For example, if a particular model calls for a machine, the natural inclination is to call the resource something like mach. If it turns out that the model or a subsequent simulation model requires more than one machine, the practitioner will have to decide what names to use. Possibilities include:

- Machb
- Mach2
- Mach02

If any of these names is selected, it would be awkward for the first machine to be called mach and the second machine to be called mach2. This means that the practitioner would probably want to rename the original machine called mach something else, probably one of the following names:

- Macha
- Mach1
- Mach01

This means that the practitioner will have to modify the previously developed program. Of course, renaming the original code could have been completely avoided if forethought had been given to naming the machine mach1 or mach01 in the first place. If either mach1 or mach01 is used in the first place, it becomes a trivial matter to copy, paste, and change the name of the machine to mach2 or mach02. This method can also be applied to other model components such as queues, transporters, and conveyors.

6.4.5.2   Program Reference Labels

Similarly, the reference labels for the program structures can use the same basic name as the structure with the letters lbl appended to the end of the structure's name. This would result in:

- Checkoutclrklbl
- Checkoutclrkqlbl

Checkoutclrklbl is the label for the clerk at the checkout counter. Similarly, checkoutclrkqlbl is the label for the queue for the clerk at the checkout counter.

6.4.6 Thinking in Advance

Most simulation models usually end up being modified in some manner in order to represent different experimental alternatives. Because the model may eventually be recycled, it is almost always of benefit to the practitioner to attempt to build as much flexibility into the model as is reasonably possible. If some forethought is given to this consideration, the additional effort that is required to enable future modification can be minimized. This approach is known in the industrial engineering world as "right the first time." In other words, the practitioner should avoid investing his time in developing a model that will have to be reworked unnecessarily some time in the future.

A typical example of this situation is one in which the practitioner knows that the experimental alternatives will require a large and varied number of resources, queues, transporters, or conveyors. In this case, it may be preferable to begin the model using an array approach. Instead of individually referring to components in the model as mach1, mach2, etc., these components are referred to as machs(i), where i is an index variable identifying a particular machine in the array of machines called machs. The array approach enables the practitioner easily to increase or decrease the number of components in the model. Furthermore, provided that the practitioner has the necessary skills, the use of array references may also reduce the amount of programming code that is needed. In place of two or more individual code sequences, the practitioner may reduce the program to one code sequence that uses the component arrays.

6.4.7 Multiple-Practitioner Program Development

Under some circumstances, the practitioner may need the assistance of other practitioners in order to complete the programming component of a simulation project. These circumstances may arise as a result of either a tight time schedule or specific skills that other practitioners may possess. In the case of a tight time schedule, the practitioner may have some latitude as to how the other practitioners are utilized. However, if the other practitioners have specific skills that the practitioner does not have, the practitioner will naturally be more limited.

Although this approach may appear to have great potential with respect to programming development, it is not without some equally significant dangers. First, the practitioner will have to maintain some level of supervision over the other simultaneously developed components. It would be most foolhardy to expect exact results at the end of the simulation project if no midterm supervision was conducted. A second potential problem with this approach is that sufficient time must be allocated toward integrating the various components well before the project is due. Just because the individual components successfully operate independently one should not assume that the merged program will not have any difficulties. In addition, the program still has to go through the verification and validation processes before experimentation can begin.

Ideally, the practitioner will have formalized the assignment of the different programming components in the existing linear responsibility chart. If this was not previously conducted, the practitioner would be well advised to revisit the linear responsibility chart and include these details. The revised linear responsibility chart should obviously be distributed to all of the other practitioners. Without some sort of formal assignment of responsibility for these work packages, the practitioner may be bitterly disappointed when attempting to collect the other practitioners' work.

6.4.8 Use of Multiple Practitioners with General Skills

When the practitioner has several other practitioners with general programming skills at their disposal, the question becomes how best to divide up the programming effort. Once again, the programming approach of "divide and conquer" can be of value. It is possible to assign specific submodels or subroutines for individual practitioners to program. Some possibilities for separate programming include:

- User-generated code
- Sequential changes in model modes
- Basic animation layout
- Machine or process failures
- Nontypical events for demonstration purposes

6.4.8.1   User-Generated Code

User-generated code would include any high-level programming code that interfaces directly with the regular simulation program. The use of high-level programming code is usually a result of a need for some sort of particularly sophisticated user rule or algorithm that the regular program does not have. The user code would be developed and compiled in some high-level programming environment like C, FORTRAN, or Visual Basic. This end code would then be linked to the simulation program.

6.4.8.2   Sequential Changes in Model Modes

Sequential changes in model modes also offer the practitioner an opportunity to model independently different parts of the model. This means that the model represents a process that flows sequentially and that there are distinct changes in the different parts of the model. This type of opportunity is particularly evident when an entity changes form or mode of movement. For example, if an entity starts out as a discrete component that undergoes a sequence of processing, each individual process could be modeled by a separate practitioner. Similarly, entities may change their mode of movement from vehicle to foot to conveyor as they proceed through an airport passenger system. The programming code corresponding to each different mode of movement is isolated. This would offer a reasonably easy means to assign separately the development of code to different practitioners.

6.4.8.3   Basic Animation

It is also possible to develop separately the basic animation components of the model. As long as the practitioner handling the model and the practitioner handling the animation agree on the model component names, significant time may be saved by simultaneously developing both of these parts. The two practitioners will have to maintain communication any time there is a change that will affect both components. For example, if additional resources are used in the model, then additional animation components will have to be added.

6.4.8.4   Machine and Process Failures

Machine and process failures may also be modeled separately from the main model. This is especially possible if the machine or process failures are not an overwhelmingly common event. In other words, if a machine failure is not the primary basis for the model, then it can usually be handled by a small segment of independently operating program code. In the meantime, the practitioner can develop the model as if the machine were not subject to failures. Although it is possible to completely debug the main model, it may not be possible to debug the failure segment completely because the failure segment must be linked to the main model in order to operate. This means that the practitioner should be prepared to spend some effort on the integration of the model and the failure submodel when the time comes to merge these two parts.

6.4.8.5   Nontypical Events for Demonstration Purposes

The practitioner may also attempt to model independently nontypical events for demonstration purposes. A nontypical event in a service-type model would include the arrival of a busload of customers. Although this would certainly not be a regular event, it could potentially be an interesting event to observe during a demonstration. The same programming and integration issues as with the machine failure situation would also apply to independently modeling nontypical events.

6.4.9 Use of Multiple Practitioners with Specific Skills

If other practitioners have specific skills, the practitioner must decide how best to use their special talents. This situation is most likely to exist when the other practitioner or practitioners have unusual graphic skills. Here, the graphic artist/practitioner can be used to enhance separately the animation component of the simulation project. This includes the development of graphics for:

- Entities
- Resources
- Transporters
- Conveyors
- Static backgrounds including buildings

The development of these items can be performed completely independently of the rest of the model. The development can take place in either a dummy model in the actual simulation program or in a graphics program. For relatively simple development it may be better to utilize the native drawing tools of the simulation program. However, complex vector-based graphics will require the use of a high-end illustrator program or even a CAD program such as AutoCAD. If some other external program is used, a new issue that must be addressed is how to get the graphic back into the simulation program. The practitioner may attempt to copy the image to the operating system buffer and paste it into the simulation program. Alternatively, most simulation software packages will allow for the importation of graphics by using a common file format, such as dxf files. In this case, the practitioner must insure that the external program has the capability to export a compatible file format.

In the meantime, the main programming components can use simplified graphics or placeholders until the time that the regular graphics are available. The attraction of this approach is that the graphic designer/practitioner actually needs to know little about the actually programming of the model, just as long as he or she is graphically skilled. Of course, the practitioner will still want to supervise the creation of the graphic components to insure that they are suitable. However, the actual merging of the graphic elements into the main model is a minor task that can be performed toward the end of the project.

6.5 Summary

The objective of the model translation phase is to translate the system into computer simulation code that can be used to test experimental alternatives. This objective requires that the simulation practitioner select a computer programming language and then actually develop the simulation model program. The practitioner can elect to utilize either a simulation-specific programming language or a general purpose programming language. This choice is dependent of the availability of computer resources, the level of practitioner proficiency, and the availability of funding.

In the second part of the model translation phase, the practitioner attempts to create an organized, functional simulation model in the program of his or her choice. If the practitioner follows a high-level flow chart, liberally comments the program, uses the graphic subroutine approach, adheres to common mnemonic naming techniques, and thinks in advance, there is a much greater chance that the practitioner will make wise use of limited program development time. The practitioner may also make use of other practitioners who are at his or her disposal. The particular skills of the other practitioners will dictate their use. The effective use of other practitioners may compress the amount of time needed for the model translation phase.

Chapter Problems

1. What is meant by mnemonic naming?

2. In order to have multiple clerks in a model, how would you name the first three clerks?

3. How can a simple animated model of a car wash be split among three practitioners so that the project is completed in the least amount of time?

4. How can programmed hot keys speed the development time of a simulation model?

5. Why is it not a good idea to keep all of your different versions of your model on the same computer?

6. What is the difference between the static and dynamic components of a simulation animation?

7. What is meant by graphic spaghetti code?

8. What two ways can simulation models be commented?

9. What is a subroutine view?

10. Why might a practitioner prefer to use a general purpose programming language over a simulation-specific software package?

Model Translation Check List

The following checklist can be used to assist the practitioner in the model translation phase to develop an organized simulation program.

_____Sequential model names
_____Backing up model files
_____Structure level program commenting
_____Program level commenting
_____Subroutine views
_____Hot keys for subroutine views
_____Mnemonic variable names
_____Mnemonic label names
_____Mnemonic resource names
_____Multiple resource naming
_____Alternative model names

References

Banks, J., Nelson, B.L., Carson, J.S., and Carson, J.S., II (2000), *Discrete-Event System Simulation*, 3rd ed., Pearson Education, New York.

Schriber, T.J. (1990), *Introduction to Simulation Using GPSS-H*, John Wiley & Sons, New York.

Chapter-Related Web Sites

Rockwell Arena: www.arenasimulation.com
Brooks-PRI Automation AutoMOD: www.automod.com
CACI: www.caci.com

7
Verification

"Building the model correctly."

7.1 Introduction

During the model translation phase, the practitioner will naturally be interested in ensuring that the simulation model has all the necessary components and that the model actually runs. In reality, we are interested in getting the model not just to run but to run the way we want it to. In other words, we are interested in ensuring that the model operates as intended. This process is known as model verification. Another way to look at the verification processes is to consider it as:

"Building the model correctly."

For some reason, even very experienced practitioners confuse model verification with model validation. Whereas verification is the continuous process of insuring that the model operates as intended, validation is the process of insuring that the model represents reality. It is pointless, perhaps impossible, to attempt to see if the model represents reality if the model does not even operate as intended. In other words, you should not attempt to validate a model that has not successfully undergone the verification process. Validation is discussed in depth in the next chapter. In the meantime, validation can be considered as:

"Building the correct model."

As we have determined, in order for the model verification process to be considered totally successful, the model must:

- Include all of the components specified under the system definition phase
- Actually be able to run without any errors or warnings

To effectively include all of the components specified under the system definition phase, the practitioner should utilize:

- A divide-and-conquer approach
- A subroutine view approach

To help insure that the program runs without errors, the practitioner can utilize the following techniques:

- Animation
- Manually advancing the simulation clock
- Writing to an output file

7.2   Divide-and-Conquer Approach

Anything but the simplest system will require a significant amount of model programming. Inexperienced practitioners frequently attempt to model the entire system without making any attempt to debug the program. These types of attempts usually result in a great deal of unnecessary frustration when the time comes to run the model. The larger and more complicated the model, the greater is the frustration. The novice practitioner will then attempt to save the program by applying band-aids to defective code. No effort is directed toward correcting the fundamental programming errors. Sometimes practitioners reach the point where things become so confusing that it is necessary to begin the program from scratch. All of the previous programming effort is lost. Obviously, this approach is not recommended for anything but the simplest models and maybe not even then.

As with any other programming language, simulation programs can benefit from the divide-and-conquer approach. This means that the practitioner should break the larger, more detailed system model into a smaller, simpler, perhaps higher-level model. The smaller, simpler model will be correspondingly easier to debug. Once the small, simple model includes all of the basic desired components, the practitioner can attempt to run the model. Any errors in syntax or variable naming can be more easily addressed. Once the model operates as intended, the practitioner can consider making a series of small enhancements to the simple model one by one. The enhancements can take the form of either:

- Enhancements to the detail of the existing components of the initial model to represent the actual system
- Expansion of the model to include other components that were not previously modeled but need to be modeled in order to represent the actual system

It is the practitioner's call as to which approach to take initially. However, it will eventually be necessary to include both the detailed enhancements and the component enhancements. Depending on the particular system to be modeled, it may be easier to begin with one or the other approach.

7.2.1   Service Model Example

Consider a service-oriented model that represents the operation of a simple bank. We should first start by making a number of simplifying assumptions about our initial model. These could include:

- All customers who wish to make a transaction enter the bank.
- No customers wish to open new accounts.
- Customers enter the teller queue immediately after entering the bank.
- Transactions are modeled as a single service distribution.

In the initial model, we can concentrate only on the walk-in customers who need to perform some sort of transaction with a teller. This simple model needs to have:

- Banking clerks
- Queues for customers who wait their turn in line

Because this model has only a very small number of components, it should be relatively easy to model and debug. Once we are able to get this simple model working, we can choose to either:

- Enhance the detail of the model
- Expand the model to include other components

If we choose to enhance the detail of the model, we could:

- Model customers stopping at the form counter
- Model different types of transactions such as deposits, withdrawals, check cashing, etc.

If we were interested in expanding the model, we could:

- Model customers wanting to start new accounts
- Model customers using the drive-up window
- Model customers using the ATM machines

7.2.2   Manufacturing Model Example

Consider a model of a small manufacturing line that assembles computers. The process consists of:

- Inserting the components into a motherboard
- Placing the motherboard into the computer case
- Inserting the drives into the case
- Testing the computer

Once this simple model is created, we can once again either:

- Enhance the detail of the model
- Expand the model to include other components

If we choose to enhance the detail of the model, we can include the following processes:

- Model the insertion of the RAM and CPU separately onto the motherboard
- Model the insertion of the drives separately for the hard drive, CD-ROM drive, and floppy drive

Similarly, we can expand the model to include:

- The percentage of computers that pass the assembly test
- The different types of failures for the assembly test
- The amount of time to rework each of assembly test failures

7.3   Animation

Animation is perhaps the most effective tool for performing basic verification (Pegden et al., 1995). Being able to visualize what the program is doing makes it easier to detect errors in the program. Animation can be used for model verification in many different ways:

- Using different entity pictures for different types of entities
- Following the entities through the system
- Changing entity pictures
- Displaying global variables or entity attribute values
- Displaying plots of global variables or entity attributes
- Displaying levels of system statistics

7.3.1   Using Different Entity Pictures for Different Types of Entities

For simulation models with different types of customers or different job priorities, the practitioner should employ different types of entity pictures. This can help insure that the different types of entities go to

FIGURE 7.1 Use of different entity pictures.

the appropriate part of the model (Figure 7.1). For example, if it turns out that the first-class air passenger customer service process is not operating properly, it will be easy to see if coach-class passengers are being served instead.

7.3.2 Following Entities

By following the animated entities through the model, it is possible to verify that the model logic is correct. The most common example of this situation is one in which practitioners discover that a decision structure is improperly sending entities to various parts of the model. One common way that this situation manifests itself is when particular queues appear to become unreasonably overloaded. Another symptom is entities disappearing in particular parts of the model and reappearing in others.

7.3.3 Changing Entity Pictures

If the entity changes form as it proceeds through the model, the practitioner should also change the picture of the entity. For example, if a roll of raw material is cut into different sections by a machine, the entities that proceed from the machine should have a different entity picture than the roll of raw material. Thus, if an entity proceeds through the model with the wrong picture, it will be obvious that the entity somehow managed to bypass the mid-stream processes. This technique is illustrated in Figure 7.2.

7.3.4 Changing Resource Pictures

A very effective means of determining what is happening to the resources in a simulation model is to change the picture corresponding to different resource states. For example, a manufacturing machine could have the following system states:

- Idle
- Busy
- Inactive
- Failed

The inactive animation picture could include the machine and the operator, without any indication that it is running. The busy machine should include the entity job in the machine with the operator still

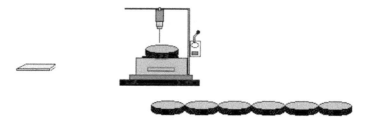

FIGURE 7.2 Changing entity pictures.

present. This approach ensures that there is no possibility that the practitioner will be confused as to the state of the machine. The resource would be modeled as inactive with the machine minus the operator. This would correspond to a period when the operator was on a break or was at a meal. The failed machine could be animated without the operator but with a mechanic bending over the machine. Different resource states are illustrated in Figure 7.3.

The animation of a transporter such as a truck or forklift should use the same concept. An empty transporter should definitely look different than a transporter which currently contains some type of cargo. It may be possible to implement a ride point where the practitioner can place the entity while being transported. When the transport process is complete, the cargo entity then dismounts the transporter and continues on through the model. The choice of different pictures for resources and transporters is really up to the practitioner. The important point to remember is that the graphics should clearly indicated which of the different states the resource is currently in. Empty versus full transporters are illustrated in Figure 7.4.

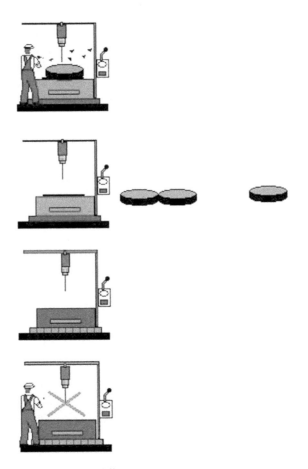

FIGURE 7.3 Different resource pictures for different states.

FIGURE 7.4 Transporter loaded and unloaded states.

Colors may also be used to indicate the state of different resources. Typically a resource that was green would mean that it is idle. Similarly, a resource that was red could represent the busy resource. The colors for inactive and failed may not be so clear cut. White could be used for inactive. Brown or black could be used for failed. Whatever color scheme is used by the practitioner, it is important to ensure that the same color methodology is used throughout the model. The practitioner should also strive to use the most vibrant colors to indicate major state differences. A large percentage of the male population is color blind with respect to red and green hues. If vibrant red and green colors are used, these people stand a greater chance to distinguish between the resource states.

7.3.5 Displaying Values

Another useful animation capability is the digital or analogue value display. This capability allows the practitioner to see either the instantaneous or the average value for any global variable or entity attribute.

Instantaneous displays are particularly useful when used to display a variety of counter values located in different parts of the model. Each time an entity enters a particular part of the model, the counter corresponding to that part of the model is incremented. When the counter for a model component is never incremented, the practitioner should check the model entry logic corresponding to that component.

Instantaneous displays are also useful for illustrating the number of entities in a queue. Because of graphic space limitations, it is not always possible to display large numbers of entities in a particular queue. A common approach to this limitation is to animate the queue so that several entities may be observed. A digital display is then used in conjunction with the animated queue to display the entire number of entities in the queue. This technique is illustrated in Figure 7.5.

Last, almost all simulation models can benefit from the use of an analogue display of the simulation time. The display of simulation time is very useful to verify that system resources are functioning correctly with respect to scheduled breaks and random failures. An analogue clock is displayed in Figure 7.6.

7.3.6 Displaying Plots

A digital or analogue animation display is limited to displaying a single value at any given moment. When the value changes rapidly, it is difficult to keep track of the past values. More importantly, the single value display cannot provide information on the value trends. When the plots display unusual or unrealistic trends, the practitioner should investigate the model logic. Figure 7.7 illustrates a system time plot for a model with errors.

As can be observed, the system time in the model is rapidly increasing. The upper level of the plot is 30 min. This indicates that a logic error, possibly with the service time, may be present in the model.

FIGURE 7.5 Digital representation of system values.

FIGURE 7.6 Analogue display of system time.

FIGURE 7.7 Plot of system time.

When using plots, the practitioner must insure that the y or vertical axis is sufficient to contain all of the variable values. If a suitable value is not used, it is likely that the plot of the values will run parallel to the top of the plot box at the maximum y value. This will prevent the practitioner from determining that some sort of problem may be occurring in the model. Figure 7.8 illustrates what may happen in this situation.

7.3.7 Displaying Levels of System Statistics

Another type of animation that can help the practitioner debug a simulation model is the level. The level is analogous to gauge. If, for example, the practitioner wanted to know the running average utilization for a resource, a level could be used. The lower or empty level would correspond to 0 utilization. Conversely, the upper or full level would correspond to 1.0 or 100% utilization. If a level is utilized in this manner, it is important to remember that the level shows the average, not instantaneous, utilization level. It can be disconcerting to novice practitioners when it appears that a resource is being used, but the level indicates otherwise. Note that it is probably inappropriate for the practitioner to utilize a level to indicate the instantaneous utilization level in any event. The level would change immediately between completely empty and completely full. It is better to illustrate this effect by changing the picture of the resource between idle and busy (Figure 7.9).

When using levels to help debug models with multiple identical resources, the practitioner must be especially careful with the equation that is used to drive the level. In this case, the practitioner must also divide the level value by the number of identical resources. If this is not performed, then the level will not necessarily provide a value that represents the average resource utilization level.

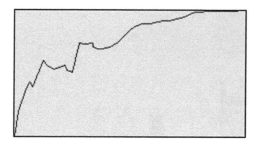

FIGURE 7.8   Use of improper plot display ranges.

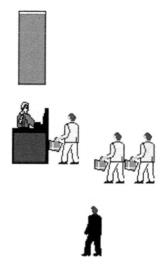

FIGURE 7.9   Level display of system utilization.

7.4   Advancing the Simulation Clock Event by Event

Simulation model animations are usually designed to run in compressed time. This sometimes makes it difficult to follow the flow of entities through the model. Although it is possible to slow significantly the animation of the model, it would be better to be able to step through the model event by event. To allow practitioners to debug their models under controlled circumstances, most simulation software packages have VCR-like controls. These allow the practitioner not only to step through the model but also to fast forward to the part of the model the practitioner is interested in viewing.

7.5   Writing to an Output File

Another potentially useful but more difficult to use verification tool is the use of output files. In contrast to the use of an animation display, output files provide a permanent record for the practitioner. These files may be written in a proprietary format or in ASCII text format. If the program writes files in a proprietary format, then the file will have to be exported by the practitioner into an ASCII file. On the other hand, if the file is in ASCII format, the file can be immediately and easily viewed using any text editor or word processor.

7.5.1 Event List File

One type of file that the practitioner may use is an event list file. Here, entry is listed according to the system time. The amount of information that is written to the file for each event can make the output very difficult to decipher. This will result in information overload on the part of the practitioner. Because not every single detail and variable value change is necessarily of interest to the practitioner, most simulation software packages will allow you to designate the level of detail that is written to the file. The following text is an example of an event list file output by a simulation software package.

Seq	# Label	Block	System Status Change
Time: 0 Entity: 2			
1.	0$	CREATE	
			arrtime set to 0.0
			Next creation scheduled at time 1.7307613
			Batch of 1 entities created
2.	1$	QUEUE	
			Entity 2 sent to next block
3.	2$	SEIZE	
			Seized 1 unit(s) of resource clerk
4.	3$	DELAY	
			Delayed by 4.279432 until time 4.279432
Time: 1.7307613 Entity: 3			
1.	0$	CREATE	
			arrtime set to 1.7307613
			Next creation scheduled at time 3.0819438
			Batch of 1 entities created
2.	1$	QUEUE	
			Entity 3 sent to next block
3.	2$	SEIZE	
			Could not seize resource clerk
			Entity 3 added to queue clerkq at rank 1

This trace file organizes the event data by model structure and time. It begins with an entity entering the system at time 0.0 with the create block. At the same time, the arrival of the second entity into the system is scheduled at 1.73. The initial entity, entity 2 is sent to the next structure, the queue block. Since there is no one already in the queue and the clerk is not busy, the entity seizes the clerk for service. The entity is then delayed for the service time until 4.27 minutes. In the meantime, the second entity, entity 3, arrives in the system at 1.73. Entity 3 enters the queue and attempts to seize the clerk. Since the clerk has already been seized by entity 2, entity 3 is placed in the queue clerkq in the first position.

By following the trace file in the above manner, the practitioner can attempt to identify where the bug in the program exists. As can been seen, the amount of data that can be sent to the trace file can be burdensome to digest. If we were interested only in the trace when queue problems occurred, we could set the trace to write events only when the queue approached some large size. Similarly, if we knew that the problem did not occur until a particular time in the model, we could set the trace to begin output at that time.

7.5.2 Variable or Attribute Output File

The practitioner may not necessarily be interested in the change of the system event by event. Perhaps only the changes or final value of a variable or attribute are of specific interest to the practitioner in order to help debug the simulation model. In this case, the practitioner may elect to output only the variable or attribute of interest to the file. By looking at the variable or attribute value, the practitioner may ensure that the entities are flowing through the model properly.

The table below is a listing of the individual entity system times from a simulation model. The first few rows of information are administrative and can be ignored. In the first column is the simulation time, tnow, when the entity left the system. The second column is the system time for each individual entity.

203 11/18/2002 Y	
1 15	7
tnow	systime
4.28	4.28
12.43	10.70
19.43	16.35
27.83	24.44
32.66	28.58
35.90	31.19
43.17	34.50
50.61	32.34
61.19	42.21
67.50	39.94
71.12	42.96
71.93	39.92
75.44	41.67
80.00	45.23

This type of output file can be used by the practitioner to ensure that there are no unusually individual system times. An unusually long system time might indicate that the entities are being inadvertently held up in the system more than they should be. If no system times at all were written to the file, there might be a complete stoppage or an infinite loop somewhere in the model.

7.6 Summary

Verification is the process of ensuring that the simulation model operates as intended. This means that the practitioner has included all of the intended components in the model and that the model is actually able to run. Verification should be considered as a continuous process rather than a one-shot effort. In order to easily develop and debug the model, the practitioner should utilize a divide-and-conquer approach. This is the development of the model in small, easily debugged components rather than an attempt to create the model all at once.

The practitioner may make good use of the animation features of simulation software packages in the verification phase. The practitioner can follow the flow of entities through the model. The different idle, busy, inactive, and failed states of resources can easily be seen in an animated model. The practitioner can also use animated displays to verify the values of different global variables and entity attributes.

If the model errors are very subtle, the practitioner may also write data to a file. This provides a static record of the changes in a particular variable or attribute. Last, the practitioner may also set up a trace file. This is the electronic equivalent of the simulation event list. The list can be set so that every event is displayed at every event time. Because this may result in information overload, the list can be activated only at a specific time or can be set to display only the information of interest.

Chapter Problems

1. What is the difference between verification and validation?

2. What are the two principal components of verification?

3. What is meant by a divide-and-conquer approach?

4. How can an output file be used to assist in the verification process?

5. How can different colors be used with resource states?

6. How can animation be used to indicate different types of customer entities?

7. An animated level is best used to display what sort of system statistic?

8. How can you use an animated plot to help you debug a model?

9. What do you have to take into consideration when specifying the values in a plot?

10. A digital display is best used for what purpose?

Reference

Pegden, D.C., Shannon, R.E., and Sadowski, R.P. (1995), *Introduction to Simulation Using Siman,* 2nd ed., McGraw-Hill, New York.

8

Validation

"Building the right model."

8.1   Introduction

In Chapter 7, "Verification," validation was defined as the process of ensuring that a model represents reality at a given confidence level. This means that the simulation practitioner will attempt to create a model that is a reasonable representation of the actual system. However, for a variety of reasons, even painstakingly constructed models may not actually represent reality. This means that no matter how well the practitioner thinks that he or she has debugged and enhanced the model, the model may still not be suitable for conducting any type of analysis. The inability of the model to represent reality may result from certain actions or omissions on the part of the practitioner with respect to any or all of the following issues:

- Assumptions
- Simplifications
- Oversights
- Limitations

8.2　Assumptions

Periodically, the practitioner will have to make modeling assumptions. These assumptions may be made because of lack of knowledge. These types of assumptions may be common when the practitioner is attempting to model a system that does not exist or a process that cannot be observed. Certain assumptions may have to be made with respect to the system components, interactions, and input data. Even if some data are available from designers or vendors, the practitioner will have to assume that the data are valid.

Another area in which the practitioner can run into trouble is with input data. It is not uncommon to fail to collect a particular type of input data. During the modeling phase, the practitioner may suddenly realize that these input data are missing. Rather than delay progress on the model development, an assumption is made with respect to the missing type of data. The assumed data distribution is inserted into the appropriate point in the model. Unfortunately, as time deadlines approach, these assumed data may be forgotten. If the model cannot be validated, the forgotten assumption is often the last place that the practitioner will look.

This situation recently occurred in a simulation model for examining the loading sequence of passengers. No data were collected for the rate at which the passengers traveled down the cabin aisle while boarding the plane. Because these data were not readily available, it was assumed that passengers moved at the same rate down the cabin aisle as they did in the jetway en route to the plane. When the model could not be validated, this assumption was reevaluated for its validity. In reality, the confined space of the cabin aisle reduced the passengers' movement rate. When a more realistic movement rate was utilized, the model was able to be validated statistically.

Each time the practitioner makes an assumption, the details should be recorded in a formal list of assumptions. Although it may take some discipline to keep this list updated, it has the potential of great benefit for the practitioner. Should the practitioner not be able to validate the model statistically, the list of assumptions is a natural place to look. This list will also ultimately be included in the simulation project presentation and report.

8.3　Simplifications

The practitioner will sometimes make deliberate simplifications in the model of the system. Some of these simplifications will be necessary in order to finish a simulation project in the allotted time. Others will be made because the internal detailed workings of the process are either too complex or thought to be insignificant.

Avoiding simplifications caused by time constraints is easier said than done. With experience, the practitioner can make better estimates for allocating time and other resources to the steps in the simulation study process. One frequent observation of teams of novice practitioners is the failure to develop the model in parallel under time constraints. Many inexperienced practitioners believe that because there is only a single model, only one individual can work on the model at a given time. There is absolutely no reason why different individuals cannot assume responsibility for different parts of the model. At a future date, the individually debugged submodels can be combined and debugged. Of course, the project team will have to maintain a robust tracking scheme to prevent version problems. If not properly handled, the team will find themselves deleting each other's work with each new submodel that is individually developed.

8.3.1 Common Simplifications

The two most common simplifications are associated with modeling complex processes as single processes or completely omitting processes. Process simplification usually results in the practitioner's collecting a single input data distribution instead of several individual input data distributions. An example of this would be to lump the checkout process in a grocery store as a single service time. The service time is actually a function of how many items are being purchased and how the customer is paying for the groceries. So in place of the single input distribution, you would actually have at least:

- Service time for scanning groceries
- Distribution of payment types
- Service times for each payment type

The second type of simplification occurs where the practitioner consciously decides that a particular process does not need to be modeled. Usually, this is because the practitioner decides that the process will not have a significant impact on the rest of the model. For example, if a particular machine's mean time between failure reliability is such that it has never been observed to fail, the practitioner could make a simplifying assumption not to model machine breakdowns.

8.4 Oversights

If there is any complexity to the system being modeled, it is extremely probable that the practitioner is going to inadvertently overlook one or more critical system components. If in fact such a component does have a significant impact on the measures of performance, the model will not be able to be validated. Because the practitioner is not aware of the potential problem in the first place, validation problems arising from oversights are much more difficult to handle. Other than conducting as detailed an orientation as possible, there are few defenses against the accidental omission of a particular event or process that may be part of the system. One thing that the practitioner can do is maintain an ever-increasing checklist of lessons learned.

8.5 Limitations

There are likely to be many limitations with respect to being able to model complex systems. The impact of each of these limitations on validating the model can vary significantly. Generally speaking, these limitations will not be under the direct control of the practitioner. As with model assumptions, the practitioner should maintain a list of limitations for the project presentation and report. Limitations may be caused by the:

- Practitioner
- Modeling software
- Data

8.5.1 Practitioner Limitations

Obviously, we cannot address the limitations inherent to the practitioner. Practitioners who feel that this is a significant problem may elect to receive additional formal simulation training. However, even with significant training, it is not uncommon for experienced practitioners to recognize errors when revisiting models from earlier in their simulation careers.

8.5.2 Modeling Software Limitations

It is theoretically possible not to be able to model some systems because of limitations in simulation-specific software. By keeping to the lowest practical level of simulation construct, it is possible to avoid this problem to a great extent. On the other hand, keeping to the lowest-level constructs increases the programming burden. However, at the same time, the flexibility of the model is also increased. Taken to an extreme, this would mean that general purpose programming languages would be the best tool. However, this would also mean a possibly unacceptable period of time to program the model. On the other hand, it is primarily when practitioners attempt to reuse old or example models or attempt to use canned models that it becomes difficult to model properly even relatively simple systems.

8.5.3   Data Limitations

It is possible that because of the nature of the system, there will be some limitations in collecting data. An obvious example of a data limitation is seen when it may take months or even years to collect the necessary quantity of input data for the system to do a robust statistical analysis. In these cases, the practitioner will simply have to do the best that he or she can under the circumstances. The lack of sufficient data may require the practitioner to accept the fact that some statistical tests will not yield as reliable results in comparison to situations where there are sufficient data.

Another manner in which insufficient data may cause the practitioner problems is with respect to modeling detail. Here, the practitioner may be forced to lump several processes together in order to collect a sufficient amount of data. For example, if there are not sufficient data for separating customers paying with cash, checks, or debit or credit cards, the practitioner is forced to combine all of the individual payment types into a single input distribution. This results in a similar situation as model simplification and possesses the same dangers.

8.6   Need for Validation

These assumptions, simplifications, oversights, and limitations cannot be so gross as to prevent the model from representing reality at a given confidence level. If any of these potential problems do prevent the model from representing reality, the practitioner may have serious problems. The validation process assists the practitioner in knowing whether it is appropriate to proceed with the simulation study or go back to the drawing board. Unfortunately, if you do have to go back to the drawing board, validation does not have the ability to tell you specifically where to look.

8.7   Two Types of Validation

There are two major types of validation of interest to the simulation practitioner. The first of these is face validity. Face validity means that the model, at least on the surface, represents reality. The second is statistical validity. Statistical validity involves a quantitative comparison between the output performance of the actual system and the model (Law and Kelton, 2000). The simulation practitioner must achieve both types of validity to have confidence that the model is accurate.

8.8   Face Validity

Face validity is normally achieved with the assistance of domain experts. A domain expert is simply an individual or group of individuals who are considered knowledgeable on the system under study. Provided that the same group of people who commissioned the simulation study in the first place are knowledgeable, it is of great benefit for the practitioner to use this group as domain experts. This approach helps:

- Instill a sense of ownership in the model
- Prevent last-minute "why didn't you…" questions
- Reduce the number of project progress inquiries

Most practitioners attempt to achieve face validity toward the end of the allocated model development period. This allows the practitioner to make minor adjustments to the model before beginning the statistical validity process. Unfortunately, the face validity process is usually considered a one-time event. Once the practitioner believes that face validity has been achieved, no further thought is given to reestablishing face validity before the final report or presentation. Instead, the practitioner should consider face validity as a continuous process. Between the time of the first face validity meeting and the final report or presentation, the practitioner should attempt to conduct additional face validity meetings. The use of additional face validity meetings will also help insure that any secondary interests or objectives for the model are identified before the final report and presentation.

8.8.1 Face Validation Animation Considerations

In order to execute properly the face validity process, the practitioner must insure that the animation of the process has sufficient visual fidelity to the actual process. This does not mean that the simulation practitioner should strive to make the animation an exact duplicate of the actual system. Attempts along this path usually involve scanning or overlaying blueprints or maps of the system. These approaches will usually yield unsatisfactory results with respect to visual detail and scaling. In other words, exactly scaled animations will rarely provide sufficient visual detail to allow the domain experts to view properly the model. It would be better to concentrate on the important visual details of the process and compromise on the exact scaling.

8.8.2 Special Animation Event Considerations

Sometimes domain experts will want to know how the model handles unusual events. This might involve:

- Breakdowns of critical pieces of machinery or vehicles
- Arrival of large numbers of customers in buses
- Late flights of air passengers seeking rescheduling help at a service counter

Although these eventualities may already be incorporated in the model, they probably are controlled with some sort of probabilistic distribution. To allow the practitioner to demonstrate these effects, most simulation packages allow the programming of hot keys to initiate these types of unusual events. For example, in Arena, the practitioner may use an arrival element to cause entities to appear at any point in the model.

8.8.3 Face Validity Is Not Individually Sufficient

It is obvious that face validity is a somewhat subjective process. The level of expertise among domain experts may vary. The level of interest in the simulation model exhibited by individual domain experts may also vary. For this reason, face validity is considered necessary but alone is insufficient for determining overall model validity. The model may look as though it represents reality on the surface, but does it really? To establish complete model validity, statistical validation analysis must also be conducted.

8.9 Statistical Validity

Statistical validity involves an objective and quantitative comparison between the actual system and the simulation model. If there is no statistically significant difference between the data sets, then the model is considered valid. Conversely, if there is a statistically significant difference, then the model is not valid and needs additional work before further analysis may be conducted. In this section, we discuss:

- Validation data collection
- Validation data analysis process

8.9.1   Validation Data Collection

Validation data collection involves collecting data from both the actual system and the base simulation model that is designed to represent the actual system. Validation data may be based on either individual observations or summary statistics. In the case of individual observations, the practitioner will perform the analysis using data from individual entity measures of performance. Conversely, the summary statistic approach involves analysis using mean data from multiple sets of observations of individual entity measures of performance. Significantly less data are required for the individual observational method than the summary method. The simpler and more rapidly executed individual observation approach is probably of more utility to the practitioner.

8.9.2   System Validation Data Collection

System validation data collection varies significantly from input data collection processes. In system validation data collection, the major concern is collecting data which reflects the overall performance of the system. A common method is to collect system or flow time. This is the time that it takes an entity to be processed or flow through the entire system under study. At this point, the practitioner can choose to use either:

- Individual entity data
- Entire system and model run data

8.9.3   Individual Entity Validation Data Approach

Individual entity validation data involve the collection of system times for individual entities going through the system and through the model. In a manufacturing order-type model, this would involve recording the time that the order was received and completed in the actual system. This type of data may be collected for as few as 30 actual orders in order to make a comparison with the orders from the simulation model. Depending on the exact system, this type of data could be collected in a number of hours or days.

The limitation to this approach is principally the issue of autocorrelation. This means, for example, if a given job has a long system time, then the job right after the given job may also have a long system time. Similarly, if a given job has a short system time, then the job right after this job may also have a short system time. The problem with the existence of autocorrelation is that it violates some statistical requirements for the appropriate use of validation tests. Whether or not the possible existence of autocorrelation is a serious issue to the practitioner is an individual decision. However, the potential problems with autocorrelation must be weighed against either not validating the model or using a much more time-consuming but more statistically robust approach.

8.9.4   Entire System and Model Run Validation Data Approach

The second approach is to use a number of entire system and model run data. This approach requires a significantly greater amount of actual system data collection than the individual entity approach. Here, we would pick a particular time period and collect all of the order-processing times observed during the time period. An average for all of the orders for that time period would be calculated. This process would be repeated many more times until perhaps a total of 30 average order-processing times were collected. The model would similarly be run 30 different times in order to generate the comparison data.

This approach does not suffer from the potential problem of autocorrelation. However, the practitioner must individually decide whether this advantage is worth the greater data collection effort over the individual entity approach.

8.9.5 Record System Validation Data Collection Conditions

A very important issue in the collection of system validation data is that the practitioner record the state of the system when the data are collected. This includes the state of all entities and resources in the system. Entities can be in queues, or they can be in transit to different parts of the system. This means, for example, that a number of individuals may already be waiting in a queue when the practitioner begins collecting data. This is a very important point because the practitioner will not know how long these entities have already been in the system. Because the practitioner needs to know when the individuals entered the system to calculate system time, the individuals who were already in the system cannot be used for validation purposes. However, the existence of the individual in the system will have an impact on how the system performs. So the practitioner needs to know how many individuals are present but must also be careful not to use those individuals for collecting system time.

Similarly, the practitioner needs to record the state of every resource in the system. Each resource may or may not be engaged by an entity. Whether or not the resources are engaged can affect the performance of the actual system data.

8.9.6 Model Validation Data Collection

Model validation data collection consists of recording the same type of output data as were collected in the system validation data collection process. In order to collect these data properly, the practitioner must load the model in the same manner that it was observed when the system validation data were collected. This means in particular that any queues that contained previously existing entities must be preloaded before the simulation model validation data are collected. This also means that the practitioner must discard any data that are generated by the previously existing entities. Thus, if the practitioner observed ten customers waiting in a queue and two customers being serviced, the validation model data from the first 12 customers must be discarded. Practitioners are advised to utilize the same number of observations from the model as the number of observations obtained from the actual system. The use of the same number of data points simplifies subsequent statistical calculations.

8.10 Validation Data Analysis Process

The validation data analysis process consists of first determining the appropriate statistical comparison of means test to execute. This is performed by determining whether or not one or both of the validation data sets are normal. If both of the data sets are normal, then a version of a *t*-test is performed. If only one or neither of the data sets is normal, then a nonparametric test is performed.

8.10.1 Examining the Validation Data for Normality

Both the system validation data and the model validation data must be checked for normality. This is usually performed by running a chi-square test on each of the two data sets individually at a particular level of statistical significance. Although the practitioner may select any reasonable level of significance, an α value of 0.05 is normally used.

To successfully perform the chi-square test, the practitioner must insure that a minimum of 20 data points are in each data set. If fewer than 20 data points are available, a chi-square test cannot be used to determine the normality of the data sets. In this event, the practitioner will be forced to utilize a nonparametric rank sum test for the comparison of means between the system and the model data sets. The chi-square test may be implemented manually in a computer spreadsheet application such as Microsoft Excel or performed automatically in a statistical package such as the ARENA Input Analyzer or AutoStat.

8.10.2   Chi-Square Procedure for Testing Normality

To test a data set for normality, the data must first be divided into a set of cells similar to a histogram. The chi-square test compares the actual number of observations in each cell to the expected number of observations in each cell if the data follow a normal distribution. The practitioner must determine both the number of cells and the boundary values of each of the cells. The number of cells that should be used in the chi-square test follows the following guidelines. The number of cells should be the maximum number, not to exceed 100, with each cell having at least five expected observations. This means that if there are 30 data points, then there should be six cells. Note that because there must be at least 20 data points, the chi-square test can never have fewer than four cells.

It is best to utilize an equiprobable approach to determine the cell boundaries. This means that the bounds of each cell are adjusted so that an equal percentage of the total number of observations should be found in each cell. Because the normal distribution is bell-shaped rather than uniformly shaped, the calculation of the cell boundaries does involve some mathematics and basic statistics.

To illustrate this process, assume that there are a total of 20 data points in the data set. Using the algorithm to determine the number of cells, we will need 20 divided by 5 or a total of 4 cells. Each of these cells is to be equiprobable in size. This means that 25% or 5 of the observed data points should be in each of the cells. To determine the cell boundaries, we will need to calculate the mean and standard deviation of the data set. Because normal or bell distributions are symmetric about the mean, 50% of the observations should fall below the mean, and 50% of the observations should be more than the mean. This also means that two of the cells will be to the left or below the mean and two of the cells will be to the right or above of the mean. It also follows that the upper boundary of the second cell will be the same as the mean and the lower boundary of the third cell will also be the same as the mean. Similarly, the left boundary of the first cell must cover the smallest value possible and the right boundary of the fourth cell must cover the highest possible value. Because we have only four cells and know the lowest possible value, the highest possible value, and the mean, the only remaining values that need to be calculated are the boundary value between the first and second cells and the boundary value between the third and fourth cells.

To calculate these two remaining values, we must utilize either a normal distribution table, which can be found in any statistics book, or the NORMINV function in a spreadsheet like Microsoft Excel. We begin the illustration using the manual method. After the practitioner has mastered the manual method, it will be an easy task to utilize the NORMINV function. The manual method begins with determining the Z or standard normal value for determining the probability value corresponding to the cell.

In our case, we can begin with finding the standard normal value for either a 0.25 or a 0.75 probability. We may need to use a 0.75 probability because some normal distribution tables begin with 0.50 probability. By looking in the body of the table, we can find this value. To convert it to a standard normal number, we need to move over to the boundary of the tables. A 0.75 probability has a corresponding standard normal value of approximately 0.67. We can then use the standard normal value to calculate the boundary values of our normal distribution. We have already determined that the lower boundary of the third cell is the mean. We can calculate the upper boundary of the third cell by taking the data mean and then adding the standard normal value multiplied by the standard deviation of the data. Remember that this value is also the lower bound of the fourth cell. We can also calculate the lower bound of the second cell by taking the data mean and subtracting the same standard normal value multiplied by the standard deviation. The completion of these calculations results in determining the boundary values for each of the four cells.

8.10.2.1   Observed Data Frequencies

The next step in the chi-square test is to count the number of observations that correspond to each of the equiprobable cells. If the data are approximately normally distributed, the number of observations will be the same as the number of expected observations. In most cases, however, there will be some differences, and the number of observations will not equal the expected number of five observations in each cell.

8.10.2.2 Calculating the Test Statistic

We now need to calculate what is known as the chi-square test statistic. This value represents the difference between the data set and how the data set would be if it were normally distributed. The test statistic is a summation of the squared difference between the actual observations and the expected observations divided by the number of expected observations for each cell.

8.10.2.3 Determining the Critical Value

The chi-square test statistic previously calculated must be compared against a critical value. The critical value is determined by consulting a chi-square distribution table or can be calculated with the CHIINV function in Excel. The chi-square critical value is determined by using a chi-square distribution table corresponding to the previously established level of significance. Chi-square tables for α of 0.05 are available in any statistics book. To find the actual critical value, it is also necessary to determine the number of degrees of freedom for the chi-square test. This is calculated by the number of cells minus the parameters used to determine the cell boundaries minus 1. In our example, we had four cells, and we used the mean and standard deviation parameters from the data set. This means that the number of degrees of freedom in our example would be 1. The corresponding chi-square critical value for an α of 0.05 and 1 degree of freedom is 3.84. Of course, the critical value will change depending on the number of cells in the chi-square test.

8.10.2.4 Comparing the Test Statistic with the Critical Value

The last step in the chi-square test is to compare the calculated test statistic with the critical value from the chi-square distribution table. In practitioner's terms, if the calculated test statistic is less than the critical value, the data set can be considered approximately normal. Conversely, if the test statistic exceeds the critical value, then the data cannot be considered approximately normally distributed. In statistical terms, if the test statistic was less than the critical value, then we could not reject the null hypotheses of the data's being approximately normally distributed. Likewise, if the test statistic were greater than the critical value, we could reject the null hypothesis of the data being approximately normally distributed.

8.10.3 Hypothesis Tests

If both the system and the model validation data sets are found to be normal, we next need to establish if the system and model data are statistically similar with a hypothesis test. Hypothesis tests involve the establishment of a null hypothesis that the means of the data sets are statistically similar. The null hypothesis, in practitioner terms, is either accepted or rejected. If it is accepted, then the simulation model is valid. If it is rejected, then the model is invalid. In this case, additional model development will be necessary.

There are a total of four different types of hypothesis tests (Johnson et al., 1999). The selection of the appropriate comparison test is dependent on whether or not the data is normally distributed, paired, or independently generated and similar or not in variance. The flow chart in Figure 8.1 can assist the practitioner in selecting the most appropriate hypothesis test.

As the flow chart indicates, the first step is to determine whether or not the two data sets are normally distributed. In order to determine this, we must either revisit the chi-square test or make an assumption that the data are either normally or not normally distributed. In the event that either one or both of the data sets are not normally distributed or that we assume that either one or both of the data sets are not normally distributed, the appropriate test is a nonparametric rank sum U test. Conversely, if we determine or assume that both of the data sets are normal, we must continue down the flow chart. If the data are naturally paired, we must perform a paired t-test. If the data are not naturally paired, then we must next determine whether or not the variance is similar. We can perform this process with an F-test. If the variance between the two data sets is similar, we use an independent t-test. If the variance is not similar, then we run a Smith–Satterthwaite test.

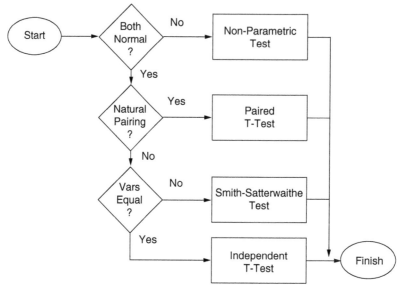

FIGURE 8.1 Hypothesis test flow chart.

8.10.4 *F*-Test

The *F*-test compares the variance of the system validation data set and that of the model validation data set. There are a number of specific implementations of the *F*-test; however, only one version is necessary for practitioner use. In this version the practitioner is interested in calculating a test statistic of the ratio of the larger variance divided by the smaller variance. It does not matter whether the larger or the smaller variance is from the system validation data set or the model validation set. This test statistic is compared to a critical *F* value at a particular level of statistical significance. A common level of statistical significance is $\alpha = 0.05$. The practitioner may acquire the critical *F* value using commonly available statistical tables or by using the *F* inverse function in a spreadsheet packet such as Microsoft Excel. In addition to the level of significance, the practitioner must also specify the number of degrees of freedom for the variance in the numerator and the number of degrees of freedom for the variance in the denominator.

There are a number of different implementations of the *F*-test. The particular version presented here requires that the data set with the larger variance be the numerator and the data set with the smaller variance be the denominator. This gives us the ability to simplify the calculations. The formula for using this version of the *F*-test is:

$$F = \frac{S_M^2}{S_m^2}$$

where

         S_M^2 is the variance of the data set with the larger variance.
         S_m^2 is the variance of the data set with the smaller variance.

The steps involved in implementing the *F*-test are:

1. Null hypothesis: variances of both groups are equal.
2. Alternative hypothesis: variances of both groups are not equal.
3. Select a level of significance.
4. Determine the critical value for *f* at the level of significance divided by 2 with degrees of freedom according to the number of samples in the data set with the larger variance: 1 for the numerator and number of samples in the data set with the smaller variance; 1 for the denominator.
5. Calculate the test statistic *f* according to the formula above.
6. Reject the null hypothesis if the test statistic exceeds the critical value.

8.10.4.1 Example of the *F*-Test

The following data sets were obtained from two different simulation models of our airport security checkpoint system. Alternative A is a model of the system with two ticket checkers, two metal detectors, and two x-ray machines. Alternative B is an alternative model: the same system with three ticket checkers, two x-ray machines, and one metal detector. The model of the system has been previously validated. Ten replications are sufficient for a desired relative precision of 0.10 for both models. The mean time in minutes for being processed through the system for 40 actual system passengers and 40 model passengers is listed in the following table:

System	System	System	System	Model	Model	Model	Model
25.92	28.44	32.04	45.00	30.92	20.11	32.35	54.13
31.68	33.48	29.52	29.52	41.39	34.26	35.05	44.20
37.44	31.68	44.28	44.28	41.72	39.08	25.38	14.71
35.28	34.20	33.48	33.12	28.23	27.57	41.86	35.81
83.52	39.60	41.40	28.44	25.44	26.85	26.77	57.31
42.12	33.48	44.28	30.60	38.32	15.46	32.18	46.65
25.92	31.32	34.20	33.12	30.83	36.28	30.43	49.68
30.96	30.24	27.36	36.72	31.04	46.69	56.46	53.49
40.32	52.20	34.20	25.92	33.40	76.24	35.19	42.64
34.56	40.68	34.92	23.40	26.23	31.50	28.71	20.71

The variance of the system is 99.36 s, and the variance of the model is 152.96 s. With this summary statistical information, we are ready to begin our *F*-test calculations.

1. H_o: The variance of alternative A is equal to the variance of alternative B; H_a: the variance of alternative A is not equal to the variance of alternative B.
2. We select a 0.90 level of significance or 0.10 as an α level.
3. The critical value for *F* at $\alpha/2$ with 39 degrees of freedom in both the numerator and denominator is 1.70.
4. The *F* statistic is calculated by:

$$F = \frac{152.96}{99.36}$$

$$= 1.54$$

5. The test statistic 1.54 is less than the critical value of 1.70, so we cannot reject the null hypothesis that the data have similar variance.

The fundamental statistics could also have been performed in Excel. The screen for this result would look like Figure 8.2.

8.10.5 Independent *t*-Test

The independent *t*-test is utilized when the data are normal and the data sets have similar variance. This test will determine if there is a statistically significant difference between two simulation models at a given level of significance.

In order to perform this test, we have to calculate the mean and sample standard deviation of both of our sets of data. The procedure is as follows:

Null hypothesis: means of both groups are equal.
Alternative hypothesis: means of both groups are not equal.
Select a level of significance.
Determine the critical value for *t* at the level of significance divided by 2 with degrees of freedom according to number of samples in both the datasets minus 2.
Calculate the test statistic *t* according to the formula above.

I	J	K
F-Test Two-Sample for Variances		
	Variable 1	Variable 2
Mean	36.13239	35.721
Variance	152.955436	99.36324
Observations	40	40
df	39	39
F	1.53935636	
P(F<=f) one-tail	0.0912285	
F Critical one-tail	1.70446413	

FIGURE 8.2   Excel *F*-test.

Reject the null hypothesis if the test statistic is either greater than the critical value or less than the negative of the critical value.

The formula for calculating the test statistic *t* is:

$$t = \frac{(\overline{x}_1 - \overline{x}_2)}{\sqrt{(n_1 - 1)s_1^2 + (n_2 - 1)s_2^2}} \sqrt{\frac{n_1 n_2 (n_1 + n_2 - 2)}{n_1 + n_2}}$$

where

t = calculated test statistic

$\overline{x}_1$ = mean of the first alternative

$\overline{x}_2$ = mean of the second alternative

s_1^2 = variance of the first alternative

s_{21}^2 = variance of the second alternative

n_1 = number of data points in the first alternative

n_2 = number of data points in the second alternative

8.10.5.1 Example of Independent *t*-Test

If we assume that the data sets from the previous example are both normal, we could perform an independent *t*-test to determine if a difference exists between the two data sets. The mean and standard deviation for the system, in seconds, are 35.72 and 9.97, respectively. The mean and standard deviation for the model, in seconds, are 36.13 and 12.37.

H_o: The mean of alternative 1 is equal to the mean of alternative 2.
H_a: The mean of alternative 1 is not equal to the mean of alternative 2.
Use a level of significance $\alpha = 0.05$.
The critical value for *t* at $\alpha/2$, 78 degrees of freedom, is 1.991.
The test statistic is:

$$t = \frac{(35.72 - 36.13)}{\sqrt{(40-1)9.97^2 + (40-1)12.37^2}} \sqrt{\frac{40 * 40(40 + 40 - 2)}{40 + 40}}$$

$$= -0.06$$

The test statistic *t* of −0.06 is between −1.991 and 1.991, so we cannot reject the null hypothesis. This means that there is no statistically significant difference between the actual system and the simulation model.

I	J	K
t-Test: Two-Sample Assuming Equal Variances		
	system	model
Mean	35.721	36.13239
Variance	99.36324	152.955436
Observations	40	40
Pooled Variance	126.159338	
Hypothesized Mean	0	
df	78	
t Stat	-0.1637982	
P(T<=t) one-tail	0.43515686	
t Critical one-tail	1.66462542	
P(T<=t) two-tail	0.87031373	
t Critical two-tail	1.99084752	

FIGURE 8.3   Excel *t*-test for equal variances.

To make life easier for ourselves, switch the order of the system and model data sets. This will result in a positive number because the mean in seconds is larger for the model data set than it is for the system data set. Thus, our test statistic is 0.06. Again, we cannot reject the null hypothesis because 0.06 is between −1.991 and 1.991.

A screen capture of the corresponding Excel *t*-test printout is illustrated in Figure 8.3.

8.10.6 Smith–Satterthwaite Test Implementation

In the event that the system and model data are both normal but the variances are dissimilar, the practitioner can utilize the Smith–Satterthwaite test to validate the simulation model. As with the small-sample *t*-test, the practitioner must reutilize the previously collected mean and standard deviation of both the system and model validation data. The Smith–Satterthwaite test accounts for the differences in variance by adjusting the degrees of freedom for the *t* critical value. The degrees of freedom estimator uses the formula below:

$$d.f. = \frac{[s_1^2/n_1 + s_2^2/n_2]^2}{[s_1^2/n_1]^2/(n_1-1) + [s_2^2/n_2]^2/(n_2-1)}$$

where

$d.f.$ = degrees of freedom

s_1^2 = sample variance of the first alternative

s_2^2 = sample variance of the second alternative

   = sample size of the first alternative

n_2 = sample size of the second alternative

In most cases the number of degrees of freedom calculated in this manner will not be an integer. The natural question again becomes whether to round up or round down. In general, as a practitioner, you want to take the most conservative approach. This means that under questionable circumstances you would rather reject the null hypothesis and conclude that the model is invalid than you would want to accept the null hypotheses and conclude that the model is valid. This means that we want a smaller rather than larger critical value. Because the *t* value increases as the number of degrees of freedom decreases, we want to round the estimated degrees of freedom upward. Once the degree of freedom estimator has been calculated, the practitioner can use the following formula to calculate the Smith–Satterthwaite test statistic:

$$t = \frac{\overline{x}_1 - \overline{x}_2}{\sqrt{\dfrac{s_1^2}{n_1} + \dfrac{s_2^2}{n_2}}}$$

where

t	=	t-test statistic for the Smith-Satterthwaite
$\overline{x}_1$	=	mean of the first alternative replications
$\overline{x}_2$	=	mean of the second alternative replications
s_1^2	=	sample variance of the first alternative
s_2^2	=	sample variance of the second alternative
n_1	=	sample size of the first alternative
n_2	=	sample size of the second alternative

The rest of the hypothesis test process would be the same as for the independent t-test.

8.10.6.1 Smith–Satterthwaite Example

Assuming that our system and model data set actually rejected the F-test null hypotheses, we would have needed to utilize the Smith–Satterthwaite hypothesis test. Using the same mean and standard deviation values for the system and model data, our process is:

H_o: The mean of the system is equal to the mean of the model; i.e., our model is valid.
H_a: The mean of the system is not equal to the mean of the model; i.e., our model is not valid.
Level of significance is 0.05.
The critical value is calculated with:

$$d.f. = \frac{[9.97^2 / 40 + 12.37^2 / 40]^2}{[9.97^2 / 40]^2 / (40-1) + [12.37^2 / 40]^2 / (40-1)}$$

$$d.f. = 74.63$$

As we discussed, the practitioner is interested in being conservative, so we round up to the nearest degree of freedom to 75. The critical value with $\alpha = 0.05$ and 75 degrees of freedom is 1.992.

The test statistic is:

$$t = \frac{35.72 - 36.13}{\sqrt{\dfrac{9.97^2}{40} + \dfrac{12.37^2}{40}}}$$

$$t = -0.16$$

Because –0.16 is between –1.991 and 1.991, we cannot reject the null hypothesis that the means of the system and model are the same. This means that the model is valid.

8.10.7 Nonparametric Test Implementation

A nonparametric test is utilized when either one or both of the validation data sets is nonnormal. An easily implemented nonparametric test is the U test or rank sum test. This test compares the sum of the ranks of the data points from each of the validation data groups.

To implement the rank sum test, the practitioner sorts the data from each group in ascending order. After sorting, the practitioner merges the data sets together while maintaining the identity of the set the data originally came from. In a spreadsheet application, the first column would correspond to the data while the second column would correspond to the data set. Typically, column 1 would represent the system validation data, while column 2 would represent the model data. A third column is used to rank all of the observations. Because the data were sorted and merged in ascending order, the first observation will have a rank of 1, and the last observation will have a rank corresponding to the total number of data points in both groups. The rank positions from group 1, the system data, are summed as variable W1. Similarly, the rank positions from group 2, the model data, are summed as variable W2.

In theory, if the model is valid, the sorted data will be approximately evenly interlaced between the system and the model validation data, and W1 and W2 will be close in value. Conversely, if the model is not valid, then the system and model data will not be approximately evenly distributed. This means that either W1 or W2 will be significantly smaller than the other.

To determine whether the difference is statistically significant, a few additional calculations must be conducted.

$$U_1 = W1 - \frac{n1(n1+1)}{2}, \quad U_2 = W2 - \frac{n2(n2+1)}{2}$$

1. U = min(U1, U2)
2. mean = n1*n2/2
3. var = n1*n2(n1+n2+1)/12
4. z = (U-mean)/std

In step 2, we retain the smaller of the two values, U1 and U2. In step 3, we are calculating the average value for all of the ranks. In step 4 we have the variance of all of the rank values. In step 5 we have the equivalent of a Z test statistic that can be used for a hypothesis test in the same manner as in the previous examples.

8.10.7.1 Nonparametric Rank Sum Test Example

In the event that either of the data sets used in our example was nonnormal, we would be most statistically robust if we performed a nonparametric rank sum test. Because we needed only ten replications of each alternative to meet our required relative precision, we really did not have enough data to perform a chi-square test on either data set.

Performing our preliminary rank sum test steps, we can sort and merge our data in ascending order.

rank	time	group	rank	time	group	rank	time	group	rank	time	group
1	14.71	2	21	29.52	1	41	33.48	1	61	41.40	1
2	15.46	2	22	30.24	1	42	34.20	1	62	41.72	2
3	20.11	2	23	30.43	2	43	34.20	1	63	41.86	2
4	20.71	2	24	30.60	1	44	34.20	1	64	42.12	1
5	23.40	1	25	30.83	2	45	34.26	2	65	42.64	2
6	25.38	2	26	30.92	2	46	34.56	1	66	44.20	2
7	25.44	2	27	30.96	1	47	34.92	1	67	44.28	1
8	25.92	1	28	31.04	2	48	35.05	2	68	44.28	1
9	25.92	1	29	31.32	1	49	35.19	2	69	44.28	1
10	25.92	1	30	31.50	2	50	35.28	1	70	45.00	1
11	26.23	2	31	31.68	1	51	35.81	2	71	46.65	2
12	26.77	2	32	31.68	1	52	36.28	2	72	46.69	2
13	26.85	2	33	32.04	1	53	36.72	1	73	49.68	2
14	27.36	1	34	32.18	2	54	37.44	1	74	52.20	1
15	27.57	2	35	32.35	2	55	38.32	2	75	53.49	2
16	28.23	2	36	33.12	1	56	39.08	2	76	54.13	2
17	28.44	1	37	33.12	1	57	39.60	1	77	56.46	2
18	28.44	1	38	33.40	2	58	40.32	1	78	57.31	2
19	28.71	2	39	33.48	1	59	40.68	1	79	76.24	2
20	29.52	1	40	33.48	1	60	41.39	2	80	83.52	1

If we add the rank totals for the system we have:

$$W1 = 5 + 8 + 9 + 10 + 14 + 17 + \ldots + 80 = 1619$$

Similarly, we sum the value of the ranks for the simulation model:

$$W2 = 1 + 2 + 3 + 4 + 6 + 7 + \ldots + 79 = 1621$$

We can now perform the rest of our rank sum test calculations.

$$U_1 = 1619 - \frac{40(40+1)}{2}, \quad U_2 = 1621 - \frac{40(40+1)}{2}$$

$$U = \min(799, 801) = 799$$

$$\text{mean} = \frac{40*40}{2} = 800$$

$$\text{var.} = \frac{40*40(40+40+1)}{12} = 10800$$

$$Z = \frac{799-800}{\sqrt{10800}} = -0.0096$$

The standard normal Z value for the two-sided 0.05 test corresponds to 0.025 probability. The Z value is plus or minus 1.96. Because –0.0096 is between –1.96 and 1.96, we cannot reject the null hypothesis that the two data groups are statistically similar. This means that the model is valid.

8.11 When a Model Cannot Be Statistically Validated and What To Do about It

Occasionally, the practitioner's first attempt at a simulation model cannot be statistically validated. This is entirely possible even though the model was successfully face validated. Any conclusions drawn from a statistically invalid model are automatically suspect. Thus, it is imperative to investigate the causes behind the inability to validate the model. There are a number of possible reasons for the inability of the model to be statistically validated. These include but are not limited to:

- Nonstationary system
- Poor input data
- Invalid assumptions
- Poor modeling

8.11.1 System is Nonstationary

With some systems it may be necessary to collect input data and validation data in separate sessions. If the input distributions such as the interarrival times and the service time differ over time, then the process is considered nonstationary. Systems may be nonstationary as a result of seasonal or cyclic changes. When the practitioner attempts to collect validation data at a later date, the system is no longer the same. This means that the practitioner is actually attempting to validate a different model than the actual system.

When the practitioner encounters a potentially nonstationary system, two different approaches may be taken. The first approach is to try to incorporate the nonstationary components of the system in the model. There are special techniques to try to incorporate input distributions that vary over time. The second approach is to attempt to gather both the input data and the validation at the same time. This way, the practitioner is guaranteed to be trying to validate the correct model. However, the model is then actually valid only for the conditions under which the input and validation data were collected. It would be inappropriate to try to draw conclusions from a model that was validated under different conditions.

8.11.2 Poor Input Data

Another potential reason for an invalid model is poor input data. If the data were collected by the practitioner, this could be a result of attempting to utilize insufficient quantities of data for proper data distribution. If the data were historical, then there is probably going to be some question as to the reliability of the data. The exact conditions under which the data were collected is likely to be unknown. If poor input data are suspected as the reason for the invalid model, the practitioner has no other choice than to collect additional real-time data.

8.11.3 Invalid Assumptions

Most simulation models are a simplified approximation of the actual or proposed system. It is simply impossible to replicate the actual system with 100% fidelity. While developing the model, the practitioner likely made some simplifying assumptions. Sometimes these simplifying assumptions are made in the interest of development speed, and sometimes they are made because some part of the process is too complex to model in detail.

When a simplifying assumption is suspect, the practitioner can either increase the level of the model detail or attempt to increase the accuracy of the simplifying assumption. Increasing the level of model detail has the potential to delay seriously the development process. Additional data collection as well as model development may be necessary to increase the level of detail. On the other hand, increasing the accuracy of a simplifying assumption can be a relatively easy task to implement with powerful effects on the statistical validation process. In one instance simply slightly increasing the fixed travel velocity of entities traveling through a system enabled a complex model to achieve statistical validity.

8.11.4 Poor Modeling

Poor modeling is an obvious possibility when a model cannot be validated. This may arise because of the level of the practitioner's modeling skills or because some part of the system was inadvertently omitted from the model. Sometimes, the animation of the system can provide clues as to possible modeling errors. In one simulation model, entities were observed disappearing from one part of the model and reappearing in another. It was obvious in this case that the practitioner needed additional model development time to produce a valid model. One related point to poor modeling is that no two models developed by different practitioners are likely to produce exactly the same results. Whether one model is necessarily better than another is debatable if both can be validated.

8.12 Summary

Validation is the process of insuring that the simulation model represents reality. The simulation model validation process consists of both face validity and statistical validity. Face validity is the continuous process of ensuring that the model, at least on the surface, represents reality. Face validity is normally performed with the assistance of domain experts. Face validity is a necessary but not individually sufficient condition for establishing complete model validity. Statistical validity involves comparison of the simulation model with the actual system. In statistical validity, some output measure of performance is collected. The same system loading conditions observed during the data collection process must be recreated in the simulation model. If statistical validity cannot be established, the model must be examined for flaws and corrected until statistical validity can be achieved.

Chapter Problems

1. What is the difference between face validity and statistical validity?

2. What are some possible causes that may prevent a model from being valid?

3. What kind of information must be observed when validation data are being collected from the system?

4. What must be done to the model data before it is compared to the system data?

5. What are the advantages and disadvantages to using individual observations versus summary statistics when validating a model?

6. A single server, single queue system is to be modeled. The following system times were obtained in minutes from the actual system and a model. When the actual system times were observed, there were two people waiting in the queue. Assuming normality for both sets of data, is the model valid at 0.10 for all tests?

   system 2, 5, 9, 6, 11, 10, 6, 4, 6, 8

   model 6, 8, 9, 5, 8, 7, 4, 7, 6, 9, 8, 4, 8, 2, 8

References

Johnson, R.A., Freund, J.E., and Miller, I. (1999), *Miller and Freund's Probability and Statistics for Engineers,* Pearson Education, New York.

Law, A.M. and Kelton, D.W. (2000), *Simulation Modeling and Analysis,* 3rd ed., McGraw-Hill, New York.

9

Experimental Design

"Do not permute yourself to death."

9.1 Introduction

Once the simulation model of the actual system has been properly validated, the practitioner can turn his or her attention to determining how to design additional models for subsequent experimental analysis. Traditional statistical references may be used to guide this process (Montgomery, 1997; Miller et al., 1990; Hildebrand and Ott, 1991). The choice of experimental design alternatives is dependent on the original simulation project objectives developed during the problem formulation phase. To assist the practitioner in the appropriate choice of experimental models, we discuss the concepts of:

- Factors and levels
- Two alternative experimental designs
- One-factor experimental designs
- Two-factor experimental designs
- Multifactor experimental designs
- 2^k experimental designs
- Interactions
- Refining the experimental alternatives

9.2   Factors and Levels

In experimental design terminology, factors are the different variables thought to have an effect on the output performance of the system. These variables are controllable in that the practitioner can vary the levels in both the actual system and the simulation models. Typical examples of factors are:

- Workers who perform specific functions
- Machines that perform specific operations
- Machine capacities
- Priority sequencing policies
- Worker schedules
- Stocking levels

Note that it is easily possible to have several different individual factors for any of the above types of factors. For example, if we have an airport security checkpoint system, there are three different types of resource factors. These include:

- Ticket checkers
- Metal detectors/operators
- X-ray machines/operators

Each of these three different types of processes can be an individual factor. Each of these factors can have different levels.

The corresponding examples of levels for the above factors are:

- Four vs. five vs. six workers performing a specific function
- An old vs. a new machine performing specific operations
- A 5-ton vs. a 2-ton capacity truck
- First-in–first-out vs. last-in–first-out priority sequence policies
- Five 8-h shifts vs. four 10-h shifts
- Restocking order levels between 10 and 25%

Note that in the above examples the number of different levels for a given factor can vary. Consider the case of the level of workers for a particular operation factor. We could be interested in the effects of:

- Four vs. five workers
- Four vs. six workers
- Five vs. six workers
- Four vs. five vs. six workers

Our experiment could be a simple comparison of any of the two-level alternative combinations, or it could be the more complex alternative involving the three-level alternative. The experimental design alternatives can be classified as either simple two-alternative designs or more than two alternative designs.

9.3   Two Alternative Experimental Designs

In the simplest experimental design, the practitioner will have only two models to analyze. We can have two situations with this type of design:

- A base system exists
- No base system exists

9.3.1   Existing System

If the actual system is currently in existence, the corresponding model will obviously be one of the other alternatives. The other alternatives could be based on:

- An alternative operating policy
- An alternative resource policy

As we have previously discussed, the alternative operating policy could include different priority scheduling policies. Similarly, in the case of an alternative resource policy, it could include a different number of resources, or it could include resources with different capacities.

9.3.1.1   Operating Policy with Existing System

Suppose our base model consists of three parallel queue clerk systems in a customer service center. This system has:

- Three clerks
- Three parallel queues

If we varied only the operating policy, the alternative might be:

- Three clerks
- A single snake queue

In this case, we altered the operating policy by changing the types of the queues from three parallel queues that feed into individual clerks to one single long snake queue that feeds into all three clerks. Any difference in the performance between the two models will be a result of the difference in the queue operating policy.

Another way that the operating policy could be changed is with the presence of express queues. In this case, our operating policy change alternative model would consist of the following arrangement:

- Two clerks with regular parallel queues
- One clerk with an express queue

Here, we are not altering the physical layout of the queues. We are, however, restricting the type of customer who can enter the queue. Obviously, our alternative model could also have contained:

- Two clerks with express parallel queues
- One clerk with a regular queue

Which of these two alternative models would be better to use is a question that must be left to the analysis phase of the project.

9.3.1.2   Resource Policy with Existing System

On the other hand, if we decided to alter the resource level, the original system would still consist of:

- Three clerks
- Three parallel queues

However, the alternative might consist of the following components:

- Four clerks
- Four parallel queues

Here we change the level of the clerk resources from three to four. We maintain the queue operating policy by keeping the same type of parallel queues as in the base model. Here the difference between the two models will be a result only of changing the level of the clerk resource levels. Note that we could just as easily have used five clerks and five parallel queues as our alternative. We would still be keeping the same factor, we would just be using a different factor level for the alternative model.

9.3.1.3 Both Operating Policy and Resource Levels

It is possible to create an alternative model with both operating policies and resource capacity factors at different levels. This means that any number of operating policies or resource capacity levels could be different. Our original model is still:

- Three clerks
- Three parallel queues

However, this now means that one possible alternative model from our original base model could be:

- Four clerks
- A single snake queue

Here we have simultaneously changed the number of clerks from three to four and the type of queue from parallel to a single snake queue in the same single alternative model. Should there be any difference in the performance of the model, we will be at a loss to determine the most influential factor. It could be the increase in the number of resources from three to four, or it could be the difference in the queue operating policy from parallel queues to a single snake queue. It could also actually be as a result of both factors being changed at the same time.

We could really confuse ourselves if we varied more than one type of factor at different levels in our single alternative model. An example of this would be the alternative with the following components.

- Two clerks with a single express line
- Two clerks with a parallel regular lines

Here we have changed the total number of clerks from three to four. We have also used both parallel and single snake queues. Finally, we have both express and regular lines. If we take this kind of mixed factor and level approach with only one alternative, the only thing that we will know is that the alternative may or may not perform differently from the base model. Obviously, this sort of approach should be discouraged.

9.3.2 No Existing System

In the case where the system does not exist, the practitioner must still start from somewhere. Here, the two different models could also differ with respect to operating or resource policies. Typical examples where the system may not exist include:

- Equipment for a new process from two different manufacturers
- Facility layout for a service facility for two different layouts

The same factor and level dangers are present when one is modeling nonexisting systems. The danger might actually be greater because even less is known about the alternative models in comparison to situations where at least one base model already exists.

9.3.3 Limitation of Two Alternative Approaches

The two alternative experimental design approach should really be of limited interest to the practitioner. The reason for this is that the identification of additional models and the analysis of those models is minor in comparison to the overall effort required for the entire project. This means that with a little bit of extra effort, the practitioner can evaluate a number of different alternatives. The result will be a much more robust analysis of the system. To determine which models to analyze, the practitioner will have to formally design an experiment.

9.3.4 Statistical Analysis of Two Alternative Systems

In any two-alternative comparison, the statistical comparison techniques are relatively simple. These comparisons primarily involve some sort of *t*-test or rank sum test. The statistical comparison of two alternative models is covered in depth in Chapter 10, "Analysis."

9.4 One-Factor Experimental Designs

The next level of sophistication is the one-factor experimental design. Here we have one specific factor that we are going to examine at three or more levels. The actual number of alternatives that we will have is simply the number of levels in the one factor.

From our previous example, we can decide to utilize a one-factor experimental design. If we select the number of clerks as our factor, our three different level alternatives could be:

- Three clerks
- Four clerks
- Five clerks

All of these models would have to use the same queue operating policy. This means that they would all have to use individual parallel queues. Any differences in the model would be expected to be a result of the different levels of clerk resources.

In the one-factor experimental design, we are not limited to the use of only three different levels, we can, in fact, use as many as we are interested in. Although there is likely to be a point of diminishing return at some point, it would be reasonable in our situation to test both slightly above and slightly below the number of clerks in the base system. This would mean that we could have a single-factor experiment with the following model configurations.

- Two clerks
- Three clerks
- Four clerks
- Five clerks
- Six clerks

We would be interested in the effect of two clerks even in the event that it appears that our base system of three clerks is insufficient for the system. In some systems, reducing the number of clerks will not necessarily reduce the performance of the system. If this is the case, the two-clerk alternative would indicate that the system could be operated at reduced cost. On the other hand, if the system performance was seriously degraded by going to two clerks, the higher number of clerks would become the focus of our attention.

We could also use an operating policy factor for our one-factor experimental design. Here we fix our number of clerks at a level of two and vary our queue operating policy. Three different one-factor model alternatives could be:

- Three individual parallel clerk queues
- One single snake queue feeding two clerks and one queue feeding one clerk
- One single snake queue feeding into all three clerks

Note that these are three completely distinct ways of operating the queues that feed into the three clerks. Any differences in the performance of these models would be attributable only to the difference in queue operating policy.

We could also evaluate the effects of having express queues. Unlike our two alternative experimental designs, we can look at the effects of as many combinations as we like at one time. This means that we can have:

- Three clerks with three regular parallel queues
- Two clerks with regular parallel queues, one clerk with an express queue
- One clerk with a regular parallel queue and two clerks with express parallel queues

The first alternative is our original base configuration. In the second alternative, two of the three clerks operate as before. However, one clerk accepts only express-type customers. The third alternative has one regular clerk with two express clerks. Each of the two express clerks has his or her own line. In these alternatives, we are maintaining the three-clerk level, and we are not mixing parallel and snake queues. The only factor that is present is the use of regular versus express queues. We would expect that any difference in the alternatives would be the result of only this factor.

9.4.1 Statistical Analysis of One-Factor Experimental Designs

One-factor experimental designs are best handled with analysis of variance and the Duncan multiple-range test. These techniques are described in detail in Chapter 10, "Analysis."

9.5 Two-Factor Experimental Designs

The next level of sophistication in selecting experimental alternatives is the two-factor experimental design. In this type of design, we are selecting two particular factors and will be examining each of these factors at a number of different levels. In this case, if we had a factor A and a factor B, the number of model alternatives that we would be generating would be easily calculated by:

$$\text{Number of levels in factor A} \times \text{number of levels in factor B}$$

Provided that we do not have too many levels in either factor A or factor B, the total number of simulation models will still be manageable. Continuing with our examples, we could have two-factor models with:

- Two different operating policies
- Two different resource policies

9.5.1 Two-Factor Operating Policy Designs

In a two-factor operating policy design, we need to have two separate operating policies that can be varied at the same time. With queue operating policies, our first factor can be the type of queue — parallel or single. Our second factor can be type of customer queue entry policy — regular or express. In a relatively simple two-factor model one combination that we could end up with is:

- Three clerks with regular parallel queues
- Three clerks with a single queue
- Two clerks with regular parallel queues and one clerk with a single express queue
- Two clerks with a single regular queue and one clerk with a single express queue

9.5.2 Two-Factor Resource Policy Design

In a two-factor operating policy design, we need to have two separate resource policies that can be varied at the same time. The obvious choice for the first resource policy is to vary the number of resources in the system. A less obvious choice for the resource policy is to use resources that are more or less capable than the normal resource. Continuing with our example, we could end up with the following set of alternatives:

- Three regular clerks with parallel queues
- Four regular clerks with parallel queues
- Three novice clerks with parallel queues
- Four novice clerks with parallel queues

In this example, our first resource factor is the number of clerks that are in the system. Our second resource factor is the skill level of the clerks, either regular or novice. Note that all of the systems still use the same type of parallel queue. There is also no difference in the type of customer — regular or express — that can be in any of the queues.

We could also have mixed the composition of the number of regular and novice clerks. This might provide a more realistic representation of what we would have in the real world, but it may also make our results a bit more difficult to interpret. This set of alternatives might include the following compositions.

- Three regular clerks with parallel queues
- Four regular clerks with parallel queues
- Two regular clerks and one novice clerk with parallel queues
- Two regular clerks and two novice clerks with parallel queues

In this example, we are not directly examining the effect of regular versus novice clerks. We are actually looking at alternatives with only regular clerks versus alternatives with mixes of both regular clerks and novice clerks.

9.5.3 Two-Factor Combination Operating Policy and Resource Policy Design

In the two-factor combination operating policy and resource policy design, we select a single operating policy factor and a single resource policy factor. Each of these can use as many levels as necessary to achieve our simulation project objectives. To illustrate the different types of model that may occur, we will continue with our example. Here we will use the type of queue as the operating policy and the number of resources as the resource policy. If we keep to two levels for each factor, one possible set of model alternatives is:

- Three clerks with parallel queues
- Four clerks with parallel queues
- Three clerks with one single snake queue
- Four clerks with one single snake queue

With this set of alternatives we can easily separate the effects of either the operating policy for the queues or the resource policy for the number of clerks.

9.6 Multifactor Experimental Designs

We are now ready to move on to a much more complicated type of experiment. Here we have multiple factors and multiple levels. The number of different model alternatives with this approach can quickly explode into an unmanageable level. Assuming that we have the same number of levels for each factor, the equation for calculating the number of alternatives is:

$$Total\ number\ of\ alternative = number\ of\ levels^{Number\ of\ factors}$$

This means that a relatively small number of levels and a relatively small number of factors can easily result in a large number of alternatives. A model must be made for each different alternative. All of the alternatives must be statistically analyzed.

For example, an analysis of the resource allocation of an airport security checkpoint system would involve the number of metal detectors/operators, x-ray machines/operators, and ticket checkers. In a small airport, the effects of 1 and 4 personnel might be examined for each component in the system. This type of experiment would result in

$$64 = 4^3$$

This is a total of 64 different alternatives. These alternatives are best represented in a chart.

Alternatives 1–16	Alternatives 17–32	Alternatives 33–48	Alternatives 49–64
1TC-1MD-1XR	1TC-1MD-2XR	1TC-1MD-3XR	1TC-1MD-4XR
1TC-2MD-1XR	1TC-2MD-2XR	1TC-2MD-3XR	1TC-2MD-4XR
1TC-3MD-1XR	1TC-3MD-2XR	1TC-3MD-3XR	1TC-3MD-4XR
1TC-4MD-1XR	1TC-4MD-2XR	1TC-4MD-3XR	1TC-4MD-4XR
2TC-1MD-1XR	2TC-1MD-2XR	2TC-1MD-3XR	2TC-1MD-4XR
2TC-2MD-1XR	2TC-2MD-2XR	2TC-2MD-3XR	2TC-2MD-4XR
2TC-3MD-1XR	2TC-3MD-1XR	2TC-3MD-3XR	2TC-3MD-4XR
2TC-4MD-1XR	2TC-4MD-2XR	2TC-4MD-3XR	2TC-4MD-4XR
3TC-1MD-1XR	3TC-1MD-2XR	3TC-1MD-3XR	3TC-1MD-4XR
3TC-2MD-1XR	3TC-2MD-2XR	3TC-2MD-3XR	3TC-2MD-4XR
3TC-3MD-1XR	3TC-3MD-2XR	3TC-3MD-3XR	3TC-3MD-4XR
3TC-4MD-1XR	3TC-4MD-2XR	3TC-4MD-3XR	3TC-4MD-4XR
4TC-1MD-1XR	4TC-1MD-2XR	4TC-1MD-3XR	4TC-1MD-4XR
4TC-2MD-1XR	4TC-2MD-2XR	4TC-2MD-3XR	4TC-2MD-4XR
4TC-3MD-1XR	4TC-3MD-2XR	4TC-3MD-3XR	4TC-3MD-4XR
4TC-4MD-1XR	4TC-4MD-2XR	4TC-4MD-3XR	4TC-4MD-4XR

where

    TC = ticket checkers

    MD = metal detectors/operators

    XR = x-ray machines/operators

This is a lot of modeling and analysis under any circumstances. The sad part about generating this large a number of alternatives is that in actual practice, most of these alternatives will not be statistically different from the others. This can be interpreted as much unnecessary effort on the part of the simulation project team. Thus, it is in the best interest of the simulation team to begin to think about different ways to reduce the number of alternatives. We can first attempt to reduce the number of alternatives by reducing either the number of levels or the number of factors.

9.6.1 Reducing the Number of Alternatives by Limiting the Number of Factors

To reduce the number of alternatives and corresponding models to a reasonable number, the practitioner has a few choices. The first choice is to reduce the number of factors. In our example, if we ignored the possible effects of the number of ticket checkers, we would have only two factors to contend with. This would result in four raised to the second power, or only 16 different alternatives and models.

Alternatives 1–4	Alternatives 5–8	Alternatives 9–12	Alternatives 13–16
1MD-1XR	2MD-1XR	3MD-1XR	4MD-1XR
1MD-2XR	2MD-2XR	3MD-2XR	4MD-2XR
1MD-3XR	2MD-3XR	3MD-3XR	4MD-3XR
1MD-4XR	2MD-4XR	3MD-4XR	4MD-4XR

where

    MD = metal detectors/operators

    XR = x-ray machines/operators

Of course, this approach prevents the examination of the effects of the ticket checkers. It could very well be that the effect of the ticket checkers is statistically significant. In other simulation projects, the use of this approach may also result in one or more important factors being ignored.

9.6.2 Reducing the Number of Alternatives by Reducing the Number of Levels

Another option for the practitioner is to limit the number of different levels for each factor. The disadvantage to this approach is that the practitioner must carefully select the appropriate levels to examine. In our security checkpoint system example, this approach would result in the three factors examined at fewer than four different levels. It may be realistic to assume that at least two personnel must man each of the system components at a given time. This would mean that the experiment could be reduced to three levels with two, three, or four personnel at each component. A total of three raised to the third power, or 27, alternatives would be needed with this approach.

Alternatives 1–9	Alternatives 10–18	Alternatives 19–27
2TC-2MD-2XR	2TC-2MD-3XR	2TC-2MD-4XR
2TC-3MD-2XR	2TC-3MD-3XR	2TC-3MD-4XR
2TC-4MD-2XR	2TC-4MD-3XR	2TC-4MD-4XR
3TC-2MD-2XR	3TC-2MD-3XR	3TC-2MD-4XR
3TC-3MD-2XR	3TC-3MD-3XR	3TC-3MD-4XR
3TC-4MD-2XR	3TC-4MD-3XR	3TC-4MD-4XR
4TC-2MD-2XR	4TC-2MD-3XR	4TC-2MD-4XR
4TC-3MD-2XR	4TC-3MD-3XR	4TC-3MD-4XR
4TC-4MD-2XR	4TC-4MD-3XR	4TC-4MD-4XR

where

    TC = ticket checkers

    MD = metal detectors/operators

    XR = x-ray machines/operators

The fact that a particular set of levels does not seem to have an effect does not necessarily mean that a different set of levels will not produce an effect. This does mean that, if it is suspected that the factor might have an effect at other levels, the other levels may need to be investigated.

9.7 2^k Experimental Designs

Yet another commonly accepted means of reducing the number of alternatives is to reduce the number of levels in each factor to two. These levels are referred to as the low level and the high level. The use of only two levels results in 2 raised to the number of factor alternatives. This approach is frequently referred to as a 2^k experimental design. In determining the alternatives for a 2^k experimental design, the concept of a binary tree can be used. This consists of a series of branches progressing from left to right. Two branches are in each set of the series. The first level will have a minus and a plus branch. The second level will have a minus and a plus branch for the first level minus branch. Similarly, the second level will also have a minus and a plus branch from the first level plus branch. This means that the second level will have a total of four branches. The third level will have a minus and a plus branch coming out of each of the four branches from the second level for a total of eight different branches. This procedure is illustrated in Figure 9.1.

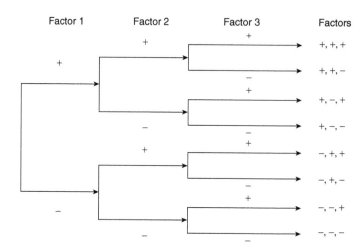

FIGURE 9.1 Experimental design branches.

In our example, we would next examine the same three factors, but only at two levels. This would result in two raised to the third power, or eight different alternatives. These alternatives are summarized in the table below:

Alternative	Ticket Checkers	Metal Detectors/Operators	X-Ray Machines/ Operators
A	−	−	−
B	−	−	+
C	−	+	−
D	−	+	+
E	+	−	−
F	+	−	+
G	+	+	−
H	+	+	+

Once the 2^k experimental design alternatives are established, the practitioner must decide on the values for the low and high levels.

For the purposes of our example, let us assume that we are interested in two to three ticket checkers, one to two metal detectors, and one to two x-ray machines. The resulting factor levels are illustrated in the following table.

Alternative	Ticket Checkers	Metal Detectors/Operators	X-Ray Machines/ Operators
A	2	1	1
B	2	1	2
C	2	2	1
D	2	2	2
E	3	1	1
F	3	1	2
G	3	2	1
H	3	2	2

The assignment of these values actually does not need to be consecutive. In other words, we do not necessarily need to examine the difference between the effects of 2 and 3 or 3 and 4 personnel assigned to each part of the security checkpoint. We could very well examine the effects between 2 and 4 personnel. A nonconsecutive level experiment might have the following low and high levels.

Alternative	Ticket Checkers	Metal Detectors/Operators	X-Ray Machines/Operators
A	2	1	1
B	2	1	2
C	2	2	1
D	2	2	2
E	4	1	1
F	4	1	2
G	4	2	1
H	4	2	2

If it turns out that there is no statistically significant difference between 2 and 4 ticket checkers, it also means that there is no difference between 2 and 3 or 3 and 4 ticket checkers. However, if there is a difference between 2 and 4 ticket checkers, we may want to redesign our levels for subsequent analysis.

9.8 Experimental Alternative Factor Interactions

One issue that the practitioner may need to address is the issue of factor interactions. This occurs when two particular factors have some sort of synergistic effect. That is, the effect of both of the factors may be larger than the sum of the effects of each of the individual factors. For example, a small number of metal detectors/operators and a small number of x-ray machines/operators may be examined in the same model. The delays in passenger processing may be much larger than that expected from analyzing either the metal detectors/operators or x-ray machines/operators individually. The effects of specific interactions can be examined through increasingly sophisticated statistical analysis techniques. However, the practitioner may need to opt for less sophisticated statistical techniques with the assumption that there are no special interactions.

9.9 Refining the Experimental Alternatives

In some instances, the initial experiment is primarily conducted to gain a feel for the performance of each of the factors that are thought to have an impact on the output measures of performance. There is no expectation that the simulation study will be limited to the initial set of factors or levels. In these circumstances, follow-up experiments are conducted that focus on the factors that were determined to be significant at the previous levels. If substantially different low and high values were used, it may be necessary to run several sets of experiments in order to bracket the point at which the levels in the factor become significant.

9.10 Summary

In this chapter we discussed the issue of simulation model experimental design. Once the practitioner has developed a valid base model, it is necessary to develop experimental alternatives. These alternatives are examined in an effort to improve the performance of the existing system. Simpler experiments will involve only two models. These types of experiments are typically the original or actual system and one alternative. More complex experiments will involve more than two models. The development of more complex experiments involves the identification of factors and levels. Factors can correspond to different types of resources in the system. Levels correspond to the number of each different type of resource to examine. The different possible factors and levels can rapidly increase. To keep the number of different alternative models under control, the practitioner can utilize a 2^k experimental design approach. This limits each factor to a high and a low level for the initial set of model alternatives. A final issue is that there may be different interactions between experimental factors. This means that particular level combinations of different factors may produce results that would not ordinarily be expected. The existence of interactions can be determined during the statistical analysis phase of the project.

Chapter Problems

1. What is the difference between a factor and a level?

2. What is meant by the term *interaction*?

3. How many alternatives will there be in a two-factor, four-level experiment?

4. Why would a practitioner be unlikely to utilize a two-alternative comparison experiment?

5. What is meant by a low level and a high level?

6. List all of the different alternatives that would be present in a 2^k-factorial experiment with 1 or 2 cashiers, 2 or 3 car washers, and 4 or 5 car dryers.

References

Hildebrand, D.K. and Ott, L. (1991), *Statistical Thinking for Managers*, 3rd ed., PWS-Kent, Boston.

Miller, I., Freund, J.E., and Johnson, R.A. (1990), *Probability and Statistics for Engineers*, Prentice-Hall, Englewood Cliffs, NJ.

Montgomery, D.C. (1997), *Design and Analysis of Experiments*, John Wiley & Sons, New York.

10

Analysis

"Avoiding analysis paralysis."

10.1 Introduction

After the simulation practitioner has determined the experimental design alternatives, attention can be directed toward the analysis phase of the simulation project. The function of this phase is to provide the simulation practitioner with the information necessary to provide decision recommendations with respect to the project objectives. The analysis phase is the most statistically challenging project phase for the simulation practitioner. A review of basic statistics before we undertake the analysis phase may be beneficial (Johnson et al., 1999). A complete but somewhat more theoretical approach to the techniques in the analysis is also covered by Law and Kelton (2000). However, the ready availability of statistical functions in Excel and the statistical analysis features found in most dedicated simulation software packages help insulate the practitioner from getting unnecessarily bogged down in this phase. The step-by-step process outlined in this chapter should provide valuable guidance for the practitioner regardless of the currency of the practitioner's statistical skills.

This chapter is divided into two main sections based on the type of system the practitioner is interested in analyzing. The first section covers the statistical analysis approach for terminating systems. The second section covers the statistical analysis approach for nonterminating systems. The difference between terminating and nonterminating systems was discussed in depth in Chapter 4, "System Definition."

In review, terminating systems have a naturally occurring event that ends the period of interest for the simulation. Terminating systems also completely clean themselves of entities between periods of interest. An example of a terminating system is the operation of a restaurant during the lunch period. The lunch period is the only time of interest, and each lunch period, the customers are new. In contrast, nonterminating systems either do not have a naturally occurring terminating event and run forever or do have a terminating event but do not clean themselves of entities between periods. An example of a nonterminating system is the manufacturing system. The system either runs continuously or shuts down after each shift. If the system shuts down after each shift, work is resumed at the beginning of the next shift on any products that are still in progress.

The analysis process for terminating systems consists of the following steps:

1. Replication analysis
2. Production simulation runs
3. Statistical analysis of the simulation run results
4. Economic analysis of statistical analysis results

Replication analysis involves the determination of the number of simulation replications or runs that are required to analyze statistically the differences between the simulation models. Production simulation runs involve the practitioner completing the number of simulation runs determined through the replication analysis. The statistical analysis of the simulation runs will interpret the level of differences between the experimental design alternatives. Last, the economic analysis puts the statistical analysis results in perspective from an engineering economy or business perspective.

In contrast to terminating systems, nonterminating system models use a single long replication run. As a result, the analysis procedure is significantly different from that of terminating systems. The non-terminating system analysis consists of:

1. Starting conditions
2. Determining steady state
3. Addressing autocorrelation
4. Length of replication
5. Batching method

Starting conditions refer to simulation run initialization procedures. Determining the steady state involves eliminating the nonrepresentative transient output performance data that are present at the beginning of the simulation run. This ensures that the remaining steady-state data are representative of the actual system performance. Autocorrelation is the correlation between successive performance observations. This must be addressed so that the variance in the output data is not underestimated. Because nonterminating systems use a single run for data analysis, we must determine how long the run should be. Finally, it is necessary to split the single long run into batches to facilitate the statistical comparison between nonterminating systems.

Even though the practitioner may be interested in analyzing a nonterminating system, practitioners are still recommended to examine the terminating system analysis section. This is because some terminating analysis procedures are still necessary to analyze statistically nonterminating system models.

10.2 Terminating System Analysis

This section includes replication analysis, production run approaches, statistical analysis, and economic analysis for terminating systems.

10.2.1 Replication Analysis for Terminating Systems

The input distributions of simulation models are usually probabilistic in nature. This input variability naturally results in some variation in the output measures of performance. Because the output measures have some variation, it is inappropriate for the simulation practitioner to recommend any given course of action based on the results from a single simulation run or replication. To reduce the chance of making a wrong recommendation, it is necessary to run a number of simulation replications and then make the recommendations based on all of the available data. A natural question is: If not one replication, then how many? This is the purpose of replication analysis.

The replication analysis process begins with selecting an initial number of replications. Summary statistics from this initial set of replications are then used to calculate whether or not additional replications are required at a particular level of confidence. If more replications are required, then the practitioner needs to run additional replications and recalculate the summary statistics and replication formulas for the process. The following examples are included in the Excel worksheet replicat.xls. The worksheets are set up so that the practitioner may also use the worksheets for conducting a simulation study.

10.2.2 Initial Number of Replications

The simulation practitioner must begin with some arbitrary number of replications to begin the replication analysis. Relatively, a smaller number of initial replications will increase the probability that the initial number is insufficient and that additional replications will be required. Conversely, a larger number of initial replications will decrease the probability that the initial number of replications is insufficient. However, a larger number of unnecessary initial replications will result in increased time on the part of the practitioner. Although this may not be a problem for relatively small simple models run on fast microcomputers, it could become a serious issue for larger, more complex models. The practitioner must also remember that it is not just one model that will need to undergo replication analysis, but all of the alternative models.

A reasonable practitioner compromise to this situation is to initially run some small number of replications. A common number of initial replications is ten. This provides a sufficient number of replications to have reasonable statistical confidence given that additional replications can always be subsequently added.

10.2.3 Replication Calculations

In order to perform the replication calculations, the practitioner must first calculate the mean and standard deviation of the ten replication means. Figure 10.1 shows the average system times for ten replications of the security checkpoint system. The mean or average of the ten values is 11.00, and the sample standard deviation is 0.42.

These summary statistical values are then used to calculate what is known as the standard error of the data using the following formula:

$$\text{Standard Error} = t_{1-\alpha/2,n-1} * s / \sqrt{n}$$

where

$t = t$ distribution for $1 - \alpha/2$ and $n - 1$ degrees of freedom

$s = $ standard deviation of the replication means

$n = $ number of observations in the sample

	C	D
	Replication Number	Mean System Time in Minutes
10		
11	1	10.80
12	2	11.96
13	3	10.47
14	4	10.70
15	5	10.80
16	6	11.35
17	7	11.04
18	8	10.68
19	9	11.15
20	10	11.00

FIGURE 10.1 Replication means.

The standard error is essentially the amount of dispersion around the mean value that data may exhibit. The first term t comes from the t probability distribution. These values are commonly available in any statistical or simulation textbook and are also available in Excel by using one of the t distribution functions. The t value depends on two parameters, the α level and the number of degrees of freedom.

The α level has to do with the level of confidence at which we wish to conduct our analysis. If we want to be 95% confident in the results of our analysis, then the α level is 1 minus the confidence level, or 0.05. We are interested in the dispersion around both sides of the mean, so we divide the α level in half. In Excel, the function that is used is:

$$= \text{TINV(probability, deg\_freedom)}$$

where

   probability = the percentage value of observations to the right of the corresponding t value

   deg_freedom = number of degrees of freedom to use

Note that the Excel implementation of the TINV function automatically assumes that you want to use a two-sided or two-tailed calculation of the t distribution value. The number of degrees of freedom is simply the number of replications minus 1.

The second term in the standard error calculation is s. This is simply the sample standard deviation of the replication averages. The mathematical formula for the sample standard deviation of the replication averages is:

$$s = \sqrt{\frac{\sum_1^{n-} x_i - \bar{\bar{x}}}{n-1}}$$

where

   s = sample standard deviation

   x_i bar = the replication average

   x bar bar = average of the replication averages

   n = number of replications

In Excel the formula for the sample standard deviation is:

$$= \text{stdev (beginning cell reference: ending cell reference)}$$

The third and last term in the standard error calculation is simply the square root of the initial number of replications.

10.2.3.1 Example

The results of the previous calculations are illustrated in Figure 10.2.

	D49	▾	*fx*	
	C			D
40	Mean			11.00
41	Standard Deviation			0.42
42				
43	t for alpha=0.025, 9 d.f.			2.26
44	square root of n			3.16
45				
46	Standard error			0.30

FIGURE 10.2 Replication calculations.

In our initial replication calculations we will have ten replication means. This means that the number of degrees of freedom will be ten minus one or nine. If we use an α value of 0.05, then the value for t with nine degrees of freedom is 2.262. We can find this value in a statistical table or use the corresponding Excel formula. The Excel implementation in Figure 10.2, cell d43, is:

$$= \text{TINV}(0.05, 9)$$

If we utilize the illustrated spreadsheet, the standard deviation of the replication means can be calculated using the formula in Figure 10.2, cell d41:

$$= \text{stdev}(d11{:}d20)$$

The square root of the n term is the easiest of all to calculate with the following formula in cell d44:

$$= 10^\wedge.5$$

All of these values can be combined in a single cell by using cell references or lumping all of the formulas together in Figure 10.2, cell d46.

$$= \text{TINV}(0.05,9)^*\text{stdev}(d11{:}d20)/10^\wedge.5$$

The resulting standard error for our data is 0.30. In practitioner terms, this means that the true mean for our security checkpoint time could vary plus or minus 0.30 min with 95% confidence. To make use of this information to determine the final number of replications that we need, we need to go one step further.

10.2.4 Selecting a Level of Precision

Once we know the standard error, we can use this information to calculate how many replications we need to run. However in order to do this, we must select a level of precision or error that is suitable for our study. There are two general approaches to determine objectively the level of precision that we are willing to accept. The first approach is based on an absolute comparison of the standard error to a particular tolerance level. The second approach is based on a relative value of the standard error in comparison to the sample mean.

10.2.4.1 Absolute Precision

In the absolute precision approach we select a tolerable level for the precision. The precision value is in the same units as the sample data. An issue with the absolute precision approach is the selection of an appropriate value for the level of precision. Unless the practitioner is very familiar with the process, the selection of an absolute precision level may appear to be somewhat arbitrary. For example, we can arbitrarily select an absolute precision level of 0.20 min. This means that we need to run enough additional replications to reduce our standard error to 0.20. Our current standard error with ten replications is 0.32 minutes. To determine the new number of replications that will be needed with this data set to achieve an absolute precision level of 0.20 min, we can utilize the previous standard error formula. In order to do this, we set the equation to our desired level of precision.

$$\text{Absolute Precision} = t_{1-\alpha/2,n-1} * s / \sqrt{n}$$

where

$t = t$ distribution for $1 - \alpha/2$ and $n - 1$ degrees of freedom

$s =$ standard deviation of the replication means

$n =$ number of observations in the sample

Next, we rearrange the terms and solve for the new *n*, the number of required replications to reach our desired level of absolute precision:

$$i = \left[\frac{t_{1-\alpha/2,n-1} * s / \sqrt{n}}{\text{Absolute Precision}} \right]^{1/2}$$

where

$t = t$ distribution for $1 - \alpha/2$ and $n - 1$ degrees of freedom

$s =$ standard deviation of the replication means

$i =$ number of replications needed to achieve the absolute precision

The Excel implementation of this formula is:

$$= ((\text{TINV}(0.05,9)*\text{STDEV}(D11:D20))/(0.2))^2$$

This is illustrated in Figure 10.3.

When the values from our data are inserted in the rearranged formula, our new number of replications is 23.04. At this point, the next question is whether to round up or to round down. Generally, it is better to be more conservative. This means that we should round up. This results in a new requirement to run a total of 24 replications to achieve an absolute precision of 0.20. The results of these replications are illustrated in Figure 10.4.

The next step in the process is to verify that 24 replications are actually sufficient for the designated absolute precision level. This means that the simulation model needs to be run for 24 replications, and the summary statistics recalculated. The new summary statistics can be inserted into the absolute precision equation and checked to make sure that the new number of replications meets the absolute precision level. If our new data set meets the absolute precision level, then we are done for the moment. However, if the new data set exceeds the absolute precision level, then we need to recalculate a new number of replications to run.

In our example, if we rerun a total of 24 replications, we obtain the summary statistics illustrated in Figure 10.5.

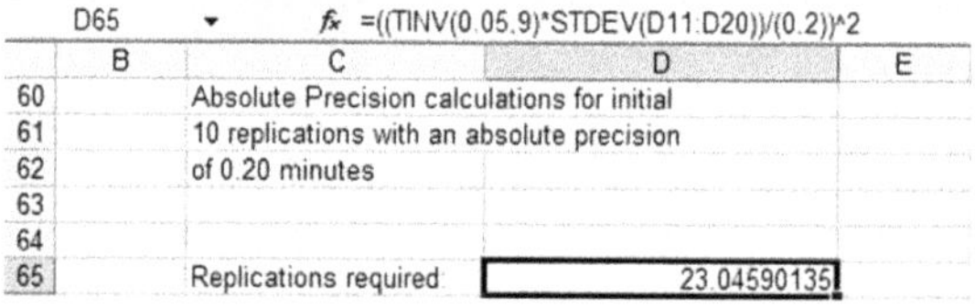

FIGURE 10.3 Replications required based on 10 replications.

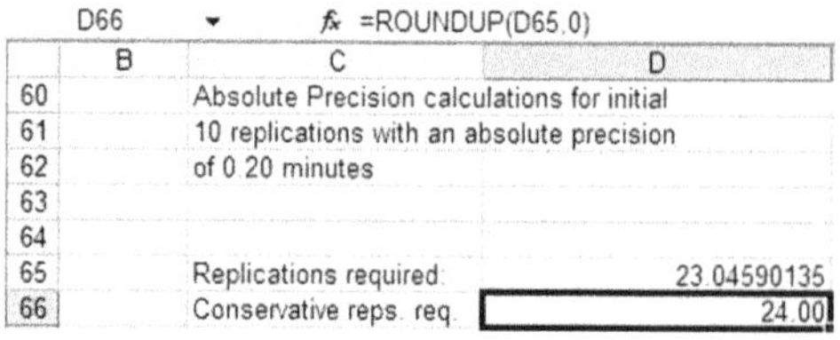

FIGURE 10.4 Conservative round-up replication approach.

	Replication Number	Mean System Time in Minutes
11		
12	1	10.80
13	2	11.96
14	3	10.47
15	4	10.70
16	5	10.80
17	6	11.35
18	7	11.04
19	8	10.68
20	9	11.15
21	10	11.00
22	11	11.02
23	12	10.66
24	13	10.94
25	14	10.88
26	15	10.78
27	16	11.04
28	17	10.68
29	18	10.99
30	19	10.99
31	20	10.57
32	21	10.92
33	22	11.04
34	23	10.91
35	24	10.90

FIGURE 10.5 New replication means.

	D46	▾	f_x =TINV(0.05,23)*STDEV(D12:D35)/24^0.5	
	B	C	D	E
40		Mean	10.93	
41		Standard Deviation	0.29	
42				
43		t for alpha=0.025, 23 d.f.	2.07	
44		square root of n	4.90	
45				
46		Standard error	0.12	

FIGURE 10.6 New replication calculations.

Our new mean is 10.93, and our new standard deviation is 0.29. Our *t* value changes somewhat because we now have a sample of 24 observations. The corresponding *t* value for 23 degrees of freedom is 2.07. If we insert these values in our absolute precision equation, the equation is equal to 0.14, as illustrated in Figure 10.6.

Because 0.14 is less than our absolute precision of 0.20, 24 replications are sufficient for a robust statistical comparison with other alternatives. As previously mentioned, if our standard error had been greater than the absolute precision of 0.20, we would have been obligated to use the modified form of our absolute precision equation to recalculate the new number of required replications.

10.2.4.2 Relative Precision

The alternative approach to absolute precision is the relative precision approach. This approach is preferred because it is not necessary to select an arbitrary absolute precision level. The relative precision approach sidesteps this issue by creating a ratio by dividing the standard error of the data by the sample mean of the data. For a robust statistical analysis, the standard error should be relatively small in comparison to the sample mean. A commonly utilized practitioner value for the desired relative precision is 0.10. This means that we want our standard error to be only 10% of the sample mean for our data. For our calculations, we actually use a value of 0.09 rather than 0.10 for a desired relative precision of 0.10. The reason for this is an obscure mathematical proof involving relative precision calculations (Law and Kelton, 2000).

To calculate the relative precision we use the following formula:

$$\text{Relative Precision} = \frac{t_{1-\alpha/2,n-1} * s / \sqrt{n}}{\overline{\overline{x}}}$$

where

$t = t$ distribution for $1 - \alpha/2$ and $n - 1$ degrees of freedom

s = standard deviation of the replication means

n = number of replications used to calculate the summary statistics

x bar bar = mean of the replication means

The Excel implementation of this formula is illustrated in Figure 10.7.

As the figure indicates, the relative precision with the replications is actually 0.03. Because this value is less than 0.09, we already have a sufficient number of replications with the minimum number of 10.

In a similar fashion to rearranging the absolute precision formula, we can also rearrange the relative precision formula to find out the required number of replications to achieve a given level of relative precision:

$$i = \left[\frac{t_{1-\alpha/2,n-1} * s}{\text{Relative Precision} * \overline{\overline{X}}} \right]^{1/2}$$

where

$t = t$ distribution for $1 - \alpha/2$ and $n - 1$ degrees of freedom

s = standard deviation of the replication means

n = number of replications used to calculate the summary statistics

i = number of replications needed to achieve the relative precision

This is implemented in Excel with the following formula:

$$= (\text{TINV}(0.05,9)*\text{STDEV}(D11:D20))/(0.09*\text{AVERAGE}(D11:D20))\wedge 2$$

10.2.4.3 Checking All of the Alternatives

Once the practitioner has an understanding of the replication analysis procedure using the base alternative, replication analysis can be performed on all of the other alternatives. Because each of the alternative models is different with respect to either resources or operating policy, there will undoubtedly be some difference in the means and standard deviations of the replication means. Thus, there will naturally be some differences in either the initial absolute precision or relative precision of the models with the initial ten replications. Although it may be mathematically possible for all of the alternatives to require only ten replications at a given relative precision, it will most likely not be the case.

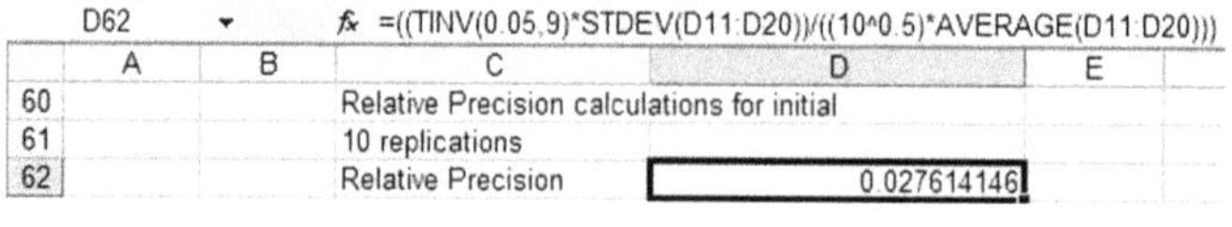

	D62		f_x =((TINV(0.05,9)*STDEV(D11:D20))/((10^0.5)*AVERAGE(D11:D20)))		
	A	B	C	D	E
60			Relative Precision calculations for initial		
61			10 replications		
62			Relative Precision	0.027614146	

FIGURE 10.7 Relative precision calculations.

A more likely situation is that one or more of the alternatives will require more than ten replications. If this is in fact the situation, then each of the individual alternatives must be rerun for the maximum number of replications required by any of the individual alternatives for the level of relative precision. The following table illustrates this process for an experiment with eight different alternatives.

Alternative	A	B	C	D	E	F	G	H
Number of ticket checkers, x-rays, and metal detectors	2, 2, 2	3, 2, 1	2, 2, 1	3, 2, 2	3, 2, 2	3, 1, 1	2, 1, 1	3, 1, 2
Mean of 10 replications: Avg. system time (min)	10.99	12.67	14.37	9.81	27.21	21.44	25.71	22.75
Std. dev. of 10 replications: Avg. system time (min)	0.43	0.66	2.17	0.26	9.72	5.30	8.71	3.83
Replications required at α = 0.05 and RP = 0.10 based on 10 replications	10	10	12	10	65	31	59	15

Note that alternative E with three ticket checkers, two x-ray machines, and two metal detectors requires a total of 65 replications to achieve a desired relative precision of 0.10. This means that all of the individual alternatives need to be rerun individually a total of 65 times in order to subsequently perform the statistical analysis. Therefore, the practitioner must next run a total of 520 simulation replications.

When the alternatives are run at the new level of replications, the practitioner can recalculate the relative precision yielded by the new mean and standard deviation of the replication means. The new level of replications will also change the t value and the n value in the relative precision equation. As long as the new calculated level of relative precision is less than or equal to 0.10, the number of replications is sufficient. In a few cases, the new calculated level of relative precision will be below the desired level of relative precision of 0.10. In these rare cases it will be necessary to repeat the process by recalculating the next number of replications required to achieve the desired level of relative precision.

10.2.5 Statistical Analysis of Terminating System Production Runs

In a very simple simulation project, the practitioner will have only two different models to compare. These will normally be the base model and the model with the alternative resource or operating policy. The statistical analysis involved in two-model comparisons is relatively simple. In a more complicated simulation project, the practitioner could easily have many more than two models to compare. With multiple models, it is possible to use the same general type of statistical analysis approaches as with only two models. However, more robust statistical methods are generally in order. To organize the statistical analysis process, we can categorize the analysis approaches into either:

- Simple two-model comparisons
- Three-model or more comparisons

10.2.5.1 Simple Two-Model Comparisons

For simple two-model comparisons, the practitioner can utilize either hypothesis test or confidence interval approaches. With the hypothesis test, we are limited to either accepting or rejecting the null hypothesis. The null hypothesis is normally that there is no difference between the two models. The corresponding alternate hypothesis is that there is a difference between the two models. Although a hypothesis approach does tell us what we need to know, many practitioners prefer to use a confidence interval approach.

The confidence interval approach is a modification of the corresponding hypothesis test. It determines the interval or range over which the difference between the two models would normally expect to be observed. Normally, practitioners utilize a 95% confidence interval. If the models are statistically similar, then the expectation is that the difference in the replication mean values will be zero. Thus, if the confidence interval covers the value 0, there is statistical evidence that the two models are the same. On the other hand, if the confidence interval does not cover 0, the two models are statistically significantly different. The confidence interval approach also has a few other advantages over the hypothesis test because it:

- Provides more information than hypothesis tests
- Graphically shows the statistical results
- Is easier to use and explain than hypothesis tests

There are two basic types of confidence interval approaches with respect to determining the difference between simulation models. The choice of the confidence interval approach is dependent on the manner in which the simulation model was designed.

10.2.5.2 Welch Confidence Interval Approach

The Welch confidence interval approach is most likely to be used by the practitioner. It is appropriate when no special effort has been made on the part of the practitioner to coordinate the entity parameters through the model as described in the model formulation section of the handbook. The Welch confidence interval approach takes a robust philosophy with respect to assumptions about the characteristics of the data. For example, with this approach, we do not need to be concerned with whether the two data sets have similar variance. If you recall, this was a major concern with conducting comparison of means tests during the validation phase. The Welch confidence interval approach assumes the worst-case scenario of having dissimilar variance between the two data sets. As you might already suspect, the Welch confidence interval approach is based on the Smith–Satterthwaite t test introduced in Chapter 8, "Validation."

After calculating the mean and standard deviation summary statistics for each data set, the practitioner must calculate the degrees of freedom estimator as with the Smith–Satterthwaite test using the formulation below:

$$d.f. = \frac{[s_1^2/n_1 + s_2^2/n_2]^2}{[s_1^2/n_1]^2/(n_1-1) + [s_2^2/n_2]^2/(n_2-1)}$$

where

$d.f.$ = degrees of freedom

s_1^2 = sample variance of the first alternative

s_2^2 = sample variance of the second alternative

n_1 = sample size of the first alternative

n_2 = sample size of the second alternative

As with the Smith–Satterthwaite test, the number of degrees of freedom calculated in this manner will most likely not be an integer. The natural question again becomes whether to round up or round down. Remember, in general, that as a practitioner, you want to take the most conservative approach. This means that under questionable circumstances, you would rather conclude that there is no difference between the two alternatives than conclude that there is. This means that we want a larger rather than smaller confidence interval to increase the probability of covering 0. Because the confidence interval is a function of the t value, we would also want as large a t value as possible. Because the t value increases as the number of degrees of freedom decreases, this time we want to round the estimated degrees of freedom downward. Note that this is the opposite of what we did with the Smith–Satterthwaite test.

The Welch confidence interval can now be calculated with the following formula:

$$\bar{x}_1 - \bar{x}_2 \pm t_{d.f.,1-\alpha/2}\sqrt{\frac{s_1^2}{n_1} + \frac{s_2^2}{n_2}}$$

where

$\bar{x}_1$ = the mean of the first alternative replications

$\overline{x}_2$ = the mean of the second alternative replications

t = the t value for the degrees of freedom previously estimated and $1 - \alpha/2$

The above equation is most commonly seen in its final form with minimum and maximum values that describe the interval at a given level of confidence. The values are normally presented with square brackets separated by a comma as shown below:

$$[\text{min value, max value}]$$

The confidence interval is sometimes graphically displayed with the mean, minimum, maximum, lower 95%, and upper 95% values. This gives a clear indication of the difference between the means of the two models.

The translation of the confidence interval approach as previously discussed involves only the relationship between the value 0 and the confidence interval. If the confidence interval covers the value 0, then there is no significant difference between the two simulation model alternatives. Conversely, if the confidence interval does not cover 0, then there is a statistically significant difference between the two simulation models.

10.2.5.3 Welch Confidence Interval Example

The following data represent the system times in seconds for two different operating policies for our airport security checkpoint system. Alternative A consists of two ticket checkers, two x-ray machines/operators, and two metal detectors/operators. Alternative B consists of three ticket checkers, two x-ray machines/operators, and one metal detector/operator. The mean and standard deviation for alternative A are 39.58 and 1.53. The mean and standard deviation for alternative B are 45.60 and 2.38.

Alt A	Alt B
38.88	42.44
43.06	44.53
37.69	46.55
38.52	48.46
38.88	48.92
40.86	48.46
39.74	45.36
38.45	44.53
40.14	42.84
39.60	43.92

To calculate the Welch confidence interval at an α level of 0.05, we begin with the degrees of freedom estimator. The degrees of freedom estimator is calculated by:

$$d.f. = \frac{[1.53^2/10 + 2.38^2/10]^2}{[1.53^2/10]^2/(10-1) + [2.38^2/10]^2/(10-1)}$$

$$d.f. = 15.34$$

Because we want to have as wide a confidence interval as possible, we round down to 15. This will result in a t value of 2.131. The actual Welch confidence interval calculation is:

$$39.58 - 45.60 \pm 2.131 \sqrt{\frac{1.53^2}{10} + \frac{2.38^2}{10}} = -6.02 \pm 1.91$$

or

$$[-7.93, -4.11]$$

The confidence interval does not cover zero with an α level of 0.05. Therefore, we can conclude that the two alternatives are statistically significantly different. Because alternative A has a lower system time of 39.58 versus the system time for alternative B with a system time of 45.60, under normal circumstances, we would recommend alternative A.

10.2.5.4　Paired *t*-Test Confidence Interval Approach

The paired *t*-test is utilized when the two models have some sort of natural pairing. This statistical comparison technique is appropriate only when the entities flowing through the system have been designed for the apples-to-apples approach described in the problem formulation chapter (Chapter 2). In this approach, we first calculate a new variable based on the difference in replication means between the two alternatives.

$$\overline{X}_{1i} - \overline{X}_{2i} = Z_i$$

where

$\overline{X}_{1i}$ = the *i*th replication mean for the first alternative

$\overline{X}_{2i}$ = the *i*th replication mean for the second alternative

Z_i = the difference in means for the *i*th replication

If you have ten replications, you end up with ten Z values from each pair of replication means. Next, we calculate the average and standard deviation of the new Z variable. Last, we utilize the following formula to calculate the confidence interval:

$$\overline{Z} \pm t_{\alpha/2,\,n-1} \frac{s}{\sqrt{n}}$$

where

$\overline{Z}$ = the mean of the Z values

$t_{\alpha/2,\,n-1}$ = the value of the t distribution for $\alpha/2$ and $n-1$ degrees of freedom

s = the standard deviation of the Z values

n = the number of pairs of replication means

10.2.5.5　Example of Paired *t*-Test Confidence Interval Approach

If we had obtained the example data from a model that used apples-to-apples pairing, we could make the following calculations at an α level of 0.05 to demonstrate the use of the paired *t*-test confidence interval technique.

Alt A	Alt B	Z
38.88	42.44	−3.56
43.06	44.53	−1.47
37.69	46.55	−8.86
38.52	48.46	−9.94
38.88	48.92	−10.04
40.86	48.46	−7.60
39.74	45.36	−5.62
38.45	44.53	−6.08
40.14	42.84	−2.70
39.60	43.92	−4.32

The mean of the Z variable is –6.02, and the standard deviation is 3.04. The t value for 0.05/2 and nine degrees of freedom is 2.262. Inserting these values into the confidence interval equation, we obtain:

$$-6.02 \pm 2.262 * \frac{3.04}{\sqrt{10}}$$

$$-6.02 \pm 2.29$$

or

$$[-8.31, -3.73]$$

Because [-8.31, -3.73] does not cover 0, there is statistical evidence to say that the two alternatives are statistically significantly different at an α level of 0.05. Because we know the the mean of alternative A was 39.58 seconds and the mean of alternative B was 45.60 seconds, we would normally want to go with alternative A if the costs for the two alternatives were the same.

To make life easier, we could also have made the equation for the Z variable the replication mean values of alternative B minus the replication mean values of alternative A. This would have resulted in the mean of the Z variable being a positive 6.02. The resulting confidence interval would have been [3.73, 8.31]. Again, the confidence interval would not have covered zero.

10.2.5.6 Three- or More Model Comparisons

Three or more comparisons involve a two-step procedure. In the first step, analysis of variance (ANOVA) is used to determine if there is a statistically significant difference among one or more of the means of the different models. If none of the means is statistically significantly different, we can stop our analysis. On the other hand, if one or more of the means are statistically significantly different, we naturally would like to know which ones they are. Unfortunately, ANOVA by itself does not provide this information. In order to determine which means are statistically significantly different from the other means, we must also run a Duncan multiple-range test.

10.2.5.7 ANOVA

Analysis of variance can be used to determine if the mean performance of one or more alternatives is statistically significantly different from the others at a given α level. One or more of the means could actually be statistically significantly better or worse than the rest of the alternatives. Analysis of variance is based on a ratio of the variance between different alternatives divided by the variance within the different alternatives. If the variation between the alternatives is large and the variance within the different alternatives is small, the ratio takes on a comparatively large value. Conversely, if the variation between the alternatives is small, and the variance within the different alternatives is large, the ratio is small. When the ratio is large, then it is more likely that one or more of the alternatives is statistically significantly different than the others.

There are a number of different implementations of ANOVA. The simplest implementation, one-way ANOVA, examines only individual factors. More complex implementations such as two-way ANOVA also examine possible interactions between individual factors. The majority of the discussion in this handbook will concentrate on one-way ANOVA. Statistical implementations of ANOVA are available in Excel as well as in most simulation software packages.

10.2.5.8 Brief Mathematical Foundation of ANOVA

This section presents a brief mathematical foundation for ANOVA. Given the software implementations available of ANOVA, it is highly unlikely that a practitioner would be called on to make these calculations by hand. However, in the interests of robustness, the following information is provided. Some practitioners may wish to bypass this entire section.

To begin the ANOVA process, the practitioner must have each of the individual replication means for each different alternative. The practitioner must also have the mean value of the replications for each alternative and the single grand mean of all replications. With these summary statistics in hand, the following calculations are required:

- Calculate the sum of squares total.
- Calculate the sum of squares between.
- Calculate the sum of squares within.
- Calculate the mean squares between.
- Calculate the mean squares within.
- Calculate the f statistic.
- Compare the f statistic to a critical f value.

10.2.5.9 Calculating the Sum of Squares Total

The sum of squares total is a summation of the difference between each individual replication mean and the grand mean squared. This term is represented by the following formula:

$$SST = \sum_{i=1}^{k} \sum_{j=1}^{n} (x_{ij} - \bar{x})^2$$

where

$\quad$ SST = sum of squares total

$\quad\quad$ k = number of different alternatives

$\quad\quad$ n = number of replications for each alternative

$\quad\quad$ x_{ij} = a single replication mean for a single alternative

$\quad\quad$ $\bar{x}$ = the grand mean of all replication means

Thus, for each individual replication mean, we subtract the grand mean and then square the difference. This is performed for each replication mean for each different alternative. Thus we will have $k \times n$ different terms that will be summed.

10.2.5.10 Calculating the Sum of Squares Between

The sum of squares between is calculated by summing the difference between the individual replication means and the alternative mean squared. SSB is calculated using the following formula:

$$SSB = \sum_{i=1}^{k} n*(\bar{x_i} - \bar{x})^2$$

where

$\quad$ SSB = sum of squares between

$\quad\quad$ k = number of different alternatives

$\quad\quad$ n = number of replications for each alternative

$\quad\quad$ $\bar{x_i}$ = the mean of the replication means for a single alternative

$\quad\quad$ $\bar{x}$ = the grand mean of all replication means

This means that for each alternative we multiply the number of replications in the alternative by the difference of the alternative mean and the grand mean squared. Next, we sum each of the k terms.

10.2.5.11 Calculating the Sum of Squares Within

Next, we must calculate the sum of squares within (SSW). We could individually calculate each of these terms by summing the squared differences of each individual replication mean and its alternative mean or we could use the following formula:

$$SST = SSB + SSW$$

By rearranging this formula we can obtain the sum of squares within with what we already know and save an enormous number of calculations:

$$SSW = SST - SSB$$

10.2.5.12 Calculating the Mean Squares Between

The mean squares between (MSB) is calculated by dividing the sum of squares between by the number of degrees of freedom associated with the alternatives. The number of degrees of freedom is the number of alternatives minus 1. The equation associated with this calculation is:

$$MSB = \frac{SSB}{k-1}$$

where

MSB = mean squares between

SSB = sum of squares between

k = number of alternatives

10.2.5.13 Calculating the Mean Squares Within

The mean squares within is calculated by dividing the sum of squares within by the number of degrees of freedom associated with the sum of squares within. The equation associated with this calculation is:

$$MSW = \frac{SSW}{k*(n-1)}$$

where

MSW = mean squares within

SSW = sum squares within

k = number of alternatives

n = number of replications for each alternative

10.2.5.14 Calculating the *F* Statistic

The next step in the analysis of variance process is to calculate the *F* statistic. This is the ratio of the mean squares between over the mean squares within. This is the same as comparing the variance between the alternatives to the variance within the alternatives that was discussed at the beginning of the ANOVA section. This equation is:

$$F = \frac{MSB}{MSW}$$

where

$F = F$ statistic

MSB = mean square between

MSW = mean square within

10.2.5.15 Compare the *F* Statistic to the Critical Value

The final step is to compare the *F* statistic to the critical *F* value. The critical *F* value is obtained through an *F* distribution table. The parameters needed to determine the *F* value are the α level, the number of degrees of freedom on the numerator, and the number of degrees of freedom on the denominator. The number of degrees of freedom for the numerator is the same number that was used to divide the SSB. This was the number of alternatives minus 1. Similarly, the number of degrees of freedom for the denominator is the same number that was used to divide the SSW. This was the number of alternatives multiplied by the number of replications in the alternative minus 1. These equations are:

$$d.f.\,numerator = k - 1$$

where k = number of different alternatives,

$$d.f.\,deno\min ator = k * (n - 1)$$

where

k = number of different alternatives

n = number of replications for an individual alternative

The calculated *F* test statistic is finally compared to the critical *F* value obtained from the *F* distribution table. If the test statistic is greater than the critical value then at least one of the alternatives is statistically different than the others. Conversely, if the test statistic is less than the critical value, then all of the alternatives are statistically similar.

10.2.5.16 ANOVA Example

As was previously discussed, it would be highly unusual for a practitioner to perform the preceding ANOVA calculations by hand. It is more likely that the practitioner will end up with the following ANOVA from a software package.

Source	df	SS	MS	F	p	
TICKET	1	906.48	906.48	4.83	0.028	*
XRAY	1	92370.60	92370.60	492.19	0.000	*
METAL	1	1944.36	1944.36	10.36	0.001	*
TICKET*XRAY	1	29.52	29.52	0.16	0.692	
TICKET*METAL	1	16.56	16.56	0.09	0.766	
XRAY*METAL	1	662.76	662.76	3.53	0.061	
TICKET*XRAY*METAL	1	15.84	15.84	0.08	0.772	
Error	528	99091.08	187.56			
Total	535	195037.20				

The table indicates that, at a 0.05 level of significance, each of the main factors (ticket, x-ray, and metal) is statistically significant. This means that altering the level of any one of these factors will have a statistically significant impact on the time it takes an air passenger to pass through the airport security checkpoint system.

10.2.6 Duncan Multiple-Range Test

In the event that the analysis of variance indicates that one or more means are statistically significantly different than the others, the practitioner will want to conduct a Duncan multiple-range test. This test will indicate which means are statistically different from the others at a particular level of confidence. Thus, the Duncan test provides additional insight over ANOVA for making informed recommendations. Obviously, running a Duncan test when the ANOVA results do not indicate one or more means are statistically significantly different would be a pointless effort.

The general concept of the Duncan test is to determine a least significant range value for a given set of adjacent means. A set of adjacent means is defined as any number of means that are adjacent when arranged in sorted order. A set of adjacent means may involve two, three, four, or more pairs of consecutive means. The idea behind the Duncan test is that any set of adjacent means with a maximum–minimum range less than the calculated least significant range critical value is not statistically significant. In other words if the spread between the maximum and minimum values in any adjacent set is less than the critical value, then all of the means in the adjacent set can be considered statistically similar.

Because adjacent sets can involve two, three, four, or more sets of adjacent means, the Duncan test can involve a fairly large number of calculations. Fortunately, if a larger set of adjacent means has been proven to be statistically similar, then all of the smaller sets of adjacent means within the larger set are also statistically similar. This means, for example, if a set of four adjacent means is found to be statistically similar then any two- or three-set combinations of adjacent means are also statistically similar.

10.2.6.1 Duncan Test Procedure

The Duncan multiple-range test consists of the following steps:

Sort the replication means for each alternative in ascending order left to right.
Calculate the least significant range value for all of the possible sets of adjacent means.
Compare each set of possible adjacent means with the corresponding least significant range value in descending order with respect to the set size.
Mark the nonsignificant ranges.

10.2.6.2 Calculating the Least Significant Range Value

The practitioner must calculate the least significant range value for each set size of adjacent means. This means that there must be $n - 1$ calculations (for $p = 2$ to n) of the least significant value. The smallest set will have two means, and the largest set will have the same number of means as the number of alternatives. The practitioner must also select an α level for calculating the least significant range value. For a typical 95% confidence level, α will be equal to 0.05.

The calculation process to determine the least significant range values is a two-step process. In the first step we calculate a value based on the mean square error of the replication means and the number of replications in a single alternative. The formula for this value is:

$$S_{\bar{x}} = \sqrt{\frac{MSE}{n}}$$

where
   MSE = the mean square error of the replication means

   n = the number of replications in a single alternative

The value for the mean square error of the replication means can easily be found in the analysis of variance statistical printout. Pay particular care that this specific n is the number of replications in a single alternative, not the number of alternatives or the number of total replications analyzed in the study. In the next step for calculating the least significant range, we will be using a different n.

Once the Duncan standard deviation of the replication means has been calculated, it is multiplied by a specific multiplier value found in Duncan multiple-range test tables to obtain the least significant range values. The formula for obtaining the least significant range value is:

$$R_p = s_{\bar{x}} * r_p$$

where

$s_{\bar{x}}$ = the Duncan standard deviation of the replication means

r_p = the Duncan multiple-range multiplier for a given level of significance, set size, and degrees of freedom

p = the size of the set of adjacent means

The table value for the Duncan multiple-range multiplier requires a given level of significance, the size of the adjacent set of means, and the number of degrees of freedom. The number of degrees of freedom is the same as the number of degrees of freedom used to calculate the mean square error of the replication alternative means. This value is readily available in the analysis of variance output.

The above calculations will result in a total of $n - 1$ values for the least significant range using p values of 2 through n. So if you have four different alternatives, you will have three values for the adjacent mean set sizes of 2, 3, and 4.

10.2.6.3 Comparison of Adjacent Means

The next step is to compare the least significant range values to each set of adjacent means. The best approach is to begin with the largest set of adjacent means and work downward to the smallest set of adjacent means. This will dramatically reduce the number of comparisons that we will have to make. The first comparison will involve the set of adjacent means that is the same as our number of alternatives. If the range or difference between the largest mean and the smallest mean is smaller than the least significant range value, then none of the means is statistically different from the others. The range for the largest set of adjacent means should always be greater than its corresponding least significant range value. The reason for this is that the analysis of variance indicated that at least one mean of all of the alternatives was different. Conversely, this means than not all of the adjacent means are statistically similar. Thus, the largest set of adjacent means must have a larger range than the least significant range value. If the range for the largest set of adjacent means is smaller than the least significant range value, then some sort of error has occurred. Either the analysis of variance result indicated that all of the means were statistically significantly the same, or there has been a calculation error.

The following step is to compare the next largest set of adjacent means. This set will consist of a total of two different sets of adjacent means. The first set will be the smallest mean through the next to largest mean. The second set will be the next to smallest mean through the largest mean. This means that if we had a total of four alternatives then the first set would encompass alternatives 1, 2, and 3. Similarly, the second set would contain alternatives 2, 3, and 4. With either of these two sets of adjacent means, we are no longer guaranteed to have the ranges larger than the least significant range. Either one or the other set must exceed the least significant range value. If a set has a range smaller than the least significant range value for that sized set, there is no significant difference among the means within the set. This means that any other smaller set of adjacent means is also statistically similar. No further analysis needs to be performed on those means.

Conversely, if a set has a range that is larger than the least significant range value for that sized set, this means that there is a statistically significant difference among the means within the set. The analysis process must continue with this set of means. The means must be divided into the next smallest sized set of adjacent means. Thus, if the range for our set of three adjacent means for alternatives 1, 2, and 3 was larger than the least significant range value, we must now examine the adjacent mean groups of size two. The first set of two adjacent means would be alternatives 1 and 2. The second set of two adjacent means would be alternatives 2 and 3. Because we know that among these three adjacent means at least one mean is different, the different mean must be from either the set with alternatives 1 and 2 or the set with alternatives 2 and 3.

If it turns out that the range for the first set with alternatives 1 and 2 exceeds the least significant range, then there is a statistically significant difference between alternatives 1 and 2. If this same set does not exceed the least significant range, then there is no statistically significant difference between alternatives 1 and 2. Similarly, the range for the set with alternatives 2 and 3 may or may not exceed the least significant range value.

10.2.6.4 Marking the Nonsignificant Ranges

To graphically illustrate which sets of adjacent means are statistically the same at a given α level, the practitioner can utilize a line. The line is placed underneath the sets of adjacent means that are not statistically significantly different. Thus, the lines can encompass two, three, or more alternative means. It is unnecessary to underline smaller sets of adjacent means within a set of statistically similar adjacent means.

10.2.6.5 Interpreting the Duncan Multiple-Range Test Results

Because several sets of means may be statistically similar and even overlap, some thought must be given to properly interpreting the results of the Duncan multiple-range test. The following table of means will illustrate this process:

Alternative	1	2	3	4
Time (min)	23.5	26.2	27.5	28.1

The following statistically significant conclusions may be stated from this table:

- There is a difference between alternative 1 and all of the other alternatives.
- There is a difference between alternative 2 and alternative 4.
- There is no difference between alternatives 2 and 3.
- There is no difference between alternatives 3 and 4.

Note how it is possible for alternative 3 to be statistically similar to both alternatives 2 and 4 while alternative 2 is statistically significantly different from alternative 4.

10.2.6.6 Duncan Multiple-Range Test Example

The airport security checkpoint system example will illustrate the process of performing the Duncan multiple-range test. First, the mean system times from each of the alternatives are listed in ascending order.

Ranking	1	2	3	4	5	6	7	8
Combination	D	A	B	C	H	F	E	G
Ticket	3	2	3	2	3	3	2	2
X-ray	2	2	2	2	1	1	1	1
Metal	2	2	1	1	2	1	2	1
Means (s)	35.07	39.10	46.50	50.56	89.51	90.59	96.97	97.70

The mean square error can be taken from the ANOVA example with the airport security checkpoint system and used to calculate the least significant range values.

$$S_{\bar{x}} = 1.67 = \sqrt{\frac{187.56}{67}}$$

The Duncan multiple-range test multiplier table values for r between 2 and 8 at 0.05 and 120 degrees of freedom are:

p	2	3	4	5	6	7	8
r	2.80	2.95	3.04	3.12	3.17	3.22	3.25

Note that the number of degrees of freedom for the error from the analysis of variance table was 528. The next nearest number available in most Duncan multiple-range tables is for 120 degrees of freedom.

The r values are next multiplied by the standard deviation of the replication means to calculate the least significant range values:

p	2	3	4	5	6	7	8
R	4.68	4.93	5.08	5.21	5.29	5.38	5.43

When the sets of adjacent means are compared to the least significant range values, our table of adjacent means becomes:

Ranking	1	2	3	4	5	6	7	8
Combination	D	A	B	C	H	F	E	G
Ticket	3	2	3	2	3	3	2	2
X-ray	2	2	2	2	1	1	1	1
Metal	2	2	1	1	2	1	2	1
Means (s)	35.07	39.10	46.50	50.56	89.51	90.59	96.97	97.70

For the set of eight adjacent means, the least significant range value is 5.43. The range between the largest and the smallest means is:

$$62.63 = 97.70 - 35.07$$

This is far greater than the least significant range value of 5.43 for eight adjacent means. This means that at least one of the means is statistically significantly different than the others. We actually expected this result because the analysis of variance indicated not all of the means were statistically similar.

The next step is to look at the sets of alternatives with seven adjacent means. We have two of these. The first is combination D through E, and the second is combination A through G.

The least significant range value for seven adjacent means is 5.38. The calculated ranges of the two sets are:

61.90 = 96.97 − 35.07 for D through E
58.60 = 97.70 − 39.10 for A through G

Both of these range values exceed 5.38, so there is at least one statistically significant different mean in both of these sets.

Following the same procedure, we have a total of three sets with six adjacent means. These are D through F, A through E, and B through G. The least significant range value is 5.29 for six adjacent means. The calculated ranges of these means are:

55.52 = 90.59 − 35.07 for D through F
57.87 = 96.97 − 39.10 for A through E
51.20 = 97.70 − 46.50 for B through G

Again, all of these ranges exceed the least significant range value of 5.29. Similarly, this means that at least one of the six means in each of these sets is statistically significantly different than the others.

Next we look at sets with five adjacent means. The least significant range value is 5.21. Our sets of five adjacent means are D to H, A to F, B to E, and C to G.

54.44.= 89.51 – 35.07 for D to H
51.49.= 90.59 – 39.10 for A to F
50.47.= 96.97 – 46.50 for B to E
47.14.= 97.70 – 50.56 for C to G

Once again, none of the four sets of five adjacent means is less than the least significant range value. As before, this means than in each of these four sets, at least one of the means is statistically significantly different than the others.

With sets of four adjacent means we have a least significant range value of 5.08. Our sets of adjacent means are D to C, A to H, B to F, C to E, and H to G. The range values for these sets are:

15.49 = 50.56 – 35.07 for D to C
50.41 = 89.51 – 39.10 for A to H
44.49 = 90.59 – 46.10 for B to F
46.41 = 96.97 – 50.56 for C to E
8.19 = 97.70 – 89.51 for H to G

Yet again, we find that all of the range values exceed the least significant range value of 5.08. Because these sets of adjacent means are not statistically similar, we must drive on to sets of three adjacent means.

With three adjacent means we have a total of six different sets. The least significant range value for three adjacent means is 4.93. The sets of three adjacent means are D to B, A to C, B to H, C to F, H to E, and F to G. The range values are:

11.43 = 46.50 – 35.07 for D to B
11.46 = 50.56 – 39.10 for A to C
43.01 = 89.51 – 46.50 for B to H
40.03 = 90.59 – 50.56 for C to F
7.46 = 96.97 – 89.51 for H to E
7.46 = 97.70 – 90.59 for F to G

Although a few of these ranges are approaching 4.93, all of the ranges still exceed the least significant range value. For at least one more level of adjacent means, we see that there is a least one mean that is statistically significantly different than the others. We now proceed to our final level of adjacent means.

The least significant range value for two adjacent means is 4.68. Because there have to be at least two adjacent means to perform the Duncan multiple-range test, we can expect some sort of results with this iteration. The sets of adjacent means are D to A, A to B, B to C, C to H, H to F, F to E, and E to G. These range values are:

4.03 = 39.10 – 35.07 for D to A
7.40 = 46.50 – 39.10 for A to B
4.06 = 50.56 – 46.50 for B to C
38.95 = 89.51 – 50.56 for C to H
1.08 = 90.59 – 89.51 for H to F
6.38 = 96.97 – 90.59 for F to E
0.73 = 97.70 – 96.97 for E to G

We finally have some results to interpret. The sets of adjacent means for D to A, B to C, H to F, and E to G have a range less than the least significant range value of 4.68. This means that there is no statistically significant difference to each of these pairs of means. Conversely, the sets of adjacent means A to B, C to H, and F to E have ranges greater than 4.68. These sets of means are statistically significantly different. We can now draw the following statistically significant conclusions:

- Alternatives D and A are the same.
- Alternatives A and B are the same.
- Alternatives B and C are the same

- Alternatives H and F are the same
- Alternatives E and G are the same

We can also draw the following additional conclusions based on pairs that are not statistically significantly the same:

- Alternative D performs statistically significantly better than all other alternatives except A
- Alternative A performs statistically significantly better than all other alternatives except D and B
- Alternative B performs statistically significantly better than all other alternatives except D, A, and B
- Alternative H performs statistically significantly better than only E and G
- Alternative F performs statistically significantly better than only E and G

On the surface, it would appear that the system should be operated with the configurations described in alternative D. Although this alternative would offer the highest level of performance, it might not be the best overall choice among the system configurations. To make appropriate recommendations, some economic analysis must usually be performed.

10.2.6.7 Economic Analysis of Statistical Results

The conclusions from the Duncan multiple-range test have more meaning when the system operational costs are considered. For example, there is no difference in the performance between alternative D with 7 individuals and alternative A with 6 individuals. Because there is no statistically significant difference between the two alternatives, the system could be operated at a lower cost with only 6 individuals rather than 7.

Another way to look at the results is to establish a particular staffing level. If, for example, only five equally trained and qualified individuals were available to operate the security checkpoint system, alternatives C, F, and E would be possible. Because alternative C is statistically significantly better than either F or E, it would be the best staffing choice.

10.3 Nonterminating System Analysis

The analysis of nonterminating systems differs significantly from that of terminating systems. This difference is a result of the following issues:

- Starting conditions
- Determining steady state
- Autocorrelation
- Length of replication
- The batch method

10.3.1 Starting Conditions

Nonterminating systems normally start only once and then continue on for an indefinite period of time. The system may actually never close, or it may close and reopen in the same state that it closed. Either way, it is considered as running for an indefinite period of time. Because the practitioner is primarily interested in how the nonterminating system performs on the average, it may not initially appear that the starting conditions are of great importance. This is because the system will first have an initial transient state before it becomes balanced and reaches the system's steady state. Even though the initial transient state may not be of the greatest importance to the practitioner, it will still need to be modeled. The practitioner can choose either of two different approaches:

1. Begin with the system empty.
2. Begin with the system loaded.

The practitioner may begin the simulation run without any entities in the system. If this approach is selected, it may take a long period of time for the system to travel through the initial transient state. The question then becomes how to identify when the system has finished with the initial transient state and started the steady state.

The practitioner may also begin the simulation run with entities in the system. At the beginning of the simulation run, the model is loaded in the same manner as the way that the model validation was conducted. This requires the practitioner to observe the system and begin the model with the same number of entities and resource states.

10.3.2 Determining Steady State

The statistical comparison between nonterminating systems must be conducted only with data from the models in steady state. So, before the practitioner can conduct any statistical comparison, the initial transient state, if one exists, must be removed. During the simulation run, the initial transient state can be typically identified as a series of observations with continuously increasing output measures of performance values. For example, the initial entities in an unloaded system will have much shorter system times than entities when the system is loaded and operating in steady state. There are a number of ways to approximate when the output measure of performance exits the initial transient state and enters the steady state. These include:

- Graphic approach
- Linear regression

Regardless of which approach is used to identify the end of the initial transient state, the practitioner may want to smooth the values of the output measure of performance. If the values are smoothed, it may be easier to identify the end of the initial transient state. The most common way to smooth the data is to take a running average of the observations. For example, a running average of 50 could be utilized. This means that each value is plotted not as its actual value but as the running average of the current value plus the value of the previous 49 observations.

10.3.2.1 Graphic Approach

In the graphic approach, the practitioner attempts to determine visually when the slope of the initial transient state approaches 0. At this point, the output measure of performance has reached steady state. Obviously, this qualitative approach is highly subjective and susceptible to influence by individual interpretation. This approach is illustrated in Figure 10.8.

10.3.2.2 Linear Regression Approach

In the linear regression approach, the practitioner is using the least-squares method to determine where the initial transient state ends. This involves testing a range of observations to see if the linear regression slope coefficient is equal to zero. If the slope for a given range of observations is not zero, then the practitioner advances the range to a later set of observations. In this manner, the practitioner will eventually have a range of data for which the coefficient of the slope is insignificant. At this point, the practitioner has reached the steady-state behavior of the system. Figure 10.9 illustrates this approach.

The first regression line clearly illustrates a situation where the slope of the observations is not zero. This means that the system is in the transient state and has not yet entered steady state. The second line includes a mix of old data plus new observations. In essence, the data window is being moved to a higher range of observations. In the second line, the regression slope is much closer to 0. This means that the observations in the data window are starting to leave the transient phase and enter the steady state.

The creation of the regression line is easily executed using an electronic worksheet such as Excel. The linear regression function is found under the Tools-Data analysis menu sequence. If it does not appear, then the practitioner must execute the Tools-Add-ins menu sequence and check Analysis Toolpak. The default installation for Excel does not automatically add-in the Toolpak. So, unless someone has already used the Toolpak, the practitioner may have to follow this sequence.

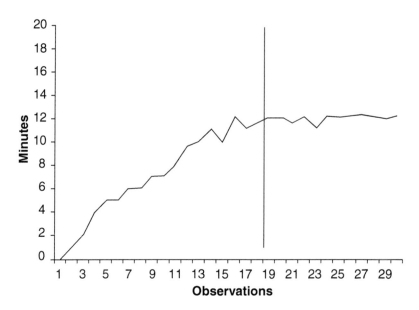

FIGURE 10.8   Graphical approach

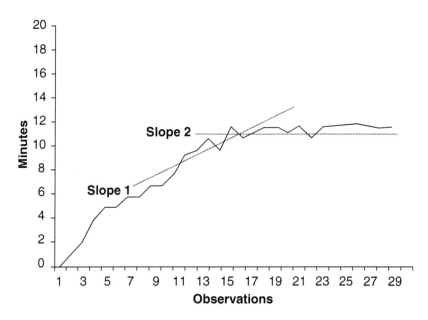

FIGURE 10.9   Regression approach.

This concept is mathematically illustrated with the following data. The data represent the first 30 system-time observations from a nonterminating system. A moving window of ten observations is used to determine when the system time moves from the transient phase to the steady-state phase.

Obs	Systime
1	0.5
2	1.0
3	2.0
4	4.0
5	5.0
6	5.0
7	6.0
8	6.0
9	7.0
10	7.0
11	8.0
12	9.5
13	10.0
14	11.0
15	10.0
16	12.0
17	11.0
18	11.5
19	12.0
20	12.0
21	11.5
22	12.0
23	11.0
24	12.0
25	12.0
26	12.1
27	12.2
28	12.0
29	11.9
30	12.0

The following regression output from Excel was obtained from the first 10 system times for this nonterminating simulation.

Summary Output:
Observations 1–10

Regression Statistics	
Multiple R	0.965
R Square	0.932
Adjusted R Square	0.924
Standard Error	0.665
Observations	10

ANOVA: Observations 1-10

	df	SS	MS	F	Significance F
Regression	1	48.492	48.492	109.792	0.000
Residual	8	3.533	0.442		
Total	9	52.025			

Intercept and Slope of Output: Observations 1–10

	Coefficients	Standard Error	t Stat	P-value
Intercept	0.133	0.454	0.294	0.776
X Variable 1	0.767	0.073	10.478	0.000

This output is based on the null hypothesis that the slope of the data is zero. The alternative hypothesis is that the slope of the data is not zero. The slope in the output is represented by the coefficient of the x variable 1. This coefficient is listed as 0.767. The P-value associated with the x variable 1 is 0.000. If we have an α level of 0.05, this value is extremely statistically significant. This means that null hypothesis must be rejected. It follows that the slope of observations 1 through 10 is not 0. Thus, this set of observations is still in the transient phase.

If we now move the window five observations forward, the window now covers observations 6 through 15. These observations result in the following Excel regression output.

Summary Output:
Observations 6–15

Regression Statistics	
Multiple R	0.966
R Square	0.933
Adjusted R Square	0.925
Standard Error	0.565
Observations	10

ANOVA: Observations 6–15

	df	SS	MS	F	Significance F
Regression	1	35.673	35.673	111.850	0.000
Residual	8	2.552	0.319		
Total	9	38.225			

Intercept and Slope of Output: Observations 6–15

	Coefficients	Standard Error	t Stat	P-value
Intercept	1.045	0.677	1.545	0.161
X Variable 1	0.658	0.062	10.576	0.000

In this output, the x variable 1 coefficient slope has been reduced to 0.658. This means that the slope is becoming more horizontal. The P-value is still 0.000. The null hypothesis must again be rejected. This means that the slope is nonzero, and the system is still in the transient phase.

Moving along another five observations to the right, we now have observations 11 through 20 in our data range. The following Excel regression output illustrates the results from this range.

Summary Output:
Observations 11–20

Regression Statistics	
Multiple R	0.878
R Square	0.771
Adjusted R Square	0.742
Standard Error	0.668
Observations	10

ANOVA: Observations 11-20

	df	SS	MS	F	Significance F
Regression	1	12.027	12.027	26.931	0.001
Residual	8	3.573	0.447		
Total	9	15.600			

Intercept and Slope of Output: Observations 11-20

	Coefficients	Standard Error	t Stat	P-value
Intercept	4.782	1.160	4.123	0.003
X Variable 1	0.382	0.074	5.190	0.001

In this regression output table the x variable 1 coefficient is 0.382. The slope continues to become closer to 0. The P-value is now 0.001. At an α level of 0.05, this value is still statistically significant. The null hypothesis is once again rejected. Although the slope is flattening, we are still in the transient phase with observations 11 to 20.

Moving on out another five data points, we are now looking at observations 16 to 25. This regression table is illustrated below:

Summary Output:
Observations 16–25

Regression Statistics	
Multiple R	0.174
R Square	0.030
Adjusted R Square	-0.091
Standard Error	0.440
Observations	10

ANOVA: Observations 16-25

	df	SS	MS	F	Significance F
Regression	1	0.048	0.048	0.250	0.631
Residual	8	1.552	0.194		
Total	9	1.6			

Intercept and Slope of Output: Observations 16-25

	Coefficients	Standard Error	t Stat	P-value
Intercept	11.203	1.004	11.162	0.000
X Variable 1	0.024	0.048	0.500	0.631

In our latest iteration, the x variable 1 coefficient is now 0.024. This is obviously close to 0. The P-value is 0.631. At an α level of 0.05, we cannot reject the null hypothesis of the slope being 0 with these results. Thus, the system has finally entered the steady-state phase.

The practitioner can now statistically claim with some certainty that the system time is out of the transient state by observation 16. Thus, observations 1 through 15 constitute the transient phase. These observations should be immediately discarded in performing any subsequent analysis.

10.3.3 Autocorrelation

An issue that the practitioner must address with nonterminating systems is autocorrelation. This is the tendency of subsequent observations of output measures of performance to be related to each other. This occurs when, for example, the system time of one entity is correlated with the following entity. If the first entity experienced a high system time, the next entity may also experience a high system time. Similarly, if the first entity has a low system time, the next entity will also have a low system time. When this occurs, the system may be suffering from autocorrelation.

The danger of autocorrelation is that the practitioner may actually underestimate the actual variance in output measures of performance, such as system time. If the practitioner underestimates the actual variance, the subsequent statistical comparisons may be affected. If the variance is underestimated, there is an increased likelihood that the practitioner will think that there is actually a difference between models when in actuality there is not. Thus, if the system has autocorrelation and the practitioner does not account for it, the practitioner may inadvertently and accidentally reject a null hypothesis of no difference between systems.

Because autocorrelation does have the potential to skew the statistical comparison between models, it is wise to attempt to address its possible effects. There are a number of statistical methods available to attempt to account for autocorrelation. Some of these are extremely mathematically challenging to implement. Because practitioners are more interested in simpler but still valid approaches, we will dispense with the more complicated techniques and utilize what is known as the batch method to account for possible autocorrelations.

10.3.4 Batch Method

Instead of running several replications of relatively short durations as with the terminating system model, we run the nonterminating system model for a single replication over a much longer time period and cut the run into separate batches. The batches are then considered as individual replications. The same comparison techniques can then be used for different alternatives. The steps in implementing this method are:

- Identify the nonsignificant correlation lag size.
- Make a batch ten times the size of the lag.
- Make the steady-state replication run length ten batches long.

10.3.4.1 Identify the Nonsignificant Correlation Lag Size

The purpose of this step is to identify the interval between observations that have little correlation to each other. If there is little correlation, then autocorrelation becomes a nonissue. This interval is known as the lag. If there is, for example, a lag of 100 observations, it means that the first and 101st, second and 102nd, etc. pairings of observations have no correlation. The mathematics for calculating the nonsignificant correlation lag size are outside of the scope of this practitioner's handbook. However, many simulation software packages have conveniently included this capability in their program. In this example, a nonterminating simulation was run for approximately 65,000 min. This resulted in a total of 26,849 system time observations. Figure 10.10 illustrates the use of the correlogram to identify the nonsignificant lag with these data. As the correlogram indicates, the nonsignificant lag occurs at an interval of approximately 600 observations.

10.3.4.2 Make a Batch Ten Times the Size of the Lag

Once the nonsignificant lag size has been determined, the practitioner can use the simulation rule of thumb of making a single batch ten times the lag size:

$$Batch\ size = 10 \times nonsignificant\ lag\ size$$

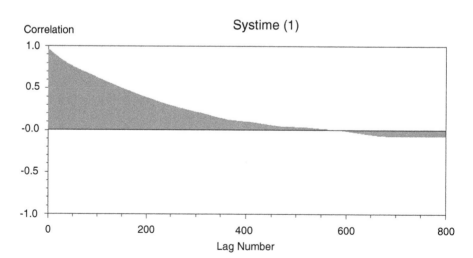

FIGURE 10.10 ARENA output analyzer correlogram.

The batch size usually has to be converted into time units for the simulation program. This can be accomplished by finding out how long it takes each system time observation to occur on the average:

$$Single\ observation\ time = \frac{Simulation\ run\ time\ length}{Observations}$$

This means that the time to obtain the number of observations for a batch size is:

$$Batch\ time = Single\ observation\ time \times batch\ size$$

In our example, this means that the batch size for each batch needs to have 6000 observations in order not to be subject to autocorrelation.

$$batch\ size = 10 \times 600$$

$$= 6000$$

Because the single observation time is:

$$single\ observation\ time = \frac{65,000}{26,849}$$

$$= 2.42$$

The nonsignificant lag batch time to produce each batch of 6000 is:

$$batch\ time = 2.42 \times 6000$$

$$= 14,520\ minutes$$

10.3.4.3 Make the Steady-State Replication Run Length Ten Batches Long

The next task is to determine how long to run the nonterminating simulation. The length of the simulation must be long enough to remove the transient phase and have at least 10 batches of the required batch size.

$$length\ of\ run = transient\ length + 10 \times nonsignificant\ lag\ batch\ time$$

In our example, our transient phase consisted of 15 observations. So, we will need to run our simulation for a total of at least:

$$length\ of\ run = 2.42 * 15 + 10 * 14,520$$

$$= 145,183\ minutes$$

This simulation run will produce at least 6015 observations of individual entity system time. We immediately discard the first 15 from the initial transient. The remaining 6000 or so will need to be divided into sequential batches of size 600. This will result in 10 batches. These batches are now considered as individual replications. We can now perform the same sequence of statistical tests as with terminating systems. This includes testing ten batches for replication analysis. In the event that ten replications are not enough, the practitioner will have to increase the length of the run. When a sufficient number of replications has been achieved, the comparison of means is done with confidence intervals or ANOVA and the Duncan multiple-range test.

10.4 Summary

In this chapter, we examined the statistical analysis of a set of validated simulation model alternatives. These models may be based on either terminating or nonterminating systems. If the system is terminating, it means that there is a naturally occurring event that ends the simulation run and that the system is cleaned of entities at the end of the run. For a terminating system analysis, the practitioner will have to perform replication analysis to determine how many production simulation runs to make, generate the production simulation runs, and statistically compare the simulation run results. The statistical comparison of simulation run results may be made with either confidence intervals or analysis of variance and the Duncan multiple-range test. Finally, the practitioner may need to perform an economic analysis of statistical analysis results.

If the system is nonterminating, it means that there is either no naturally occurring terminating event and the system runs forever, or the system does end but then starts up with entities from the previous period still in the system. Nonterminating systems utilize a single long simulation run for analysis purposes. For a nonterminating system, the practitioner needs to be concerned with the starting conditions, determining when the steady state in performance occurs, addressing autocorrelation, the length of replication, and the batching method to divide the simulation run. The practitioner then needs to follow the same statistical comparison techniques as described for terminating systems.

Chapter Problems

1. A pilot run of two simulation model alternatives was conducted using 10 replications. Based on the replication means, how many replications are necessary to achieve a relative precision of 0.10 at an α level of 0.05. What do you have to do after calculating the intial number of replications required?

   Alternative 16, 8, 7, 5, 8, 7, 6, 7, 6, 8

   Alternative 25, 9, 6, 11, 10, 6, 4, 9, 12, 10

2. Two similarly priced, completely new flexible manufacturing systems are considered for purchase. The following replication average processing time data in minutes were obtained for components being manufactured on each system. Assuming that you have a sufficient number of replications and both data sets are normal, determine which system should be purchased. Use 0.10 for any F-test and 0.05 for any other tests.

   System 1: 13, 15, 10, 7, 18, 15, 13, 18, 19, 25, 28, 21

   System 2: 10, 14, 13, 9, 19, 19, 16, 14, 21, 21, 29, 26

3. Three different operating policies are considered for a warehouse distribution system. The previous analyst (who no longer works there) ran only five replications of each model. The replication average output data are in hours to complete the orders. Given only the following data, what conclusions can you draw at an α level of 0.05?

Policy A	39	51	58	61	65
Policy B	22	38	43	47	49
Policy C	18	31	41	43	44

4. The following partial ANOVA output was obtained from running 67 replications of eight alternatives for different staffing and operating policies of a customer service center.

Source	DF	SS	F	P
Ticket	1	251.8	4.83	0.028
X-ray	1	25658.5	492.19	0.000
Metal	1	540.1	10.36	0.001
	528	27525.3		

Alternative	A	B	C	D	E	F	G	H
Employees	6	6	5	7	5	5	4	6
Mean (min)	10.86	12.92	14.04	9.74	25.94	25.15	27.14	24.68

Analyze the results and make recommendations at an α level of 0.05.

References

Johnson, R.A., Freund, J.E., and Miller, I. (1999), *Miller and Freund's Probability and Statistics for Engineers,* Pearson Education, New York.

Law, A.M. and Kelton, D.W. (2000), *Simulation Modeling and Analysis,* 3rd ed., McGraw-Hill, New York.

11

Project Reports and Presentations

"Readiness is all."
Shakespeare

11.1 Introduction

The final step in the simulation process is to write a report and create a presentation on the simulation project recommendations and conclusions. This chapter covers:

- Written report guidelines
- Executive summaries
- Equations
- Screen captures and other graphics
- Presentation guidelines
- Presentation media
- Electronic presentation guidelines
- Electronic software presentation issues
- Actual presentation

To assist the practitioner in the report and presentation process, a number of example reports and PowerPoint presentation files have been included in the appendix. In addition, template documents for both reports and presentations are also available on the CD-ROM accompanying the handbook.

11.2 Written Report Guidelines

Generally, both the report and the presentation should follow the same format that was used as a guide for the simulation project. These include:

- Problem statement
- Project planning
- System definition
- Input data collection and analysis
- Model formulation
- Model translation
- Verification
- Validation
- Experimentation and analysis
- Recommendations and conclusions

The level of detail within each of those steps is at the discretion of the practitioner. However, the assembly of a professionally prepared report will need to include items such as an executive summary, equations, and worksheet and simulation software screen captures.

11.3 Executive Summary

It is a fact of simulation analysis that a good number of the individuals who need to be briefed on the project will not have as technical a background as most practitioners. Others will not necessarily be interested in the level of detail available in a comprehensive report. For these reasons, the practitioner should consider including an executive summary. The executive summary should not be more than two or three pages and should contain condensed descriptions from the report, including:

- Project objectives
- Results
- Recommendations and conclusions

11.4 Equations

A professionally prepared report should include the equations that were used during the analysis phase of the report. The existence of equation-editing software like that included in the Microsoft Office software suite make this previously painful process much less tedious. It is actually to the advantage of the practitioner to use an equation editor for the report because the equations can be then copied and pasted into the electronic presentation. Furthermore, once the practitioner has created all of the basic simulation equations, they can be reused in future simulation studies. The use of the Microsoft Equation Editor is beyond the scope of this book. However, to facilitate this process, the practitioner can copy and paste the equations included on the CD-ROM provided with this book.

11.5 Importing Screen Captures

Another characteristic of a professional report is the inclusion of system photographs and simulation software screen captures. Photographs are invaluable in describing the operation of specific system processes. Simulation software screen captures can include the simulation model flow charts, the animation, and statistical plots or graphics.

Photographs are usually imported into word processor documents, including those in Microsoft Word, by importing the graphic file. However, two distinct methods may be used to bring screen captures into the document.

11.5.1 File Import Method for Digital Photographs

This method is used for importing digital photographs into the document. Most digital cameras and scanners will allow the user to save the image in jpg format. This format is significantly more space efficient than other formats. There is some loss of resolution because of the encoding algorithm used by this format. However, for report purposes, this loss is not significant. The sequence for this method, assuming that the image already exists, is:

- Switch to the word processor such as Microsoft Word.
- Create a text box.
- Import the jpg or other graphics file into the text box.
- Size and position as necessary.

11.5.2 Buffer Method for Screen Captures

The second method is to copy the screen images to the operating system buffer and then paste the buffer directly into the document. This sequence is:

- Acquire the desired screen image.
- Press the alt and print screen keys simultaneously.
- Switch to the document.
- Paste the buffer directly into the document or into a text frame in the document.

The use of a text frame provides the practitioner with additional control over the graphic. This includes sizing, cropping, and positioning. The buffer method does have the disadvantage that the captured image cannot be significantly edited before being placed into the document. Because of this limitation, the image must be relatively free from defects. A second disadvantage to the buffer method is that the image may dramatically increase the size of the document. When this occurs, it may be because the image is a bitmap with high color resolution.

11.5.3 File Import Method for Screen Captures

This method is also utilized for importing graphic files of screen captures into the document. It is a combination of the buffer method for screen captures and the file import method. This method offers the most image editing control, minimal memory size, and flexibility to easily reuse the screen capture image. The sequence for this method is:

- Acquire the desired screen image.
- Press the alt and print screen keys simultaneously.
- Open an image-editing software package such as Microsoft Photoeditor.
- Paste the buffer directly into the image-editing software.
- Edit the image as necessary.
- Save the image as a jpg graphics file.
- Switch to the word processor such as Microsoft Word.
- Create a text box.
- Import the jpg file into the text box.
- Size and position as necessary.

With this method, the practitioner can use the same graphics in both the report and the presentation. This can significantly reduce the amount of time needed to create both a professionally appearing report and a presentation.

11.6 Presentation Guidelines

Although the electronic presentation should follow the same basic simulation project process as the report, the practitioner will have to individually determine the level of detail that should be provided during the presentation. Factors that will contribute to this decision include:

- Objective of the presentation
- Time for the presentation
- Technical level of the audience

11.6.1 Objective of the Presentation

The objective of the presentation is of paramount importance in deciding on the level of detail to be provided in the presentation. Is the presentation for solving a problem, or is it for making a financial decision? If solving a problem is of primary importance, then the analysis section would probably need to be more detailed. This would specifically include a detailed discussion of the potential benefits of each alternative. Conversely, if the presentation is for making a financial decision, then additional detail will need to be included on the economic analysis of each alternative.

11.6.2 Time for the Presentation

The amount of time available for the presentation is also an important factor in deciding the level of detail to be provided in the presentation. If sufficient time is available, the practitioner can explain both the high-level flow chart and the interworkings of the model. If time is limited, the practitioner may be able to present only the high-level flow chart. Similarly, a demonstration of the model can easily require a significant amount of presentation time. Displaying a few screen captures of the running model may be advantageous when time is short.

11.6.3 Technical Level of the Audience

If the audience has limited technical knowledge, it may be beneficial to bypass or greatly reduce the presentation of the modeling and statistical details. These types of technical details may be either beyond the comprehension of or of little interest to the audience. The time that is available would be better spent demonstrating the animation and concentrating on the conclusions and recommendations.

When demonstrating the animation, the practitioner should take particular care that the simulation run does not produce any abnormal situations. The audience will latch onto the fact that the animation may illustrate a situation that has not necessarily been observed. If this occurs, then the validity of the simulation model will be in question. This can easily lead to a situation where any other results generated from the model will be questioned. It is better not to have to explain to the audience that the fact that a particular situation occurs in the single simulation run does not necessary mean that it will occur in every simulation run.

11.7 Presentation Media

The practitioner has several choices of presentation media to select from. With the increasing availability of LCD computer projectors, the electronic presentation has become the preferred choice with major corporations and professional conferences. The electronic presentation holds the distinct advantage of making last minute corrections or updates possible. However, the use of LCD computer projectors does involve a few disadvantages including increased preparation time and computer compatibility issues.

11.7.1 LCD Projector–Based Presentations

In this section, we discuss the peculiarities of presenting an LCD projector electronic presentation. These include:

- Allocation of additional preparation time for LCD presentations
- LCD projector and notebook compatibility issues
- Other LCD projector presentation issues

11.7.1.1 Allocation of Additional Preparation Time for LCD Presentations

An LCD projector presentation requires significant preparation time over other presentation methods. This is principally because of the need to set up both the LCD projector and the notebook computer. If the presentation is conducted with someone else's LCD projector, the practitioner must allow sufficient time to become familiar with the video input connectors and the warmup sequence required by the projector. In one professional conference, the presenter connected the output video from the notebook computer to the output connector on the the LCD projector. It was several minutes into the presenter's allocated time before it was discovered that the cable needed to be connected to the LCD projector's input connector not the output connector. The additional stress created by this problem obviously did not help with the delivery of the presentation.

Another requirement for additional electronic presentation preparation time occurs when more than one presenter is involved. This is particularly evident with professional conferences. Not only does the next presenter have to set up equipment, but the previous presenter must also remove or detach his or her equipment. Significant time must be allocated to allow for this breakdown and setup sequence.

11.7.1.2 LCD Projector and Notebook Compatibility Issues

The use of an LCD projector presentation presents several computer compatibility issues. Some computers will not display images through both the external video port and the screen at the same time. Those that do will usually require the user to toggle through different modes that will alternatively display only the external port, only the notebook display, both at the same time, or neither. There is also the issue of notebook screen resolution and refresh rate versus the LCD projector screen resolution and refresh rate. Periodically the notebook screen resolution and refresh rate must be adjusted so that the LCD projector can handle the resolution and refresh rates. If the image flashes on the projector or screen for a brief moment and then reverts to the default projector image, this problem may be occurring.

Yet another compatibility problem is that the LCD projector may attempt to interpolate the image if it does not have the same native screen resolution as the notebook computer. This problem usually presents itself by "greeking" or distorting the text and images on the LCD projector image. In extreme cases, this can prevent the presenter from properly communicating project results. In mild cases, this is an unnecessary annoyance.

11.7.1.3 Other LCD Projector Presentation Issues

A few other issues that the electronic presenter should consider are plug multipliers and extension cords. There may be only one power outlet near the presenter's area of the room. If the jack is occupied by the electronic projector, the presenter will have to find an alternative power source. If the presenter has a plug multiplier with a male plug on one side and three female sockets on the other side, there will be no need to find another power source. The other issue is whether or not the power outlet is close enough for the presenter's notebook computer. Some setups require the presenter to be in one location while the electronic projector is in another location. Under these circumstances it may be helpful to bring your own extension cord. The lack of a plug multiplier and an extension cord has caused many embarrassing delays until a support technician could be contacted.

11.7.2 Transparency-Based Presentation Issues

Transparencies are an alternative means of presenting the results of the simulation study. One advantage to using transparencies is that transparency projectors are significantly more commonly available than LCD computer projectors. Another advantage is that transparencies are obviously less susceptible to technology-related problems. Unless the practitioner has access to a color printer, the presentation will be in old fashioned black-and-white laser print.

If the practitioner has a color inkjet printer, he or she should be aware that inkjet printers cannot use regular transparencies. The slick surface of normal transparencies does not provide an adequate base to absorb the liquid inkjet ink. Inkjet printers must use special transparency paper that has a coating to capture the ink. This special transparency paper may not be readily available. So, if any possibility exists that transparencies will have to be printed with an inkjet printer, the special transparency paper should be acquired in advance.

Last, corrections to transparencies are more likely to be difficult to execute. Despite the limitations of hard copy transparencies, the practitioner should still prepare a backup set of transparencies to augment the electronic media for any important presentation.

11.8 Electronic Presentation Software Issues

The proper use of electronic presentation software can make a significant difference in the successful communication of the simulation project results. In this section, we discuss:

- Presentation masters
- Use of colors
- Use of multimedia effects
- Speaker's notes
- Use of presentation handouts

11.8.1 Presentation Masters

Most presentation software packages such as PowerPoint include a dizzying array of presentation masters or templates to assist the practitioner with the rapid development of a presentation. These presentation masters typically include different backgrounds from which the practitioner may choose. If the practitioner does decide to use one of these canned masters, careful consideration should be given to using one that has some applicability to the simulation study. All too often, a practitioner will use a master that is very colorful or has an attractive pattern. A better course of action would be to develop a new master specifically for the project. This could include a copy of the company logo or a muted background image of something significant to the project. For example, the company logo could appear at the top left of each slide. Similarly, a presentation given to a group of professional manufacturing engineers could include a muted background of a manufacturing facility. If either of these approaches is taken, care must be given to insure that the color or density of the logo or the background does not interfere with the presentation of any text or graphics.

11.8.2 Use of Colors

If the practitioner wisely decides to develop a simulation project specific master, one of the fundamental issues that must be resolved is the color scheme. The selection of the color scheme must always insure the readability and viewability of the presentation. This dictates the use of contrasting colors for the background, main text, and bullets.

11.8.2.1 Background Colors

The choice of the most appropriate background slide colors for a particular type of presentation has been the focus of many research efforts. Without delving into the psychology or validity of these efforts, some commonly accepted concepts will be presented. First, there is some evidence that supports the use of reddish color backgrounds when the purpose of the presentation study is to excite or even incite the audience. Other evidence supports the use of blue hues to create a calming effect. There is also some thought that the color of money, green, should be used when discussing financial matters.

Because the practitioner will probably be presenting to a diverse audience with a wide background, the most conservative approach would be to promote a calming approach. This means that we would generally want to use a medium to dark blue background.

Presentations also may or may not use gradients in the slide background. If gradients are utilized, consideration should be given to insuring that the lighter color is on the upper half of the screen and the darker color is on the lower half of the screen. The reason why some professional presenters prefer this color gradient scheme is that it mimics the nighttime skyline horizon.

11.8.2.2 Text Colors

As with title colors, the text colors should contrast with both the background and the title text. With a blue background, an effective color is white. If more than one level of bulleted line is used, the practitioner must also decide if the second- and lower-level text should keep the same color or change to another. To avoid unnecessary distraction, it may be more advantageous to retain white as lower level text when using a blue background.

11.8.2.3 Title Colors

The two considerations for the title color are that it contrasts with both the background and the main body text. If a blue background and a white main body text are used, yellow is one candidate for the title color.

11.8.2.4 Bullet Colors and Shapes

One color combination that appears to be effective is the use of a medium to dark blue background, yellow title, white text, and red first-level bullets. The blue background ensures that glare from the electronic projector is minimized. The yellow title and white text contrast well against the blue background. The red bullets call attention to themselves as well as contrasting well with the white text and blue background.

11.8.3 Use of Multimedia Effects

A dizzying array of multimedia effects is available to the practitioner in an electronic presentation. These include:

- Screen transitions
- Text and graphics animations
- Sound effects

Screen transitions involve the method for removing the previous screen and displaying the next screen. The default type effect is a direct replacement in which the previous screen is nearly instantaneously replaced by the new screen. Other transitions include wipes from one side to another, boxing in or out, or pixilation between the two screens. The practitioner should carefully consider the use of any screen transitions other than a direct replacement. The novelty of more sophisticated screen transitions quickly wears off and in fact can become an unnecessary distraction.

Another common multimedia screen effect is the animation of text lines or graphics on the slide. This is commonly implemented by having the text or graphics appear from one side or another. It is also possible for the text or graphics to follow some sort of simple path. As with screen transitions, the practitioner should be careful in selecting possible text or graphics animations to make sure that they are not counterproductive. The use of animated text lines should not be confused with the use of text builds. These are successive slides that display progressive lines on a particular slide. The use of text builds is an effective presentation technique that should be used by the practitioner. Figures 11.1 and 11.2 illustrate the use of a text build.

WHY SIMULATE?

• Gain insight into the operation of a system

FIGURE 11.1 Initial bullet.

WHY SIMULATE?

• Gain insight into the operation of a system
• Develop operating policies to improve system performance

FIGURE 11.2 Bullet build.

Sound effects constitute another possible multimedia effect in an electronic presentation. More common uses of sound effects include:

- Opening presentation sound effect
- Slide clicking noises
- Closing applause

An opening presentation sound effect can be used to great effect to obtain the audience's attention. However, the use of any kind of peculiar types of sound effect should be avoided. If an opening sound effect is to be used, it should have some applicability to the subject material. For example, the opening of a presentation to an audience of airlines managers could include the sound of a jet passing overhead. If the organization has a company jingle, it might also be an acceptable opening sound effect.

Some practitioners use an electronic slide sound effect when advancing slides. This is more of a neutral sound effect. If the sound level can be output at an appropriate level, it may be useful for signaling the display of the next slide to audience. Members of the audience who are concentrating on taking notes or reading the handouts will be able to direct their attention to the screen. The use of any type of sound effect is at the discretion of the presenter.

Obviously, the use of electronic applause at the end of the presentation should be avoided at all costs. Not only does this sound effect appear unprofessional, it also indicates a lack of self-confidence in the presenter.

11.8.4 Speaker's Notes

Most software presentation packages will enable the practitioner to print copies of speaker notes. This form of output includes a less than full-size image of the original slide plus an area where the practitioner can record additional information beyond that on the slide. Because one of the fundamental presentation rules does not allow more than eight lines and eight words per line, the speaker note format can be a lifesaver during a particularly stressful presentation. Figure 11.3 illustrates a typical speaker's note format for a presentation.

11.8.5 Presentation Slide Handouts

A professional touch to a practitioner's presentation is the distribution of handouts to the presentation attendees. If the presentation is of any length, the handouts should be printed with multiple slides per page. With most presentation software packages such as Microsoft Powerpoint, it is also possible to print the slide images on only one side of the page. The other side of the page is reserved for viewer notes. Figure 11.4 illustrates the use of presentation handouts.

```
┌─────────────────────────────────────────────┐
│           EXAMPLES OF SYSTEMS THAT            │
│               CAN BE SIMULATED                │
│                                               │
│   •   Manufacturing Systems                   │
│                                               │
│   •   Service Systems                         │
│                                               │
│   •   Transportation Systems                  │
│                                               │
│                                               │
│                                               │
│                                               │
└─────────────────────────────────────────────┘
```

Production and assembly lines, job shops, and flexible manufacturing systems

Service systems include retail establishments, restaurants, and repair facilities

Transportation systems include airports, train stations, and bus stations

FIGURE 11.3   Speaker's notes page.

11.8.6 Handout Output Issues

An issue that is often overlooked with presentation handouts is that the practitioner may have access to only a black-and-white printer or a black-and white-copier. This becomes an issue if the practitioner attempts to print color slides converted to gray scale. The original presentation colors and the presentation software's color to black-and-white translation scheme may seriously reduce the viewability or resolution of the handouts.

The most easily implemented solution to this problem is to attempt to set the printer output to pure black and white. This solution is particularly effective when no photographs or high-color images are present in the original presentation. If photographs are present in the presentation, they may result in an obviously useless solid black image. The solution in this case is to temporarily omit the photographs or high-color images from the presentation before printing the handouts.

11.9 Actual Presentation

This section addresses issues associated with the actual presentation that will help the presenter deliver a professional standard presentation. These issues include:

- Rehearsal
- Dress
- Positioning
- Posture
- Presentation insurance

11.9.1 Rehearsals

For any important presentation, the practitioner should make a concerted effort to perform a number of rehearsals. The rehearsals should continue until at least one totally satisfactory rehearsal has been performed. The practitioner should not just go through the presentation in his or her mind while standing in the office, but actually perform a rehearsal:

INDE6370 Digital Simulation Comparison of More Than Two Systems University of Houston Department of Engineering Houston, Texas	AGENDA • ANOVA • Duncan multiple range test

ONE WAY ANALYSIS OF VARIANCE

- Preferred tool for more than two groups (alternative)
- Compares variability between and within groups
- Indicates if one or more means are significantly different
- Assumptions
  - Samples are normal
  - Samples are independent
  - Variables are equal

ONE WAY ANALYSIS OF VARIANCE

- Calculate means
- Calculate sums of squares
- Calculate mean sums of squares
- Calculate F statistic
- Calculate F statistic to critical F value
- Reject or cannot reject Ho

ONE WAY ANOVA CALCULATE MEANS

- Grand mean
- Group means

ONE WAY ANOVA CALCULATE SUMS OF SQUARES

- Sums of Squares Total (SST)
- Sums of Squares Between Groups (SSB)
- Sums of Squares Within Groups (SSW)

FIGURE 11.4 Presentation handouts.

- In the same room
- With the same equipment
- With the same presentation and the same model

There are no shortages of horror stories about individuals who performed many rehearsals but in a different room with different equipment. The same room is necessary so that the practitioner is totally familiar with the lights, the screen controls, and even where the electrical sockets are placed.

Similarly, the practitioner should be familiar with the electronic projection equipment. This includes how to connect the practitioner's notebook computer to the electronic projector. It also includes knowing what toggle controls are necessary to output the image to the electronic projector.

Last, the practitioner should insure that the final presentation and model are actually available on the notebook computer. Sometimes there is an overwhelming temptation to make last minute changes. These last minute changes can easily cause the model to malfunction or otherwise perform in an unexpected manner.

When the electronic presentation or the simulation model malfunctions, there is no one to blame but yourself.

11.9.2 Dress

For a formal presentation the practitioner should be dressed formally. Formal dress preferably includes a suit and tie. However, under some circumstances it may be acceptable to wear a dress shirt, tie, and coat. Do not be led astray by the corporate casual Friday dress policy. If you dress casually, the audience may also think that you conducted the study in a casual manner. This is particularly important in the corporate environment when individuals other than the immediate stakeholders will be attending. Dressing in an appropriate manner allows you to concentrate on other difficulties that may arise.

11.9.3 Positioning

Positioning refers to the physical location of the presenter to the audience. The preferred position for the presenter is in the front to the left of the screen. Many conference rooms and auditoriums are set up in this manner with a podium in this position. With this layout it is difficult for the presenter to look at the audience and see the screen at the same time. The most common result is for the inexperienced presenter to turn away from the audience and concentrate on the projector screen. This inevitably results in loss of rapport with the audience.

This situation is actually easily resolvable by making use of the notebook computer screen. This enables the presenter to both face the audience and have the opportunity to keep track of the presentation. Utilizing the computer in this manner does require some discipline to constantly avoid looking back at the screen to insure that the projector is still operating correctly.

One combination method that the presenter can use to great effect is to use the notebook computer screen but point to the projector screen for periodic emphasis. With this method, it is important that the presenter use the left arm and hand to point to the screen. This enables the presenter to maintain a forward facing position with eye contact with the audience.

11.9.4 Posture

Little can be said about presentation posture. The presenter should avoid continuously leaning on the podium or keeping hands in pockets. The audience derives energy or lack of it from the presenter. If the presenter appears lethargic or excessively nervous, the audience may fall asleep or become restless.

11.9.5 Presentation Insurance

The practitioner cannot purchase an insurance policy to cover accidents during the course of the presentation. However, the practitioner can take certain measures to increase the probability of being able to successfully make the presentation. Some steps that may be taken to help insure a success presentation include acquiring:

- Presentation and model on both floppy and CD-ROM media
- Simulation software package CD-ROM
- A multiplexer plug
- An extension cord

In the event that the practitioners' notebook computer malfunctions, it may be necessary to conduct the presentation and demonstration on another system. Many current notebook computers have only one bay for both the floppy disk and CD-ROM. To preclude problems should either the floppy disk or CD-ROM not be available, the practitioner can bring copies of the presentation and model on both floppy and CD-ROM disks.

As previously discussed, a multiplexer plug is a device that expands the number of electrical outlets that are available for use. This can become a critical issue because many built-in outlets have only two sockets. If a socket is used by the LCD projector, then without a multiplexer, only one other socket will be available for the practitioner.

Likewise, a short lightweight extension can also be an invaluable aid to the practitioner. This will allow the practitioner to place the computer in a better presentation position away from the LCD projector. Although an extension cord may be available from the host organization, it is better not to have to wait until one can be located or to wait until it is convenient for a technician to bring one.

11.10 Summary

In this chapter, we discussed the standards and practices that the practitioner may utilize in order to develop a comprehensive and professional project report and presentation. All too often, a great amount of effort is expended during the technical phases of the project, but too little effort or too little time is available to communicate properly the work and results of the project.

In general, the project report and presentation should follow the same sequence as the actual project. The project report should also include an executive summary in nontechnical language. To enhance the report and presentation, the practitioner should make use of resources such as automatic table of contents generation, electronic equation editors, digital photographs, and screen captures. The practitioner should avoid the use of meaningless multimedia effects and inappropriate slide backgrounds.

When utilizing electronic presentation equipment, the practitioner should be aware of the additional time and technical demands inherent with the use of this technology. It is always a good idea to keep a set of transparencies as a backup. Obviously, any important presentation should be sufficiently rehearsed. However, an effort should be made to rehearse the presentation in the same location and with the same equipment as the actual presentation.

Chapter Questions

1. If the notebook computer and the LCD projector are properly connected, and both are powered on, what should be checked if no image is being projected?

2. What is meant by greeking?

3. What is a presentation slide build?

4. What type of information should be included in an executive summary or abstract?

5. Why is it necessary to allot additional time for an electronic presentation?

6. What is a multiplexer?

7. What is the safest color to use for slide backgrounds?

8. If you are forced to use someone else's notebook computer, why would you want to bring the presentation and model on both floppy and CD-ROM?

9. What technique can you use to help avoid looking at the projected image on the screen?

10. Why must you have special paper to make transparencies with inkjet printers?

12

Training Simulators

"I hear and I forget."

"I see and I remember."

"I do and I understand."

Aesop

12.1 Introduction

Advances in microcomputer technology in the last 10 years have expanded the horizons of computer simulation beyond discrete, continuous, and combined simulation models. One area that has greatly benefited from these advancements is the area of computer training simulators. A fundamental distinction between simulation and simulators can be found with respect to their ultimate use. Traditional simulation is typically used to analyze systems and to make operating or resource policy decisions. In contrast, simulators are generally used for training users to make better decisions or for improving individual process performance.

For the remainder of the introduction section, we discuss the advantages and disadvantages of simulators, simulator applications, and the simulator development process.

12.1.1 Simulator Advantages

Simulators are frequently compared to other types of training techniques that are commonly utilized. These include classroom and practical exercise training. Although it is the mainstay of the educational system, classroom training is typically limited to theory-based lecture-type instruction. The effectiveness of the training is assessed through written exams. Frequently this lecture-type instruction is limited to an initial course and then periodic refresher classes.

If the necessary resources are available, it may be possible to augment the classroom training with some sort of practical exercise. The practical exercise is used to reinforce the theory presented in the classroom. Practical exercises are particularly important when the application involves a combination of mental and physical skills. Unfortunately, for many applications, the use of practical exercises may require an inordinate amount of time or other resources to be properly implemented. Sometimes conducting a practical exercise will require that the subject system be shut down for the exercise. In many situations, it is operationally difficult or even impossible to shut down the system. As a result, even if practical exercises are used, it is unlikely that many practical exercises will be performed for training purposes.

The limitations and disadvantages of classroom and practical training methods give birth to the advantages of using computer training simulators. These types of training simulators utilize the multimedia capabilities of the current generation of personal computers to provide realistic training that might not otherwise be possible because of:

- Time limitations
- Cost limitations
- Operational limitations

The first advantage is associated with time limitations. These come in two forms. First, computerized training simulators do not require special scheduling of training sessions. If a personal computer is available for use, then individual training sessions can be conducted. In addition, an unlimited number of sessions may be scheduled for refresher training. During operational lulls, system managers and equipment operators can easily receive training during what might otherwise be nonproductive time.

The second time advantage can be found in the fact that many processes involve a significant time lag between when an action is taken and when the results of the action can be observed. As a result, it is difficult for individuals to establish any sort of cause-and-effect relationships. Simulators can compress what would ordinarily be lengthy periods of time into much smaller time segments. Thus, there is a greater probability that cause-and-effect relationships can be established by the individual.

The second simulator advantage is associated with training costs. Because the training is simulated on a computer, additional resources are not required to perform the training. This means that training can be obtained with reduced travel and meeting facility costs. In addition, any mistakes that are made with the training simulator do not translate into capital loss, as might occur with the actual system. Similarly, a training simulator means reduced costs associated with any consumable training resources.

The last simulator advantage involves operational limitations. Because the training does not interfere with ongoing organizational operations, the training can be performed in parallel, with little or no respect to operational considerations. Clearly, it would be operationally impossible to shut down many systems, but it would be a relatively simple task to perform simulated training on the same system.

The differences between conventional classroom, practical exercise, and training simulators can be summed up in this short passage frequently credited to Aesop:

"I hear and I forget."

"I see and I remember."

"I do and I understand."

Many studies have indicated that most people quickly forget or distort the content of verbal communications. However, if the verbal instruction is augmented visually, there is a greater probability that the communication will be remembered. However, to obtain a true understanding of the message, it is necessary to actually experience the content. Clearly, Aesop would have approved of the use of multimedia interactive training simulators for mission-critical applications.

12.1.2 Simulator Disadvantages

As with traditional simulation, computerized training simulators are not without their disadvantages. The primary disadvantages are:

- Availability of high-end multimedia personal computers
- Increased initial development costs

The first potential disadvantage to the use of training simulators is that the programs must generally be run on high-end multimedia personal computers. In order to maximize the training effectiveness, the instructional material must be presented in as many forms of media as possible. This includes text, graphics, sound, and motion. In addition, training simulators are designed to require a high degree of interactivity. These requirements place a heavy load on the graphic and computational capabilities of the personal computer. For this reason, it is necessary that the organization have higher-end multimedia personal computers to properly host training simulators. However, with the recent increase in personal computer capabilities and the accompanying reduction in purchase costs, this disadvantage is not nearly as onerous as it was a few years ago.

The second disadvantage of the computerized training simulator is associated with increased development costs. As with traditional simulators, it may be necessary either to contract out the development of the training simulator or personally to acquire the necessary skills to develop a training simulator in house. The in-house development of a training simulator has become much easier in recent years. The same advances in computer technology that leverage the capabilities of multimedia training simulators also make it easier to develop multimedia training simulators. This means that the practitioner should not consider the task of developing an in-house training simulator as an impossible task but one that may require as much effort as a traditional simulation modeling and analysis project.

12.1.3 Simulator Applications

The advantages and disadvantages of computerized multimedia training simulators lend themselves to many organizational mission-critical operations. A few applications include:

- Operating sophisticated machinery
- Troubleshooting malfunctioning equipment
- Responding to emergency situations

12.1.3.1 Operating Sophisticated Machinery

Training simulators can be utilized for providing orientation and refresher training on a variety of equipment and machinery. The value of training simulators increases as the probability and cost of possible damage to the equipment or machinery by inexperienced operators increases. Consider the operation of a computer numerically controlled lathe. To use this piece of equipment, the operator must:

- Insert a piece of raw stock
- Prepare the correct tooling to turn the raw stock

- Determine the set point between the tooling and the raw stock
- Load the computer program
- Run the computer program
- Monitor the operation
- Remove the finished stock

In this relatively simple operation, may things can go wrong. For example, the operator can insert the wrong stock or insert the correct stock incorrectly. The operator could also improperly position the tooling in the tool turret. The orientation set point between the tooling and the raw stock could be recorded incorrectly. The operator could also load the wrong computer program. Last, if some problem occurs during the turning cycle, the operator needs to know how to stop the equipment safely.

Any of these errors would at best prevent the finished part from being produced correctly. In this case, considerable previously performed work plus the cost of the raw stock could be lost. At worst, the lathe tooling could unexpectedly collide with the raw stock. This could create a dangerous situation for the operator as well as possibly damage the expensive computer numerically controlled lathe. If there is only one of this piece of equipment, the entire production may be forced to halt until repairs or replacement can be made.

A training simulator can be developed both to instruct and test novice operators on the correct procedures for using the computer numerically controlled lathe. So instead of wasting time and resources by making learning mistakes on the actual equipment, the novice operator can safely and inexpensively make errors on the simulator. Even a relatively simple process such as removing the key from the stock chuck can be caught by the training simulator. Not removing the key from the actual machine could result in its becoming a missile when the machine is put in the cycle mode.

12.1.3.2 Troubleshooting Malfunctioning Equipment

Another area where training simulators can be of great benefit is in troubleshooting malfunctioning equipment. The more complicated the piece of equipment, the more difficult it may be to identify the malfunctioning component. Even when the malfunctioning component is identified, it may be very difficult to repair or replace the component. An example of a troubleshooting training simulator was one recently developed for marine diesel engines. This training simulator possessed the capability to handle:

- Starting failures
- Overheating
- Power loss

Starting failures could be the result of inadequate battery capacity, engine starter problems, poor lubrication, or fuel problems. Overheating could be the result of a closed or blocked raw water circuit, loss of cooling fluid, poor lubrication, belt damage, or ruptured hoses. Power loss could result from fuel problems, electrical problems, lost propeller or shaft, damaged transmission, and so on. It may be apparent from the previous discussion that the symptoms can be a result of different problems, and the original problems can generate a new batch of other different problems.

In this example, to prevent damage to the engine, the operator must first be able to assess the problem rapidly and correctly. The operator must then execute the appropriate actions in a timely manner. The consequences of failure could lead to the loss of the vessel as well as create a potentially life-threatening situation.

A great advantage to using a simulator in this capacity is the flexibility to create an enormous number of different troubleshooting scenarios for training purposes. It may even be possible to create trouble-shooting scenarios that would not ordinarily be possible to recreate in the training environment. This forces the user to employ logic instead of memorization in order to respond to the problem. The result is a more effectively trained engineer, technician, or operator.

12.1.3.3 Responding to Emergency Situations

Training simulators can be adapted to help organizations respond to emergency situations. This is a particularly effective use of training simulators because it can be particularly difficult to receive realistic training on responding to a wide variety of emergencies. Most organizations do provide some training on emergencies such as bomb threats. However, few conduct enough training to have an effective response when the emergency actually occurs. In the case of a bomb threat, the organization will actually have to:

- Receive the threat
- Gather additional information
- Decide whether the threat is credible
- Search the area
- Evacuate the area to a safe area
- Search the safe area
- Reenter the area

Clearly the simple process of responding to an emergency bomb threat is not as simple a process as it might seem on the surface. By studying the process of receiving and responding to a bomb threat, the practitioner can develop a training simulator that can be used to provide realistic training in this process. Obviously, emergency response managers within the organization can use this kind of simulator to receive training that might not otherwise be possible because of time, cost, or operational limitations. There is little question that few organizations will allow or can afford its operations to be disrupted regularly to practice emergency response procedures of this type.

12.1.4 Training Simulator Development Process

The training simulator development process is very similar to that of traditional simulation modeling and analysis projects. This should not really be surprising because the two are closely related. The training simulator development process consists of:

- Problem formulation
- Project planning
- System definition
- Input data collection
- Model translation
- Verification
- Validation
- Implementation

The first three steps are directly equivalent to a traditional simulation project. The input data collection phase differs somewhat in that it also includes the acquisition and editing of graphic and sound data. The model translation and verification phases are also similar to traditional simulation. We will still be translating our model into a computer model, and the model will still have to be debugged. The validation phase, however, is significantly different. Here we are attempting to validate the model for both face and training validity. Face validity is the same. However, training validity means that the simulator represents the actual model sufficiently to impart improved learning. Note that there are no experimental design or analysis phases. This is because we are not trying to analyze the system by designing and evaluating different system alternatives. Last, we have the implementation phase in which the simulator is distributed for individual use. This is also significantly different from a traditional simulation model, which is normally retained by the practitioner. In the implementation phase, we are concerned with publishing the simulator on a CD-ROM, distributing the CD-ROM, user installation, and support. To insure that the proper development process is followed, we discuss each of these steps in turn.

12.2 Problem Formulation

As we previously discussed, the problem formulation phase is similar in concept to that of a traditional simulation model. We already know that this phase will consist of:

- A formal problem statement
- An orientation of the system or equipment
- The establishment of specific project objectives

12.2.1 A Formal Problem Statement

The formal problem statement for a training simulator would involve one of the following issues:

- Improving management decision-making process skills
- Improving equipment-operating skills

Consider a management decision application involving the implementation of advanced manufacturing technology. This is a complex process involving conceptual, planning, installation, and startup phases. If an organization does not properly manage this process, the consequences could be increased project cost, delayed operational startup, or even complete project failure. These types of projects require numerous decisions throughout the project. There is usually a significant delay before the results of these decisions take effect. As a result, it may not be possible to establish cause-and-effect relationships to help improve the performance of the implementation project.

Similarly, the operation of equipment can present complex training problems. For example, the preoperational checks, operation, and troubleshooting of a marine diesel engine can present a significant number of different tasks. In this type of application, the operator must be able to perform a large number of tasks correctly and in the proper order. In addition, the operator must be able to apply problem-solving skills when troubleshooting operational malfunctions. These malfunctions may occur with such speed that it is impossible for the operator to consult with a technical manual before irreparable damage occurs. Thus, the operator must also have sufficient on-hand knowledge and skills to properly and rapidly solve the problem.

12.2.2 An Orientation of the Management Process or Equipment

It will be necessary for the practitioner to become oriented to the system, process, or equipment. Although it may not be necessary for the practitioner to be a domain expert, he or she will have to have ready and repeatable access to domain experts. It is the skills, knowledge, and attitudes that the practitioner is attempting to incorporate in the simulator and develop in the simulator users. The practitioner may also elect to utilize similar types of orientation visits as with projects involving traditional simulation models:

- Initial orientation visit
- Detailed orientation visit
- Review orientation visit

During the initial orientation visit, the practitioner will attempt to gain a high-level understanding of the management process or equipment. The actual activities during the detailed orientation visits will differ somewhat depending on whether the training simulator is management process or equipment oriented. If the subject is a management process, it is more likely that the detailed orientation visits will consist of interviews with the expert domain managers. On the other hand, if the subject is equipment oriented, the practitioner needs to be introduced to the basic functioning of the equipment by a knowledgeable engineer or equipment operator. The review orientation visit is used to help insure that the practitioner's understanding of the management process or equipment is accurate.

If, for example, the training simulator is for managing the implementation of advanced manufacturing technology, the orientation visits should cover high-level details such as:

- What types of advanced manufacturing technology are to be implemented
- What application is the technology to be used for
- Who is involved in the decision-making processes
- What resources are involved in the implementation process
- How much the resources cost to allocate
- What decisions need to be made with respect to the resources
- How successful the companies are that implement the technology

Similarly, for an equipment simulator designed to train operators or users, the orientation visits might cover high-level details such as:

- What type and dimension raw materials are used
- What operations or processes the equipment performs
- What controls the equipment possesses
- What product is produced by the equipment
- What types of problems may occur in the process

12.2.3 The Establishment of Specific Project Objectives

When the orientation visits are complete, the practitioner should establish specific project objectives. In contrast to traditional simulation model projects, the project objectives for a training simulator project may be somewhat more ambiguous. Examples of training simulator project objectives could include producing a training simulator to provide realistic training on:

- Implementing advanced manufacturing technology projects
- Responding to bomb threats, which might not be possible because of time, cost, or operational limitations
- Performing preoperational checks, operation, and troubleshooting of marine diesel engines
- Setting up and machining a computer numerically controlled turning center

12.3 Project Planning

The project-planning phase is somewhat similar to that of traditional simulation model projects. However, because the steps in the development of a training simulator are different, the work breakdown structure, linear responsibility chart, and Gantt chart will also be somewhat different.

12.3.1 Work Breakdown Structure for Simulator Project

The following table illustrates a two-level WBS modified for a simulator project.

WBS Activity

1.0 Problem Formulation
    A. 1.1 Orientation
    B. 1.2 Problem Statement
    C. 1.3 Objectives
2.0 Project Planning
    A. 2.1 Work Breakdown Structure
    B. 2.2 Linear Responsibility Chart
    C. 2.3 Gantt Chart
3.0 System Definition
    A. 3.1 Identify System Components to Model
    B. 3.2 Identify Input and Output Variables

4.0　Input Data Collection
  A.　4.1　Collect Data
  B.　4.2　Analyze Data
  C.　4.3　Edit Data
5.0　Model Translation
  A.　5.1　Select Modeling Language
  B.　5.2　High-Level Flow Chart of Model
  C.　5.3　Develop Model in Selected Language
6.0　Verification
  A.　6.1　Debug
7.0　Validation
  A.　7.1　Face Validity
  B.　7.2　Training Validity
8.0　Implementation
  A.　8.1　Setup Disks
  B.　8.2　Distribution
  C.　8.3　Installation
  D.　8.4　Support

12.3.2　Linear Responsibility Chart for Simulator Project

Similarly, the corresponding LRC for a simulator project could be organized as follows:

WBS	Activity	Eng. Mgr.	Analyst 1	Analyst 2	Analyst 3
1.0	Problem formulation				
1.1	1.1 Orientation	A	P		
	1.2 Problem statement	A	P		
	1.3 Objectives	A	P		
2.0	Project Planning				
	2.1 WBS	A	P	S	
	2.2 Linear Resp. Chart	A	P	S	
	2.3 Gantt Chart	A	P	S	
3.0	System Definition				
	3.1 ID System Comp.	A	P	S	
	3.2 ID I/O Vars.	A	P	S	
4.0	Input Data Collection				
	4.1 Collect Data	A	P	S	W
	4.2 Analyze Data	A	S	P	W
	4.3 Edit Data	A	S	P	W
5.0	Model Translation				
	5.1 Select Language	A	P		
	5.2 High Level Flow Chart	A	P		
	5.3 Develop Model	A	P	S	P
6.0	Verification				
	6.1 Debug	A	W	S	P
7.0	Validation				
	7.1 Face Validity	A	P	S	
	7.2 Training Validity	A	P	S	
8.0	Implementation				
	8.1 Publishing	A	P	S	
	8.2 Distribution		P	S	
	8.3 Installation	A	P	S	W
	8.4 Support		P	S	

Note: P = primary responsibility; S = secondary responsibility; W = worker; A = approval; R = review.

12.3.3 Gantt Chart for Simulator Project

The corresponding Gantt chart for a simulator project could be as follows:

WBS	Activity	1	2	3	4	5	6	7	8	9	10	11	12
1.0	Problem formulation												
1.1	Orientation												
1.2	Prob. stat.												
1.3	Objectives												
2.0	Project planning												
2.1	WBS												
2.2	LRC												
2.3	Gantt chart												
3.0	System definition												
3.1	Id. comp. to model												
3.2	Id. I/O variables												
4.0	Input data collection												
4.1	Collect data												
4.2	Analyze data												
4.3	Edit data												
5.0	Model translation												
5.1	Select language												
5.2	Flow chart of model												
5.3	Develop model												
6.0	Verification												
6.1	Debug												
7.0	Validation												
7.1	Face validity												
7.2	Statistical validity												
8.0	Implementation												
8.1	Publishing												
8.2	Distribution												
8.3	Installation												
8.4	Support												

12.4 System Definition

The system definition phase of the simulator model consists of identification of the following:

- What system components to model
- What input data to collect
- What output data to generate with the model

12.4.1 Identification of System Components to Model

The specific system components to model will vary depending on both the general type of training simulator and the specific application itself. In particular, the needs for management-type training simulators and equipment-type training simulators are quite diverse.

12.4.1.1 Management Training Simulators

For management training simulators we would consider modeling:

- The input decisions that can be made by the user
- The mathematical model of the process that translates inputs into outputs
- The output resulting from the input decisions made by the user

12.4.1.2 Equipment-Type Simulators

For equipment-type simulators the practitioner would need to model:

- The equipment itself
- Normal operation of the equipment
- Abnormal operation of the equipment

Regardless of whether the simulator is based on management decision or equipment, it will be necessary to program the simulator components through the following operating modes:

- Instructional mode
- Training mode
- Test mode

The instructional mode is intended to provide the user with all of the necessary information to understand the system modeled in the simulator. The instructional mode can consist of a set of either static or interactive screens. If static screens are employed, it may be possible for the practitioner to recycle existing instructional material. A menu system or a set of buttons can be used to navigate through the instructional material pages. Figure 12.1 is an example of a static instructional screen.

Interactive screens may also be used, particularly with equipment-based training simulators. With equipment-based simulators, the practitioner can present a graphic image of different sides of the equipment. The images should be programmed so that when the cursor is run over the different components, the component names and functions will appear on the simulator screen.

For example, with a marine diesel simulator, the user will need to know the location of the following components:

- Raw water pump
- Primary fuel filter
- Secondary fuel filter
- Oil filter
- Coolant tank

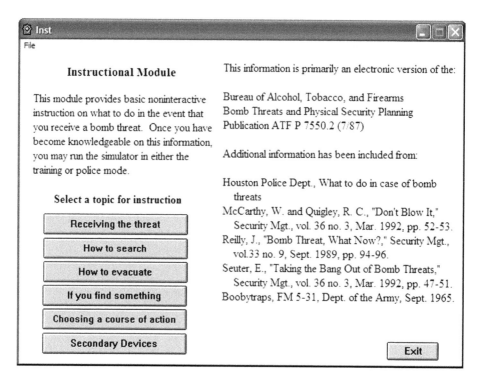

FIGURE 12.1 Instructional screen.

These components are located on both the front and the side of the engine. This means that the practitioner will need to have graphic images of both of these views. The user would first have to navigate to the correct view (Figure 12.2). Suppose we are interested in the raw water pump. When the cursor is run over the graphic image of the raw water pump, the simulator can be programmed to change the shape of the cursor. As this occurs, a small window can appear identifying the highlighted component as the water pump. The simulator can also be programmed to inform the user that the raw water pump pulls water from the sea to cool both the heat exchanger and the exhaust gasses.

The second operating mode is the training mode. This mode is the primary operating mode of the simulator. In this mode, the practitioner will attempt to replicate the actual functioning of the management process or equipment. For management process simulators, the training mode will need to accept input decisions made by the user. The simulator will then need to take the input decisions and run them through the mathematical simulator model. The mathematical simulator model will then display the output results of the user's input decisions.

For equipment simulators, the training mode will accept user input in the form of manipulating the operating controls. Users should be able to power the equipment, operate the equipment, and power down the equipment. The training mode should also include the capability to generate randomly different types of malfunctions that may occur with the subject equipment. The simulator will monitor the user's actions and determine whether or not the actions are appropriate. In Figure 12.3, the engine oil light has come on. The user can choose what actions to take.

The last mode, the test mode, can be used to present textual or graphic questions to the user. In the case of textual questions, the user should be presented with a single question and multiple answer responses (Figure 12.4). After a response is given for the current question, the following question can be presented. The textual questions for the test mode can be the same questions that are used for the test instrument that is discussed in the validation section of this chapter.

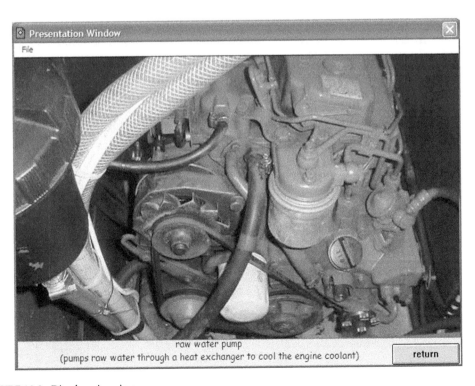

FIGURE 12.2 Diesel engine view.

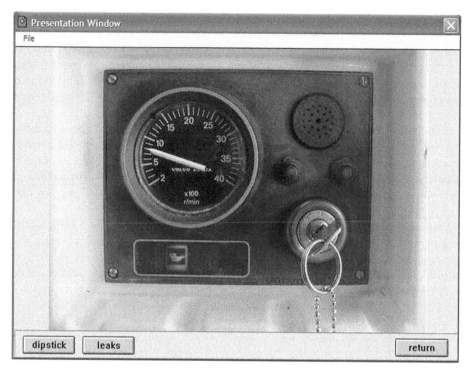

FIGURE 12.3   Diesel engine insrument panel view.

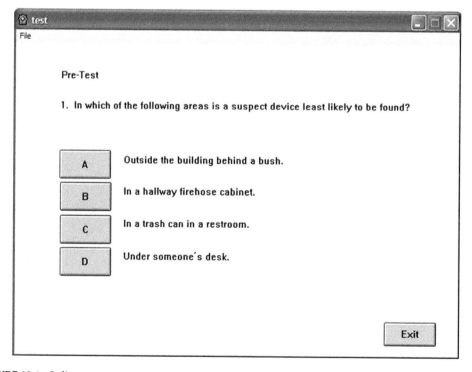

FIGURE 12.4   Online test.

The graphic version of the test mode can be designed to present the name of a particular component of the equipment. The practitioner can also reinforce the identification of the different components by testing the user. The simulator can be programmed so that the names of different components are generated randomly. The user must then place the cursor over the selected component. If the user correctly identifies the component, the simulator will provide positive feedback. Conversely, if the user places the cursor over the wrong component, the simulator will provide negative feedback.

12.4.2 What Input Data to Collect

The type of training simulator that is being developed will dictate what type of input data is needed to be collected. For management training-type simulators, the practitioner will need to collect data on:

- Different types of scenarios that may be encountered
- Different types of responses to the different types of scenarios
- Different types of outcomes resulting from the different types of responses

A management training simulator must be able to recreate most if not all of the scenarios that the user may encounter in real life. This means that the practitioner must be able to include historical data and other parameters that may be relevant to each particular scenario. Figure 12.5 illustrates the presentation of historical data at a scenario briefing.

The practitioner must also collect data concerning the possible responses that are available to users in each of the various training scenarios. For a management scenario, the user will probably be making decisions concerning the allocation of resources to respond to the scenario. The management decisions that are made in one particular scenario may be completely different from those of another, but very similar, scenario. Similarly, the management decisions that the practitioner may observe can be very similar to each other even though they may be in response to two entirely different scenarios.

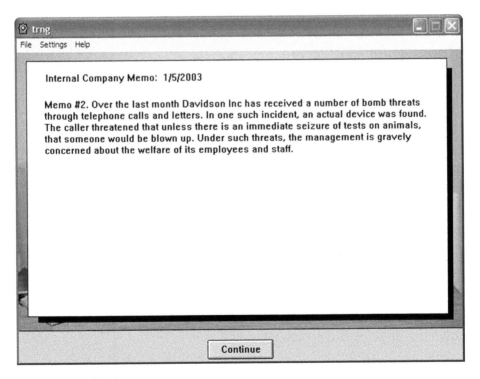

FIGURE 12.5 Scenario briefing.

Last, the practitioner must collect data on the effectiveness of different management decisions with respect to individual scenarios. For a given scenario, one set of management decisions may be very effective, while another set of management decisions may be very ineffective. Similarly, the same course of action may be very effective in one scenario but very ineffective in another because of differences in the individual scenario parameters.

For operational training-type simulators, the practitioner will need to collect data that represent the equipment in different functional states. This may include:

- Graphic images of the static component of the equipment
- Graphic images of the dynamic components of the equipment
- Equipment sounds during normal operation
- Equipment sounds during abnormal operation

12.4.3 What Output Data to Generate with the Model

The type of output data to generate also depends on the type of training simulator being developed. For management training-type simulators this may include:

- The utility or success of the user's management decisions
- The speed with which the user makes decisions
- The resources that the user consumes in making the decisions
- The consequences of the user's decisions

For operational training-type simulators, this may include:

- How well the user can identify the equipment components or controls
- How correctly the user sets up or operates the equipment
- How long it takes the user to set up or operate the equipment
- How well the operator can respond to different malfunctions

12.5 Input Data Collection

The input data collection process for training simulators differs significantly from that of traditional simulation models. The primary difference is that the type of data that are collected also involves graphic images, movie clips, and sound files. Below is a discussion of:

- Collecting data
- Analyzing data
- Editing data

12.5.1 Collecting Data

As previously stated, the development of a multimedia training simulator will require that the practitioner collect mathematical, graphic, movie, and sound-type data. Each of these different types of data presents its own data collection issues.

12.5.1.1 Mathematical Data

Mathematical data pertain to creating the mathematical model on which the simulator is based. This type of approach is generally used only for management simulators. Depending on the exact subject material, the data can be obtained through:

- Assumption
- Expert opinion
- Focus group

- Survey
- Observation
- Experimentation

Of these methods, the survey approach can most effectively be used to obtain numeric data for use in a management training simulator. The general idea is to obtain a set of independent input decision variables that are thought to affect the dependent output performance variables. The data for both the independent and dependent variables can be obtained by administering a Likert-style agree/disagree survey. The results of the survey are analyzed as discussed in the input data analysis section of this chapter.

12.5.1.2 Graphic Data

The practitioner will generally have to use vector and bitmap graphic files in the training simulator. Vector files are mathematical representations of the graphic. Typical vector files are graphics generated by a computer-aided drawing package such as AutoCAD or a presentation package such as Microsoft PowerPoint. Because vector graphics are mathematical representations, the file sizes are relatively small. Bitmap graphic files, on the other hand, are pixel-by-pixel representations of the graphic. Theses types of graphics are typically generated in a paint or graphics program, captured from a computer screen window, or acquired with a digital camera. As a result of pixel-by-pixel representation, these files are generally large in size. Because the majority of the graphic data used in a training simulator will be of the photographic bitmap type, we will concentrate on the issues associated with collecting this type of graphic data.

There are two common types of bitmap formats that the practitioner should be familiar with. These are:

- BMP
- JPG

BMP stands for bitmap. The BMP file format maintains the integrity of the original image. However, the integrity comes at the cost of extremely large file sizes. Bitmaps do have the advantage of being readily editable by a wide variety of image or graphics programs.

In contrast, the JPG or Joint Photograph Experts Group format involves a lossy algorithm to reduce the size of the file. As a result, JPG files are much smaller than BMP files. However, sometimes BMP files are not completely accurately translated into JPG files. This is most notable when a BMP file is translated to a JPG file and then back to a BMP file. You may notice that some of the colors have been altered. Once the BMP file has been translated to JPG, there is no way to recover the lost data. For most applications, this minor loss of detail is of little consequence. The smaller size of JPG files versus BMP bitmap files, however, is a major issue. This enables the simulator program to be smaller and run faster.

A second issue with bitmap graphics is the bit color depth. The bit color depth determines how many different colors the file may display at one time. Two different levels are of interest to the practitioner:

- 24-bit color
- 16-bit color

Twenty-four-bit color corresponds to 16.7 million colors. This is more colors than the human eye can detect. This means that any file with 24-bit color will appear photorealistic. This is the effect that is most desirable with a training simulator. You want to keep the graphics as photorealistic as possible to avoid cartooning or distorting the graphical images. The downside to 24-bit color is that large images such as full screen captures can require significant space.

It is possible, but generally undesirable, to use 16-bit color in some cases. A 16-bit color depth results in 65,000 colors. In some applications, the reduction in colors does not necessarily degrade the images, and 16-bit color does save a significant amount of image space. Some older computers have display cards capable of displaying only 16-bit color. If this is expected, it may be advantageous to use only 16-bit color images.

To collect bitmap graphic data, the practitioner will have to acquire either a digital camera and/or a scanner. It is best to acquire the type of digital camera that uses either floppy disks or CD-ROMs to record the images. This will allow the practitioner to take a greater number of pictures than would be possible with memory module-type cameras. The practitioner should also ensure that the camera has the ability to save the images in 24-bit color JPG format files. When collecting graphic data with a digital camera, the practitioner is advised to take a large number of photographs for every individual image that is required. The need for multiple images is apparent when it is necessary to obtain an image that is perfectly level and square to the camera. Slight variations in positioning or lighting can produce images with significant difference in quality. The use of a tripod for the camera may help in obtaining high-quality images. Because it is usually more difficult to obtain access to the actual system or equipment for multiple visits, it makes practical sense to attempt to acquire as many images as possible in as few sessions as possible.

In the event that the simulator will need to use a large number of previously existing hardcopy images, it may be beneficial to acquire a digital scanner. This device is used to convert the images into electronic format, which may be edited. Scanners are usually bundled with image-to-text conversion or optical character recognition (OCR) software. This software allows the practitioner to scan hardcopy text and convert it into an electronic file. During this process some minor conversion errors may occur. Spelling errors can easily be corrected by running a spell check on the text file. The electronic text file can then be easily imported in the simulator for instructional purposes.

12.5.1.3 Movie Data

Live motion movie data add to the quality of the training simulator. This is particularly important when training someone on a process that requires a sequence of actions. The movie allows the practitioner to demonstrate repeatedly how to perform the operation.

Significant movie clips are beyond the current capability of most training simulators. However, short movie clips in conjunction with other interactive multimedia can be very effective. The most usable format for training simulators is the Motion Pictures Expert Group or MPEG format. Many high-end digital cameras have the capability to record limited length MPEG movie files. Even 15 to 60 s of movie data can be quite useful. It is also possible to string several MPEG movie files together for longer sequences.

The use of MPEG files introduces a few performance and installation issues. First, the practitioner must decide what size window to use to record the movie. The smaller the window, the smoother the playback. However, the window must be large enough to communicate the message that the practitioner is attempting to send. Two common sizes are 320 × 240 pixels and 160 × 112 pixels. The practitioner must also be aware of the fact that the end user must have the appropriate software drivers installed to use MPEG files. It is possible for the practitioner to insert an error routine in the simulator that will instruct the user either to install the drivers or to bypass the use of the MPEG files altogether.

MPEG files also have the capability of simultaneously recording sound. This means that the practitioner can not only visually demonstrate processes but explain them at the same time. This combination can make the use of MPEG files a very effective training tool. Unfortunately, it is relatively difficult to edit MPEG files. This means that several takes may be necessary before the desired visual and audio clip is obtained.

12.5.1.4 Sound Data

Sound data can be used for instructing simulator users and for providing feedback for the users' actions. There are two different sound formats that are normally used in training simulators. These are:

- MIDI
- WAV

Musical instrument digital interface or MIDI files are an electronic instrumental representation of sound. This type of file is generally used in training simulators for introduction and background music. MIDI files have the advantage of being very compact; however, MIDI files cannot be used for verbal messages.

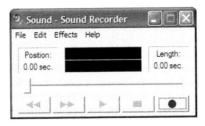

FIGURE 12.6 Sound recorder window.

The creation of MIDI files is beyond the capability of the average engineering or operations research analyst. As a result, MIDI files generally originate from sources such as the Internet or shareware files.

Wave files are a digital representation of sound. In training simulators, WAV files are used for providing:

- Verbal instructions
- Feedback
- Sound effects

Most notebook computers have the ability to record sounds directly into WAV files using the Windows accessory sound recorder and a microphone. Figure 12.6 illustrates the sound recorder accessory.

Synthesized verbal instructions may also be used in training simulators. These types of programs use ASCII text to generate the WAV files. The use of synthesized verbal instructions can actually be advantageous in comparison to human speech. This is because it can be very difficult to record acceptable WAV files. During this activity, both the practitioner and the subject must be present to record the files. The computer never becomes impatient at having to record a large number of messages or at having to record the same message a large number of times. Another advantage to the use of computerized speech is that it is completely politically neutral. The voices can also be changed between male and female and different tenors.

The use of WAV files requires that the practitioner make several decisions with respect to the specifications of the files. These include:

- Format
- Frequency
- Bits
- Mode

The WAV format specifies the algorithm that is used to represent the sound. There are actually a number of different WAV formats available. Common formats include PCM, ADPCM, and GSM. The differences among these WAV formats would be of little interest except for the fact that the practitioner may encounter compatibility issues with different software programs. For this reason, it is best to stay with an established format such as pulse code modulation or PCM.

The frequency of the file is an important issue with respect to sound quality. The higher the sampling frequency, the better or less distorted will be the sound. Common frequencies are 44 kHz, 22 kHz, 11 kHz, and 8 kHz. For example, 44 kHz corresponds to CD quality sound. The downside to higher frequencies is the amount of disk space required by the file for a given number of seconds. If the file is large, playback performance may also be affected.

The number of bits associated with a WAV file corresponds to the potential accuracy of the hardware and software to playback the sound. Bits can be 8, 16, 32, 64, or even more. The higher the number of bits, the more accurate the sound; 16 bits corresponds to CD quality sound. The number of bits can present a compatibility issue with some sound cards. Higher-level bit sound cards can play lower-level bit sounds. The opposite is not necessarily true. There is no way to control the type of sound card or the installation setup in a user's computer. The file size issue is also a problem with higher-level bit sounds. The higher the level bit, the greater the space required for a given sound duration. These issues mean that the practitioner should probably not record WAV files with greater than 16 bit sound.

A final WAV file issue is whether to record in mono or stereo. For the practitioner this has little significance. Recording in stereo does require more file space. This means that it is better for the practitioner to record in mono for most training simulators.

Because we are likely to have a large number of WAV files in a training simulator, we will be interested in saving disk space. In addition, we are interested in keeping our files small to enhance the program performance. We also want the files to be as compatible as possible with the potential users' computers. For these reasons, it is best to use PCM either 8 or 16 bits in mono with a sampling rate of 11 or 22 kHz. This provides acceptable sound quality and loading performance for training purposes while conserving disk space. You can determine the current parameter values of a sound file by executing the file, property menu sequence in the Sound Recorder. Figure 12.7 illustrates this Sound Recorder window.

It is also possible to change the current format of a WAV file to a more appropriate format using the sound recorder. This can be performed by pressing the Convert Now button in the properties for sound window. The sound selection window that appears is illustrated in Figure 12.8.

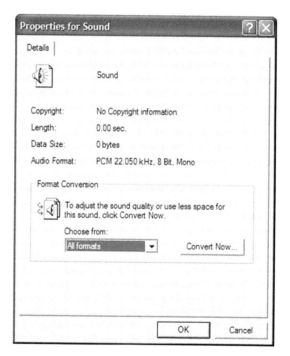

FIGURE 12.7 Properties for sound window.

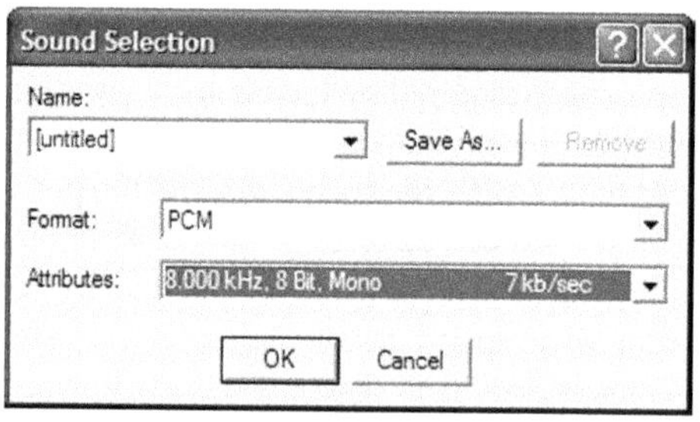

FIGURE 12.8 Sound selection window.

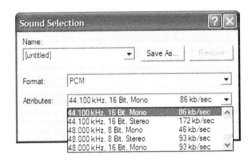

FIGURE 12.9 File attributes drop-down list.

The desired format can be selected by dropping down the attributes box and scrolling through the different options. Note in Figure 12.9 that the Sound Recorder also displays the file space required for that format option.

12.5.2 Analyzing Input Data

In management decision-training simulators, the practitioner may have to perform both statistical analysis of the input data and suitability analysis of the image, movie, and sound files.

12.5.2.1 Analyzing Input Survey Data

If the practitioner is developing a management process-training simulator, input data may have been obtained through the use of a qualitative survey. If this approach is taken, the data will have to be analyzed with some statistical technique such as multiple linear regression. With multiple linear regression, the basic concept is to develop a set of equations that will predict the output performance of the system based on the input decisions that are made by the user. The use of multiple linear regression in a management-training simulator is discussed in the model translation section of this chapter. For a more in-depth discussion of multiple linear regression, practitioners are directed to the references at the end of this chapter.

12.5.2.2 Suitability Analysis

The suitability analysis will primarily be concerned with identifying the best images, movies, and sound files for the training simulator. This is actually a very time-consuming process, as the practitioner might have to view and select among tens or even hundreds of images in the case of graphic data before a suitable image is found. The selection of a suitable image means that minimal editing will be required in order to use the image in the training simulator. Images that are most suitable are:

- Equipment images with horizontal edges parallel to the horizontal image frame
- Equipment images with vertical edges parallel to the vertical image frame
- Equipment images centered with respect to the vertical axis of the image frame
- Equipment images centered with respect to the horizontal axis of the image frame
- Images without distracting reflections

These requirements basically mean that the objects are square with respect to the image frame. This can be performed only if the position of the camera is centered on the center of mass of the equipment that is being photographed. Careful positioning in this manner will prevent keystoning or perspective errors, which are difficult or impossible to remove by editing.

For surfaces with reflections, it may be necessary to photograph the item from underneath a dark cover or blanket. This will prevent the image of the picture taker from appearing on the image. In other situations, it may be necessary to erect a temporary screen to eliminate unwanted reflections.

12.5.3 Editing Input Data

It will probably be necessary for the practitioner to edit both the graphic and sound data to achieve the desired effects. Because of the different nature of these types of data, we will examine them individually.

12.5.3.1 Editing Graphic Images

The practitioner may want to edit graphic images for a variety of reasons. Common reasons include:

- The contrast, brightness, or size needs to be adjusted.
- Some defect exists in the image.
- A single object in the image is needed.
- The background in the image is needed.

It is highly unlikely that the original images that were photographed will be completely acceptable in terms of contrast or brightness. Contrast or brightness adjustments can be made using image-editing tools such as Microsoft Photo Editor. Photo Editor can also be used to cut down the original image to a size more appropriate for use in the simulator.

It is possible that an otherwise usable image contains some sort of defect. It may not be possible to edit out the defect with an image-level editing tool such as the Photo Editor. In this case, it may be necessary to utilize a bitmap-level editor such as Microsoft Paint. These programs allow the practitioner to remove or change the colors of individual pixels in the image. A minor disadvantage to using Paint is that the image must be in BMP format before the file is opened. This means that Photo Editor may be needed to convert the original JPG file to a BMP file for editing. Once the editing is complete, the Photo Editor can be used to convert the edited BMP file into the finished JPG file.

A very common need for graphics editing is to capture the image of a specific object in a picture into a different file. For example, we may need the image of a specific engine component that is located in the engine compartment of a car. The original image will be the entire engine compartment. This image will need to be edited, so that we end up with the image of the single component. Suppose we need an image of the battery installed in the engine compartment. A logical sequence of steps to acquire this item is as follows:

- Take a digital photo of the engine compartment in JPG format.
- Open the file in Photo Editor and cut out the region that contains the battery.
- Paste the battery region as a new image in Photo Editor.
- Save the battery region image as a BMP file.
- Open the battery region image in Paint.
- Use Paint to remove the excess pixels that are not part of the battery.
- Save the edited picture as a BMP file in Paint.
- Open the edited picture BMP file in Photo Editor.
- Save the edited picture as a JPG file in Photo Editor.

The longest part of the process will be removing the excess pixels that are not part of the desired image. This can be slow, difficult work. If the image is not geometrically solid, the editing process can be grueling. An example of this is if an image of a plant is needed. The areas between the leaves must be removed from the plant in order for the plant image to be used properly.

It may be possible to use a more sophisticated graphics package than either Photo Editor or Paint. This will enable the practitioner to perform all the editing and saving needs in one program. Corel Paint is one example of this type of software package. Of course, the additional expense of purchasing a high-end graphics package may make Photo Editor and Paint appear attractive.

12.5.3.2 Editing Sound Files

The practitioner may also need to edit sound files. Editing is usually needed when the sound file involves human speech instruction. When the file is recorded, the sound may have any one of the following problems:

- Too soft
- Too loud
- Starting delay too long
- Ending delay too long

The practitioner can use the Microsoft Sound Recorder for doing minor editing jobs. It is an easy matter to attempt either to boost or reduce the volume of the WAV file. In addition, the practitioner can easily delete the opening or closing dead sound spaces before and after the usable text.

The practitioner may also be interested in increasing the length of the sound file when the original file is not of sufficient length. For example, the practitioner may obtain a short WAV file of a diesel engine running. In the simulator, it may be necessary for the sound to be of greater length. In order to increase the length of the sound, the practitioner can use the sound recorder to copy and paste the same original segment of sound several times until the desired length is obtained.

Special effects may also be needed for sounds in the simulator. The sound recorder allows the user to add an echo to the original file as well as to reverse the entire file altogether. Though these and other special effects are available, the practitioner should resist the use of inappropriate and insignificant distractions during the development of the simulator.

12.6 Model Translation

The model translation phase for training simulators consists of:

- Selecting the modeling language
- High-level flow chart of model
- Develop model in selected language

12.6.1 Selecting the Modeling Language

As with traditional simulation models, the practitioner has the choice of using either a high-level programming language or a specialized software package. Specialized software packages that are used for developing training simulators are known as authoring software. Authoring software is specifically designed for facilitating the use of interactive sound, music, movie, and graphic components. Because of these sophisticated multimedia capabilities, authoring software packages are generally preferred over general purpose programming languages. However, this does not preclude the possibility of simulator development with a graphically based programming language such as Microsoft's Visual Basic.

The most popular authoring software package is currently Macromedia's Authorware development platform. This particular authoring package utilizes icons to represent graphics, sound, movies, and interactions such as buttons. By assembling different icons on the program's flowline, the practitioner can control how the program operates.

12.6.2 High-Level Flow Chart of the Simulator Model

To facilitate the model development process, it is necessary to create a high-level flow chart of the simulator model. The same flow chart symbols that are used for developing a high-level flow chart for traditional simulator model are also used for developing a high-level flow chart for a training simulator. As you may recall, this includes:

- Start oval
- Process rectangle
- Decision diamond
- Input/output tilted parallelogram
- Stop oval

A major difference between simulators and traditional simulation models is the need to account for real-time user interaction. This means that the practitioner must take into account the different possible actions that the user may take with the actual system. Figure 12.10 illustrates a high-level flow chart for a training simulator.

This particular high-level flow chart is for a simulator designed to train users to respond to bomb threats. We will use this flowchart to illustrate how user decisions can affect the flow of the simulator. When the program starts, a credits screen is first illustrated. When the credits screen is cleared, the user has the opportunity to run the simulator in the instructional, training, or police mode. Note how the mode selection is represented by a series of decision diamonds. If a particular mode is not chosen, the flow chart moves to the next mode.

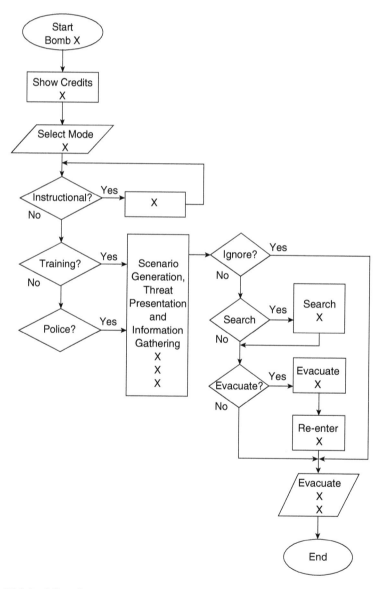

FIGURE 12.10 High level flow chart.

If the instructional mode is selected, the program will run the inst.exe executable program. This covers basic bomb threat search and evacuation procedures. It is designed to be used by individuals who have little or no knowledge of bomb threat response procedures. When the user is sufficiently knowledgeable on these procedures, he or she can proceed to either the training or police simulator components.

Experienced users may select either the training or police mode. The primary difference between these two modes is the target user. The training mode is for civilians, whereas the police mode is for law enforcement officers. If the training mode is selected, the program will run the executable trng.exe. This executable component generates a bomb threat scenario, provides a scenario briefing to the user, presents the threat to the user, and allows the user to gather information from the caller. Once the caller hangs up, the user can decide whether to ignore the bomb threat, search the facility for the bomb, or immediately evacuate. These user choices are also represented by a series of decision diamonds.

If the user decides to ignore the threat, the program will go directly to the evaluation subroutine in trng.exe. On the other hand, if the user elects to search the facility, the program will run the executable sofc.exe. The executable sofc.exe is modeled as a process. If the user elects to evacuate the facility immediately, the program will run the module eofc.exe, which is also modeled as a process. Note that if the user did elect to search the facility first, the facility can also be evacuated by running the evac.exe module. This is modeled in the flow line by having the search process positioned before the evacuate process. Thus, immediately after searching, the user can decide whether or not to evacuate. If the facility is evacuated, the user will have to decide when to reenter the facility. At the end of the simulation run, the program will return to trng.exe, where the user's search and evacuation decisions will be evaluated. After receiving the evaluation feedback, the simulator run ends.

Note that this flow chart represents the operation of the simulator at a very high level. The practitioner might also find it useful to create a flow chart for each of the individual sofc.exe, eofc.exe, and rofc.exe processes.

12.6.3 Develop Model in Selected Language

During this component of the model translation phase, the practitioner will be programming the high-level flow chart in the selected authoring language. Although the actually programming of the simulator is an application-dependent process, some general guidelines can be provided for the practitioner. A number of particularly important guidelines involve the following subjects:

- Operating modes
- Help screens
- Subdirectory organization

12.6.3.1 Operating Modes

As previously discussed, all training simulators should have the capability to operate in multiple modes. Strong consideration should be given to programming some form of each of the following modes:

- Instructional mode
- Training mode
- Test mode

12.6.3.2 Credit Screens

The practitioner should insure that the simulator has an appropriate number of credit screens to recognize all of the individuals involved in the process. The credits screen should also include contact information for acquiring the simulator or receiving technical support. The credits screen will normally appear only once, when the program is first run.

12.6.3.3 Help Screens

Because the simulator will be used by other individuals, it may be necessary to include a variety of help screens. These screens should be of two different types:

- Orientation
- Operational help

The orientation screen is needed to provide the users with basic information about the purpose and capabilities of the program. In addition, it should instruct the user on how to obtain additional program help. This screen should be presented the first time that the program runs. The orientation screen should also be available at any time during the simulation run though the program menu.

Operational help screens also need to be available to the user throughout the program. It is best to provide the help screens in a context-sensitive manner. This means that the program will display help screens appropriate for the part of the program that is currently running. By including only context-sensitive help, the practitioner can reduce the number of help screens that are available at any given time. This also helps reduce the possibility of the user misunderstanding the help instructions in various sections of the program.

12.6.3.4 Menu System

It is essential that the practitioner include some sort of menu system in the simulator. This menu system should be available regardless of the mode or section of the simulator the user is currently using.

12.7 Verification

The verification process for training simulators consists of the same general processes as for traditional simulation models. As you may recall, verification is the process of insuring that the model operates as intended. This phase generally consists of debugging. As with traditional simulator programs, the debugging process can be facilitated by:

- Using variable displays in the simulator screens
- Stepping through the program
- Writing data to an external file

12.8 Validation

The validation process for training simulators involves the process of ensuring that the simulator model represents reality to a sufficient level to have a training effect. The validation process for training simulators is similar but substantially more complex than the validation process for traditional simulation models. The validation process includes achieving:

- Face validity
- Training validity

12.8.1 Face Validity

Face validity means that the simulator on the surface appears to represent the training simulator subject matter and would be useful for training purposes. The face validation process for training simulators requires the use of domain experts. These are individuals who have a high degree of knowledge of the training simulator subject material. During the development phases, the domain experts should regularly be solicited for suggestions to improve the reality and usability of the training simulator. By the time the practitioner is ready to undertake the training validity process, the domain experts should be satisfied with the operation of the training simulator.

12.8.2 Training Validity

Training validity means that the simulator has the ability to increase the knowledge of the user in the subject area. Assessing training validity involves the following steps:

- Development of a test instrument
- Administering the test instrument
- Analyzing the test instrument results

12.8.3 Development of a Test Instrument

A test instrument is essential to determine whether or not the user has increased his level of knowledge as a result of using the training simulator. If the subject material is well developed, it is possible that some sort of subject test is already in existence. However, if no test currently exists, or the reliability of the existing test is suspect, the practitioner will need to develop a new test.

When developing a new test, the practitioner must insure that the test is statistically reliable and has the ability to discriminate between levels of performance. Reliability means that if the same test is taken with the same level of knowledge, the test scores will be consistent. The ability to discriminate between levels of performance means that individuals with less knowledge must consistently score lower than individuals with higher levels of knowledge. The ability of the test to discriminate in this manner is frequently referred to as known group validity.

12.8.3.1 Test Format

The simplest format for developing a training validity test instrument involves multiple-choice answers. This format involves asking a series of questions on the subject matter and providing a limited number of defined responses. For our purposes, it is sufficient to make the exam a total of 20 questions with four possible answers for each question. This will allow a possible range of test scores between 0 and 100% in increments of 5%.

In developing the questions, we will need to take into consideration our method for determining the reliability of the test. One method to obtain a reliability value is to create odd and even halves of the test with equivalent questions. This means that the odd-numbered questions test the user on different elements of the subject matter. The even numbered questions test the user on the same ten different elements of the subject matter. In other words, there is an equivalent odd and even question. If the user can provide a correct response to a given odd question, the user should also be able to provide a correct response to the corresponding even question. Similarly, if the user cannot provide a correct response to a given odd question, he or she should not be able to provide a correct response to the corresponding even question. This means that the score on the odd set questions should be similar to the score on the even set of questions.

The practitioner can develop the even version of each question by either rewording or inverting the original odd question. Significant care must be taken by the practitioner to insure that it is not possible for the test taker to apply a process of deductive reasoning to determine the correct response by merely looking at the odd and even versions of the questions. It is also a good idea to randomize the order of the odd and even versions of the questions. This ensures that the same odd and even halves of the questions are not adjacent to each other. Here is an example of an odd question:

What is the most important question the recipient of a bomb threat should ask the caller?
- a. What does the bomb look like?
- b. Where is the bomb?
- c. When will the bomb go off?
- d. Why are you doing this?

An example of a corresponding even question would be:

What is the least important question the recipient of a bomb threat should ask the caller?
- a. What does the bomb look like?
- b. Where is the bomb?
- c. When will the bomb go off?
- d. Why are you doing this?

This means that if the user has the knowledge needed to answer questions about the bomb threat correctly, he or she will be able to respond correctly to both the odd and even questions. There is really no way to determine the correct answer to the even question either by the way the odd question is worded or by the possible responses to the questions.

12.8.3.2 Administering the Test Instrument to Knowledgeable Individuals for Reliability Testing

Once the initial version of the test instrument has been developed, the practitioner must administer the test to a sufficient number of individuals to determine the reliability of the test. In this process, it is very important that the practitioner select a set of individuals who are assumed to have knowledge of the process. Although this group of individuals will not be expected to answer every question correctly, it will be expected that the equivalent forms of each question will be answered consistently. If a group of unknowledgeable individuals are used for reliability testing, it is likely that the responses will involve a great deal of random guessing. This will prevent the practitioner from properly analyzing the test results to see if the test is reliable.

The larger the number of knowledgeable individuals who are used for the reliability testing the better. In situations with a large number of domain experts, the practitioner should strive to obtain a minimum of 30 individuals. However, in most cases there is usually some practical limitation in terms of the number of available domain experts. In one extreme case, a domain expert stated that only three individuals in the United States possessed suitable knowledge in the subject matter area.

Sometimes it is possible to obtain a large number of knowledgeable individuals at a professional conference. The testing session can be scheduled during an individual conference session. Knowledgeable individuals can be encouraged to participate by offering a complimentary copy of the training simulator on its completion.

12.8.3.3 Reliability Testing

When the practitioner has obtained a suitable number of initial test instrument scores, it is time to determine if the test instrument is reliable. We will use what is known as the split-half method to determine the test reliability. To perform this analysis, the practitioner needs to follow the following steps:

- Score the even and odd halves of each test separately.
- Insert the even and odd half-scores in a worksheet.
- Calculate the correlation coefficient between all of the even and odd halves.
- Calculate the reliability of the overall test.

When the practitioner scores the even and odd halves of the test, the result should be a set of odd half-scores and a set of even half-scores for all of the individuals who took the test. It is important to maintain the identity of each odd and even half-score. This is because we will be analyzing how the odd half-scores relate to the even half-scores for all of the tests.

The practitioner should create a worksheet with three columns. The first column should have an index number starting with 1 and ending with the total number of individuals who took the test. The second column should contain the half-score for the odd half of each individual's test. The third column should contain the half-score for the even half of each individual's test. This is illustrated in Figure 12.11.

Once all of the data are entered in the spreadsheet, the practitioner should run a correlation test between the odd and even halves of the test. The mathematical equation for correlation is:

$$correlation = \sqrt{\frac{\sum (x - \bar{x})(y - \bar{y})}{\sum (xy - \bar{x}\bar{y})}}$$

	A	B	C
10		odd	even
11	1	45	45
12	2	40	35
13	3	30	35
14	4	25	30
15	5	40	40
16	6	45	45
17	7	35	30
18	8	35	40
19	9	40	35
20	10	50	45
21	11	35	40
22	12	40	35
23	13	45	45
24	14	40	45
25	15	35	30
26	16	35	35
27	17	40	45
28	18	40	40
29	19	50	45
30	20	30	30

FIGURE 12.11 Split half reliability scores.

In Excel, this test is accessed by the tools–data analysis–correlation menu sequence. If the data analysis option does not come up under the tools menu, the data analysis option will need to be installed. When this menu sequence is executed, the window shown in Figure 12.12 will appear.

If our data are arranged in columns B and C between rows 11 and 30, we would insert this range in the input range text box. Because we have the words odd and even in row 10, we also need to check the labels in the first row check box. This window would appear as in Figure 12.13.

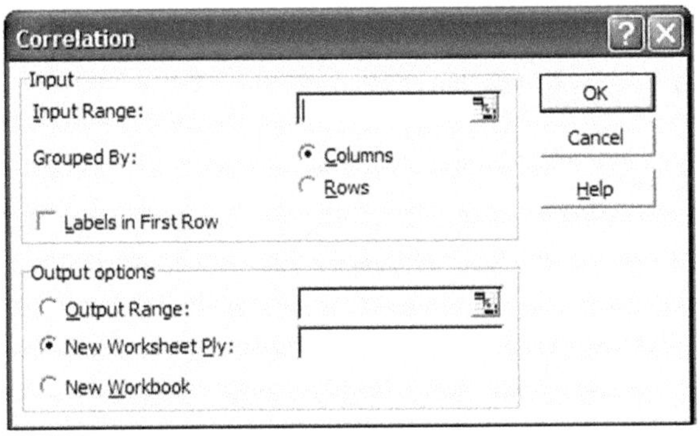

FIGURE 12.12 Excel correlation window.

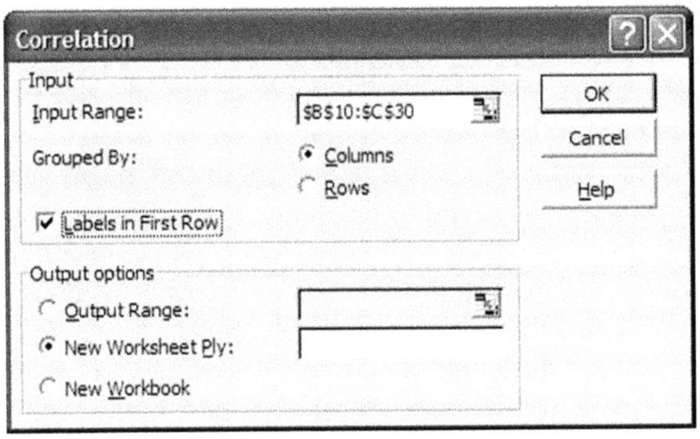

FIGURE 12.13 Completed Excel correlation window.

	A	B	C
1		*odd*	*even*
2	odd	1	
3	even	0.780117	1

FIGURE 12.14 Excel correlation results.

When the OK button is pressed, Excel will calculate the correlation coefficient between the odd and even halves of the test instrument. The results from Excel are illustrated in Figure 12.14.

As the worksheet indicates, the correlation coefficient between the odd and even halves is 0.78. This is actually a favorable correlation. However, what we are really interested in is calculating the test reliability between the odd and even halves. The equation for calculating the overall test reliability is given in the following equation:

$$reliability = \frac{2 \times correlation}{1 + correlation}$$

A frequently referenced acceptable standard for test reliability is 0.70. In our particular example, the test reliability is:

$$reliability = \frac{2 \times 0.78}{1 + 0.78}$$

This is equal to 0.88. Because this easily exceeds the 0.70 standard, we can safely assume that our test instrument is reliable. Once we have established that the test is consistent or reliable, we can move on to the known group validity testing of the test instrument.

12.8.3.4 Known Group Validity Testing

Known group validity testing determines whether or not the test instrument has the capacity to distinguish different levels of knowledge. Even though the test may be reliable, it will be useless for determining training validity if it cannot tell if there has been an increase in the user's level of knowledge. The general concept behind known group validity testing is that a group of individuals who are known to possess subject knowledge should score higher on a valid test than groups of individuals who are expected not to possess subject knowledge. In other words, a test with known group validity will result in higher scores for knowledgeable individuals and lower scores for unknowledgeable individuals.

Fortunately, we already have scores on the test instrument for individuals who are considered knowledgeable through our reliability testing process. We can simply add the odd and even halves of the test to come up with our total score for each individual. We will still have to acquire test scores for our unknowledgeable group. If we have access to the same individuals whom we will ultimately be using for our training validity, we can save ourselves significant effort. These same unknowledgeable individuals can be given the test instrument during the known group validity phase and be compared to the known group. Once we administer the test to the unknown group, we can also enter their scores in the spreadsheet. This is illustrated in Figure 12.15.

We can now use the same statistical comparison of mean techniques as presented in Chapter 8, "Validation." As you will recall, we first need to determine if the data sets are normally distributed. If both data sets are normal, then we can use an independent *t*-test. To determine which independent *t*-test to use, we must perform an *F*-test on the data sets to determine if the variances are equal. If the variances are equal, then we can perform the regular independent *t*-test. However, if the variances are not equal, we will have to perform the Smith–Satterthwaite *t*-test. Conversely, if one data set, or both are not normally distributed, we will have to use a nonparametric rank sum test.

For our comparison test, we will have the following hypotheses:

	A	B	C	D	E
10		odd	even	known	unknown
11	1	45	45	90	50
12	2	40	35	75	45
13	3	30	35	65	70
14	4	25	30	55	65
15	5	40	40	80	60
16	6	45	45	90	70
17	7	35	30	65	50
18	8	35	40	75	65
19	9	40	35	75	75
20	10	50	45	95	40
21	11	35	40	75	35
22	12	40	35	75	60
23	13	45	45	90	55
24	14	40	45	85	50
25	15	35	30	65	35
26	16	35	35	70	60
27	17	40	45	85	40
28	18	40	40	80	35
29	19	50	45	95	40
30	20	30	30	60	55

FIGURE 12.15 Known group validity test scores.

H_o: No difference between the known group scores and the unknown group scores.
H_a: There is a difference between the known group scores and the unknown group scores.

If the H_o is not rejected, then there is no statistically significant difference between the known and unknown groups. In this situation, our test does not have the ability to distinguish between different levels of knowledge. This means that we will have to redesign our test instrument from the beginning and go back through the reliability and known group validity testing.

On the other hand, if the H_o is rejected, it means that there is a statistically significant difference between the known and unknown group test scores. This means that the test does have the ability to distinguish between different levels of knowledge. If this is the case, we are ready to move on.

Consider our example. Assuming that both data sets are normal, we can first perform the F-test. Our hypotheses for the F-test are as follows:

H_o: The variances are equal between the known group and the unknown group
H_a: The variances are not equal between the known group and the unknown group

The F-test is executed by following the tools, data analysis, F-test menu sequence. When this sequence is followed, the window shown in Figure 12.16 appears.

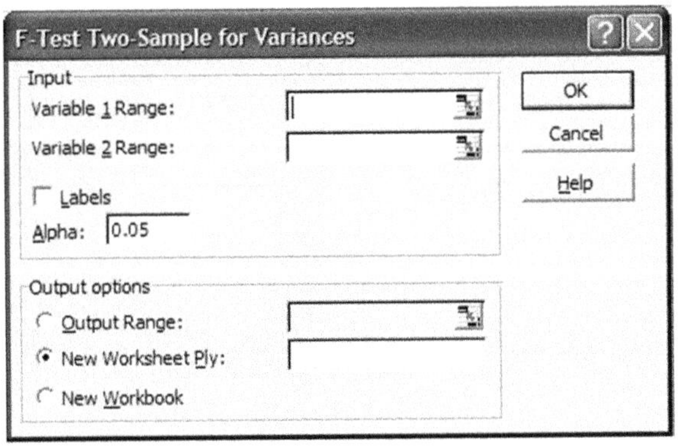

FIGURE 12.16 Excel F-test window.

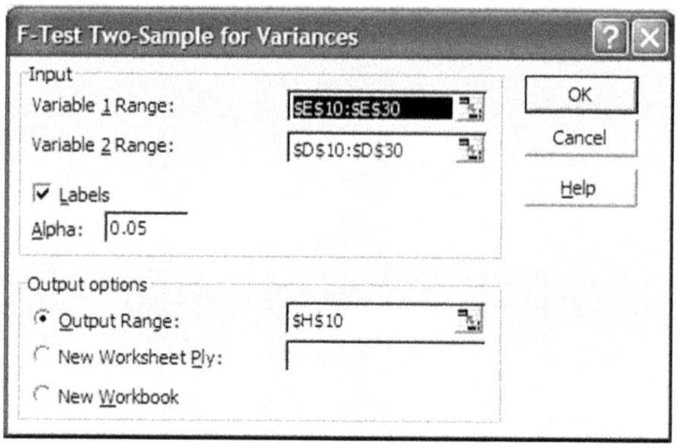

FIGURE 12.17 Completed Excel *F*-test window.

F-Test Two-Sample for Variances

	unknown	known
Mean	52.75	77.25
Variance	161.7763	135.4605
Observations	20	20
df	19	19
F	1.194269	
P(F<=f) one-tail	0.35137	
F Critical one-tail	2.16825	

FIGURE 12.18 Excel *F*-test results.

We can now designate our input ranges in the *F*-test window. It is helpful to put the data range with the higher variance in the input variable 1 range and the data range with the lower variance in the input 2 variable range. This is illustrated in Figure 12.17.

When we execute the *F*-test, Excel provides the results illustrated in Figure 12.18. As the results indicate, the *F*-test statistic is 1.19. The critical *F* value for $\alpha = 0.05$ is 2.17. Because 1.19 is less than 2.17, our null hypotheses of the variances being equal cannot be rejected. Because the variances are assumed to be equal, we can perform a traditional independent *t*-test. Our hypotheses will be:

H_o: The means are equal between the known group and the unknown group.
H_a: The means are not equal between the known group and the unknown group.

This test is executed by following the tools, data analysis, *t*-test: Two-Sample Assuming Equal Variances menu sequence. When this is performed, the window shown in Figure 12.19 appears.

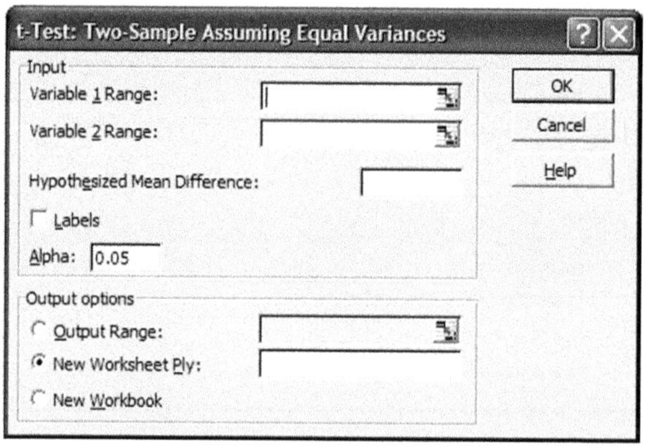

FIGURE 12.19 Excel *t*-test for equal variances window.

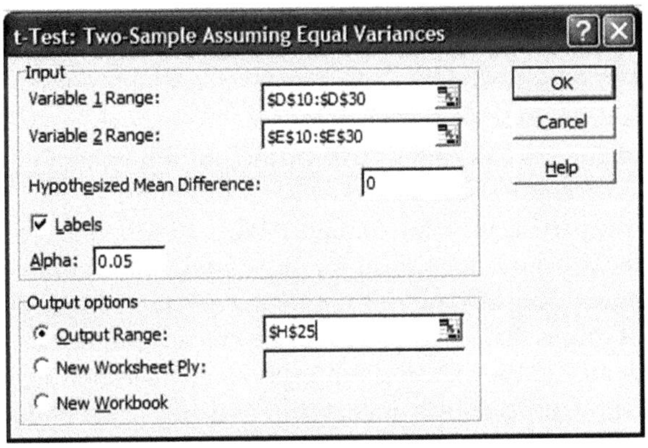

FIGURE 12.20 Completed Excel *t*-test for equal variances window.

t-Test: Two-Sample Assuming Equal Variances

	known	unknown
Mean	77.25	52.75
Variance	135.4605	161.7763
Observations	20	20
Pooled Variance	148.6184	
Hypothesized Mear	0	
df	38	
t Stat	6.355208	
P(T<=t) one-tail	9.28E-08	
t Critical one-tail	1.685953	
P(T<=t) two-tail	1.86E-07	
t Critical two-tail	2.024394	

FIGURE 12.21 Excel *t*-test results.

We can now fill in the input variable ranges with our data. We can use our known group as the variable 1 range and our unknown group as our variable 2 range. Note that we are going to include the row with the data labels. Our hypothesized mean difference is 0. We will now have the screen shown in Figure 12.20.

When we execute the *t*-test, Excel will provide the results in Figure 12.21. As that figure indicates, the test statistic is 6.35. The critical *t* value is 2.02. Because 6.35 is greater than 2.20, the null hypothesis of no difference in the means must be rejected. This means that the test instrument does in fact have the ability to distinguish between the levels of knowledge of the known and unknown groups.

12.8.4 Administering the Test Instrument

Once the test instrument has been determined to be both reliable and known-group valid, we can begin with our testing procedure to determine if the simulator itself has training validity. In general, there are two experimental design methods of determining training validity, depending on the availability of test subjects.

The simpler of the two methods is suitable when the number of available test subjects is limited and the subject matter was not previously presented in other forms. This means that we will be able to determine only the effect of using the training simulator by itself. We will determine the effect through gains in test scores as a result of using the simulator. In this method we will perform the following actions:

- Administer a pretest version of the test instrument.
- Train individuals with the simulator.
- Administer a posttest version of the test instrument.

If the practitioner was careful in selecting the individuals for the known group validity testing, the scores from these same individuals can be used as the basis for the pretest. This means that the practitioner can bypass the first step and go directly to the training on the simulator. If this method is used, it is critical that the practitioner be able to associate the scores for the unknown group data with the scores for the posttest. One method of doing this is assigning a participant control number. For confidentiality purposes, the practitioner does not necessarily need to know the actual identity of each individual. However, the practitioner must know the participant control number. When distributing the posttest, the practitioner will need to instruct each participant to mark down the control number on the posttest.

When the posttest data are compiled, we will analyze the difference in the pretest and posttest scores using a paired *t*-test. This means that we will focus on the average gain by each individual as a result of being trained with the simulator. We will cover this method in a following section of this chapter.

Some theoretically based statisticians will undoubtedly criticize the use of this single group gain score pretest–posttest approach to validating the training effectiveness of the simulator. However, real-world limitations frequently necessitate this approach.

In the more complicated of the two methods, we will:

- Administer a pretest version of the test instrument.
- Divide the original group into two equivalent control and simulator groups.
- Train the control group with existing instructional methods.
- Train the simulator group with the training simulator.
- Administer a posttest version of the training simulator.

This method requires a larger number of test subjects because the original group must now be split into a control group and a simulator group. This method does have the advantage of determining the difference in performance between the existing instructional methods and the training simulator. When this method is used, analysis of covariance (ANCOVA) is employed to perform the statistical analysis.

ANCOVA is somewhat more difficult to execute that the more common ANOVA. However, for training simulator validation, it is worth the extra effort to perform the ANCOVA. The reason for this is that ANCOVA allows us to separate the effects of a covariate. In our case, this is the prior knowledge of the subject as demonstrated by the pretest scores. This is important because a high pretest score has the potential to be interpreted as learning if we ignore the effect of the covariate as in an ANOVA analysis. With ANCOVA, we can separate the effect of the covariate pretest score. This will allow us to properly interpret the posttest scores with respect to the group treatment.

With either method, it is important that an equivalent version of the pretest be used as the posttest. This will help insure that the test individuals are not merely memorizing the answers to specific questions. One simple method of developing the posttest version is to randomize the order of all of the questions on the test. Because we are no longer concerned with the reliability analysis at this point, it does not matter if the odd and even halves are not maintained. Some care, however, should be given to insuring that the equivalent questions are not adjacent to or even near their corresponding questions.

12.8.5 Analyzing the Test Instrument Results

The statistical approach that needs to be used to analyze the test instrument results depends on the type of experimental design used by the practitioner. If the practitioner used the simple single-group before-and-after approach, a paired *t*-test should be used. Conversely, if the control and simulator group approach was used, an analysis of covariance test should be employed.

12.8.5.1 Gain Score Analysis

Once the posttest has been completed and scored, we can enter the data into our worksheet. In this case, we will reuse our unknown group data for our pretest data in column F. The posttest data are listed in column G. Figure 12.22 illustrates our current worksheet.

	A	B	C	D	E	F	G
10		odd	even	known	unknown	pretest	posttest
11	1	45	45	90	50	50	65
12	2	40	35	75	45	45	50
13	3	30	35	65	70	70	65
14	4	25	30	55	65	65	70
15	5	40	40	80	60	60	65
16	6	45	45	90	70	70	80
17	7	35	30	65	50	50	55
18	8	35	40	75	65	65	60
19	9	40	35	75	75	75	80
20	10	50	45	95	40	40	60
21	11	35	40	75	35	35	75
22	12	40	35	75	60	60	70
23	13	45	45	90	55	55	80
24	14	40	45	85	50	50	90
25	15	35	30	65	35	35	80
26	16	35	35	70	60	60	95
27	17	40	45	85	40	40	65
28	18	40	40	80	35	35	85
29	19	50	45	95	40	40	65
30	20	30	30	60	55	55	80

FIGURE 12.22 Gain score analysis data.

To analyze the difference between the pretest and the posttest scores, we will utilize either a paired *t*-test or a nonparametric rank sum test. We will use the paired *t*-test if the data are normal. If the data are nonnormal, we will use the nonparametric rank sum test.

To determine if the paired *t*-test data are normal, we will first need to calculate the difference between the pretest and the posttest. Because the posttest scores are generally greater than the pretest scores, we can facilitate the process if we actually calculate our difference in scores by calculating for each set of data:

$$\text{gain score} = \text{posttest score} - \text{pretest score}$$

These calculations will result in the worksheet seen in Figure 12.23.

	A	B	C	D	E	F	G	H
10		odd	even	known	unknown	pretest	posttest	gain
11	1	45	45	90	50	50	65	15
12	2	40	35	75	45	45	50	5
13	3	30	35	65	70	70	65	-5
14	4	25	30	55	65	65	70	5
15	5	40	40	80	60	60	65	5
16	6	45	45	90	70	70	80	10
17	7	35	30	65	50	50	55	5
18	8	35	40	75	65	65	60	-5
19	9	40	35	75	75	75	80	5
20	10	50	45	95	40	40	60	20
21	11	35	40	75	35	35	75	40
22	12	40	35	75	60	60	70	10
23	13	45	45	90	55	55	80	25
24	14	40	45	85	50	50	90	40
25	15	35	30	65	35	35	80	45
26	16	35	35	70	60	60	95	35
27	17	40	45	85	40	40	65	25
28	18	40	40	80	35	35	85	50
29	19	50	45	95	40	40	65	25
30	20	30	30	60	55	55	80	25

FIGURE 12.23 Gain score analysis gains.

The mean value for the gain score is 19.0, and the standard deviation is 16.59. We can perform a chi-square goodness of fit test for a normal distribution as demonstrated in Chapter 5, "Input Data Collection and Analysis":

1. Our hypotheses are H_o: normal (19.00, 16.59); H_a: not normal (19.00, 16.59).
2. Level of significance is 0.05.
3. The critical value for alpha = 0.05 and one degree of freedom is 3.84.
4. The test statistic is 1.6.
5. The test statistic of 1.6 is less than the critical value of 3.84. This means that we cannot reject the null hypothesis of the data being normal (19.00, 16.59).

These results are illustrated in the Excel worksheet shown in Figure 12.24. Because the data are normally distributed, we can proceed to the paired *t*-test.

12.8.5.2 Paired *t*-Test Analysis

Because the gain score data were found to be normally distributed, we can utilize the paired *t*-test approach to determine the training validity of the simulator. The hypotheses for our test are as follows:

H_o: There is no difference in means between the pretest and posttest scores.
H_a: There is a difference in means between the pretest and posttest scores.

The paired *t*-test is executed in Excel by following the tools, data analysis, *t*-test: Paired Two Sample for Means menu sequence. This results in the window shown in Figure 12.25.

We can designate our posttest data as the input variable range 1 and our pretest data as input variable range 2. We designate our data in this manner because we expect the posttest data to have higher values than the pretest data. This facilitates our subsequent statistical calculations. We have also designated the

	J	K	L	M	N	O	P	Q
10	cell	lower %	upper %	lower x	upper x	obs	exp	((o-e)^2)/e
11	1	0.00	0.25	0.00	7.81	7	5	0.8
12	2	0.25	0.50	7.81	19.00	3	5	0.8
13	3	0.50	0.75	19.00	30.19	5	5	0
14	4	0.75	1.00	30.19	999.00	5	5	0
15								1.6
16								
17							critical	3.84

FIGURE 12.24 Chi-square results.

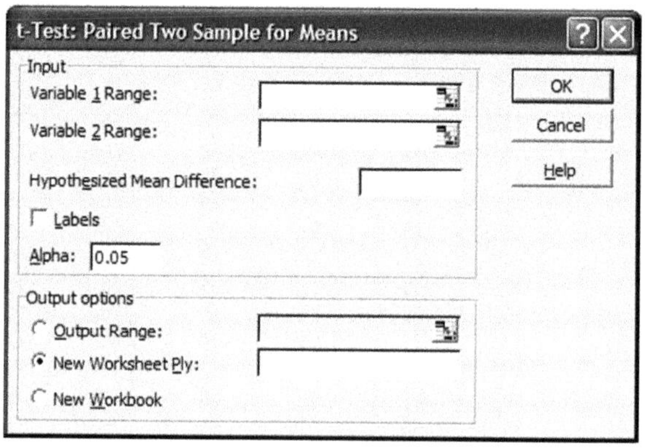

FIGURE 12.25 Excel paired *t*-test window.

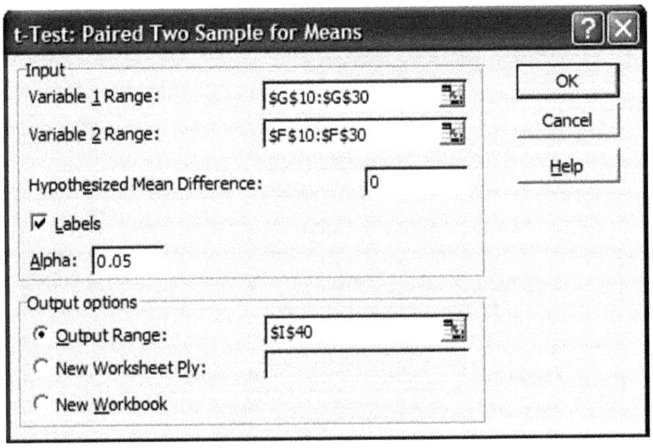

FIGURE 12.26 Completed Excel paired *t*-test window.

t-Test: Paired Two Sample for Means

	posttest	pretest
Mean	71.75	52.75
Variance	140.1974	161.7763
Observations	20	20
Pearson Correlation	0.08868	
Hypothesized Mean	0	
df	19	
t Stat	5.121469	
P(T<=t) one-tail	3.03E-05	
t Critical one-tail	1.729131	
P(T<=t) two-tail	6.07E-05	
t Critical two-tail	2.093025	

FIGURE 12.27 Paired *t*-test results.

first row of the data for labels. Next, we will have to insert a 0 in the hypothesized mean difference text box. Last, we will designate the output to go to cell I40. The resulting window is illustrated in Figure 12.26.

When we execute the test, Excel presents the window of Figure 12.27.

The Excel results indicate a test statistic of 5.12. The two-tailed critical value for $\alpha = 0.05$ is ±2.09. Because 5.12 is greater than 2.09, the null hypothesis of no difference between the pretest and the posttest is rejected. This means that there is a statistically significant difference in the pretest and posttest at an α level of 0.05. In other words, there is evidence that the simulator has training validity.

12.8.5.3 Nonparametric Rank Sum Test

If the gain score distribution had been nonnormal, we would be obligated to validate the simulator using the nonparametric rank sum test. We will utilize the same data for demonstration purposes. The nonparametric rank sum technique that we will utilize is the same as was introduced in Chapter 5, "Input Data Collection and Analysis."

We begin by sorting our pretest and posttest scores in ascending order. We will designate the pretest scores as group 1 and the posttest scores as group 2. This results in the following values.

Score	Group	Rank	Score	Group	Rank
35	1	1	65	1	21
35	1	2	65	2	22
35	1	3	65	2	23
40	1	4	65	2	24
40	1	5	65	2	25
40	1	6	65	2	26
45	1	7	70	1	27
50	1	8	70	1	28
50	1	9	70	2	29
50	1	10	70	2	30
50	2	11	75	1	31
55	1	12	75	2	32
55	1	13	80	2	33
55	2	14	80	2	34
60	1	15	80	2	35
60	1	16	80	2	36
60	1	17	80	2	37
60	2	18	85	2	38
60	2	19	90	2	39
65	1	20	95	2	40

The hypotheses are:

H_o: There is no difference in means between the pretest and posttest scores.
H_a: There is a difference in means between the pretest and posttest scores.

We will examine the hypothesis at an α level of 0.05. The sum of the ranks are equal to:

$$W1 = 255$$

$$W2 = 565$$

$$U_1 = 255 - \frac{20(20+1)}{2}, \quad U_2 = 565 - \frac{20(20+1)}{2}$$

$$U = min\ (45, 355) = 45$$

$$Mean = 20*20/2 = 200$$

$$Var. = \frac{20*20(20+20+1)}{12} = 1366.66$$

$$Z = \frac{45-200}{\sqrt{1366.67}} = -8.09$$

The critical Z value for $\alpha = 0.05$ is ±1.96. Because -8.09 is less than -1.96, the null hypothesis of no difference between the pretest and the posttest is rejected. This means that the simulator has training validity.

12.8.5.4 Preparing for the Analysis of Covariance Method

As you will recall, an initial step in using the pretest and posttest method with both a control group and a simulator group is to insure that the two treatment groups have equivalent pretest knowledge. This is accomplished by assigning and reassigning individual pretest scores into the two groups until the means and the variances of the two groups are similar.

Number	Control Pretest Group	Simulator Pretest Group
1	70	85
2	65	60
3	75	55
4	80	55
5	65	45
6	50	65
7	80	60
8	55	85
9	45	80
10	90	65
11	65	85
12	70	85
13	55	55
14	75	70
15	75	45
16	75	55
17	50	75
18	70	45
19	65	60
20	85	60

The mean and variance of the control group in this table are 68.0 and 148.42, respectively. The mean and variance of the simulator group are, respectively, 64.5 and 191.84. To insure that the means and variances are statistically similar, the practitioner may decide to perform the usual comparison of variance and means tests.

A comparison of variances *F*-test can be performed on the pretest data:

1. Our hypotheses are: H_o: no difference between the control and simulator group variances; H_a: difference between the control and simulator group variances.
2. Use a 0.05 level of significance.
3. The critical value for the *F* distribution with an α value of 0.05 and 19 degrees of freedom in the numerator and 19 degrees of freedom in the denominator is 2.17.
4. The *F* statistic is 1.29.
5. Because the *F* statistic is less than the critical value of 2.17, we cannot reject the null hypothesis of no difference between the variances of the control and simulator test groups.

The Excel results are illustrated in Figure 12.28.

We can also perform an independent *t*-test assuming equal variances to convince ourselves that the means are also statistically similar. For this test our hypotheses are:

1. H_o: no difference between the means of the control and simulator group pretests; H_a: difference between the means of the control and simulator group pretests.
2. Use a 0.05 level of significance.
3. The critical *T* distribution value for 0.05 and 38 degrees of freedom is ±2.02.
4. The calculated *T* statistic is 0.85.
5. The *t* statistic of 0.85 is less than the critical value of 2.02, so we cannot reject the null hypothesis of no statistical difference between the means at an α level of 0.05.

F-Test Two-Sample for Variances

	Variable 1	Variable 2
Mean	64.5	68
Variance	191.8421053	148.4211
Observations	20	20
df	19	19
F	1.292553191	
P(F<=f) one-tail	0.290735492	
F Critical one-tail	2.16824958	

FIGURE 12.28 Excel *F*-test results.

t-Test: Two-Sample Assuming Equal Variances

	Variable 1	Variable 2
Mean	68	64.5
Variance	148.4210526	191.8421
Observations	20	20
Pooled Variance	170.1315789	
Hypothesized Mean	0	
df	38	
t Stat	0.848546366	
P(T<=t) one-tail	0.200723781	
t Critical one-tail	1.685953066	
P(T<=t) two-tail	0.401447562	
t Critical two-tail	2.024394234	

FIGURE 12.29   Excel *t*-test results.

This information is summarized in the Excel worksheet in Figure 12.29.

The results of these two tests indicate that we can consider our control group and simulator groups to be equivalent. Because we have decided which individuals will be in the control group and the simulator group, each group can receive its respective training. After the training treatment, both groups are given the posttest. The following table represents the pretest and posttest scores.

Number	Control Group Pretest	Control Group Posttest	Simulator Group Pretest	Simulator Group Posttest
1	70	80	85	75
2	65	60	60	65
3	75	55	55	70
4	80	65	55	75
5	65	65	45	55
6	50	55	65	70
7	80	80	60	75
8	55	70	85	90
9	45	55	80	90
10	90	70	65	80
11	65	70	85	90
12	70	65	85	85
13	55	65	55	90
14	75	80	70	90
15	75	80	45	75
16	75	65	55	55
17	50	45	75	90
18	70	60	45	70
19	65	70	60	70
20	85	80	60	80

The hypotheses for the analysis of covariance are:

H_o: No difference between using existing training methods and the training simulator
H_a: Difference between using existing training methods and the training simulator

In contrast to most of the statistical analysis that we have performed so far, ANCOVA cannot be performed by the data analysis tools available in Excel. ANCOVA is a more specialized statistical technique, and it will be necessary to acquire a copy of a more sophisticated software package. Common statistical packages include SPSS (Statistical Package for the Social Sciences), MiniTAB, and SAS. As with the common simulation packages, it is possible to obtain student versions of these statistical packages under certain circumstances.

Regardless of which statistical package we choose, we will designate the posttest scores as the dependent variable, the group treatment as the model factor, and the pretest scores as the covariate. Our data will have the following ANCOVA results.

Source of Variation	Sum of Squares	DF	MS	F	Signif.
Covariate pretest	1699.803	1	1699.803	23.878	0.000
Group treatment	1423.762	1	1423.762	20.000	0.000
Explained	2750.428	2	1375.214	19.318	0.000
Residual	2633.947	37	71.188		
Total	5384.375	39	138.061		

These results indicate that the covariate of the pretest did indeed have a statistically significant effect with an F statistic of 23.878. This means first that an individual's pretest score did have an effect on his or her final score. Second, it means that the ANCOVA technique was necessary to remove the effect of the covariation to obtain accurate group factor results. If we had used the more common ANOVA method to analyze the data, we might have misinterpreted the statistical results.

The group treatment with the covariate effect removed was also statistically significant with an F statistic of 20.000. This means that the simulator had a statistically significant effect in comparison to the classroom/practical exercise training method. From these results, we can conclude that there is evidence that the simulator has training validity.

12.9 Implementation

Once the training simulator has been determined to have training validity, it is time for implementation. This will generally include activities involving:

- Publishing
- Distribution
- Installation
- Support

12.9.1 Publishing

Publishing includes all of the activities necessary to take the simulator code and supporting files and place them on a CD-ROM disk suitable for distribution. This process is actually significantly more complicated than first appears. The publishing process includes the following activities.

- Compiling the source code into executable code
- Collecting supporting files
- Creating a setup distribution disk
- Testing the distribution disk on various platforms

12.9.1.1 Compiling the Source Code into Executable Code

Practitioners should avoid distributing the source code to the users at all costs. There are two primary reasons for this. First, it would be undesirable to allow the users to modify and even take credit for the original source code. Second, if others have the original source code, there is nothing preventing them from distributing the source code to third parties. The situation can quickly go out of control.

To help avoid these problems, the practitioner should compile or package the source code into executable program files. With the executable version of the program files, users do not need the original authoring program in order to use a simulator. It is also very difficult to reverse engineer executable program files back into source code. This helps protect the integrity of the simulator program.

12.9.1.2   Collecting Supporting Files

Typically, the final program will operate correctly when running on the practitioner's development computer but will not function correctly when copied to other computers. This is because simulator projects normally require a large number of supporting files in order for the program to function correctly. If these supporting files are not present on the user's computer or are present in the wrong subdirectory, the program will malfunction. Aside from the executable files, the following files must also be prepared for creating a setup distribution disk.

- Sound files
- Movie files
- Driver files

12.9.1.3   Creating a Distribution Disk

The practitioner has the choice of directly copying the program files to the distribution CD-ROM or using an installation program to create a setup CD-ROM. The first approach is relatively easy and inexpensive for the practitioner to implement. No additional software is required, and the practitioner simply copies the executable and support files from a hard drive to the CD-ROM. The user installation process consists of copying the executable and support files from the CD-ROM onto the user's local computer. Despite explicit instructions to this effect, all sorts of installation problems may occur. Users may copy only part of the simulator files or even try to run the simulator from the CD-ROM drive. This method of distribution is less professional and is inherently unreliable.

The second approach involves using an installation program such as the Wise Installation System. This type of installation program allows the practitioner to identify easily the various executable and supporting files required by the simulator. The installation program combines all of the files into a single large setup.exe installation file. It is much easier for the practitioner to create a CD with this single setup.exe file than a large number of individual executable and supporting files. The practitioner will also have control over how the program is installed on the user's computer. This helps reduce the probability of poor installation practices by users that can lead to other problems. The only disadvantage to the use of a specific installation package is the additional cost of the software.

The practitioner should also give some thought to designing and printing a disk label. Round disk labels are readily available through computer and office supply stores. The design for the disk label can be produced in a graphics presentation package such as Microsoft PowerPoint. Some disk labels come in a kit with a disk label creation program. These can also be used to produce the label. The kit may also include a device to help affix the label to the CD. These are small circular plugs that fit into the center hole of a CD-ROM case. The use of one of these plugs greatly facilitates the correct positioning of the label on the CD.

The disk label should include the name and version of the program. It should also include minimal installation instructions. Other information such as the name and the contact method for the practitioner are optional. They may be helpful, however, in the case of installation difficulties.

12.9.1.4   Testing the Distribution Disk on Various Platforms

The practitioner cannot assume that just because the simulator operates on the host computer, it will operate on the user's computer. It is likely that the simulator will be installed on a wide variety of computers with respect to:

- Operating systems
- Hardware
- Drivers

At the time this handbook was assembled, there were four commonly used operating systems. These were Windows 98, Windows NT 4.0, Windows 2000, and Windows XP. Some incompatibilities have been discovered with certain authoring software versions. Although it will be impossible to predict which operating system the user will have, it is possible to check the functioning of the program on the different operating systems. In some cases compatibility problems may be circumvented by running the newer operating systems in more compatible modes.

The users' computers are also likely to vary widely in computer hardware. The practitioner may have to include a list of minimum hardware requirements for the program in a readme.text installation program. The heavy use of multimedia components is likely to stress a lower performing computer system. This may result in unacceptable graphic screen displays or unsynchronized WAV files.

Last, special drivers may be required for different parts of the simulator to function. For example, in order to play properly MPEG movie clips, the correct MPEG driver must be installed on the host computer. Some error trapping is possible to program into the simulator. If this is performed, a suitable bypass subroutine can be developed that prevents the simulator from crashing if the appropriate driver is not present.

12.9.2 Distribution

The second component to the distribution process is the actual distribution of the simulator. Distribution of the simulator can be performed either with a CD-ROM or through downloading from an Internet web site. It is likely that there will be a relatively large number of sizable files associated with the simulator. For this reason, distribution by CD-ROM may be the only feasible alternative.

12.9.3 Installation

The third component to the distribution process is the installation of the simulator program onto the user's computer. During the installation procedure, two issues may arise.

- Hardware issues
- Network issues

12.9.3.1 Hardware Issues

Even though the practitioner may have attempted to prevent any hardware installation problems during the software publication phase, it is still possible that problems may arise. For example, some early DVD CD-ROM drives would not properly read regular CD-ROM disks. The operating system would not report any type of error. Instead, the installation program would simply pretend to be going though the normal installation process.

12.9.3.2 Network Issues

Most training simulators are likely to be programmed for local computer installation rather than network installation. In this type of installation, the executable and support files will normally reside in the same program directory or subdirectory. In addition, if the simulator needs to generate temporary data files, these files will also most likely reside in the same subdirectory as the executable or support files. In a local computer installation, the simulator will expect all of these subdirectories and files to be located on a physical drive in the user's computer.

Unfortunately, some organizations do not permit the installation of applications on a user's local computer. In this situation, all applications are installed on a centrally maintained file server. Unless special consideration is given to the possibility of network installation, it is possible that the simulator will look for certain files in the wrong drive or subdirectory. These issues mean that the practitioner may be contacted by the user's information technology group to discuss network installation issues.

12.9.4 Support

The practitioner should anticipate providing support for any simulator that has been distributed. Support questions may arise from either the simulator installation process or the operation of the simulator. Even for a program of the limited applicability of a training simulator, it may be beneficial to establish a web site that provides basic frequently asked questions (FAQ) support. These can include answers to:

- Installation problems
- Hardware requirements
- Necessary multimedia drivers
- Network issues

12.10 Summary

In this chapter, we discussed a new type of simulation model, the training simulator. These types of simulators enable users to receive realistic training that might not otherwise be possible because of time, cost, or operational limitations. In addition, this training may be received in compressed time and without consequence.

Training simulators may be based on either management processes or complex equipment. In the case of management processes, the practitioner will be concerned with the input decisions, the underlying simulator model, and the output measures that result from the mathematical model. In the case of a training simulator for complex equipment, the practitioner will be concerned with the equipment components, operating controls, normal equipment operation, and abnormal equipment operation. These activities are normally presented in the simulator in the form of instructional, training, and test modes. In the instructional mode, the simulator presents all of the information necessary to understand the subject process or equipment. The training mode allows the user to experience the process or operate the equipment. Last, the test mode allows both the practitioner and the user to determine the extent of learning.

Many of the simulator development processes are similar to the traditional simulation model study process. Major differences can be found in the input data collection and validation phases. The input data collection phase includes the collection, analysis, and editing of not only mathematical data but graphic and sound data as well. The validation phase includes the normal face validation. However, statistical validation is replaced with training validation activities.

In addition, the simulator development process does not include an experimental design and analysis phase or a project recommendations and conclusions section. Instead, the simulator development process includes an implementation phase. In the implementation phase, the practitioner is concerned with software publishing, distribution, installation, and support activities.

Chapter Problems

1, What is the difference between traditional simulation models and training simulators?

2. What type of format should be used for a simulator test instrument?

3. What is the difference between face validity and known group validity?

4. Why is ANCOVA preferred over ANOVA for determining the training validity of simulators?

5. What consideration must be given when creating a distribution disk?

References

Hildebrand, D.K. and Ott, L. (1991), *Statistical Thinking for Managers,* PWS-Kent, Boston.
Johnson, R.A., Freund, J.E., and Miller, I. (1999), *Miller and Freund's Probability and Statistics for Engineers,* Pearson Education, New York.

13

Examples

Abu M. Huda
Continental Airlines

Christopher A. Chung
University of Houston

Somasundaram Gopalakrishnan
University of Houston

Erick C. Jones
University of Houston

The purpose of this chapter is to provide the practitioner with some specific examples of studies involving simulation modeling and analysis. These papers can be utilized by the practitioner to provide guidance for future technical reports and manuscripts. A copy of a master's thesis has also been included for graduate students contemplating a thesis in the area of simulation.

13.1 Combined Continuous and Discrete Simulation Models in the High-Speed Food Industry

Abu M. Huda and Christopher A. Chung

13.1.1 Introduction

The food-processing industry is a typical manufacturing sector in which the system must first be modeled with a continuous event approach and then later with a discrete event approach. This combination-type model is necessary because the subject material must frequently be first represented as a fluid and is later transformed into individual packaged goods.

For a combined discrete and continuous event system, the complexity stems from the different elements that control the two systems. The continuous variable is monitored as it passes a threshold value, which in turn triggers a discrete change. The threshold value is defined as a range because the step change within a time advance may not necessarily match an exact defined value, unlike a discrete change. Therefore, a combined system must be carefully designed to maintain appropriate balance between the continuous and discrete parts.

13.1.2 Objective

The objective of this report is to explore the application of simulation modeling and analysis to a high-speed combined continuous and discrete food industry manufacturing process. Also, the minimal coverage and limited literature in this area constitute a significant difficulty, particularly for simulation modelers faced with high-speed food-processing systems. By addressing the issues that may be encountered with a real world example of a high-speed combined system, this process may assist analysts in the development of more valid simulation models.

13.1.3 System Background

The issues involved in developing and analyzing a combined discrete and continuous event simulation model are presented in the context of an actual high-speed coffee-manufacturing facility located in Houston, Texas. The facility is representative of many high-speed manufacturing processes that include a number of production lines that produce a variety of food products that must be modeled first as a fluid and later as individual packages. For the purposes of this report, attention is focused on a single production line that produces different blends of bagged coffee from specific types of roasted coffee beans. This manufacturing system includes storage bins, a screw conveyer, whole roast coffee bins, roasters, grinders, ground coffee hoppers, surge hoppers, and a packaging machine. The basic manufacturing process is illustrated in Figure 13.1.

The system shown in Figure 13.1 follows a continuous process from the storage bins to the surge hopper. From the surge hopper, coffee is packaged in individual bags for shipping, following a discrete process. Because coffee beans are measured in terms of weight, and not individual beans, the flow process of the coffee beans through different hoppers must be treated as a continuous process. Once the coffee has been packaged, the batching of individual packages follows a discrete process and therefore needs to be modeled so.

Four storage bins store whole roast coffee beans. The primary purpose of these bins is to allow the roasted coffee beans to cool and degas for 4 to 5 h. A screw conveyer transports the coffee from the storage bins to the whole roast coffee bins. Five whole roast coffee bins are used as buffers between the grinder and the storage bins. The whole roast coffee bins also feed the grinders, once the grinders are in operation. The production line has two variable-speed grinders that feed the ground coffee hoppers. Six ground coffee hoppers receive coffee from the grinders and feed the packaging machine surge hopper.

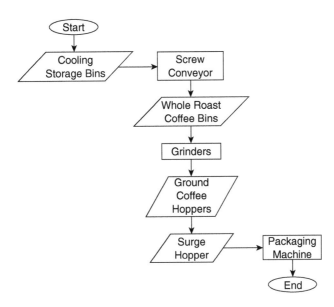

FIGURE 13.1 High level flow chart.

A single packaging machine surge hopper feeds the ground coffee directly to the packaging machine. The packaging machine contains a weighing scale and a bagging apparatus.

The daily manufacturing operation involves two shifts of 10 h. Each shift has a line operator and a grinder operator. The first shift normally begins the daily production schedule, and the second shift completes the schedule. A typical daily production schedule consists of a number of separate orders with varying blend, grind, package sizes, and materials. Orders with the same type of blend and grind are grouped together in the production schedule to reduce the number of changeovers.

The process begins when the line operator sends orders to the grinder operator. The grinder operator pulls different types of coffee beans from the storage bins into the whole roast coffee bins for each specific blend. The grinder operator then sends the coffee through the screw conveyer system. When the whole roast coffee bin gets filled, grinding begins. The grinding rate depends on the type of grind. After grinding, the coffee flows into the ground coffee hopper.

The line operator opens the gate between the ground coffee hopper and the surge hopper. When the coffee flows into the surge hopper, the line operator starts the packaging machine. As the packaging machine fills bags, the line operator puts the required number of bags into cases and sends the cases to the shipping area on a conveyer. The line operator runs the packaging process continuously until a changeover is required.

13.1.4 Combined Continuous and Discrete Model Issues

In a continuous model, the state of a system changes with time. Even though the change can be represented by a state equation, the development of such an equation is not very straightforward because in most cases there is not enough insight into the complex interactions within the system's components. However, an equation involving the rate of change of the state of system can be devised. That rate of change is called the derivative of the state of the system. Such equations involving one or more derivatives are called differential equations. Therefore, the continuous process components in a combined model are represented by a set of differential equations. These equations characterize the behavior of the system over the simulated period of time. They generate the output for the time-dependent state variables and control the behavior of the model accordingly. A state variable does not remain constant between events, and its value changes according to the differential equations. An example of a state variable would be the level of coffee in the surge hopper. Because the status of the state variables changes with time based on the differential equations, the analyst must carefully monitor the system to insure that impossible state variable values do not arise. For example, the demand for coffee by the packaging machine cannot result in a negative level of coffee in the surge hopper. Although this issue may not present much of a problem with larger-scale chemical plant–type systems, it is far more difficult to address with the higher-speed, smaller-capacity components typically found in food-processing systems.

The integration process used to solve the differential equations in the continuous component can also present problems for the analyst. The time advance involved in the integration process must be carefully specified. Too large a time advance can also result in completely impossible negative state variable values, such as a negative amount of coffee in a storage bin.

An issue unique to the food-processing industry versus other combined model manufacturing applications involves the multiple repackaging of the product. More commonly modeled liquid product combined models, such as those found in the oil refinery sector, involve one continuous-to-discrete component interface as the liquid is transferred to individual tankers of finite capacity. In the food-processing industry, many products are first packaged individually into bags or cans and then repackaged into cases for transportation.

A final issue involved in combined models is how the discretely and continuously changing components of a combined model can affect each other. This can occur in three ways (Prisker, 1986). First, a continuous state variable achieving a threshold value may cause a discrete event to occur. Second, a discrete event can affect the value of a continuous state variable. Last, a discrete event can cause the relationship governing a continuous state variable to change at a particular time. The following sections illustrate these issues.

13.1.5   Combined Model Translation

The manufacturing process was modeled in the simulation software package ARENA. The complete model consisted of a total of over 500 model block and experiment element structures. Although it is beyond the scope of this chapter to individually describe each of these structures, a discussion of the most significant continuous and combined model components follows. These include the creation of orders, supplying raw coffee beans, packaging the coffee, changing the blend, changing the package size, and changing the packaging film. Figure 13.2 shows a part of the ARENA model with different blocks.

At the beginning of a shift, orders are created according to the production schedule. The orders are described in terms of order size, blend type, grind, package size, case size, and packaging film type in an

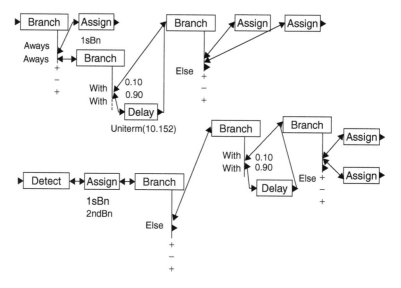

FIGURE 13.2 Combined model translation.

ASCII file. This information is imported into the model as read from the ASCII file. The file contains the individual attributes of each order.

Before each shift, the designated ground coffee hoppers are supplied with normally used blends of coffee. When the conveyer becomes available, the model allows coffee to flow into the designated whole roast coffee bins for feeding the grinder by changing the rates. The model detects the total amount of coffee ordered for each blend compared to the amount available in the ground coffee hopper. For each ground coffee hopper there are three detection parameters: one when the bin becomes full, one to check whether there is a need for more coffee to complete the order, and one to detect when the bin becomes empty. For order sizes larger than the capacity of the hoppers, the supply of coffee is switched back and forth between the designated hopper and one of the extra hoppers by varying the rate of flow.

In the event that either of the hoppers' level approaches 0, the model checks the status of the total order size and signals the ordering of more coffee. An important continuous modeling issue encountered at this point is that the threshold value of the level of the coffee in the bin must be closely monitored. Should the actual level become negative, a run time error may result. This is of particular concern because coffee flows through the system in a continuous manner, and the different bins have a variety of inflow and outflow rates.

The model starts the coffee-packaging process by changing the "in" and "out" rate values for the surge hoppers and the ground coffee hoppers. In this stage, the program becomes a combined model, as coffee flows continuously from the surge hopper into the weighing scales and then is packaged first into individual bags and then later into cases. When the level of the coffee scale indicates that the required package weight is present, the program triggers a count mechanism, which tracks the number of packages in the case. This is an example of a continuous state variable causing a discrete event to occur. Once the required number of packages is prepared, the model engages the seizure of the line operator. There is an associated delay period for completing the packaging of the case. The number of completed cases is recorded with another counter element. This is an illustration of a discrete-to-discrete component interface primarily found in food-processing industry combined models.

Once the model has determined that the packaging of one particular type of blend is complete, it starts the changeover to next blend. The model accomplishes this by setting the bin outflow rate values. This is an example of a discrete event discretely changing a continuous state variable. On certain changeovers, there is also a change in the package size. A delay period represents the time to change the scale setting on the packaging machine. This change might occur after every single order, even with the same type of blend. Changeovers may also require a change in the aluminum film packages. This requires another delay element to attach the new aluminum film roll to the back of the packaging machine. After the completion of each order, the model triggers the start of the next order. Finally, once the new order has been initiated, the program checks the new order's film type for packaging to determine whether a film changeover is required. The high-level flow chart in Figure 13.1 provides the basic model logic.

13.1.6 Input Data Collection and Analysis

The collection of input data is of obvious importance in a simulation analysis. The preferred method of input data analysis is to obtain empiric data and fit the data to theoretical probability distributions. Unfortunately, this approach may not always be possible because of system complexity and operating uncertainty. This is particularly evident in the continuous event components of combined systems. For example, the progression of minute material such as coffee beans can be described in terms of flow rates but cannot be individually recorded as single coffee beans or ground coffee particles. In this situation, some types of system input data must be obtained through equipment operational specifications and operator or management estimates.

For this model, equipment operational specifications were utilized for conveyor transportation rates, grinding rates, and packaging rates. Similarly, operator estimates were utilized for blend changeover delays, packaging film type changeover delays, packaging weight changeover delays, shift orientation delays, the probability of conveyor availability on demand, and the delay after a convey request if it is

TABLE 13.1 Input Data Parameters

Operation	Data Parameters
Conveyor rate	400 lb/min
Grinding rates	40 lb/min for URN type
	35 lb/min for S/C type
	30 lb/min for GL type
Packaging machine rate	35 packages/min for 1-lb bags
	15 packages/min for over 1-lb bags
Blend changeover	Uniform (10,15) min
Packaging film type changeover	Uniform (7.5, 20) min
Packaging weight changeover	Uniform (2, 5) min
Shift orientation	Uniform (10,15) min
Conveyor availability on demand	Binomial (0.10)
Delay after convey request if not available	Uniform (10,15) min
Box preparation, box filling, packaging labeling, and sealing	Normal (0.432, 0.07) min

not available. The preferred empiric data collection and theoretical distribution-fitting approach was utilized for the manual operation of box preparation, box filling, packaging labeling, and sealing. The values for the input data parameters are listed in Table 13.1.

13.1.7   Performance Measures

A variety of measures may be used to evaluate the performance of this type of combined system. Traditionally, simulation models use system times or resource utilization levels. For this system, the primary performance measure was the total time required to complete orders for a given blend. This measure was chosen because the objective of the model was to determine how to complete orders in the minimum possible time. This measure is also a function of both the continuous and discrete event components in the model.

13.1.8   Model Verification and Validation

Verification is the process of ensuring that a model operates as intended. Validation is the process of ensuring that a model represents reality. Verification was performed using ARENA's debugging tools and system animation. Validation was performed using both face validation and statistical validation. Face validation was achieved through a cyclic model review and improvement process with the assistance of a number of plant managers and engineers. For statistical validation, 20 historical data points were collected from the records at the facility. The data included the time to complete individual orders and their corresponding parameters of package size, bag size, case size, and amount of coffee for those orders. The base model was run for the required number of replications with the associated parameters of each of those 20 orders. The measure of performance was the time to complete the individual orders. A nonparametric rank sum test was conducted to determine whether there was a significant difference between the two data sets. The resulting Z value of -0.40 was between the critical values of -1.96 and $+1.96$. Because the model did not statistically significantly differ from the actual system, it was considered to be valid.

13.1.9   Experimental Design

The selection of experimental factors is a critical one for combined continuous and discrete systems, similar to any entirely discrete or continuous system. The selection of factors becomes more complicated because the analyst may select continuous related factors, discrete related factors, or factors of both types. Discussions with the plant management indicated that a total of three factor variables in the continuous component of the process should be investigated to determine their effects on improving production output.

The first continuous variable concerned the storage capacity of the coffee bins. The bins were thought to have insufficient capacity for the line operator to run the system without repeatedly asking the grinder operator for additional coffee. The second continuous variable thought to affect system performance was conveyor availability. There was only one conveyer available at the facility, which supplied coffee to all the production lines. Because the particular production line being modeled did not have any priority over other production lines, the line operator often had to wait for the conveyer to be available to supply the coffee. This affected the production time and therefore impacted timely production of the orders. Once the conveyer was available, it was able to supply the coffee within a very short period of time. The third continuous variable was the grind rate. The current grind rate was 40 lb/min. That rate was suspected of slowing down the process of supplying ground coffee to the ground coffee hoppers.

Each of these factors was tested at two levels according to management specifications with a 2^3 full factorial design. The bin sizes were set at 1200- and 1600-lb capacity levels. The grinding rates were 40 lb/min and 60 lb/min. The experimental levels for the conveyer availability were 10% and 60%. These alternatives are summarized in Table 13.2. The basic simulation model was modified to incorporate these factor variations. Each of the eight alternatives was run to determining the number of replications required for statistical comparison. Different random number streams were used for each alternative to avoid the possibility of correlation.

13.1.10 Results

A total of 30 replications were required to obtain a relative precision of 0.10 at an α value of 0.05. A 95% confidence interval of the expected average order completion time based on the 30 replications is summarized in Table 13.3.

ANOVA and a Duncan multiple-range test were performed on the alternative means. ANOVA was performed to determine whether there was a difference among the means of the different factor combinations. The summarized results are shown in Table 13.4.

The ANOVA results at an α value of 0.05 indicated that the main effects for the bin size and the grind rate were statistically significant. Since at least one mean among the alternatives was statistically significantly different, a Duncan multiple-range test was conducted to determine which means differed from the rest. The Duncan test was conducted at an α value of 0.05. The results of the Duncan test are summarized in Table 13.5.

13.1.11 Discussion

The ANOVA results indicate that bin size and grind rate are the only statistically significant effects. The Duncan multiple-range test indicates that there was no statistically significant difference between alternatives D and C; H, B, and C; B, H, A, and G; or A, G, F, and E as shown in Table 13.5.

Note that no statistically significant difference exists between the means at an α level of 0.05 for 1–4, 3–6, 5–7, or 7–8.

TABLE 13.2 Experimental Design

Factor Combination	Bin Size (lb)	Grind Rate (lb/min)	Conveyor Avail.
A	1200	40	10%
B	1200	40	60%
C	1200	60	10%
D	1200	60	60%
E	1600	40	10%
F	1600	40	60%
G	1600	60	10%
H	1600	60	60%

TABLE 13.3 Output of Different Experimental Designs

Factor Combination	Grand Mean (min)	95% Confidence Interval
A	411	[400, 417]
B	406	[400, 412]
C	398	[392, 403]
D	394	[389, 399]
E	417	[408, 425]
F	418	[411, 426]
G	410	[403, 418]
H	405	[400, 410]

TABLE 13.4 ANOVA Analysis

Source	DF	SS	MS	F	P
Bin size	1	6458.4	6458.4	21.16	0.000*
Grind rate	1	7470.5	7470.5	24.48	0.000*
Conv. Avail	1	592.2	592.2	1.94	0.165
Bin size and grind rate	1	100.1	100.1	0.33	0.567
Bin size and conv avail.	1	87.6	87.6	0.29	0.593
Grind rate and conv avail.	1	199.8	199.8	0.65	0.419
Bin size, grind rate, and conv avail.	1	266.7	266.7	0.87	0.351
Error	232	70813.1	305.2		
Total	239	85988.5			

*Significant for an α value of 0.05.

TABLE 13.5 Duncan Multiple-Range Test Output

Order	1	2	3	4	5	6	7	8
Factor Combination	E	F	G	A	H	B	C	D
Bin Size	1600	1600	1600	1200	1600	1200	1200	1200
Grind Rate	40	40	60	40	60	40	60	60
Conveyer Avail	10%	60%	10%	10%	60%	60%	10%	60%
Means	417	418	410	411	405	406	398	394

Because there is no statistically significant difference between D and C, alternative C with the higher grind rate but lower bin size and conveyor availability offers the best alternative from a performance standpoint. Although the ANOVA results indicated that the effect of bin size was statistically significant, the Duncan test indicated that its effect was overshadowed by the effect of the grind rate. This would allow the organization to keep the existing bins and conveyor system and only concentrate capital investment on increasing the grind rate. The saving of cycle time for a set of orders for option C compared to option F is 20 min (398 vs. 418). The manufacturing facility being modeled runs two 10-h shifts per day. Therefore, over a longer period of time, this would entail considerable savings of labor hours, increased production, and consequently improved productivity. Performance measurement of all these alternatives would not have been possible without the availability of this simulation model. The model provides a cost-effective means to measure different manufacturing options without actual implementation. Therefore, there are improvements in terms of cycle time, labor hours, and operating and capital cost.

13.1.12 Conclusions

The primary purpose of this report was to illustrate some of the issues associated with a real-world example of a combined continuous- and discrete-event high-speed food-processing simulation model. During the course of this type of combined-event model study, the modeler can expect to encounter unique data collection, continuous-to-continuous component, continuous-to-discrete component, discrete-to-continuous component, and factor selection modeling and analysis issues. By providing insight to ways to address these issues, this section should assist simulation analysts in developing more valid combined continuous and discrete event simulation models.

The model acts as an efficient tool to test different production scenarios as well as to provide an understanding of the production process. This will help identify any bottleneck with any production schedules as well as determine the completion time for the schedule. Therefore, better forecasts can be made for completion time of different orders. Changes can be made in the model to alleviate the bottleneck, and the corresponding impact can be easily observed. Also, for future changes in the system and the process, the model will act as an effective tool for testing feasibility before actual implementation. This should help investing for proven improvements.

Reference

Pritsker, A.B. (1986), *Introduction to Simulation and SLAM II.* Systems Publishing Corporation, West Lafayette, IN.

13.2 Operation of Airport Security Checkpoints under Increased Threat Conditions

The following example is an illustration of a technical report for a service system. In this case, the service system is a simulation model of an airport security checkpoint system. The format of this report can be used as a guideline for preparing technical reports to audiences with simulation backgrounds.

ABSTRACT: Many foreign and domestic events increase the concern of terrorist actions directed toward the U.S. commercial air transportation system. Government and airport officials respond to these threats by increasing security measures. However, even under increased levels of security, airport officials are still responsible for insuring the effective movement of air passengers. To address this problem, a simulation analysis was conducted on the operation of an airport security checkpoint under increased threat conditions. The results provide guidance for airport officials in maintaining effective passenger movement under these conditions.

13.2.1   Introduction

Many foreign and domestic events increase concern about terrorist actions directed toward U.S. commercial airports. Recent examples of such events include the Gulf War, the Unibomber, and the Oklahoma City bombing. Government and airport officials respond to these threats by increasing security measures. These have included the restriction of curbside baggage check-in and the refusal of passage of individuals without tickets to the loading gates. However, the single most significant security measure continues to be the x-ray and metal detection screening of ticketed individuals proceeding to the loading gates.

Although the detection of restricted items remains paramount, real world considerations dictate the rapid processing of individuals through the security checkpoint. This report examines the processing of individuals through a security checkpoint at a major metropolitan airport. The results provide guidance for airport officials in maintaining effective passenger movement under these conditions.

13.2.2   Previous Relevant Work

Both the air- and land-side operations of airport terminals have been the subject of simulaton analysis. Air-side simulation analysis includes runway exits (Sherali et al., 1992; Hobeika et al., 1993) and aircraft arrival gates (Hamzawi and Salah, 1986; Sokkar et al., 1990; Hassounah and Steuart, 1993). Land-side simulation analysis includes the high level system modeling of passenger and luggage flow (Hassan et al.,1989). Other land-side simulation includes the design of passenger transportation interchanges within a terminal (Codd, 1990) and space standards for check-in areas (Seneviratne and Martel, 1995). This study adds to this existing work through a tactical approach to the operation of security checkpoints under increased threat conditions.

13.2.3   Background

Processing of each individual through the security checkpoint begins with ticket inspection. This consists of one of the security personnel verifying that the individual's ticket is current and valid. Provided that the individual passes this examination, the individual may proceed to the x-ray/metal detector area of the checkpoint.

The next process is determined by the presence of luggage. If the individual does not have luggage, he or she proceeds directly to the metal detector. If the individual possesses luggage, he or she must first place these items on the front of the x-ray machine. Once this is accomplished, the individual may then proceed to the metal detector. Luggage that is placed on the x-ray machine is examined by the x-ray operator for the presence of restricted items. Provided that none exists, the operator moves the luggage out of the x-ray machine to await retrieval by the owner. The individual proceeds through the metal detector, which is supervised by a third security operator. This operator insures that if the metal detector indicates the presence of suspect metallic objects, that the individual removes any possible problem objects before proceeding through the metal detector a second time. If the individual successfully negotiates the metal detector, he or she is free to retrieve any luggage and then exit the system.

Exceptions to the normal flow of this process are addressed by the checkpoint supervisor. These include manual baggage inspection and individuals in wheel chairs. An exception to the ticketed individual rule occurs in the case of airline crew members and airport employees. However, these individuals must still proceed through the system in a manner similar to the ticketed individuals.

Under normal operating conditions, the security checkpoint will be manned by one supervisor, two x-ray operators, two metal detector operators, and two ticket checkers. These numbers may fluctuate up or down according to equipment serviceability and personnel availability. The checkpoint is located in the main concourse and services a total of 21 gates. The layout of the security checkpoint is illustrated in Figure 13.3.

13.2.4   Objective

The objective of this simulation analysis is to operate the security checkpoint in the most effective manner. The most effective manner is defined as operation with the minimum average individual time in the system with the minimum number of personnel while maintaining the current level of security. Though

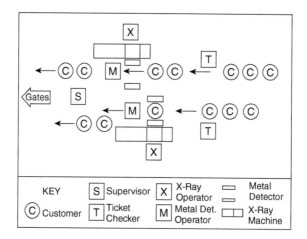

FIGURE 13.3   System layout.

the checkpoint operates throughout the day, effective operation is critical during periods of high arrival and departure activity. These periods are a result of the hub-and-spoke system used by the major airline in the airport involved in this study. These periods of high activity dictate the use of a terminating system simulation approach.

Given the objective of this analysis, the following questions need to be addressed:

1. For a given number of security personnel, which configuration will result in the minimum average customer service time?
2. If the checkpoint experiences an equipment malfunction, how should the personnel be reassigned to result in minimum average customer system time?

13.2.5   Data Collection and Fitting

A total of eight data distributions were collected and fitted. These included the interarrival times of customer batches, the customer arrival batch size, ticket-checking service time, travel time from the ticket checker to the metal detector, whether or not the customer has luggage, the x-ray service time, the metal detector service time, and the customer metal detector failure rate.

13.2.5.1   Interarrival Times of Customers

The interarrival times of customers were obtained by observing the interval between the arrival of one batch of customers and the arrival of the next batch of customers. Arrival was defined as either the physical location within arm's length distance of the ticket checker or cessation of forward movement upon reaching the end of the ticket-checking queue.

13.2.5.2   Customer Arrival Batch Size

A batch of customers consisted of one or more individuals who arrived at the first point of the system simultaneously. Batch sizes larger than one typically occurred with husband and wife passengers, families of passengers, groups of businessmen, and air crew. The size of the arrival batch was recorded with the batch interarrival time.

13.2.5.3   Ticket-Checking Service Time

This period begins with the arrival of the customer within arm's reach of the ticket checker. The service time consists of checking either the ticket or identification of the customer for validity. The ticket-checking service time ends when the customer begins forward movement and allows the ticket checker to service the next customer.

13.2.5.4   The Presence of Luggage

The presence of luggage was defined as any object that the customer was required to have x-rayed. This included luggage, boxes, and radios.

13.2.5.5   Travel Time

This is defined as the time required after the end of the ticket-checking time and before arrival at either the x-ray machine, the metal detector, or the queue of either.

13.2.5.6   X-Ray Service Time

X-ray service time begins with the customer first placing objects on the end of the x-ray machine. The service time includes the time required by the x-ray operator to inspect the objects and ends when the individual removes the luggage. Data for this specific service time were recorded so that they were not dependent on the time the passenger was being processed through the metal detector. Thus, these data reflect the system time that can be attributed to the presence of baggage.

13.2.5.7   Metal Detector Service Time

The collection of metal detector service times depended on whether the customer possessed luggage. In the case of no luggage, metal detector service time began with passenger passing the front of the x-ray machine. In the case of the presence of luggage, the time began when the passenger released the luggage onto the front of the x-ray machine and began forward movement toward the metal detector. Service time includes the period during which the customer has exclusive control over the metal detector and operator. Time includes the customer passing through the metal detector. Time ends when either the customer is not stopped by the operator and physically passes the operator or fails the inspection and passes backwards through the metal detector. Observations of customers who failed the metal detector indicated that other customers passed through the metal detector while the failed customer searched for possible causes of failure.

13.2.5.8   Customer Metal Detection Failure

This occurred when the customer failed to remove suspect metal objects from his/her person. These data were collected on the basis of the number of customers who failed out of the total number of customers observed.

13.2.6   Data Fitting

This section consists of fitting the observed data to theoretical probability distributions and performing hypothesis tests to determine if the empirical data could have come from the theoretical probability distribution. If the observed data can be fit to a theoretical distribution, sampling for simulation purposes will be taken from these theoretical distributions.

Theoretical distributions are preferred because observed data may contain irregular values; values outside of the observed data that could possibly have occurred may be obtained, and theoretical distributions are more readily utilized with simulation languages. In the event that observed data cannot be fit to a theoretical distribution, the observed empirical data are used.

Distribution fitting was conducted with the chi-square and Kolmogorov–Smirnov goodness of fit tests. The personal computer software package Statgraphics, version 2.6, was utilized to fit data to theoretical distributions. In all cases, an α value of 0.05 was utilized for the hypothesis test.

13.2.6.1   Interarrival Times of Customer Batches to the Checkpoint

This data were fitted to a lognormal distribution with a mean of 3.14 and a standard deviation of 3.07. The null hypothesis was that these data could have come from a lognormal distribution with a mean of 3.14 and a standard deviation of 3.07. A chi-square goodness of fit test resulted in a calculated statistic of 12.8. This yielded a significance level of 0.46. Because this value is not less than 0.05, the null hypothesis cannot be rejected.

13.2.6.2   Customer Arrival Batch Size

These data were fit to a geometric distribution. In this case, the empirical distribution was modified by subtracting one from each observation. Thus, the batch size minus one was fit to a geometric distribution with $p = 0.552695$.

The null hypothesis was that these data could have come from a geometric distribution with $p = 0.552695$. The chi-square goodness of fit test resulted in a calculated statistic of 8.96158. This yielded a significance level of 0.110606. Since this value is not less than 0.05, the null hypothesis cannot be rejected. Although the batch sizes minus one were fit to a geometric distribution, it is necessary in the simulation to add back the subtracted one after generating the random number for the geometric distribution.

13.2.6.3   Ticket-Checking Service Time

These data were fit to a γ distribution with an α value = 1.91 and a β value of 0.99. The null hypothesis for these data was that it could have come from a γ distribution with an α value = 1.90784 and a β value of 0.989. The chi-square goodness of fit test resulted in a calculated statistic of 14.089. This yielded a significance level of 0.079. Because this value is not less than 0.05, the null hypothesis cannot be rejected.

13.2.6.4 Travel Time

Travel time to the x-ray/metal detector group I was fit for coming from a uniform distribution [0.8, 2.2] thousandths of hours. Travel time to the x-ray/metal detector group 2 was fit for coming from a uniform distribution [1.1, 3.2] thousandths of hours. The null hypothesis for group 1 was uniform [0.8, 2.2] and yielded a Kolmogorov–Smirnov D_n value of 0.271. This results in a significance level of 0.124, which is larger than 0.05 and therefore cannot reject the null hypothesis. The null hypothesis for group 2 was uniform [1.1, 3.2] and yielded a Kolmogorov–Smirnov D_n value of 0.238095. This results in a significance level of 0.185, which is larger than 0.05 and therefore cannot reject the null hypothesis.

13.2.6.5 The Presence of Luggage

These data consisted of the observation that, of the customers observed, 10% did not possess objects that required being x-rayed. This is a Bernoulli trial with only one of two outcomes: the customer has objects or the customer does not.

13.2.6.6 X-Ray Service Time

These data could not be fit to a theoretical distribution with $\alpha = 0.05$. Thus, the empirical distribution is utilized in the simulation.

13.2.6.7 Metal Detector Service Time

These data could not be fit to a theoretical distribution with $\alpha = 0.05$. Thus, the empirical distribution is utilized in the simulation.

13.2.6.8 Customer Metal Inspection Failure

These data consisted of the observation that 3% of the customers observed failed the inspection. This is a Bernoulli trial with one of two possible outcomes: the customer fails the inspection, or the customer does not.

13.2.7 Model Building

This system was modeled using the personal computer simulation package SIMAN IV. Modeling components and issues are presented in this section.

13.2.7.1 Assumptions

A total of eleven assumptions were incorporated in the modeling of the security checkpoint system. The first assumption is that passengers cannot balk. This means that in order to obtain access to the loading gate area, all customers must proceed through the security checkpoint. The second assumption is that passengers do not jockey. Jockeying is jumping between queues if another queue becomes shorter in length. Actual observation indicated that 90% of the customers who possessed luggage appeared unwilling to jockey between queues.

13.2.7.2 Queue Selection

When presented with more than one queue, it is assumed that the customer will first select the shortest queue. If queues exist of similar length, it is assumed that the customer will select the nearest queue.

13.2.7.3 Size of Queue

It is assumed that the security personnel will not shut down the system regardless of how long a system queue becomes. In the case of the ticket-checking queue, the queue extends down the main concourse hall. When the x-ray queue exceeded a small number of customers, the ticket checker himself moved down the main concourse hall.

13.2.7.4   Customers Who Fail the Metal Detector Go to the End

It is assumed that customers who fail the metal detector inspection go to the end of the queue. Observation of this situation indicated that while the failed customer attempted to determine the cause of failure, other customers in the queue bypassed the failed customer.

13.2.7.5   Service Rates Are the Same among Checkers and Operators

It is assumed that the service rate between ticket checkers and machine operator is the same for a given type of operation. This is a simplifying assumption.

13.2.7.6   Service Rate Is Independent of Queue Size

It is assumed that thorough inspection is paramount and that the number of customers waiting in a queue will not result in a ticket checker or machine operator changing work method so that the customer system time is reduced.

13.2.7.7   Customers Do Not Leave without Their Luggage

It is assumed that customers will not leave the system without their luggage.

13.2.7.8   No Restricted Items Are Discovered

The discovery of a restricted item such as a firearm or explosive device is not a regular occurrence. Thus, this situation is outside the scope of this model. In order to model this event correctly, knowledge and observation of such an incident would be necessary.

13.2.7.9   No Breaks

It is assumed that, during the period of concern, all security operators will remain on duty without taking breaks.

13.2.7.10   First In–First Out

It is assumed that the operation of all queues is performed in a first in–first out manner.

13.2.8   Modeling Using SIMAN

The simulation of this system consisted of one basic model. Alternative simulations were developed from the basic model. This model consists of the following components.

13.2.8.1   Generation of Customers

The generation of customers included the creation according to interarrival times and batch sizes. As each customer is created, ticket service time and whether or not the customer has luggage are first determined. If the customer does not have baggage, the customer's metal detector service time is generated. If the customer has luggage, then both an x-ray and metal detector time is determined. Because SIMAN IV does not possess a geometric distribution generator, the batch size values were determined by obtaining the integer of Alog[uniform(0,1)]/[Alog(q)] + 1, where q is 1 minus the probability for the geometric distribution. This yields batch sizes of one or greater.

   This method of determining the customer profile provides flexibility in the use or nonuse of common random numbers to reduce or not to reduce variance. If common random numbers are used, similar customers will be processed through the system under different alternatives. Without common random numbers, the customers will be independent. Common random numbers were not used because this analysis utilizes ANOVA, which requires independence.

   Immediately after the creation of the customers, the SIMAN program invokes a FORTRAN event. The purpose of this is to determine that the period of concern between 7:00 A.M. and 9:15 A.M. has expired. The event is utilized to stop customers from entering the system after 9:15 but allows all customers in the system to be processed.

13.2.8.2   Ticket Checker

The first element of the system that the customer encounters is the ticket checker. In the event of more than one ticket checker, the customer selects the shortest queue using the PICKQ statement. In this section, customers queue if necessary, seize the checker when available, are delayed the service time, and finally release the checker.

13.2.8.3   Does the Customer Have Bags?

Based on a customer attribute, the next block determines whether or not the customer possesses luggage. If the customer does not have luggage, he or she picks the metal detector with the smallest total number in the queue and in transit toward that queue. If the customer has luggage, he or she then proceeds toward the x-ray machine that has the smallest total number in the queue and in transit toward that queue.

13.2.8.4   Travel Delay

The customer then undergoes the appropriate travel delay to the chosen metal detector or x-ray.

13.2.8.5   X-Ray Machine

If the customer has luggage, he or she must proceed to an x-ray machine. If a queue exists, the customer joins the queue. Once the customer reaches the x-ray machine, there is a delay while the customer loads the luggage on the x-ray machine. As soon as this delay is over, the customer proceeds to the metal detector adjacent to the x-ray machine. The separation of the customer and the luggage is executed by a BRANCH statement. From this point, the luggage is held for the x-ray machine service time while the customer either queues for or is serviced by the metal detector and operator. The identity of the customer is preserved by a luggage attribute. This is necessary to later match the passenger with the luggage.

13.2.8.6   Metal Detector

All customers must go through the metal detector process. If the metal detector is being utilized, the customer will queue. A 3% probability exists that the customer passing through the metal detector will fail the inspection. This is accomplished using a BRANCH/WITH statement, which sends the customer back to be reinspected.

13.2.8.7   Does the Customer Have Bags?

If the customer does not have luggage, clearance of the metal detector constitutes exit from the system. At this point, the system time of the customer is tallied. In the event that the customer has bags, the customer proceeds to the end of the x-ray machine to attempt to collect the luggage.

13.2.8.8   Collecting Luggage

Customers who have luggage proceed to the end of the x-ray machine and attempt to collect their luggage. If the customer arrives before his or her luggage is inspected by the x-ray operator, the customer waits. If the luggage has already been inspected, the customer undergoes a delay while the customer unloads the luggage from the end of the x-ray machine. This process requires the utilization of one queue each for the customers and the luggage, and the presence of the MATCH statement where the customer entity number and the luggage passenger attribute number are matched. When the correct passenger and luggage are matched, and the customer has unloaded the luggage, the customer exits the system. The system time for the customer is then tallied.

13.2.9   Tactical Considerations

Arrival and departure of flights occur during the specified period of concern, 7:00 A.M. to 9:15 A.M. Outside of this period and until the next period of activity approximately three hours later, little activity occurs at this security checkpoint. The nature of this operation dictates that it be modeled using terminating simulation techniques. The tactical considerations of this terminating simulation are addressed below.

13.2.9.1   Starting Conditions

This issue concerns the state of the system when the simulation starts. Because, in this simulation, flights do not depart until approximately 7:25 A.M. to 7:50 A.M., the system is empty until approximately 7:00 A.M., the beginning of the period of concern. Thus, the decision is to begin the simulation with the system in an empty state.

13.2.9.2   Terminating Event

This issue concerns the existence of a natural terminating event in the process. In this situation, the last departure occurs at 9:15 A.M., and no other flights depart until the next cycle. Thus, the last departure at 9:15 represents the natural terminating event. Should any customers be in the system at this time, they will be allowed to depart. However, no new customers are expected after 9:15, so for simulation purposes, any customers arriving past that time will be disposed of with a FORTRAN event routine.

13.2.9.3   Measure of Performance

A single measure of performance was selected for this simulation analysis. This measure is the total time the customer spends in the system. Total time in the system was selected in contrast to queue waiting time because a long total system time and not a long queue waiting time would cause a customer to miss a flight. An additional issue existed in that SIMAN IV limitations restricted the number of attributes that could be assigned to entities. This encouraged the use of as few as possible total attributes, so that the model could handle a larger number of entities.

13.2.9.4   Number of Replications

The number of replications for each alternative simulation was based on the sequential stopping procedure specified in *Simulation Modeling and Analysis* (Law and Kelton, 1991), Section 9.4.1. This procedure utilizes the equation:

$$Relative\ Precision = \frac{t_{n-1,\,1-\alpha/2} * s}{\overline{X} * \sqrt{n}} \qquad (13.1)$$

where

$\overline{X}$ = grand mean of individual replication means

n = number of replications

s = standard deviation of individual replication means

α = 0.05

t = t value for n - 1 degrees of freedom and α = 0.025

Replications were conducted until the desired relative precision of 0.10 was obtained. A 95% confidence interval of the expected average customer time in system can then be calculated based on the following equation:

$$95\%\ C.I. = \overline{X} \pm t_{n-1,\,0.0975} * s / \sqrt{n} \qquad (13.2)$$

where all variables are the same as above.

13.2.10   Validation

The definitive test of model validity is determining if the simulation times in the system closely resemble that from the actual system. For this purpose, forty random system times were taken from the actual two ticket checker, two x-ray, two metal detector system and compared with forty random system times for the identical configuration simulation model.

The samples from each group were first tested for a null hypothesis of coming from a normal distribution. The system time null hypothesis was rejected, while the simulation null hypothesis could not be rejected. Based on those results, a nonparametric rank sum test was conducted with Statgraphics to determine if the null hypothesis that there was no difference between the two sets of data was true. The test yielded a Z value of 0.530931. Because this value is between -1.96 and $+1.96$ for $\alpha = 0.05$, the null hypothesis cannot be rejected. Thus, any difference between the system and this particular model is not statistically significant, and the conclusion can be drawn that the model is valid.

13.2.11 Experimental Design

In order to meet the objectives of this analysis, a 2 full factorial design was utilized. With this method, the three main factors are the number of ticket checkers, the number of x-ray machines/operators, and the number of metal detectors/operators. Each of these three main factors is examined at two levels.

In the case of the number of ticket checkers, the two levels consist of two and three checkers. The two levels for the x-ray machines are one and two. The two levels of the metal detector are one and two. These levels were determined by the objectives of the simulation, physical space limitations, equipment availability, and observation of the system. The physical space limitation was a particularly important determination because the width of the hall that the checkpoint is located in would have difficulty holding more than two x-rays and two metal detector machines. Likewise, more than three ticket checkers may impede the flow of traffic in the opposite direction. The factor combinations are summarized in Table 3.6.

The basic simulation model was modified to accommodate these factor combinations. This resulted in eight separate sets of programs. The experimental design was then simulated and examined using analysis of variance and the Duncan multiple-range test.

13.2.12 Simulation Run Results

13.2.12.1 Replications

An initial set of runs was conducted to determine the number of replications required for each factor combination to obtain an expected average system time with a desired relative precision of 0.10. The maximum number of replications necessary to obtain the 0.10 desired relative precision was 67 with factor combination E. The simulation for each factor combination was then performed for 67 replications. The means and 95% confidence intervals for system times for each of the factor combinations is listed in Table 3.7.

TABLE 13.6 Factor Combinations for Personnel

Factor Combination	Ticket Checkers	X-Ray Machine	Metal Detectors
A	2	2	2
B	3	2	1
C	2	2	1
D	3	2	2
E	2	1	2
F	3	1	1
G	2	1	1
H	3	1	2

TABLE 13.7 Mean and 95% Confidence Interval Results

Factor Combination	Grand Mean (s)	95% Conf. Int. (s)
A	39.10	[38.78, 39.41]
B	46.50	[45.45, 47.54]
C	50.56	[49.24, 51.88]
D	35.07	[34.86, 35.28]
E	93.37	[84.18, 102.57]
F	90.55	[82.39, 98.72]
G	97.70	[89.22, 106.17]
H	88.86	[79.71, 98.02]

13.2.12.2 Analysis of Variance

For each of the factor combination simulation runs, different random number streams were utilized. This specifically avoided the possibility of correlation which is sometimes induced using common random numbers so that paired t-test comparisons may be performed. The data obtained with the 67 replications for each of the eight factor combinations was examined using the analysis of variance routine in the VAX software package MINITAB. The analysis of variance results are listed in Table 13.8.

The results indicate that the effects of the three main factors are all significant at $\alpha = 0.05$. None of the interactions are significant at $\alpha = 0.05$.

13.2.12.3 Duncan Multiple-Range Test

A Duncan multiple-range test was performed on each of the main factor combinations. This test was performed at $\alpha = 0.05$, and Duncan test table r values for 120 total replications were used as an approximation for the total 536 replications. This was necessary because of table limitations. Table 13.9 summarizes the results of the Duncan multiple-range test.

TABLE 13.8 Analysis of Variance Results

Source	DF	SS	MS	F	p	
TICKET	1	906.48	906.48	4.83	0.028	*
XRAY	1	92370.60	92370.60	492.19	0.000	*
METAL	1	1944.36	1944.36	10.36	0.001	*
TICKET*XRAY	1	29.52	29.52	0.16	0.692	
TICKET*METAL	1	16.56	16.56	0.09	0.766	
XRAY*METAL	1	662.76	662.76	3.53	0.061	
TICKET*XRAY*METAL	1	15.84	15.84	0.08	0.772	
Error	528	99091.08	187.56			
Total	535	195037.20				

*Indicates significant at $\alpha = 0.05$

TABLE 13.9 Duncan Multiple-Range Test Results

RANKING	1	2	3	4	5	6	7	8
COMBINATION	D	A	B	C	H	F	E	G
TICKET	3	2	3	2	3	3	2	2
XRAY	2	2	2	2	1	1	1	1
METAL	2	2	1	1	2	1	2	1
MEANS (s)	35.07	39.10	46.50	50.56	89.51	90.59	96.97	97.70

Differences between 1 and 2, 2 and 3, 3 and 4, and among 5–8 are not significant at $\alpha = 0.05$.

The results of the Duncan test may be interpreted by stating that factor combinations D, A, B, and C are all superior to H, F, E, and G. D is also superior to B and C, and A is also superior to C.

13.2.13 Conclusions

13.2.13.1 Selection of the Best Alternative

Interpretation of the Duncan multiple-range test indicates that the existing system of two ticket checkers, two x-ray machines/operators, and two metal detectors/operators (2,2,2) performs well in comparison to the other alternatives. The single alternative of 3,2,2 resulted in a lower mean time in system. However, the Duncan test indicates that the difference between this alternative and the present system is not significant. Because the alternative 3,2,2 possesses the disadvantage of requiring the economic burden of additional personnel without significant statistical reduction in customer system time, it cannot be considered the best alternative.

13.2.13.2 Responses to Simulation Objectives

With respect to the objectives of the simulation, the following answers can be provided:

1. The addition of a third ticket checker will not significantly reduce customer time in the system.
2. The lowest customer time in the system with fewer than six security personnel other than the supervisor is obtained with two ticket checkers, two x-ray machines/operators, and one metal detector/operator (2,2,1). However, this time is significantly more than the existing configuration of (2,2,2). Thus, the reduction in personnel must be seen as enough incentive to accept the increased time in system if this option is utilized. Because the paramount concern is security, and the second most important concern is system throughput, an economic-performance comparison might be inappropriate to this analysis.
3. If only five security personnel other than the supervisor are available, the configuration should be (2,2,1). For only four available personnel, configuration (2,1,1) can be utilized and can be expected to perform as well as any of the six-person configurations with the exception of (2,2,1).
4. If an x-ray machine malfunctions, it does not matter how the other personnel are assigned as long as at least two ticket checkers are utilized. in this case, the individual who was operating the metal detector might be better utilized at other security checkpoints.

Overall, it would appear that the physical limitations of the system restrict the reduction of customer system time. Thus, other avenues of approach should be considered. Possible areas include work method and ticket modification. In the case of work method modification, alternative techniques may be developed that result in reduced service times. With ticket alterations, changing the design of the boarding pass envelope or the ticket itself may enable the ticket checker to more readily determine the validity of the ticket.

13.2.13.3 Final Conclusion

The use of simulation modeling and analysis techniques enabled the examination of the operation of one of the principal Greater Pittsburgh Airport security checkpoints. This process consisted of data collection, data fitting, model building, tactical considerations, experimental design, validation and verification, and statistical analysis.

With the proper execution of these steps, insight was obtained concerning the most effective mode of operation under a variety of alternative configurations. Because real world security demands operate on this system, it is unlikely that many of these configurations could be physically examined through the reconfiguration of the system. Thus, it appears that simulation is the only effective analytic technique that can be used to examine the operation of this system without disruption or compromise of security requirements.

References

Bouisset, J.F. (1994), Security technologies and techniques: airport security systems. *J. Test. Eval.* 22(3), 247–250.

Codd, J.A. (1990), Planning for interchange. *Highways Transport.* 37(8), 25–30.

Hamzawi, S.G. (1986), Microcomputer simulation aids gate assignment. *Airport Forum* 16(5), 66–70.

Hassan, M.F., Younis, M.I., and Sultan, M.A. (1989), Management and control of a complex airport terminal. *1989 IEEE International Conference on Systems, Man, and Cybernetics,* Cambridge, MA, pp. 274–279.

Hassounah, M.I. and Steuart, G.N. (1993), Demand for aircraft gates. *Transport. Res. Rec.* 1423, 26–33.

Hobeika, A.G., Trani, A.A., Sherali, H.D., and Kim, B.J. (1993), Microcomputer model for design and location of runway exits. *J. Transport. Eng.* 119(3), 385–401.

Law, A.M. and Kelton, W.D. (1991), *Simulation Modeling and Analysis.* McGraw-Hill, New York.

Miller, I., Freund, J.E., and Johnson, R.A. (1990), *Probability and Statistics for Engineers.* Prentice-Hall, Englewood Cliffs, NJ.

Polski, P.A. (1994), International aviation security research and development. *J. Test. Eval.* 22(3), 267–274.

Seneviratne, P.N. and Martel, N. (1995), Space standards for sizing air-terminal check-in areas. *J. Transport. Eng. ASCE* 121(2), 141–149.

Sherali, H.D., Hanif, D., Hobeika, A.G., Trani, A.A., and Kim, B.J. (1992), Integrated simulation and dynamic programming approach for determining optimal runway exit locations. *Management Sci.* 38(7), 1049–1062.

Sokkar, F., Harjanto, A., and Nelson, S.V. (1990), Examination of air traffic flow at a major airport, Proc. 1990 Winter Simulation Conf. IEEE, Piscataway, NJ, pp. 784–792.

13.3 Modeling and Analysis of Commercial Aircraft Passenger Loading Sequencing

Somasundaram Gopalakrishnan (Christopher A. Chung, Advisor)

The following example document is part of an actual fall 2002 master's thesis. This example has been included in the handbook to illustrate the content that is required to complete a master's thesis in the area of discrete event simulation.

ABSTRACT: Efficiency in terms of flight operations involves many characteristics among which the "on time" arrival and departure of flights play an important role. There have been constant efforts to reduce the time from the arrival of an aircraft for passenger loading to the point where the flight is ready for departure. Passenger loading of aircrafts is one of the important phases of commercial airline operations and it is one of the prime events that act toward the delaying of flights. The possibility of reducing the loading time has always existed, and the sequence in which the passenger loading could be carried out holds the solution to passenger loading delays.

The basic purpose of this research is to analyze the various loading and unloading sequences of the passengers in an aircraft and to suggest ways to improve the existing systems so that flight delays caused by passenger loading and unloading can be reduced.

The loading of passengers typically starts with the loading of physically handicapped people followed by first class, family/children, back third coach, middle third coach, and in the end front third coach passengers. This general sequence typically occupies a major portion of the waiting time. This work examined the process by testing various feasible alternative loading sequences. It was discovered that by altering the loading sequence, there was an opportunity to statistically significantly reduce the plane loading time.

13.3.1 Introduction

13.3.1.1 Overview

According to Air Transport Action Group (ATAG), over 1,600 million passengers per year rely on the world's airlines for business and vacation travel, and around 40% of the world's manufactured exports are transported by air [1]. Passenger and freight traffic is also expected to increase at an average annual rate of around 4–5% between the years 1998 and 2010. By 2010 the number of people traveling by air could exceed 2.3 billion each year with an economic impact that could exceed $1,800 billion. Last, with this increase in activity, many carriers are trying to establish their own hold in the commercial aircraft market, but not many are able to exercise complete superiority. Even established carriers sometimes face losses that in extreme cases lead to the complete disruption of their business. There are a variety of problems and challenges faced by the carriers, and their success or even their existence in the market depends on their ability to solve these problems.

Customer satisfaction is one of the important criteria that decides the success of a carrier. Carriers spend a substantial amount of time and money to satisfy their customers in every possible aspect. Some of the factors in determining customer satisfaction levels are airfare, comfort provided, baggage-handling efficiency, and to an important extent the "on time" departure and arrival of flights.

According to the Air Transport Association of America (ATA), air traffic delays cost businesses $825 million in lost productivity during the year 2000 [3]. Another ATA report predicted that by the year 2008, additional traffic would cause a 250% rise in delays [2].

The Department of Transportation believes there are a variety of reasons that contribute to flight delays; these include severe weather, aircraft maintenance, runway closures, air traffic control system decisions, equipment failures, and to an important extent on passenger loading and unloading [4]. One Department of Transportation (DOT) report also found that 74% of gate departure delays were caused by airline operations issues such as passenger loading and unloading, aircraft refueling, and baggage handling [5]. This means that it is imperative that carriers monitor these criteria so as to improve their performance with respect to reducing flight delays.

13.3.1.2 Thesis Objective

The basic purpose of this research was to analyze the various loading and unloading sequences of the passengers in an aircraft and to suggest ways to improve the existing systems so that the flight delays caused by passenger loading and unloading can be reduced. Different operating strategies were tested with respect to the loading sequence of the passengers who are characterized as physically handicapped/old passengers, children/family, first class, back third coach passengers, middle third coach passengers, and front third coach passengers using simulation modeling and analysis.

13.3.1.3 Limitation of Applicability

This study analyzed the loading and unloading of passengers only, which is one of the factors contributing for flight delays. It does not necessary solve flight delay or gate problems, which may not be directly related to loading and unloading problems.

13.3.1.4 Thesis Organization

The remainder of this proposal is divided into literature review, problem statement, proposed research methodology, and potential contribution to the body of knowledge sections. The literature review section discusses previous work that has been done to reduce flight delays. The problem statement section discusses the problem that is solved by this research. The proposed research methodology section starts with an introduction about simulation and its advantages. Then the operating algorithm, parameters involved, model assumptions, and output statistics are discussed. It then presents the modeling approach that is being adopted, a high-level flow chart, an introduction to ARENA simulation software, and the actual system to be represented in ARENA. It then discusses system verification and validation. This section also presents experimentation and analysis approaches for the various operating policies and changes that can be made to the existing system so as to improve the current operating efficiency of the system. Last, the potential contribution to the body of knowledge section discusses the potential contribution of this research work to the existing body of knowledge.

13.3.2 Literature Review

13.3.2.1 Previous Relevant Work

Customer service has always been one of the most important criteria and driving forces for the commercial aircraft industry. There have been numerous previous efforts in the field of customer service for the airline industry. This began when the Air Travel Consumer Report was published in 1987 to help reduce customer dissatisfaction and to help airlines improve their service. The primary purpose of this report was to evaluate the performance of different commercial airlines on a periodic basis.

Lee and Cunnigham (1996) worked on customer loyalty in the airline industry where the important goal was the measurement and management of service quality and customer loyalty [6]. They developed a framework that accounts for the process underlying customer evaluation of service quality and the formation of loyalty. Then, based on the framework, this work reviewed and evaluated various perspectives regarding determinants of service quality and customer loyalty in the airline industry.

Dhir and Wallace (1996) tried to assess airline service quality by passenger, travel agents and airline employees by applying social judgment theory [7]. It provided means to reduce the gap in the expectations among the airline employees, travel agents, and passengers.

Snowdon et al. (1998) analyzed ways of improving the customer service [8]. Their objective was to improve asset utility, handle more flights in shorter periods of time, coordinate schedules with alliance partners, and quickly respond to irregularities, such as weather and malfunctioning equipment delays. They suggested improvement in operating level by use of advanced information technology to get passengers through check-in, security, and boarding faster and to improve the baggage-handling system, thus improving the passenger experience.

An additional major criterion contributing to the dissatisfaction of customers is flight delays. There have been many previous efforts to reduce flight delays.

Cheng (1997) proposed a knowledge-based airport gate assignment system integrated with mathematical programming techniques to provide a solution that satisfies both static and dynamic situations within a reasonable computing time [9]. A partial parallel assignment was introduced that considers a group of aircraft and looks at all the available gates and then does the gate assignments by optimizing a multiobjective function.

Haghani and Chen (1998) assessed ways of optimizing gate assignment at airport terminals [10]. They gave a new integer programming formulation for the gate assignment problem, and their main aim was to reduce the distance traveled by the passenger to a flight departure gate within the airport, which in turn reduces the delay of the flight.

Bolat (1999) worked on assigning arriving flights at airports to available gates [11]. Here a mathematical model was developed to assign the flights with the minimum range of utilized time periods of gates, subject to the level of service offered to passengers and other physical and managerial considerations.

Barros and Wirasinghe (2000) analyzed the location of the new large aircraft gate positions in pier terminals, and their important consideration was to reduce the passenger walking and the baggage transfer distances [12]. Based on the proportions of hub transfer passengers, a mathematical formulation was derived for pier terminals to help in the choice of the specific location of the gates. Wald (2001) worked on the growth of passenger loading time as the size and quantity of jumbo jets increase [13].

Vandebona and Allen (1993) determined passenger flow distributions at airports [14]. They proposed a theoretical framework. Analogous models in other areas of engineering applications were investigated, and they attempted to take advantage of the prevailing experience in those fields. The proposed model provided a sound starting point for queuing theory applications leading to the analysis of congestion, delay, and travel times. The passenger behavioral aspects, such as the desire to reach the terminal early and apprehension related to missing flights, were also taken into account.

Chang and Schonfeld (1995) found ways to sequence flights in airport hub operations [15]. Their main aim was to reduce the aircraft costs and passenger transfer times at hubs through the efficient sequencing of flights. They suggested sequencing bigger aircraft as last in first out (LIFO) and smaller aircraft as first in first out (FIFO) to maximize gate utilization and terminal capacity. Therefore, they suggest LIFO for larger aircraft when airports are not busy and gate utilization is not important and FIFO for smaller aircraft when airports are very busy.

Ng (1992) proposed a multicriteria optimization approach to aircraft loading of materials. It provided timely planning and improved airlift support [16]. It introduced the goal programming approach to loading problems. This model reduced the number of aircraft loads required to complete an airlift by 9%.

The literature review yielded a number of previous usual efforts directed at improving customer service through gate assignment mathematical models, passenger flow, and aircraft sequencing. With the exception of Ng (1992), no useful references were identified that addressed aircraft loading. No references were found that specifically identified passenger loading sequences.

13.3.3   Problem Statement

The efficient running of a system is very important in having a competitive edge over others in today's industry. This philosophy applies very well to commercial aircraft operations. Efficiency in terms of flight operations involves many characteristics, among which the "on time" arrival and departure of flights play an important role. Passenger loading of aircrafts is one of the primary contributing causes of flight delays. From the literature review section it is seen that there has been little work previously done on passenger loading, and the scope of improvement in this particular aspect is large with huge benefits.

Typically, the passenger loading process consumes a lot of time. There have been many efforts to reduce the time from the arrival of an aircraft for passenger loading to the point where the flight is ready for departure. Passengers who are traveling in a commercial aircraft are generally classified as physically handicapped/old passengers, children/family passengers, first class, back third coach, middle third coach, and front third coach passengers. The loading of these passengers also follows a typical pattern. The typical sequence generally followed for loading the passengers starts with the physically handicapped passengers/old passengers followed by first class, family/children, back third coach, middle third coach, and in the end front third coach passengers. This general sequence typically occupies a major portion of the waiting time. This work caters to the reduction of this waiting time and suggests improvements in the loading sequence of passengers.

13.3.4   Research Methodology

This section describes the method adopted to conduct the research. This includes a brief description of simulation and the different steps involved in the research study. The major steps involved in the proposed methodology are system definition, model formulation, model translation, verification and validation, and experimentation and analysis. Each step is discussed below with respect to this research work.

13.3.4.1   Simulation

13.3.4.1.1   *Definition*

Simulation is the process of designing a model of a real system and conducting experiments with this model for the purpose of understanding the behavior of the system and evaluating various strategies for the operation of the system (Pegden, Shannon, and Sadowski, 1995).

13.3.4.1.2. Advantages of Simulation

Simulation has become one of the widely used study techniques these days because of its powerful problem-solving capabilities for complex systems that cannot be easily analyzed using the traditional study techniques available. The important advantages of the simulation study are (Pegden, Shannon, and Sadowski, 1995):

- New policies, operating procedures, decision rules, organizational structures, information flow, etc. can be explored without disrupting ongoing operations.
- Hypotheses about how and why certain phenomena occur can be tested for feasibility.
- Bottlenecks in material, information, and product flow can be identified.
- A simulation study can prove invaluable to understanding how the system really operates as opposed to how everyone thinks it operates.
- New hardware designs, physical layouts, software programs, transportation systems, etc., can be tested before committing resources to their acquisition and/or implementation.

Though simulation has many advantages it also has some disadvantages. These include:

- Model building requires specialized training. The quality of the analysis depends on the quality of the model and the skill of the modeler.
- Simulation results can be sometimes difficult to interpret.
- Simulation analysis can be time consuming and expensive.

13.3.4.2 System Definition

The system definition includes the following components: aircraft, passengers, and the output measures of performance. These are the important elements of the system that completely define the system to the level that the simulation study can be conducted satisfactorily. Each element is incorporated into the model to conduct the simulation study. The following section defines each of these.

13.3.4.2.1 Aircraft

It is the aircraft that is under study. The simulation model includes the loading process including the gate area, jetway, and plane fuselage.

13.3.4.2.2 Passengers

Passengers form the entities of the system. This research tries to reduce the loading time of the passengers in the aircraft loading process. As already mentioned, the passengers are classified into six major categories, i.e., handicapped/old passengers, children/family passengers, first class, back third coach passengers, middle third coach passengers, front third coach passengers and they might be categorized into further classes depending upon the air carrier. Each class has a different loading time, and the time for loading follows a particular distribution. These distributions were ascertained by using the input analyzer feature of ARENA 5.0.

13.3.4.2.3 Output Measures of Performance

These are the indicators of performance of the system. Overall time is the important output measure of performance for this study.

Overall time is the time taken for the complete loading of all the classes of passengers in the aircraft. The primary aim of this work is to reduce this overall time of loading. A variety loading sequences were studied to suggest the best alternative for reducing the overall time.

13.3.4.3 Input Data Collection and Analysis

New input data were collected, and no old or historical data was used. Each input data set was fit using the ARENA input analyzer. Once the input distributions were ascertained, a model representing the original system was developed after analyzing the actual working algorithm of the system. Then the input distributions were fed into the model so that the system might represent the actual system.The following input data were collected. The theoretical distribution of each data set was determined using the ARENA Input Analyzer.

13.3.4.3.1 *Interarrival Time*

This is the time between the arrival of consecutive passengers. Each passenger category will have a particular interarrival distribution.

The passengers are assumed to be waiting for boarding as soon as the aircraft is ready for boarding.

13.3.4.3.2 *Batch Size and Maximum Batch Size*

The passengers arrive in batches, and hence the batch size was necessary for modeling the system. Generally the batch size is assumed to be 1. The maximum batch size controls the entity quantity in the system. The following quantities are the values for the maximum batch size for the various classes of passengers.

First class	16
Physically handicapped	2
Family	6 family batches with a uniform (3,6) distribution of the remaining family members
Back third	36*loadfactor-familybacktotal
Middle third	36*loadfactor-familymiddletotal
Front third	36*loadfactor-familyfronttotal

Here the load factor is assumed to be 0.8; i.e., the flight is assumed to be 80% full.

Thirty-six passengers can sit in total in any section of the aircraft, i.e., the back third, the middle third, and the front third.

The variables familyfronttotal, familymiddletotal, and familybacktotal indicate the total number of family members sitting in front third, middle third, and back third, respectively.

13.3.4.3.3 *Load Time*

This is the time taken by a passenger to load bags into the overhead compartment and to seat him or herself. The loading time is going to change depending on the class of the passenger. The following are the distributions obtained for the various classes of passengers after the raw data were fed into the input analyzer feature in the ARENA. For more details into the raw data detail and the distribution, see Appendix.

The passengers with no bags have been assumed to be taking 1 s to seat themselves, and it is applicable to all the classes of passengers.

Distribution Summary: Ordinary Passengers 1 Bag

Distribution: Lognormal
Expression: 1 + LOGN(12.7, 9.17)
Number of data points = 48

Distribution Summary: Ordinary Passengers 2 bags

Distribution: Beta
Expression: 4 + 35 * BETA(1.04, 1.54)
Number of data points = 20

Distribution Summary: Family with 1 Bag

Distribution: Gamma
Expression: 8 + GAMM(18.6, 0.196)
Number of data points = 2

Distribution Summary: Family with 2 Bags

> Distribution: Normal
> Expression: NORM(19.1, 6.15)
> Number of data points = 14

Distribution Summary: Physically Challenged with 1 or 2 Bags

> Distribution: Uniform
> Expression: UNIF(33, 88)
> Number of Data Points = 6

13.3.4.3.4   *Route Time*

This is the time the passengers took to travel from the starting of the ramp to the entrance at the aircraft door. The ramp has a length of 100 feet, but each category of passenger has a different speed of travel (walk), and hence they have different route time. The speeds at which the passengers walk were measured for the various categories of passengers by observing the walking distance of the people for a particular passenger type and the time that they take to cover that distance. Thus, the speed with which the passengers walked were measured.

Passenger Type	Rate
Ordinary passengers	3.4 ft/s
Physically challenged	2 ft/s
Family members	2 ft/s

13.3.4.3.5   *Row Travel Time*

It is the time taken by passengers to travel from one row to another. It also again depends on the type of passenger.

Passenger Type	Time
Ordinary passengers	2 s
Physically challenged	2 s (Does not play an important role)
Family members	3.4 s

13.3.4.4   Model Formulation

The existing system is represented in the flow chart in Figure 13.4, which represents the working of the system in sequence. Each stage contributes to the loading time of the aircraft.

13.3.4.5   Model Translation

Simulation models can be represented in many simulation languages. ARENA 5.0 has been chosen in this research for representing the system. ARENA is a hierarchical SIMAN/Cinema based modeling system that can be tailored for use in specific application areas such as transportation, communication, business process reengineering, etc. It also provides an interactive graphical environment for building SIMAN/ Cinema models.

The model was created in the ARENA 5.0. A general ARENA model consists of three major portions. They are the blocks, elements and animation components. Blocks define the general logical flow of the system. The logic of the system is implemented using the components of the blocks. Blocks are used to define the way the entities flow in the system. The picture of the blocks shown in the next page depicts the blocks used in the above system. Each sequence of blocks depicts the creation and the movement of the entity for a particular class of passengers. The diagram of the blocks is shown in the next page.

Elements are used to define the general parameters, variables and attributes of the system and its components. They also define the characteristics of the entities. The picture depicts the elements that are used in the model for implementing the system.

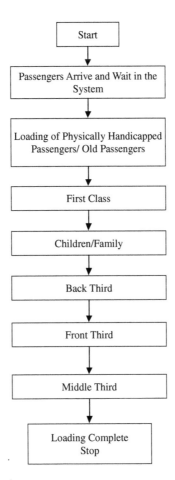

FIGURE 13.4 High level system flow chart.

Animation is the third component, and it consists of pictorial representation of the system and its entities and it helps to see the working of the system in the form of an animation. Animation helps in debugging and better understanding of the working of the simulation model. The picture (Figure 13.5) depicts the ramp and the aircraft. The two counters (digital displays) show the tnow (the system time at any present moment) and the time the last entity enters the aircraft and sits in the aircraft. The picture shows the queues in the planes where they stand before they occupy the seat assigned to them. Once they reach the seat they are put in storages where they are delayed for a very long time.

The blocks of the existing model that defines the logic consists of separate sequence of blocks for each class of passengers. For each category of passengers there is a "Create" block that creates the entities. Once the entities are created they are assigned a picture and a passenger category using the "Assign" block. Then they are assigned the number of bags that they carry along with the particular row and seat as their seat assignment using "Assign" block. Based on the number of bags that have been assigned to each passenger, a particular delay for that passenger is assigned. The passenger type is identified using a "Branch" block. If a particular seat has already been assigned, then the entity is diverted back to the assign block where the entity is assigned a new seat, and the entity continues through the model. The sequence of blocks in Figure 13.6 explains the before-mentioned sequence.

In the next stage the entities are assigned a particular jet-rate and row travel time as previously discussed in the input data section. Again, a "Branch" block is used to identify the type of passenger and assign the particular values for jet rate and row travel time. Then the entities are assigned various amounts of delays using the "Delay" block and this block helps in controlling the sequence especially for the family

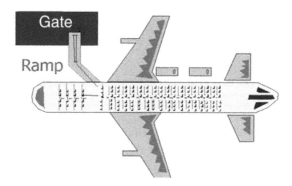

FIGURE 13.5 Model animation of ramp and aircraft.

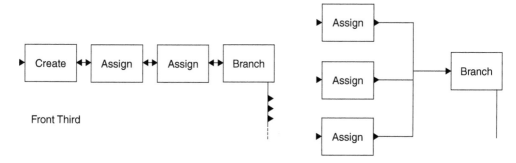

FIGURE 13.6 Model block sequence for seating passengers.

members and the first class passengers. After the delay has been assigned the entities reach the station called "gatesta1," which is the entrance to the ramp. The entities then reaches the door, which is a station called "planesta1." The entity travels from the "gatesta1" to "planesta1" via a "Route" block, which defines the characteristics of the entity when it travels along the ramp. A particular duration is assigned in the "Route" block based on the passenger type. This duration is governed by the expression $(100-2*[nq(aisleq)])/jetrate$ where 100 denotes the length of the ramp, "nq(aisleq)" denotes the number of passengers standing in the aisleq queue, and the jetrate denotes the speed with which a particular type of passenger travels in the ramp.

Once the entity reaches the door, it enters the queue called the aisleq, which is the queue to the entrance of the plane. Once inside the queue, the entity seizes the first row of the section of the plane that it is going to sit in. Then a logical comparison operation is done to find whether the assigned row number matches the row number in the plane. If it matches, the entity is allowed to sit in the row; otherwise, the entity is allowed to pass on to the next row. The comparison between the seat number in the plane and the actual seat number assigned is carried out by a Branch block. The sequence mentioned is shown in Figure 13.7 in the form of actual blocks here.

There are a variety of pictures assigned to various types of passengers, and the pictures are assigned to them based on whether they are walking in or putting their bags overhead or whether they are seated. These activities are carried out on the following sequence of blocks. The blocks are separated into two sequences. One sequence takes care of entities if they find their seats and the other takes care if the entities are not sitting in a particular row. If they find their particular seat, then the particular row delay assigned previously in the model for loading the bags on the overhead bins is actually incorporated in the system and the entities get delayed. After the delay the entity gets seated in their respective rows. Pictures of passengers sitting in the row are assigned based on the passenger type using a Branch block. The entities

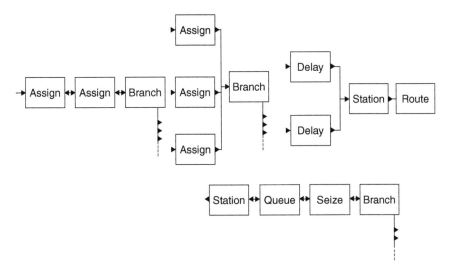

FIGURE 13.7 Model code for assigning seats.

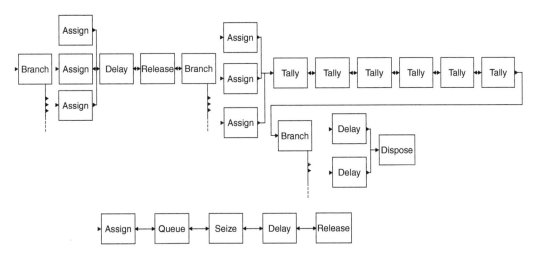

FIGURE 13.8 Model code for seating passengers.

then are stored in an ARENA feature called storages, and they are delayed for a very long period of time to complete the loading (Figure 13.8).

If the passenger's row and the actual row did not match, then they are made to go to a different sequence of blocks where they move into the next row available, and they are delayed by a value called row travel that has been previously defined in input data section. After they pass through the particular row, then there is again a comparison operation done to check whether the passenger has reached a particular row. If they have reached, they are made to sit; otherwise, they are allowed to carry on to the next row.

The loading is carried out in a particular sequence as previously stated. The main flow sequence is the same for each class of passenger except for a few minor changes in the flow while seating family members because the entity is duplicated based on the family size, and they are made to sit in the same row (Figure 13.9).

The creation sequence is the same as that discussed before. After the creation, the population is divided into three categories, and they are made to go to the three sections of the aircraft i.e., the back third, middle third, and the front third. After the division of the entities, a particular entity is duplicated with

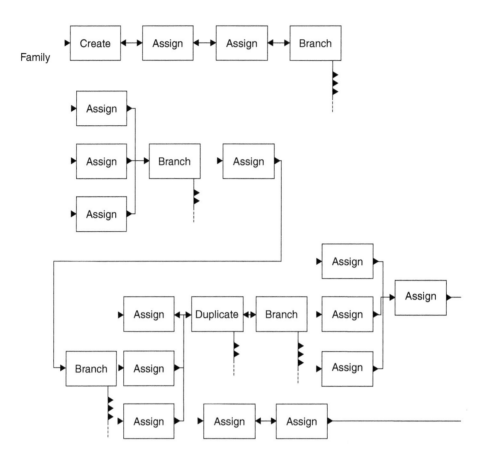

FIGURE 13.9 Passenger loading code for families.

a quantity to duplicate familysize-1, where the family size is equal to uniform (3,6). After the duplication, again the population is divided into different categories based on whether they have no bags, a single bag, or two bags.

The loading sequence for the first class passengers is essentially the same as that discussed in the ordinary class passengers, except that the first class passengers are considered a different set of entities, and they get seated in a different section of the plane (the front portion of the plane), and then the other set of passengers get seated.

The animation section shows the actual loading of the sequence of the system in the form of moving pictures and hence makes it possible to visualize the system, which in turn helps in the face validity of the model. The animation section of the aircraft when the model is not run was shown before. The following picture shows the animation when the model is running. The picture shows the passengers coming down the ramp and waiting in queue while the other passengers seat themselves in their seats. The physically handicapped passengers sit in the first row. The screen capture of the system shows the existing model, where the passenger sequence is physically challenged, first class, family members, back third, middle third, and the front third. Each category of passenger is represented with a particular color. Ordinary passengers are represented with a blue color, physically challenged with red, family members with brown, and first class passengers with black color. The screen capture of the animation while the model is running is shown in Figure 13.10, here depicted in blank and white. Figure 13.11 shows the animation when the model has run to completion.

Once the model has been created, the model requires validation to check whether the model represents the actual system. The process to verify and validate the model is presented in the next section.

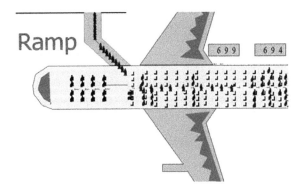

FIGURE 13.10 Model animation during loading.

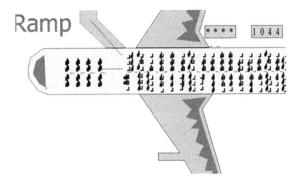

FIGURE 13.11 Model animation after loading.

13.3.4.6 Verification and Validation

Verification is a process that confirms whether the model behaves as it was intended to behave. It checks at a very basic level whether the model developed actually functions properly. It also checks the logical flow of the system.

Validation checks whether the model behaves the same as the real system.

Face validity was carried out by showing the model to an expert. The expert gave his opinion on whether the model could be taken to represent the actual system. In this research, face validity was carried out by experts from Continental Airlines.

Statistical validity was carried out by statistically comparing the outputs from the actual system with the data from the model that was developed using ARENA. Eight data points were available from the actual data of the system time. The simulation model was also run for eight replications and the system time (loading time) was obtained. The following standard algorithm was followed while carrying out the statistical validity. The model data and real data were limited to quantity and assumed to be non-normal.

	System Data and Validation Data	
	System	Model Alt 01c
Average	20.08	19.51
Standard deviation	2.98	1.86

Because the model and the actual system data were assumed not to follow normal distribution, a "nonparametric U test" or "the rank sum test" was carried out to check the validity of the model. The process starts with mixing the validation data and the model data, arranging them in ascending order, and finding the ranks of the two sets. Once the ranks were found, they were added for the validation data and the model data.

The sums of the ranks for the two sets of data were called W_1 and W_2, respectively. Then the following formulas were used to conduct the test:

H_o = The two sets represent similar systems: Null Hypothesis.
H_1 = The two sets represent different systems: Alternate Hypothesis.

$$U_1 = W_1 - \frac{n_1(n_1+1)}{2}$$

$$U_2 = W_2 - \frac{n_2(n_2+1)}{2}$$

and the parameter U is equal to the smaller of the two. For the given data:

$$W_1 = 71, W_2 = 65$$

$$n_1 = n_1 = 10$$

and

$$U = U_2 = 29$$

Mean and variance for the parameter U are found using the following formulas:

$$\text{Mean} \quad = \mu_{U_1} = \frac{n_1 n_2}{2}$$

$$\text{Variance} \quad = \sigma^2_{U_1} = \frac{n_1 n_2(n_1+n_2+1)}{12}$$

$$Z = \frac{U_1 - \mu_{U_1}}{\sigma_{U_1}}$$

In this case, mean = 32, variance = 53.33, and $Z = -0.41$.

Because $Z = -0.41$, which is between 2.575 and –2.575, the confidence intervals for $\alpha = 0.01$ level of significance, the null hypothesis cannot be rejected. Hence, the model is validated.

Once the model had been verified and validated, then experiments were carried out after finding the necessary number of replications. The number of replications was determined using the relative precision method. Once the number of replications that the model needs to be run was determined, the model was ready for experimentation.

13.3.4.7   Experimentation

Experimentation is the process of designing alternative systems to the existing system so as to improve the performance of the existing system. The various alternatives in this research focused on changing the loading sequence of the passengers so as to reduce the loading time. There were six major categories of passengers existing in the system and changing their sequence leaves about 6! = 720 alternatives to the existing system including the present system. Of these alternatives, most were not feasible in effect. The majority of alternatives conflict with the general rules that are currently in effect for the loading of passengers. The general rules were: The first class passengers cannot be loaded before the general class passengers. No other sequences were accepted in practicality. Hence, each alternative was analyzed in view of the practical feasibility of each loading technique and the best alternative to reduce the overall time for loading the passengers. The following eight alternatives were suggested, including the existing system, so as to improve the system with respect to the loading sequence.

13.3.4.7.1   *Alternative Notations*

Physically challenged–PC
Children/family–Family
First class–F
Back third–BT
Middle third–MT
Front third–FT

13.3.4.7.2   *Loading Sequence Alternatives*

Alternative 1 PC, F, Family, BT, MT, FT (existing system)
Alternative 2 Family, PC, F, BT, MT, FT
Alternative 3 F, BT, MT, FT, PC, Family
Alternative 4 F, BT, MT, FT, Family, PC
Alternative 5 PC, F, BT, MT, FT, Family
Alternative 6 Family, F, BT, MT, FT, PC
Alternative 7 F, PC, Family, BT, MT, FT
Alternative 8 F, Family, PC, BT, MT, FT

The seven alternatives were created in ARENA 5.0. There are no major differences in the seven alternative models when compared to the base model with respect to the blocks used. The differences arise from the difference in the times when the entities are created in the models and the difference in the delay time for the physically challenged and the family members. These changes make the required changes to the model, and thus the required alternatives are created. Once the alternatives were created, they were run, and the results for the loading times for various alternatives were analyzed.

13.3.4.8   Analysis

The analysis stage involves the process of evaluating the number of replications for the alternatives, conducting ANOVA for the various alternatives, and, based on its outcome, whether or not a Duncan multiple-range test is carried out. The number of replication is necessary for the model and its alternatives so that the models results could be void of any errors. ANOVA is used to compare the means initially and Duncan multiple-range test is used to check whether which alternative is better among the rest.

13.3.4.8.1   *Replication Analysis*

After the models were generated for the alternatives, it was necessary to find the number of replications that each alternative should be run so that the model and the alternatives could be compared. The number of replications needed was found using the relative precision method in the following manner.

   Each alternative is run at first for 10 replications. Then the half-width confidence interval for the alternative is found, and then it is divided by the mean, and the resulting value is checked to be less than or more than 0.09. If it is less than 0.09 then the models are run at 10 replications. Otherwise, the replications are increased till the half-width confidence interval divided by the mean is less than 0.09. The process is carried out for each of the alternatives, and the highest value for the number of replication required is taken to be the number of replications required to compare the alternatives.

   The formula for half-width confidence interval (C.I.) is:

$$C.I. = t_{\alpha/2} \cdot \frac{s}{\sqrt{n}}$$

where, s is the standard deviation of any particular model run and n is the number of replications.

Once the number of replication is determined, the models are run for the required number of replications, and the results are then utilized to see if there are any major differences in the means of the various alternatives. ANOVA is used to check whether there is any significant difference in the means.

13.3.4.8.2   *Analysis of Variance (ANOVA)*

ANOVA was used first to find whether there were any significant differences in the alternatives that had been analyzed. The following formulas are used to carry out ANOVA.

$$\text{Total Sum of Squares (TSS)} = \sum_{i=1}^{k} \sum_{j=1}^{ni} (y_{ij} - \bar{y})2$$

$$\text{Between Sum of Square (BSS)} = \sum_{i=1}^{k} ni\ (y_{ij} - \bar{y})2$$

$$\text{Error Sum of Square (ESS)} = \sum_{i=1}^{k} \sum_{j=1}^{ni} (y_{ij} - \bar{y}_i)2$$

and it is known that TSS = BSS + ESS, where y represents the data points and i the respective columns for the data for the various alternatives.

H_o: (Null Hypothesis) There is no significant differences between the alternatives.

H_1: (Alternative Hypothesis) There is significant difference in the means of the alternatives.

The ratio shown below has an F distribution with $k - 1$ and $N - k$ degrees of freedom where k is the number of alternatives or number of groups in general and N is the total number of data points available from all the alternatives.

$$F = \frac{BSS\,/\,(k-1)}{ESS\,/\,(N-k)}$$

The F value above is compared to the critical value of F from the F table with $(k - 1)$ and $(N - k)$ degrees of freedom.

If, F crit < F, Reject H_o then, there is significant difference in the alternatives and if it turns out otherwise then cannot reject H_o and there is not much difference between the alternatives. If it turns out that there is significant difference between the means then the analysis is continued to Duncan multiple-range test to find which alternative is better than the other as ANOVA only tells whether there is a statistical difference between the alternatives. It does not tell which alternative is better and which is not.

13.3.4.8.3   *Duncan Multiple-Range Test*

The Duncan multiple-range test involves the process of comparing the means of the various alternatives to find which alternative is better than the other statistically at a particular level of significance.

The test compares the range of any set of p means with an appropriate least significance range, R_p, given by

$$R_p = S_x r_p$$

Here S_x is an estimate for the standard deviation and is computed by the formula:

$$S_x = \sqrt{\frac{MSE}{n}}$$

where MSE is the mean square of error from ANOVA. The value of r_p depends on the desired level of significance and the number of degrees of freedom corresponding to MSE and it obtained from tables for a particular α value (generally 0.05).

13.3.5 Results and Conclusions

13.3.5.1 Replication Analysis

Replication analysis was done to find the number of replications. The simulation models were run for 10 replications for all the alternatives and the value for the relative precision was found to be less than 0.09. Hence, 10 replications is sufficient for analyzing the alternatives further.

System	Alt 01d	Alt 02d	Alt 03d	Alt 04d	Alt 05d	Alt 06d	Alt 07d	Alt0 8d
Average	19.85	18.21	21.64	20.81	21.22	19.03	20.95	22.56
Stdev	1.79	1.19	1.66	1.73	1.97	1.35	1.53	1.56
T Value	2.69	2.69	2.69	2.69	2.69	2.69	2.69	2.69
Half Width Conf-int	1.52	1.01	1.41	1.47	1.68	1.14	1.30	1.32
Relative precision value	0.08	0.06	0.07	0.07	0.08	0.06	0.06	0.06

13.3.5.2 ANOVA

After conducting ANOVA on the alternatives it is seen that there were significant differences in the alternatives.

ANOVA

Source of Variation	SS	df	MS	F	P-value	F crit
Between groups	141.79	7.00	20.26	7.77	0.00	2.14
Within groups	187.73	72.00	2.61			
Total	329.52	79.00				

Because *F crit* < *F*, there is significant difference between the means.

13.3.5.3 Duncan Multiple-Range Test

Duncan multiple-range test was carried out to check which alternative was best among the rest. The results were as obtained in this table.

P	2	3	4	5	6	7	8
r_p	2.82	2.97	3.06	3.13	3.19	3.23	3.27
R_p	1.43	1.51	1.55	1.59	1.62	1.64	1.66

$$R_p = S_x . r_p$$

$$S_x = \sqrt{\frac{MSE}{n}} = \sqrt{\frac{2.6}{10}} = 0.509$$

where:

    p = The number of the mean to be compared

    r_p = Multiplier with the standard deviation

    R_p = Least significant range

Comparing the ranges with the critical values we find the following means to have significant differences:

Means

Alt1	Alt2	Alt3	Alt4	Alt5	Alt6	Alt7	Alt8
19.85	18.21	21.64	20.81	21.22	19.03	20.95	22.56

Means in Ascending Order

Alt2	Alt6	Alt1	Alt4	Alt7	Alt5	Alt3	Alt8
18.21	19.03	19.85	20.81	20.95	21.22	21.64	22.56

Interpretations of the above results are:

- There is no statistical difference between alternative 2 and alternative 6.
- There is no statistical difference between alternative 6 and alternative 1.
- There is no statistical difference among alternative 1, alternative 4, alternative 7, alternative 5.
- There is no statistical difference among alternative 4, alternative 7, alternative 5, alternative 3.
- There is no statistical difference among alternative 5, alternative 3, alternative 8.

Hence, alternative 2 is statistically significantly better than alternative 1 (base model) at an α level of 0.05. This means that alternative 2 comprising of the loading sequence "Family, PC, F, BT, MT, FT" has a faster loading time than the base alternative "PC, F, Family, BT, MT, FT" model or any other alternatives. Hence it is seen that if the family members along with the small children were allowed to load before the other group of passengers, the overall loading time is significantly reduced. The reason for the reduction could be that the family members tend to sit together, and if one family is in the queue waiting to get to their row then the whole group of people belonging to the family is delayed, as all the members tend to go to the same row. Hence, it seems to be a better idea to send the family members first to occupy their seats and then send all the other categories of passengers. Also, the current sequence of loading operated by the airline industry is the second best loading process, and the industry need not resort to any other loading sequence because all the rest of them have been proved in this research to be increasing the loading time.

13.3.6 Contribution to the Body of Knowledge

The body of knowledge is benefited by the knowledge of the correct loading sequence required for reducing the loading time for passengers in aircraft. It allowed the elimination of faulty and inappropriate loading sequences that seemed to be contributing to huge delays in loading passengers and the loss of millions of dollars. The major advantage of this work is that it helps to change the existing situation, taking into consideration the logical and economic feasibility of the loading sequence. Logical feasibility deals with analyzing the various alternatives available for loading without violating the set standards of sequence that need to be always followed while loading the passengers in an aircraft. For example, the ordinary class passengers cannot be loaded before the first class. Economic feasibility deals with changing the system without any major change in the resources available or without any major cost involved with the change. To implement the suggested alternative, the carrier need not spend any additional money to improve its existing system. It would just need to change its sequence of loading so as to improve the loading time. Thus, this work will help in the reduction of flight delays that are crippling the airline industry without any major changes to their operating policy and without major involvement of cost.

13.3.7 Future Work

In this research there is the basic assumption that all the passengers are ready to board once the aircraft is ready for boarding. But, that might not be the case in real scenarios because there are always some passengers who come in late for boarding the flight. Hence, if the factor of passengers coming in late to board the aircraft is also included in the model, then the model would be more robust and better combinations of loading sequences could be found so as to reduce the loading time for the passengers.

References

1. Air Transport Action Group report, URL: *http://www.atag.org/ECO/default.htm.*
2. Travel Guide News, URL: *http://europe.cnn.com/TRAVEL/NEWS/9910/15/flight.delays/.*
3. Thibodeau, P., Airport Delays Cost $825M, *Computer World*, 33(42), October 18, 1999, p. 16.
4. Department of Transportation, Air Travel Consumer Report, August 2001, URL: *http://www.dot.gov/airconsumer/0108atcr.DOC.*
5. Department of Transportation (DOT) Report No. CR-2000–112, July 25,2000.
6. Lee, M. and Cunnigham, L.F., Customer Loyalty in Airline Industry, *Transport. Q.*, 50(2), 1996.
7. Dhir, K.S. and Wallace, I.L., Assessment of airline service quality by passengers, travel agents and airline employees: an application of social judgment theory, Annual Meeting of the Decision Sciences Institute, November 24–26, 1996.
8. Snowdon, J.L., El-Taji, S., Montevacchi, M., Macnair, E., and Miller, S., Avoiding the blues for airline travelers, Winter Simulation Conference Proceedings, 1998, IEEE pp. 1105–1112.9.
9. Cheng, Y., Knowledge-based airport gate assignment system integrated with mathematical programming, *Comput. Indust. Eng.*, 32, 837–852, 1997.
10. Haghani, A. and Chen, M.C., Optimizing gate assignments at airport terminals, *Transport. Res.*, 32, 437–454, 1998.
11. Bolat, A., Assigning arriving flights at an airport to the available gates, *J. Oper. Res. Soc.*, 50, 23–34, 1999.
12. Barros, A.G. and Wirasinghe, S.C., Location of new large aircraft gate positions in pier terminals," Proceedings-International Air Transportation Conference, June 18–21, 2000.
13. Wald, M.L., As jumbo jets grow so does loading time, *New York Times*, June 10, 2001, p. 3.
14. Vandebona, U. and Allen, D., Passenger flow distributions at airports, *Transport. Res. Rec.*, 1423, 40–46, 1993.
15. Chang, C. and Schonfeld, P., Flight sequencing in aircraft hub operations, *Transport. Res. Rec.*, n1506, July 1995.
16. Ng, K.Y.K., A multi criteria optimization approach to aircraft loading, *Op. Res.*, 40(6), 1200–1205, 1992.
17. Pegden, D.C., Shannon, R.E., and Sadowski, R.P., *Introduction to Simulation Using SIMAN*, 2nd ed., McGraw-Hill, New York, 1995.

13.3.8 APPENDIX

13.3.8.1 Input Data–Fitting Distributions

The following pages show the fit distributions for the input data collected.

The passengers with no bags have been assumed to be taking 1 sec to seat themselves for all the classes of passengers.

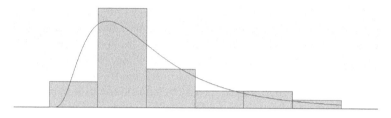

FIGURE 13.12 Distribution summary (ordinary passengers, 1 bag).

Distribution: Lognormal
Expression: 1 + LOGN(12.7, 9.17)
Square Error: 0.015500
Chi-square test
    Number of intervals   = 4
    Degrees of freedom   = 1
    Test statistic   = 2.63
    Corresponding p-value   = 0.108
Kolmogorov–Smirnov test
    Test statistic   = 0.0883
    Corresponding p-value   > 0.15
Data summary
    Number of data points   = 48
    Min data value   = 1.88
    Max data value   = 34.2
    Sample mean   = 13.4
    Sample std dev   = 7.75
Histogram summary
    Histogram range   = 1 to 35
    Number of intervals   = 6

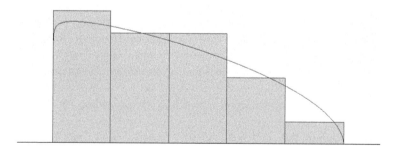

FIGURE 13.13 Distribution summary.

Distribution: beta
Expression: 4 + 35 * beta(1.04, 1.54)
Square error: 0.003651
Chi-square test
    Number of intervals   = 3
    Degrees of freedom   = 0
    Test statistic   = 0.0982
    Corresponding p-value   < 0.005

Kolmogorov–Smirnov test
    Test statistic   = 0.154
    Corresponding p-value > 0.15
Data summary
    Number of data points = 20
    Min data value   = 4.63
    Max data value   = 38
    Sample mean   = 18.1
    Sample std dev   = 9.06
Histogram summary
    Histogram range   = 4 to 39
    Number of intervals   = 5

FIGURE 13.14 Distribution summary (family, 1 bag).

Distribution: Gamma
Expression: 8 + GAMM(18.6, 0.196)
Square Error: 0.259761
Kolmogorov–Smirnov test
    Test Statistic   = 7.42
    Corresponding p-value < 0.01
Data summary
    Number of data points = 2
    Min data value   = 8
    Max data value   = 15.3
    Sample mean   = 11.7
    Sample std dev   = 5.17
Histogram summary
    Histogram range   = 8 to 16
    Number of intervals   = 5

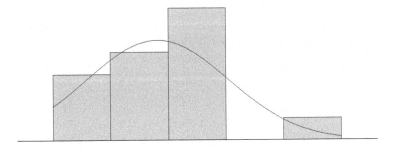

FIGURE 13.15 Distribution summary (family, 2 bags).

Distribution:Normal
Expression: NORM(19.1, 6.15)
Square error: 0.044003
Kolmogorov–Smirnov test
    Test statistic   = 0.123
    Corresponding p-value   > 0.15
Data summary
    Number of data points   = 14
    Min data value   = 10.2
    Max data value   = 35
    Sample mean   = 19.1
    Sample std dev   = 6.38
Histogram summary
    Histogram range   = 10 to 35
    Number of intervals   = 5

FIGURE 13.16 Distribution summary (physically challenged, 1 or 2 bags).

Distribution: Uniform
Expression: UNIF(33, 88)
Square error: 0.133333
Kolmogorov–Smirnov Test
    Test statistic   = 0.289
    Corresponding p-value   > 0.15
Data summary
    Number of data points   = 6
    Min data value   = 33
    Max data value   = 87.3
    Sample mean   = 61.9
    Sample std dev   = 23
Histogram summary
    Histogram range   = 33 to 88
    Number of intervals   = 5

13.3.8.2 Duncan Multiple-Range Test

P	2	3	4	5	6	7	8
rp	2.82	2.97	3.06	3.13	3.19	3.23	3.27
Rp	1.43	1.51	1.55	1.59	1.62	1.64	1.66

$$R_p = r_p{}^* S_x$$

$$S_x = [\text{mean square error (MSE)}/n]^{\wedge}0.5 = = (2.60/10)^{\wedge}0.5 = 0.509$$

Comparing the ranges with the critical values we get the following means to be significantly different:

Means

Alt1	Alt2	Alt3	Alt4	Alt5	Alt6	Alt7	Alt8
19.85	18.21	21.64	20.81	21.22	19.03	20.95	22.56

Means in Ascending Order

Alt2	Alt6	Alt1	Alt4	Alt7	Alt5	Alt3	Alt8
18.21	19.03	19.85	20.81	20.95	21.22	21.64	22.56

Hence, alternative 2 is significantly better than alternative 1 (base model).

13.4   Multitheater Movie Complex

The following example report is based on a simulation modeling and analysis project conducted in the entertainment industry. In this project, the operation of a multitheater movie complex was modeled and analyzed.

13.4.1   Problem Definition

A current trend in the movie theater industry is to build large complexes that show several movies simultaneously. With a number of miniature theaters, these complexes allow the theater management to screen movies that appeal to larger sectors of the public. The close physical proximity of the miniature theaters provides management with the additional advantage of being able to pool ticket- and concession-selling resources.

As with any system that seeks to minimize operating costs and utilizes multiple resources, two questions immediately arise. First, at what level should the resources be staffed to attain a given level of performance? Second, if the desired number of employees are not available, how should the remaining employees be configured? With multitheater movie complexes, the first question translates into how many ticket-selling and concession-selling employees should be utilized to operate the theater so that customers pass through the system in a given amount of time. Similarly, the second question translates into configuring any less than normal number of ticket- and concession-selling employees so that the degradation of system time is minimized.

In order to seek answers to these questions, the theater management may physically configure the ticket and concession sellers and then attempt to ascertain the effect of the configuration on the customer. Operation in this manner will have the effect on one extreme of increasing operating costs, and on the other extreme of alienating customers. An alternative to physically configuring the system is to conduct a computer simulation and analysis. Because this approach simulates the operation of the system, it eliminates disadvantages commonly associated with investigating physical configurations.

Although this simulation and analysis may be applicable to all multitheater movie complexes, a multitheater complex with ten theaters was utilized for this model. It should be noted that access to this complex was graciously provided by the complex owner, and invaluable assistance was obtained from the manager and assistant managers.

13.4.2   Project Planning

The activities for this simulation and analysis were planned according to the 10-step simulation study process. These steps directly correspond to the chapters listed in the Table of Contents.

The project-planning sequence began with the development of a timetable for the procedure steps. An initial 3-day period was allocated for system observation and data collection functions. A period of 3 weeks with three 4-h sessions per week was designated for model formulation, data preparation, model translation, verification and validation, experimental design, and experimentation. One week was allocated for analysis, interpretation, and report generation. A final week was allocated for presentation preparation.

13.4.3   System Definition

13.4.3.1   Basic System Definition

During the initial visit, on day 1 of the system observation and data collection period, the theater management provided an informal briefing on the operation of the complex and a tour of the facilities. During the briefing, certain system parameters during peak operating hours were determined. During the high-activity periods of interest, Friday and Saturday nights, 7 and 9 P.M. showings, a typical config-uration would include four ticket sellers, six concession sellers, and two ticket collectors. A basic schematic of this system is illustrated in Figure 13.17.

The basic flow of movie viewers consists of purchasing a ticket from one of the four ticket sellers, possibly purchasing concessions, and finally having the ticket torn by the ticket collector before entering the theater area.

The flow time for this system is defined as the interval between the arrival time of the customer in the system and the time that the customer is given back a torn ticket and passes to the theater area. The layout schematic does not include some features of the actual layout. These include a game room located on the side of the lobby and the men's and women's restrooms. The absence of these components is addressed in the assumptions section of this report.

13.4.3.2   Period of Interest

The period of interest for this simulation occurs during the 7 and 9 P.M. screenings on Friday and Saturday nights. At each one of these screening periods, high activity occurs from approximately 15 min before the hour until the end of the hour. After the end of the hour, the system empties out until the next period of high activity.

13.4.3.3   Measures of Performance

For this simulation and analysis, the primary measure of performance will be customer flow time. This is defined as the interval of time beginning when the customer queues for buying a ticket and ending when the customer is done with the ticket collector. Secondary measures of performance include the time-average queue size and the worker utilization for the ticket sale, concession sale, and ticket collection processes.

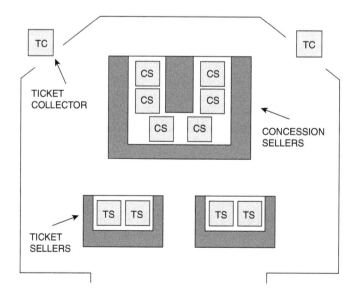

FIGURE 13.17 Movie theater layout.

13.4.4   Model Formulation

After defining the basic system as illustrated in Figure 13.17, a preliminary model of the complex was developed. This model was utilized to determine the components in the system and the types of data that would be required. The significant components in the model are the arrival, ticket-selling, concession-selling, and ticket-collecting processes. For each block in the model, the upper lines of text indicate the action that is being performed. The lower lines of text specify what type of data must be collected.

13.4.4.1   Input Data

Input data consist of interarrival data, ticket-selling data, and concession-selling data. Arrival data include interarrival times, batch arrival sizes, and whether the batches split into smaller groups. Ticket-selling data include ticket seller selection and service time. The concession purchase decision is whether the customer will buy concessions or proceed directly to the ticket collector. The concession and ticket collection data include travel time to the process, server selection, and service time.

13.4.4.2   Assumptions

The initial system model is subject to a number of simplifying assumptions. These are necessary in order to reduce the model to a feasible level of complexity.

13.4.4.2.1   *Game Room*

The lobby of the complex contains a small game room with four arcade-type video games. This subsystem was not modeled.

13.4.4.2.2   *Restrooms*

Restroom activity was not modeled.

13.4.4.2.3   *Balking and Reneging*

It is assumed that the movie viewers will not leave the complex immediately after arrival or after entering one of the system queues.

13.4.4.2.4   *Breaks*

It is assumed that the system servers will not take breaks during the high-activity periods of interest.

13.4.4.3   Grouping

Batch splitting into groups is defined as the division of an arrival batch into two or more groups, each of which operate through the system independently until proceeding to the ticket collector. This specifically assumes that each group of one or more customers will pay for all the members' tickets at the same time, decide whether to purchase concessions as a group, and will wait until all groups are ready to proceed to the ticket collector before proceeding to the ticket collector as a reassembled batch.

13.4.4.4   Entry to the Theater

It is assumed that customers are allowed in the theater after processing through the system.

13.4.4.5   Data to Collect

As indicated in the conceptual model section, the arrival, ticket-selling, concession-selling, and ticket-collection data must be collected. These include interarrival times, batch sizes, group splitting, travel times, server selection, and service times.

13.4.5   Input Data Collection and Analysis

13.4.5.1   Input Data Collection

Data were collected over a 3-day (Friday, Saturday, Sunday) period. A decimal hour stopwatch was utilized to time interarrival, travel, and service times. All time measurements unless specifically stated are expressed in thousandths of hours.

13.4.5.2   Input Data Analysis

The following data-fitting results were obtained from the raw data. The SIMAN IV input processor was utilized for the fitting of continuous-type data. Discrete data fitting for batch sizes was attempted with UNIFIT II. Bernoulli-type data were tabulated in LOTUS 123. Finally, some data could not be successfully fit. In these cases, empiric values were utilized for that parameter.

13.4.5.2.1   *Interarrival Times of Batches*
This is defined as the time in thousandths of hours between the arrival of batches of movie viewers to the ticket seller or ticket seller queue. The summary statistics for this data set were:

Data pts = 321
Mean   = 2.59
Std dev   = 2.76
Min   = 0.1
Max   = 24.6

The data were determined to be best fit with an Erlang distribution with parameters of 2.59 and 1.

13.4.5.2.2   *Batch Size*
This is defined as the number of individual customers who arrive together as a batch. This usually occurs when a family or group of people arrive in the same vehicle. A data fit for the batch sizes was attempted with DATAFIT II. The table below summarizes the result of the fit. As can be determined, each of the four discrete distributions offered a very poor chi-square test fit. As a result, empiric data were used in the SIMAN model to generate batch sizes.

13.4.5.2.3   *Group Splitting*
Splitting is defined as a batch splitting into two or more groups, each of which paid for its tickets independently. Assumptions concerning group splitting are addressed in the conceptual model formulation section. These data were tabulated in LOTUS 123 with the following results.

Batch size of 2
0.814 Stay together
0.186 Split into two groups of one individual each.

Batch size of 3
0.676 Stay together.
0.270 Split into one group of 2 and one group of 1
0.054 Sprit into three groups of 1

Batch size of 4
0.778S tay together.
0.222 Split into four groups of 1.

All other batches stay together.

13.4.5.2.4   *Ticket Seller Service Time*
This is the time required for the ticket seller to serve the group. This distribution was fit with the SIMAN IV Input Processor. This process had the following summary statistics:

Data pts = 166
Mean   = 4.22
Std dev   = 1. 64
Min   = 1.4
Max   = 19

The best fit for this model was 1 + Erlang (9.896, 4)

13.4.5.2.5 *Buy Concessions?*

This is a Bernoulli distribution based on whether the independently functioning group buys concessions. These data were tabulated and summed in an electronic spreadsheet.

0.544 Buys concessions
0.456 Does not buy concessions

13.4.5.2.6 *Travel Time to the Concession Stand*

This is defined as the time required to travel between the ticket seller and the concession stand. These data were fit with the SIMAN Input Processor. This distribution had the following summary data:

Data pts = 42
Mean   = 2.12
Std dev   = 9.91
Min   = 1.4
Max   = 5

This distribution was best fit as a Weibell distribution with 1.93 + WEIB(1.91, 2.24).

13.4.5.2.7 *Concession Seller Service Time*

This is defined as the time required for the concession seller to service the group at the concession stand. This distribution was fit with the SIMAN IV Input Processor. The result is illustrated in Figure 13.18.

13.4.5.2.8 *Travel Time to the Ticket Collector*

This is defined as the time required to travel between the concession stand and the ticket collector. Note that if the group does not purchase concessions, then the travel time between the ticket seller and the ticket collector is the sum of the time required to travel between the ticket seller and the concession stand and the time required to travel between the concession stand and the ticket collector. These data were fit with the SIMAN IV Input Processor. The result is illustrated in Figure 13.19.

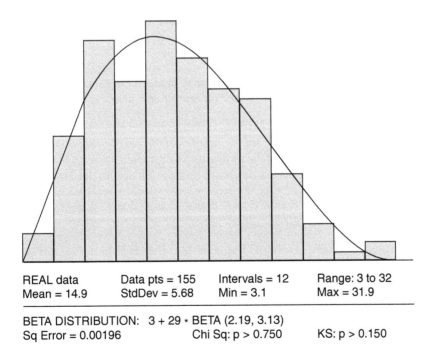

REAL data	Data pts = 155	Intervals = 12	Range: 3 to 32
Mean = 14.9	StdDev = 5.68	Min = 3.1	Max = 31.9

BETA DISTRIBUTION: 3 + 29 * BETA (2.19, 3.13)
Sq Error = 0.00196   Chi Sq: p > 0.750   KS: p > 0.150

FIGURE 13.18 Distribution fitting for concession seller service time.

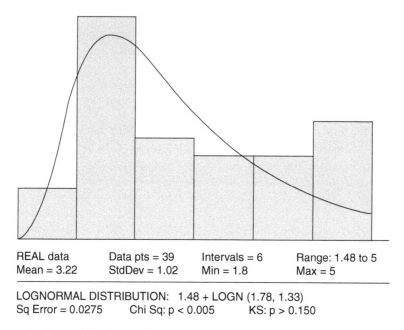

REAL data   Data pts = 39   Intervals = 6   Range: 1.48 to 5
Mean = 3.22   StdDev = 1.02   Min = 1.8   Max = 5

LOGNORMAL DISTRIBUTION: 1.48 + LOGN (1.78, 1.33)
Sq Error = 0.0275   Chi Sq: p < 0.005   KS: p > 0.150

FIGURE 13.19   Distribution fitting for travel time to ticket collector.

13.4.5.2.9   *Ticket Collector Service Time*

This is defined as the time required for the ticket collector to accept, tear, and return the tickets to the group and allow the group to pass. This distribution was fit using the SIMAN IV Input Processor.

13.4.6   Model Translation

The model translation phase consists of selecting a simulation software package to model the system and to perform the actual process of programming the model.

13.4.6.1   Software Packages

The simulation software package SIMAN IV was used to model the system. The simulation software packages CSIMAN IV and CINEMA IV were used for generating a system animation.

13.4.6.2   Modeling Translation Approach

During the modeling translation phase, the system modeling and the animation layout development were conducted simultaneously. This permitted the verification of the programming code with the animation, at each stage of development, as the project progressed from the initial to the final model. This approach minimized the development time while having each successive version provide a more accurate representation of the system.

13.4.6.3   Model Code

This section consists of a description of the operation of the model. Throughout the model, generic stations were employed. This permitted the system to be modeled within the Microlab version limit of 100 blocks. In other cases, the use of normal functions such as PICKQ were replaced with FINDJ to permit further block reductions or allow the model to be properly animated.

13.4.6.4   Arrival of Customers

The arrival of customers is modeled using a CREATE block with the interarrival and offset time as specified in Section 13D.5.2.1. The time of arrival is assigned to the entity attribute arrtime using the mark modifier. Also at the time of creation, the attributes collside, corrside, symbol, and travcoll time are assigned.

13.4.6.4.1 *Collside Attribute*

This attribute is used only by batches that split into groups. This designates which ticket collector the group will proceed to after reassembling group following ticket and concession purchases. Groups have a 50% chance of proceeding to one collector or the other. A value of 0 instructs the group to proceed to the left collector. A value of 1 instructs the group to proceed to the right collector.

13.4.6.4.2 *Corrside Attribute*

This attribute is used to determine if the group has proceeded to the ticket collector on the correct side of the complex. If the group has chosen incorrectly, this means that the desired movie is being shown in the part of the complex serviced by the other collector. Groups have a 50% chance of having made the correct selection. A valueof 0 signifies an incorrect selection. A value of 1 signifies a correct selection.

13.4.6.4.3 *Symbol Attribute*

This attribute performs the dual purpose of being an animation symbol and determining the number of customers in the batch before splitting into groups. The number of customers in the batch is determined by the empiric distribution parameter specified in Section 13.4.5.2.2 and the experiment file.

13.4.6.4.4 *Travcoll Attribute*

This attribute is the time required for the entity to travel between the concession stand and the ticket collector. If a batch splits into groups, it is assumed that the group will travel together from the assembly point to the ticket collector.

13.4.6.5 Batch Splitting into Groups

All batches less than or equal to 4 are subject to splitting into two or more groups. This process is modeled through a series of BRANCH, ASSIGN, and DUPLICATE blocks. Depending on the entities' batch size, the entity will be sent a specific sequence of blocks. For a given sequence of blocks, the batch will be split into two or more groups. This is performed by assigning the entity the values of btchgrps, grpindex, and a new symbol. After receiving the new attributes, each group/entity proceeds through the ticket and concession selling processes independently. If the batch entity does not split into groups, it proceeds as a group with the original batch size directly to the ticket-selling process.

13.4.6.5.1 *Btchgrps Attribute*

This attribute contains the value of the number of groups that the batch splits into. This is necessary for reassembling the batch before proceeding to the ticket collector.

13.4.6.5.2 *Grpindex Attribute*

This attribute contains the number of this entity's group in the original batch. Thus, for a batch that splits into two groups, one entity/group will have a grpindex of 1, and the other will have a grpindex of 2.

13.4.6.5.3 *New Symbol Attribute*

If the batch splits, the entities will represent a different number of customers. A batch entity that originally had a symbol of two may now have two groups of one, each with a symbol of 1, representing one customer.

13.4.6.6 Buying the Ticket

Before actually purchasing the tickets, some administrative functions are performed including the assignment of values to the global variable total and the entity attributes travconc and gotoconc. After this ASSIGN block, the entity chooses a ticket seller and then goes through the normal QUEUE-SEIZE-DELAY-RELEASE sequence.

13.4.6.6.1   Total Global Variable
This variable is incremented by the number of customers represented by the entity. It is used to determine the number of customers in the system.

13.4.6.6.2   Travconc Attribute
This attribute holds the travel time for the group to go from the ticket selling station to the concession stand. Travel time is determined by the distribution specified in the data collection section.

13.4.6.6.3   Gotoconc Attribute
This attribute is a logical that determines if the group will purchase concessions. A value of 1 instructs the group to proceed to the concession stand.

13.4.6.6.4   Which Ticket Seller?
The groups pick from among many ticket sellers. The server is selected who is either idle or has the smallest number of groups waiting in line. This is modeled with a FINDJ block. It was necessary to use this block in order for the animation to function correctly.

13.4.6.6.5   Generic Stations for Ticket Selling: QUEUE-SEIZE-DELAY-RELEASE Sequence
A generic station sequence was utilized for the QUEUE-SEIZE-DELAY-RELEASE procedure for each of the ticket sellers. This allowed an overall reduction in the number of blocks for the model.

13.4.6.6.6   Does the Group Want Concessions?
This question is modeled with a BRANCH block. If the group wants concessions, it proceeds to the concession area. If the group does not want concessions and it was part of a batch that split into two or more groups, it proceeds to an assembly area to await the arrival of all the other groups in the original batch. If the group is the same as the original batch, it proceeds to either the left or right ticket collector.

13.4.6.6.7   Traveling Directly to the Ticket Collector
If the group does not want concessions, it will travel directly to the ticket collector. This is modeled using the ROUTE block with a travel delay equal to the sum of the attributes travconc and travcoll.

13.4.6.6.8   If the Group Wants Concessions
If the group wants concessions, the concession seller is selected who is either idle or has the smallest number of customers in line or en route to the line. This is modeled using a FINDJ block in a manner similar to the ticket seller FINDJ block.

13.4.6.6.9   Generic Stations for the Concession QUEUE-SEIZE-DELAY-RELEASE Sequence
Generic stations are utilized for this process in a manner similar to the ticket-selling sequence.

13.4.6.6.10   Was the Group Split?
If the group finished with the concession stand was part of a batch that split, the group proceeds the assembly area to await the arrival of all other groups in the batch. If the group was not split, it travels to the nearest ticket collector with the ROUTE block.

13.4.6.6.11   Assembly Area for Split Group Batches
All batches that were split into two or more groups must assemble here before proceeding to the ticket collector. This process is modeled with a set of detached QUEUES and a MATCH block for each batch size. As each group is finished at either the ticket selling station or the concession stand, the groups wait in a QUEUE for the other groups from the batch. When the correct number of groups arrive with a common ARRTIME attribute, the groups are MATCHed and released to go to the ticket collector.

13.4.6.6.12   Split Groups Traveling to the Ticket Collector
When split groups are reassembled to travel to the ticket collector, a slight delay is imposed on all groups after the group with the grpindex of 1. This is to represent the succeeding groups following the preceding group and to preserve the identity of each group in the animation. Because the delay is imposed only on succeeding groups, which must queue for the ticket collector, there is little impact on the groups' flow time.

13.4.6.6.13 *Is the Correct Theater on the Chosen Side of the Complex?*

If the group proceeded to the correct side of the complex, the group proceeds to the ticket collector. If the group chose the incorrect side, the group proceeds across the complex to the other ticket collector.

13.4.6.6.14 *Generic Station for Ticket-Collecting QUEUE-SEIZE-DELAY-RELEASE Sequence*

When the group arrives at the correct side of the complex, the group goes through the QUEUE-SEIZE-DELAY-RELEASE sequence for the ticket collector.

13.4.6.6.15 *Exiting the System*

After RELEASEing the ticket collector, the flow time for the group is TALLYed, and the global variable total is decremented by the number of customers in the group. At this point, the group ceases to be of interest. The group, however, does proceeds to the generic STATION theater for animation purposes and then is destroyed.

13.4.6.7   Experiment File

The experiment file specifies the components of the system. These include the attributes, variables, resources, queues, stations, distribution parameters, and statistical expressions. Specific points of interest are the use of the parameter element, index resources, and the output element.

13.4.6.7.1 *Indexed Resources*

Indexed resources are utilized for the ticket sellers, concession sellers, and the ticket collectors. The use of indexed resources enables the modeling of generic stations and the use of the FINDJ block for resource selection.

13.4.6.7.2 *Parameters Element*

The majority of the probability distributions in the model file utilize the parameters element in the experiment file. This permits additional flexibility in the event that the parameters for these distributions require changing.

13.4.6.7.3 *Output Element*

At the end of each replication, the primary measure of performance, average flow time, is written to a file for later analysis. This file is eventually loaded into the output processor for determining confidence intervals on the measure of performance.

13.4.7 Verification and Validation

13.4.7.1   Verification

Verification is the process of ensuring that a model performs as intended. Verification was performed utilizing the TRACE element and the system CSIMAN/CINEMA animation. Use of the animation for verification was possible throughout the development of the model because the animation layout was developed simultaneously. In a number of instances, previous models were also utilized to verify the next animation layout.

13.4.7.2   Validation

Validation is the process of insuring that the model represents reality. Both face and statistical validation was performed with the base configuration model. Face validation was performed by demonstrating the model animation to members of the movie theater management. Statistical validity was performed by collecting the system times for customers entering the theater and comparing those times to the system times generated for customers in the model. The model was found to possess both face and statistical validity.

13.4.8 Experimental Design

13.4.8.1 Base and Alternative Configurations

The two primary questions for this simulation were earlier stated as involving the number of employees necessary to maintain a given level of performance and the configuration of the system with a reduced number of employees. These questions may be answered by investigating the performance of the base configuration versus alternate configurations. These configurations may be determined by utilizing a 2 to the 2nd factorial experiment. With this approach, the number of ticket and concession sellers will function as the factors, and different numbers of ticket and concession sellers will utilized for each of the two factor levels.

The base configuration of four ticket sellers and six concession sellers will establish the high factor levels. The configuration with the low factor levels will utilize three ticket sellers and five concession sellers. The two remaining configurations will be four ticket sellers and five concession sellers; and three ticket sellers and six concession sellers.

These configurations will allow the investigation of system performance with other numbers of employees as well as indicate which configuration should be used if it is necessary to operate with a reduced number of employees.

The final experimental design utilizes the same alternatives specified in Section 13D.5.. Because the period of interest clearly specifies a terminating simulation approach, the REPLICATION element of each alternative was set to run for 1250 thousandths of hours with an initial 10 pilot runs.

For each alternative, a 0.95 confidence interval with a relative precision of 0.05 was to be obtained for the group flow times with the SIMAN IV Output Processor. If necessary, additional replications were to be run so that each alternative had the same number of replications with the required 0.95 C.I. at 0.05 relative precision. These data were then to be subjected to an analysis of variance and a Duncan multiple-range test.

13.4.8.2 Experimentation

Each of the four alternatives was run on an IBM PC. The results for each of the 10 pilot runs for each alternative are shown in the following table. These values are measured in thousandths of hours.

Identifier	Average	SD	0.950 C.L Half-Width	Min Value	Max Value	Number of Obs.
CNFG352	26.9	1.41	1. 01	24.1	28.8	10
CNFG362	25.0	1.25	0.89	23.2	27.8	10
CNFG452	25.5	2.39	1.71	21.5	29.9	10
CNFG462	23.9	1.07	0.76	22.5	26.2	10

Using the relative precision method with a 0.95 confidence interval and 0.10 relative precision, configuration 452 requires an additional eight replications.

The REPLICATE element in each one of the alternative experiments was then set to run for 18 replications, with the results shown in the following table.

Flow Times for Alternative Final Runs in Thousandths of Hours

Identifier	Average	SD	0.950 C.L Half-Width	Min Value	Max Value	Number of Obs
CNFG352	26.1	1. 78	0.885	23.1	28.8	18
CNFG362	24.8	1.16	0.575	23.2	27.8	18
CNFG452	24.9	2.01	0.998	21.5	29.9	18
CNFG462	23.6	0.94	0.469	22.3	26.2	18

13.4.8.3 Analysis and Interpretation

13.4.8.3.1 *Analysis of Variance*

The flow time replication data for each alternative was exported to an ASCII file using the SIMAN Output Processor. The data were then imported into an electronic spreadsheet and reformatted for use with MINITAB. The data were reformatted so that the flow time for each replication was in column 1 of the worksheet, and the numbers of ticket and concession sellers were in columns 2 and 3. This resulted in a worksheet that was three columns wide and 72 rows long. Each column was then exported to an ASCII file. These ASCII files were then READ into MINITAB columns, and the ANOVA procedure was executed. The results are summarized below:

Source	DF	SS	MS	F	P
TICKSELL	1	25.19	25.19	10.70	0.002*
CONCSELL	1	28.69	28.69	12.18	0.001*
TICKSELL*CONCSELL	1	0.005	0.005	0.00	0.965
Error	68	160.2	2.355		
Total	71	214.053			

13.4.8.3.2 *Duncan Multiple-Range Test*

The Duncan multiple-range test permits comparisons to be performed on data that have been analyzed with analysis of variance. The Duncan test yielded the following results:

$$Sxbar = SQRT(\text{mean square error/replications}) = SQRT(2.355/18)$$

$$= 0.362$$

The least significant range values are obtained by multiplying Sxbar by the Duncan least significant range values r_p at an α level of 0.05:

$$\text{Least significant range} = Sxbar \times r_p$$

p	r_p	R_p
2	2.83	1.02
3	2.98	1.08
4	3.07	1.11

Configuration Flow Time in Thousands of Hours

CNFG462	CNFG362	CNFG452	CNFG352
23.6	24.8	24.9	26.1

Because the range between CNFG352 and CNFG462 is greater than the R4 value of 1.11, there is a significant difference among the four means. We actually expected this because the ANOVA null hypotheses was rejected. Because the range between CNFG452 and CNFG462 and that between CNFG352 and CNFG362 are both greater than the RJ value of 1.08, there also exists a significant difference among each of these sets of three means. Finally, among the pairs of CNFG352 and CNFG452; CNFG452 and CNFG362; and CNFG362 and CNFG462, there exists a significant difference for the pairs CNFG352 and CNFG452; and CNFG362 and CNFG462. There does not exist a significant difference between the pair CNFG452 and CNFG362. This is represented by the line under this pair of means.

13.4.8.3.3 *Interpretation of Results*

The analysis of variance and the Duncan test provide statistically significant results from the simulation analysis. These results can be utilized to provide answers to the two primary questions of the analysis.

What is the Effect on the System Performance with other Configurations?
The Duncan tests indicate that if the base configuration of four ticket sellers, six concession sellers, and two ticket collectors is reduced to any other number of ticket or concession sellers, then the flow time of the customer groups will be statistically significantly lengthened. In other words, the customer will have to spend more time in the system if the number of employees is reduced in any manner.

If the Number of Available Employees is Reduced, What Should the Configuration Be?
If an employee is absent, flow time will be degraded. A configuration should be chosen so that the flow time degradation is minimized. Because the Duncan test indicated that there is no significant difference in flow time between the three ticket, six concession seller configuration and the four ticket, five concession seller configuration, it does not matter which of these configurations is chosen. This result also indicates that if an employee must leave early, that it is not necessary to replace this employee from another area to minimize the degradation of the flow time.

13.4.9   Conclusions and Recommendations

As with most commercial organizations, multitheater movie complexes seek to offer maximum service to the customer at minimum operating expense. Because these complexes utilize a number of parallel resources for ticket selling, concession selling, and ticket collecting operations, staffing level and configuration impact on the performance of the system. Through the utilization of discrete event simulation and analysis, it is possible for the system to be analyzed without physical alteration. This permits information to be obtained without increasing operating expenses or alienating customers, events that might occur if the alternatives were physically implemented.

In the model, it was determined that any configuration utilizing fewer employees than the base configuration would statistically significantly lengthen the flow time of the groups of movie viewers. However, whether the differences are of a practical level is another issue. The other significant result was that if the base configuration of employees is reduced because of an absence, it does not matter whether the employee was a ticket or a concession seller. This means that it will not be necessary to reconfigure the system to offer the best service to the customers.

References

Miller, I.R., Freund, J.E., and Johnson, R., *Probability and Statistics for Enqineers,* Prentice-Hall, Englewood Cliffs, NJ, 1990, pp. 404–406.
Pegden, C.D., Shannon, R.E., and Sadowski, R.P., *Introduction to Simulation Using SIMAN,* McGraw-Hill, New York, 1990, pp. 12–13.

13.5   Will-Call Operations Simulation Modeling and Analysis

Erick C. Jones

13.5.1   Project Definition

This project was defined by a client who wanted improve his will-call operations. The will-call operation consists of clients who have placed orders to be filled. When the order is ready, the client is called to pick up the order. The clients arrive throughout the day to pick up their orders.

The will-call operation is for a building supplies wholesale company that has approximately 10,000 square feet of area. The client is considering whether he should invest money to automate his facility. He is entertaining a proposal for adding a power conveyor to his operation. The only way that this investment can be justified is by reducing some of his personnel who move material from the clerk area to the loading docks.

Other relevant information and assumptions are listed below:

- The cost of a FTE is $42,500; this is the burdened rate.
- The cost of a powered conveyor is $300/square foot.
- Material handlers travel approximately 700 feet one way to pick up materials from the will-call counter.
- The proposal for conveyor services is 1500 feet of power conveyor.
- The second option for conveyor is 1000 feet of power conveyor, which would require the material handlers to transport goods approximately 500 feet.
- The client does not want to slow his people loading the docks.

13.5.2 Project Planning

During this phase a Gantt chart, work breakdown structure, and a linear responsibility chart were developed using the Microsoft Project software. The Gantt chart is illustrated in Figure 13.20.

13.5.3 Model Planning

Information that needs to be collected for the model were determined to include:

Arrival times
Clerk service times
Cart speeds
Conveyor speeds
Loader loading time

All this information should be modeled in order to make the model work.

13.5.4 Input Data

Time was scheduled with the operations manager for the wholesale distribution company, and the input data that were identified for the system were collected. The data were then imported into the Arena Input Analyzer to determine the summary statistics and best fit distributions for each type of data.

13.5.4.1 Input Distributions

Data were collected from the operation, saved as an Excel spreadsheet, and then imported into the Arena simulation software. The Arena simulation software was utilized to create the following distribution information. Based on the data collected and imported from the Excel spreadsheet, the following distributions were created.

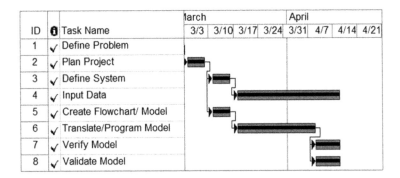

FIGURE 13.20 Gantt chart.

13.5.4.2   Arrival Time

A total of 36 data points were utilized to fit the customer interarrival time data to a theoretical distribution. The analysis of these data resulted in the following summary statistics and best fit distribution.

Min data value = 4
Max data value = 32
Sample mean   = 16.9
Sample std dev = 6.59

The arrival time was best fit with a triangular distribution with a minimum value of 3.5, a mode of 14.0, and a maximum value of 32.5.

13.5.4.3   Clerk Service Time

A total of 39 data points were utilized to fit the clerk service time data to a theoretical distribution. The analysis of these data resulted in the following summary statistics and best fit distribution.

Number of data points = 39
Min data value   = 3
Max data value   = 17
Sample mean   = 9.18
Sample std dev   = 3.39

The clerk service time was best fit with a Weibull distribution offset by 2.5.

13.5.4.4   Loader Service Time

A total of 34 data points were utilized to fit the loader service time data to a theoretical distribution. The analysis of these data resulted in the following summary statistics and best fit distribution:

Min data value = 4
Max data value = 8
Sample mean   = 5.47
Sample std dev = 1.08

The loaders time was best fit with an Erlang distribution of 3.5 + ERLA (0.657, 3).

13.5.5   Model Formulation

A flow chart of how the system would capture information was utilized to make sure the process flow was logical before the actual coding of the system was done.

13.5.6   Model Translation

The Arena software was utilized to create a model, with both blocks and elements. The given software creates both experiment and model code, in the SIMAN programming language. A screen capture of the model is illustrated in Figure 13.21.

13.5.7   Verification and Validation

13.5.7.1   Validation

Face validation was agreed to initially by the operations manager in reference to the current operations; this is our baseline model (Scenario 1).

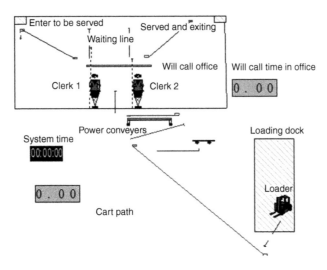

FIGURE 13.21 Model translation.

Next, we utilize the statistical validation methodology: we first use the *F*-test to determine the data type, which then in turn would indicate which *t*-test to use to determine statistical significance.

Classical C.I. Intervals Summary

ID	Avg.	St.Dev.	0.900 C.I.	Min.	Max.	Number
SYSTIME	45.9	8.38	5.2	31.4	51.8	9

13.5.7.1.1 *F-Test*

Comparison of Variances

Identifier	Var	0.900 C.I.	Value	Number
SYSTIME	0	5.2	1.46e-316	20
SYSTIME	0	5.2	1.46e-316	9
Reject H_0: Variances are not equal at 0.1 level.				

13.5.7.1.2 *t-Test*

Paired-t Means Comparison

Identifier	Est.Mean	Std Dev	0.900 C.I.	Min	Max	Number
PSYSTM	45.9	8.38	5.2	31.4	51.8	9
Reject H_0: Means are not equal at 0.1 level.						

13.5.7.2 Verification

We seek to have verification that there are enough replications to have a statistical significance that the model will produce the same or similar results. The chosen level is a statistical significance of 90%. The following information was utilized to determine the information.

Classical C.I. Intervals Summary

Identifier	Average	Std Dev	0.900 C.I.	Min	Max	Number
tavg(systime)	44.5	1.32	0.765	42.5	46.1	10

Then I (number of verifications) was recalculated to determine if the number of replications were sufficient and the number of replications that it was determined to give the sufficient precision at a 90% confidence level was I = 0.3651. Because we ran 10 replications it can be concluded that the data yielded the appropriate precision.

13.5.8   Experimentation and Analysis

After validating the baseline model, we ran the baseline model to compare the appropriate statistics that represent improved customer service and worker utilization. The following changes were compared:

1. Scenario 1 (baseline): Orders are picked up from the will-call service center, and then four material handlers pick up the orders and travel 700 feet and take them to the loader.
2. Scenario 2 (all conveyors): Orders are placed on a conveyor system that takes packages 1500 feet to the loader.
3. Scenario 3 (combination): Orders are placed on 1000-foot conveyor system, where two material handlers pick up packages from the conveyor and travel 300 feet.

The statistics of importance were:

1. Total order time
2. Loader utilization

The methodology utilized in the class is used below to determine correlations and variance differences.

13.5.8.1   Analysis

13.5.8.1.1   Analysis of Variance
Mean information from output analyzer using the confidence interval function.

Classical C.I. Intervals Summary

Identifier	Average	Std Dev	0.900 C.I.	Min	Max	Number
Baseline	44.5	1.32	0.765	42.5	46.1	10
Scenario 2	34.4	0.671	0.389	33.5	35.8	10
Scenario 3	39.3	0.909	0.527	37.8	40.4	10

Analysis of variance information from output analyzer:

One-Way ANOVA Table
L1 tavg(systime)
L2 tavg(systime)
L3 tavg(systime)

Source of Variation	Sum Squares	DF	Mean Squares	F-Exp
Between Treatments	467.600	2	233.800	1243.898
Error (W/inTreatments)	4.511	24	0.188	
Total	472.100	26	F-Crit	2.538

13.5.8.1.2   Duncan Test

$$Sx = sqrt\ (MSE/n) = sqrt(0.188/10) = 0.1371$$

Use Duncan chart for values for L1, L2, L3.
L2 to L3 = 2 pairings, 2 means, the value at 0.10 with MSE degrees of freedom of 26 (from chart) = 2.92
L1 to L3 = 3 pairings, 3 means, the value at 0.10 with MSE degrees of freedom of 26 (from chart) = 3.07

Given: $R_p = Sx * r_p$

R2 = Sx * r2 = 0.1371 * 2.92 = 0.4004
R3 = Sx * r3 = 0.1371* 3.07 = 0.4209

Sorted means into ascending order:

L3(Combo)	L2(All conveyors)	L1(Baseline)
34.4	39.3	44.5

Ranges:

  L3 to L2 = 4.90
  L3 to L1 = 10.10
  L1 to L2 = 5.20

Correlation:

  Because 4.90 is greater than 0.4004, there is no correlation between L3 and L2
  Because 5.20 is greater than 0.4209, there is no correlation between L1 and L2

The recommendation therefore would be to stay with the current scenario for efficiency unless there is a relevant cost impact.

13.5.9 Conclusions and Recommendations

13.5.9.1 Cost Analysis

The driving factors for the decisions on whether to buy the conveyor are cost and payback period. The following information is given as current information. The cost of the power conveyor is $300 per square foot. The cost of a full-time equivalent (FTE) is $ 42,500 per year. Considering this information, the following information was created.

Category	Baseline	Scenario I	Scenario II
Order time	44.50	34.40	39.30
Loader utilization	12.74%	13.52%	13.48%
FTE cost	$170,000	$0	$85,000
Conveyor cost	$0	$450,000	$300,000
FTE savings	$0	$170,000	$85,000
Payback period	NA	2.65	4.53

13.5.9.2 Recommendations

Considering the payback period and the increase in efficiency, the following conclusions and recommendations can be offered:

  1. Go with scenario II, the all-conveyor option.
  2. Reduce the loader to a part-time schedule.
  3. Consider evaluating the need for two clerks.

13.5.10 Appendix

13.5.10.1 Model Code

```
5$       STATION,     conveyofsta1; exitseg1.lbl EXIT: conveyseg1,1;
21$      QUEUE,       cartque;
34$      GROUP,       Temporary:1,Last:NEXT(22$);
22$      REQUEST,     1,cartsto:cart(por),1,startsta;
25$      STATION,     startsta;
24$      TRANSPORT:   cart,stopsta,1;
6$       CREATE,      1,0:TRIA(3.5, 14, 32.5): MARK(ARRTIME): NEXT(15$);
15$      BRANCH,      1:   With,.5,custype1.lbl,Yes;
                      custype1.lbl ASSIGN: custype = 1:
                      PICTURE = CUST1;
9$       STATION,     insta;
10$      ROUTE:       1,questa;
11$      STATION,     questa;
```

18$	PICKQ,	LRC:
		CLERKQ2.LBL:
		CLERKQ1.LBL;
		CLERKQ2.LBL QUEUE, clerk2que;
		CLERKSZ2.LBL SEIZE, 1,Other:
		clerk2,1:NEXT(16$);
16$	DELAY:	2.5 + WEIB(0, 0),,Other:NEXT(17$);
17$	RELEASE:	clerk2,1;
12$	STATION,	releasesta;
13$	ROUTE:	1,disposesta;
		CLERKQ1.LBL QUEUE, CLERK1QUE;
		CLERKSZ1.LBL SEIZE, 1,Other:
		clerk1,1:NEXT(7$);
7$	DELAY:	2.5 + WEIB(0, 0),,Other:NEXT(8$);
8$	RELEASE:	clerk1,1:NEXT(12$);
14$	STATION,	disposesta;
20$	TALLY:	willcalltime,TNOW-ARRTIME,1;
0$	ASSIGN:	picture = sku1:
		picture = box;
4$	STATION,	conveyonsta1;
1$	QUEUE,	conveyorque1;
2$	ACCESS:	conveyseg1,1;
3$	CONVEY:	conveyseg1,conveyofsta1;
		custype2.lbl ASSIGN: custype = 2:
		PICTURE = CUST2:NEXT(9$);
19$	DISPOSE:	No;
27$	STATION,	stopsta;
26$	FREE:	cart;
32$	ASSIGN:	PICTURE = box;
33$	SPLIT:	:NEXT(28$);
28$	QUEUE,	LOADERQUE;
29$	SEIZE,	1,Other:
	LOADER,	1:NEXT(30$);
30$	DELAY:	3.5 + ERLA(0.657, 3),,Other:NEXT(31$);
31$	RELEASE:	LOADER,1;
35$	TALLY:	systime,TNOW-ARRTIME,1;
23$	DISPOSE:	No;

13.5.10.2 Experiment Code

PROJECT, "Unnamed Project","Erick Jones",,,No,Yes,Yes,Yes,No,No,No;
ATTRIBUTES:
    skutype:
    custype:
    ARRTIME:
    sku1a:
    sku2a:
    customer1:
    customer2;
STORAGES:
    skucartsto:
    cartsto;

VARIABLES:
    seqtype,CLEAR(System),CATEGORY("None-None");
QUEUES:
    1,CLERK1QUE,FirstInFirstOut,,AUTOSTATS(No,,):
    sortconveyque2,FirstInFirstOut,,AUTOSTATS(Yes,,):
    sortconveyque3,FirstInFirstOut,,AUTOSTATS(Yes,,):
    sortconveyque4,FirstInFirstOut,,AUTOSTATS(Yes,,):
    clerk2que,FirstInFirstOut,,AUTOSTATS(Yes,,):
    skucartque,FirstInFirstOut,,AUTOSTATS(Yes,,):
    cartque,FirstInFirstOut,,AUTOSTATS(Yes,,):
    LOADERQUE,FirstInFirstOut,,AUTOSTATS(Yes,,):
    conveyorque1,FirstInFirstOut,,AUTOSTATS(Yes,,):
    conveyorque2,FirstInFirstOut,,AUTOSTATS(Yes,,):
    conveyorque3,FirstInFirstOut,,AUTOSTATS(Yes,,):
    conveyorque4,FirstInFirstOut,,AUTOSTATS(Yes,,);
PICTURES:
    cust1:
    cust2:
    box:
    sku1:
    sku2;
RESOURCES:
    1,clerk1,Capacity(1),,Stationary,COST(0.0,0.0,0.0),,AUTOSTATS(Yes,,):
    2,CLERK2,Capacity(1),,Stationary,COST(0.0,0.0,0.0),,AUTOSTATS(Yes,,):
    LOADER,Capacity(1),,Stationary,COST(0.0,0.0,0.0),,AUTOSTATS(Yes,,);
STATIONS:
    conveyonsta1:
    conveyonsta2:
    conveyonsta3:
    conveyonsta4:
    conveyofsta1:
    conveyofsta2:
    conveyofsta3:
    conveyofsta4:
    skustopsta:
    stopsta:
    skustartsta:
    disposesta:
    questa:
    startsta:
    insta:
    releasesta;
DISTANCES:
    skucartmap,skustartsta-skustopsta-10:
    cartmap,startsta-stopsta-20;
TRANSPORTERS:
    skucart,1,Distance(skucartmap),264 — -,AUTOSTATS(Yes,,):
    cart,4,Distance(cartmap),1.0 — -,AUTOSTATS(Yes,,);
SEGMENTS:
    conveyorseg1,conveyonsta1,conveyofsta1–20:
    conveyorseg2,conveyonsta2,conveyofsta2–100:

conveyorseg3,conveyonsta3,conveyofsta3–50:
conveyorseg4,conveyonsta4,conveyofsta4–100;
CONVEYORS:
conveyseg1,conveyorseg1,60,1,Active,1,Nonaccumulating,,AUTOSTATS(Yes,,):
conveyseg2,conveyorseg2,60,1,Active,1,Nonaccumulating,,AUTOSTATS(Yes,,):
conveyseg3,conveyorseg3,60,1,Active,1,Nonaccumulating,,AUTOSTATS(Yes,,):
conveyseg4,conveyorseg4,60,1,Active,1,Nonaccumulating,,AUTOSTATS(Yes,,);
TALLIES:
willcalltime,"willcalltime.dat":
SYSTIME,"systimep2a.dat";
DSTATS:
NQ(CLERK1QUE),AVGNUMQUE1:
NR(CLERK1),CLERK1UTIL:
NR(CLERK2),CLERK2UTIL:
NQ(CLERK2),AVERAGENUMQ2;
OUTPUTS:
tavg(systime),"Aprojectreplicate.dat";
REPLICATE,
10,0.0,MinutesToBaseTime(480),Yes,Yes,0.0,,,480,Minutes,No,No;

14

ARENA User's Minimanual

14.1 Introduction

This is one of a number of introductory user manuals for some of the more popular simulation software packages. In this manual, the practitioner is introduced to the simulation software package ARENA. ARENA actually consists of three different components. The primary component of ARENA is a graphically based software package that utilizes the SIMAN simulation programming code. The graphic basis of ARENA allows models to be quickly developed and easily animated. The second component of ARENA is the Input Analyzer. This software package is used to fit input data. The last component of ARENA is the Output Analyzer. This component is used to statistically analyze and compare output measures of performance. In further discussion, the simulation programming component of the software is referred to as ARENA. The Input Analyzer and the Output Analyzer are referred to separately by their respective names. This minimanual begins with a tutorial on ARENA and ends with tutorials on the Input Analyzer and the Output Analyzer.

ARENA can be utilized by the practitioner at a variety of different levels. Toward the lower level, the practitioner is responsible for designating individual structures known as model blocks and experiment elements. Model blocks control the flow of entities though the model. Experiment elements define the characteristics of the different model components. The lower-level blocks and elements are initially more difficult to use; however, they offer increased flexibility over high-level structures.

Toward the higher level, the practitioner can use process or custom modules. These modules consist of a number of the lower-end modules combined together for commonly encountered processes. The higher-level components are somewhat easier to use but may lack flexibility for specific applications.

The approach taken in this chapter is to introduce the practitioner to the lower-level model blocks and experiment elements. Using these lower-level structures helps the practitioner obtain a greater understanding of how the system operates. At a later date, the practitioner may decide that the higher-level structures are more appropriate for specific uses. Until then, we concentrate on the lower-level model blocks and experiment elements. In this minimanual, we cover:

- ARENA user interface
- Frequently used model blocks
- Frequently used experiment elements
- Basic animation
- Modeling transporters
- Modeling conveyors
- Outputting data for comparison purposes
- Input analyzer
- Output analyzer

14.2 ARENA User Interface

ARENA operates under the various Windows operating systems. The ARENA interface contains the normal menu bar and icon tool bars, a project bar on the left side of the screen, a model view in the center of the screen, and a data view on the bottom of the screen. Figure 14.1 illustrates the opening ARENA screen.

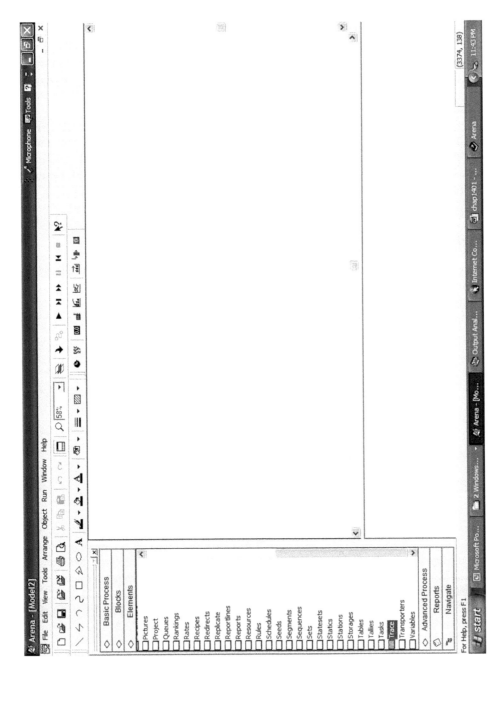

FIGURE 14.1 ARENA opening window.

The menu bar contains the usual file, edit, view, and help menus. Some of the more common functions are duplicated on the icon bar. In addition to the normal functions, the menu bar also contains ARENA-specific functions for controlling simulation runs and links to launch other ARENA-related programs.

The project tool bar on the left side of the screen contains the different structures that the practitioner will use to build the simulation project. There are a number of panels that correspond to the different types of structures that are available. If a particular panel that is needed does not appear, the practitioner can right-mouse-click above the project tool bar to load the correct panels.

The center model view screen is where the practitioner will spend most of the model-building time. This area is used to build the model structure by structure. The model is built by dragging structures from the project tool bar on the left side of the screen into the model view. The model flow is established by connecting the individual structures together. Because different types of structure are available, it is essential that the practitioner attempt to keep this area as organized as possible.

The bottom data view screen can be used by the practitioner to change the values of different components that are used in the model. For our introductory purposes, we will dispense with the use of the data view screen for the moment. To increase our modeling space it will be useful to minimize the size of the data view screen or eliminate it altogether. However, it will still be available for use by the practitioner in the event that it is deemed to be helpful.

14.3 Frequently Used Model Blocks

There are an enormous number of model blocks available to the practitioner in ARENA. It is likely that over the course of the practitioner's career, only a fraction of the available model blocks will ever be utilized. The number of frequently used model blocks is correspondingly smaller. In this introductory user's manual we will cover the most common of the frequently used model blocks. These include:

- Create
- Assign
- Queue
- Seize
- Delay
- Release
- Tally
- Dispose

14.3.1 Create

The create block is used to generate entity batches in the system. As discussed in the Introduction (Chapter 1) of the handbook, an entity batch is the number of individuals in an arrival batch. This corresponds to a family batch of a particular size arriving at one time. The create block allows the practitioner to specify when the first creation is to occur and when every subsequent creation will occur thereafter. The practitioner can also limit the maximum number of batches that can arrive from a particular create block. The create block icon and the create block dialogue box are illustrated in Figure 14.2.

14.3.2 Assign

The assign block is used to set global variable and entity attributes to particular values. In the case of a simple simulation model, the most common use of the assign block is to record when each entity arrives in the system. This is performed by pressing the add button in the variable frame of the assign block window. An entity attribute such as *arrtime* can be used to record this time by placing its name in the new variable or attribute text box. Likewise, the simulation time variable tnow value is assigned to the

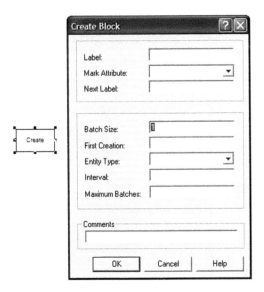

FIGURE 14.2 Create block.

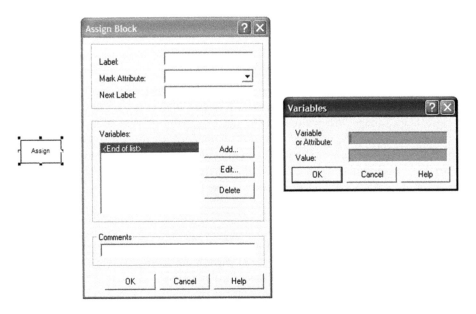

FIGURE 14.3 Assign block.

attribute *arrtime* by placing tnow in the assign value text box. The assign icon, assign block window, and add variable window are illustrated in Figure 14.3.

14.3.3 Queue

The queue block (Figure 14.4) is used to represent a system waiting line or queue. The practitioner must specify the name of the queue in the queue ID text box. The practitioner may also limit the number of entities that can be in the queue at any given time. If any entities arrive in the queue when the maximum number has been reached, they can also be diverted to a program structure with the specified label identified in the balk label text box.

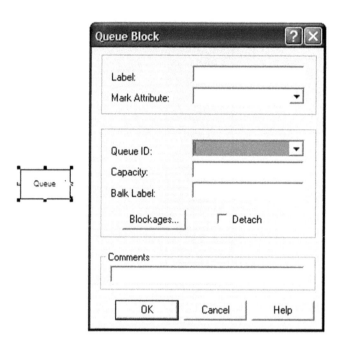

FIGURE 14.4   Queue block.

14.3.4 Seize

The seize block is used by the entity to seize control of a particular resource. This corresponds to a customer engaging the services of a checkout clerk. The seize block requires that the resource be added from the seize block window. When a resource is added, the resource window appears. In this window, the practitioner can specify the resource ID and the number of resources that the entity should seize. Normally, entities will seize only one resource at a time. However, it is possible to have more than one resource serving a single entity. As an example, more than one bagger could be bagging groceries for a single customer in a grocery store. The seize windows are illustrated in Figure 14.5.

14.3.5 Delay

The delay block is used to model service times during which the entity has seized the resource. The only text box of importance to the practitioner for the moment is the duration text box. The practitioner can insert either a deterministic value or a probability distribution in this text box. Alternatively, the practitioner may also insert a variable in this text box that has previously been set to a deterministic value or a probability distribution value. The delay icon and the delay block window are illustrated in Figure 14.6.

14.3.6 Release

The release block frees the resource from the entity at the end of the previous delay block time. The practitioner must specify a resource to release by first pressing the resource add button on the release block window. In the resource window, the practitioner can specify the resource ID to release. The release icon, release block window, and the resource window are illustrated in Figure 14.7.

14.3.7 Tally

The tally block is used to compile observational statistics generated by individual entities. A typical use of the tally block is to compile the system time for each individual entity. This block requires the practitioner to identify the tally in the tally ID block and also provide an entry in the value text box. If

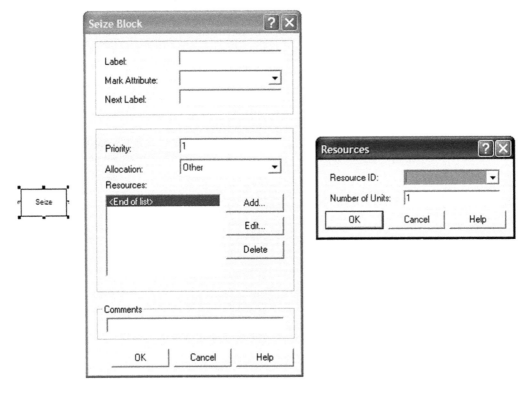

FIGURE 14.5 Sieze block.

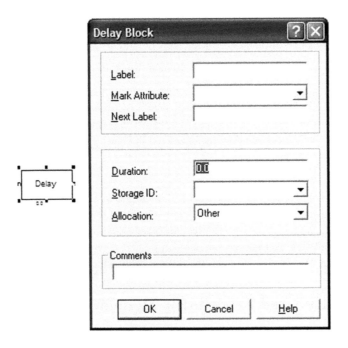

FIGURE 14.6 Delay block.

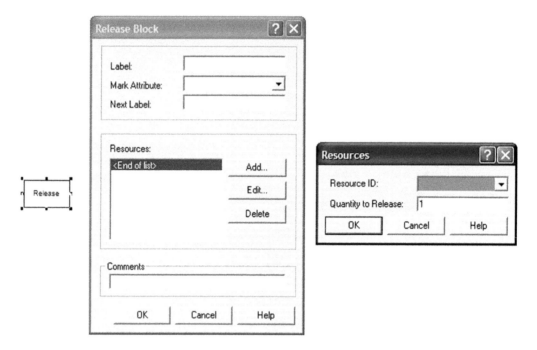

FIGURE 14.7 Release block.

the tally block is used to compile system time, the practitioner can insert the following equation into the value text box:

$$Tnow - Arrtime$$

where

$Tnow$ = Built-in global variable for the current simulation time.

$Arrtime$ = A user-named attribute that records what time the entity arrived in the system.

The result of this equation is the interval of time that the entity has spent in the system from the time that it arrived to the time that it reached the tally block. Because the current time is tnow and time advances, the earlier *arrtime* value is subtracted from *tnow*. This yields the total system time for the entity. The tally icon and the tally block window are illustrated in Figure 14.8.

14.3.8 Dispose

The dispose block simply disposes of the system entity. Once the entity has traveled though the system, and the necessary statistics have been recorded, there is no need to continue maintaining the entity. By disposing of the entity, we free program resources for other uses. In addition, the training and academic versions of some simulation packages do not allow more than a certain number of entities to be in the model at a given time. The dispose icon and the dispose block window are illustrated in Figure 14.9.

14.4 Frequently Used Experiment Elements

There are also an enormous number of experiment elements available to the practitioner. These elements specify the characteristics of the model components. In many cases, there is a direct relationship between the experiment elements and the model blocks. The most common experiment elements include:

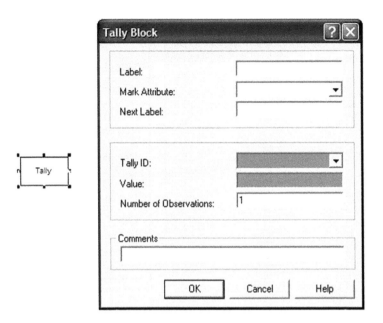

FIGURE 14.8 Tally block.

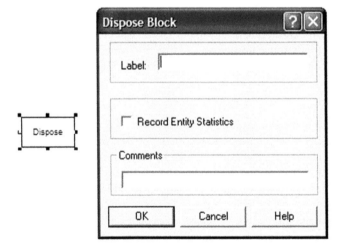

FIGURE 14.9 Dispose block.

- Attributes
- Resources
- Queues
- Tallies
- Dstats
- Replicate

14.4.1 Attributes

Attributes are variables that are associated with individual entities. If an attribute is specified in the attributes element, then all entities that pass through the system will have an attribute with the same name. The value of the attribute for each individual entity, however, can be different. There can be as

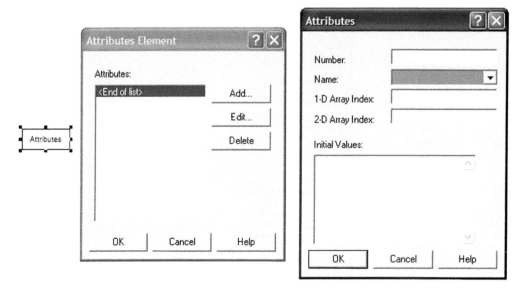

FIGURE 14.10 Attributes element.

many different values for this attribute as there are entities. In our simple model, we have particular use for one attribute. This attribute needs to hold the individual arrival time in the system for each individual entity. As previously discussed, this information is necessary to calculate the entity's system time. In order to assign an attribute, the practitioner must first press the add button in the attribute element window. In the new attributes window, the practitioner can specify a name as well as starting values. The attribute icon, attribute element window, and the attribute window are illustrated in Figure 14.10.

14.4.2 Resources

The resources element is used to specify the characteristics of the resources in the model. To define a resource, the practitioner must press the add button in the resource element window. The new resources window that appears allows the practitioner to identify the name of the resource. The next text box can specify the use of a capacity or schedule to define the availability of the resource. If capacity definition is used, the practitioner can set the level of available identical resources for the model. Normally, the capacity of the resource is one. However, if the practitioner is interested in modeling multiple identical resources, the capacity resource can be increased as necessary. Alternatively, the practitioner can define the resource to operate under a schedule. This option would be used if the capacity of the resource needs to vary over time. If the resource needs to become unavailable for breaks and meal periods, then the schedule option should be specified. The resources icon, resources element window, and resources window are illustrated in Figures 14.11 and 14.12.

14.4.3 Queues

The queues element is used to identify and specify the characteristics of the queues in the model. Queues are added by pressing the add button in the queues element window. In the new queues window, the practitioner can specify the name of the queue and the ranking criterion for the queue. The ranking criterion is a type of sequencing for the entities in the queue. Common sequencing rules include first in–first out, last in–first out, highest value first, and lowest value first. For systems that have rush jobs versus regular jobs, an attribute can be used to hold the priority. If the value of 1 is used to represent the rush jobs and 2 for the regular jobs, the practitioner could use lowest value first for the ranking criterion. The queues icon, queues element window, and queues window are illustrated in Figure 14.13.

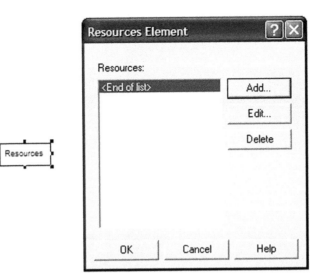

FIGURE 14.11 Resources element.

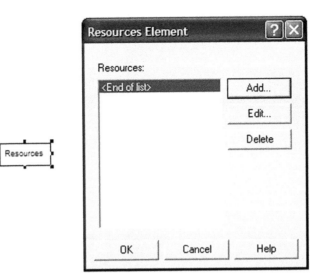

FIGURE 14.12 Resources element details.

14.4.4 Tallies

The tallies element is used to define the observations statistics for the model. A typical use of the tallies element is to record system time statistics for the entities. The tally will automatically keep a running average for the observations. To add a tally, the practitioner needs to press the add button in the tallies element window. In the new tallies window, the name of the tally can be specified. The practitioner may also specify that the results of the tally be written to an output file. The tallies icon, tallies element window, and tallies window are illustrated in Figure 14.14.

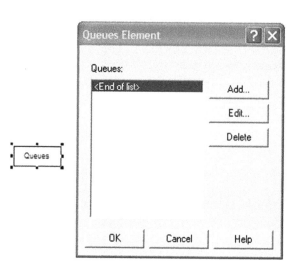
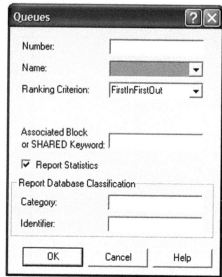

FIGURE 14.13   Queues element.

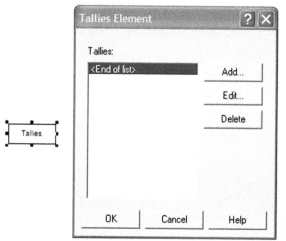

FIGURE 14.14   Tallies element.

14.4.5 Dstats

The dstats element is used to define the discrete event statistics for the model. These are the same as the time-dependent statistics discussed in the Introduction (Chapter 1) of the handbook. To create a dstat, the practitioner needs to press the add button in the dstats element window. In the new dstat window, the practitioner can name the dstat. The practitioner must also provide an equation for the dstat expression text box. The expression is a function that operates on a specific component of the model. The two different types of dstats that have been discussed are:

- Time average number in queue
- Resource utilization

To generate a dstat for the time average number in queue, the practitioner must use the nq function on a particular previously defined queue. For example, to obtain the time average number in queue for a queue defined as clerkq, the practitioner would need to enter the following formula in the dstat expression text box:

$$nq(clerkq)$$

Similarly, to generate a dstat for the average utilization, the practitioner would need to use the nr function on a previously defined resource. So to obtain the average utilization for a resource defined as clerk, the practitioner would need to enter the following formula in the dstats expression text box:

$$nr(clerk)$$

The practitioner may also elect to write the utilization values to an output file. This might be useful for debugging purposes while developing the model. In order to use the output file text box, the practitioner must enclose the name of the file in double quotes. For example, to write data in this manner, the practitioner might enter:

$$c:\backslash files\backslash simproj\backslash util.dat$$

This would write the data to a file called *util.dat* in the *files\simproj* subdirectory of the practitioners hard drive. The dstats icon, dstats element window, and dstat window are illustrated in Figure 14.15.

14.4.6 Replicate

The replicate element is used to control the length and number of the simulation runs that the practitioner desires. If necessary, the practitioner can also start the simulation time clock at a time other than 0.0. The practitioner may also specify the number of hours per day and the base time units. All of the other options should be left at their default values. The replicate element window is illustrated in Figure 14.16.

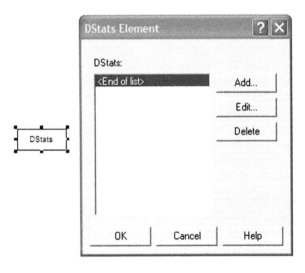
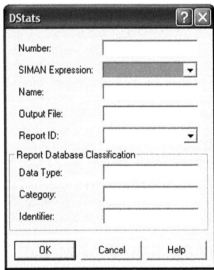

FIGURE 14.15 DStats element.

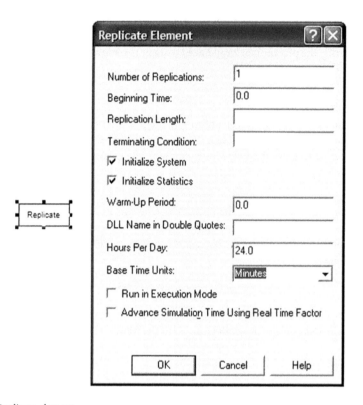

FIGURE 14.16 Replicate element.

14.5 General Approach for Developing a Simple Nonanimated Model

Now that the practitioner is familiar with the most frequently utilized model blocks and experiment elements, it is time to assemble a simple nonanimated model. The best approach to assembling a model is first to define the block flow for the model. The data for the text boxes for the model blocks do not yet need to be entered. The reason for this is that many of the model blocks require component names that will be subsequently identified in the experiment elements. The blocks contain drop-down boxes with all of the defined component names. So instead of misspelling the names of the components in the model text boxes, it is better to accurately enter the data with the drop-down boxes. When the practitioner has assembled and connected all of the model blocks, attention can be turned to the experiment. Each experiment element can be added, and the names and characteristics of the model components can be defined. When the practitioner is finished with defining the experiment elements, he or she can return to the model and enter the defined data through the drop-down boxes. This procedure is summarized as:

- Select and connect all model blocks
- Select and define all experiment elements
- Return to the model blocks and identify the components

14.6 Building the Model

We are now ready to begin building our first model. The model is created by dragging individual model blocks from the left-hand project tool box to the center flow chart screen window. If a model block is previously highlighted, a new model block will be automatically connected to the highlighted block. This can save a significant amount of time while developing the model. If the practitioner forgets to highlight the last block, the last block and the new block can still be connected by clicking on the connector symbol in the menu tool bar. The connector icon has a small white square connected to a small white diamond with a S-shaped line.

Our first model will consist of a single-queue, single-server system. Fortunately, you are already familiar with the appropriate sequence of model blocks from the previous model block section of this chapter. To refresh your memory, you need to drag and connect the following blocks in this exact sequence:

- Create
- Assign
- Queue
- Seize
- Delay
- Release
- Tally
- Dispose

Remember not to bother with entering data in any of the blocks. If any problems occur connecting the blocks, just delete the offending block and drag a new block into place. Similarly, if the connector becomes damaged, simply highlight the connector and delete it. With all of the model blocks in place, the screen should look like the one shown in Figure 14.17.

Once all of the blocks are in place, it is time to place the experiment elements. The best place to position the experiment elements is in a part of the flow chart window that is directly below the model blocks. The experiment elements are not connected to each other, so the sequence is not essential. Although it is possible to fill in each element one at a time, it is better to drag all of the elements over to the flow window first and fill out the elements afterward at the same time. The practitioner needs to drag the following elements over to the flow chart screen:

- Attributes
- Resources
- Queues
- Tallies
- Dstats
- Replicates

Some thought should be given to keeping the experiment area organized. One approach is to keep all of the related elements in the same column. This means that one column would have entity-related elements such as the attributes element. Another column would have the system component elements including the resource and queues elements. The statistics-related tallies and dstats elements would be in another column. Finally, the project-related replicate element would be in its own column. When complete, the experiment should look like Figure 14.18.

14.7 Filling in the Experimental Data

It is now time to begin filling the experiment elements with data. As with all the elements, it is necessary to double click on the icon. When each icon is double clicked, the corresponding elements screen appears.

FIGURE 14.17 Simple model block sequence.

FIGURE 14.18 Simple experiment elements.

14.7.1 Attributes Element Entry

We will begin with the attributes element. Here, we will need to press the add button. When the new attributes screen appears, insert the attribute name *arrtime*. We will use this attribute to record the time that the individual entity enters in the system. There is no need to insert any other data in the attribute screen. Click the OK button until you are out of the attributes element.

14.7.2 Resources Element Entry

Next, we will specify the resource in the resources element. Like the attributes element, it is first necessary to press the add button in the resources element window. In the new resources window enter the resource name *clerk*. The other important data can be defaulted to capacity and 1. Exit all of the resources windows.

14.7.3 Queues Element Entry

In the queues element window, you will need to press the add button. In the new queues window, enter the name of the queue *clerkq*. The name of the queue uses the mnemonic convention of the resource name plus the character q. This convention is described in detail in the model translation section of the handbook. In our simple model it is not necessary to add in a queue capacity or the balk label. Exit all of the queues windows.

14.7.4 Tallies Element Entry

In the tallies element window it is necessary to press the add button. When the tallies window opens, insert the tally name *systime* in the name text box. It is not necessary to insert any other data in the tallies window. Exit all of the tallies windows.

14.7.5 Dstats Element Entry

In the dstats element window, we will have to enter two different dstats. To enter the first dstat, press the add button. In the new dstats window, enter the SIMAN expression:

$$nq(clerkq)$$

This will create a dstat that will eventually generate the time average number in queue for the queue clerkq. Next name the dstat that you just created. It is best to name this dstat something descriptive like *avgnuminq*. This stands for average number in queue. Exit the current dstat window.

In the main dstats element window, press the add button once again. This time we will create a time-dependent statistic for resource utilization. In the SIMAN expression text box enter:

$$nr(clerk)$$

Name this dstat *avgutil* for average utilization. Exit completely all of the dstats windows.

14.7.6 Replicate Element Entry

A final set of element entries is needed in the replicate element. In the replicate element window enter the length of the simulation run as 480. In the base time units text box change the units to minutes. This will result in running the simulation for a total of 480 min. The number of replications should be defaulted at 1. The starting time should also be defaulted at 0.0.

14.8 Filling in the Model Data

The practitioner should fill in the model block data only after the experiment element data entry has been completed. The major reason for this is that the model block drop-down boxes can be used to enter data without a need to enter manually the names of different components. Generally, it is also a good idea to fill in the model data information in the same order as the entity flow.

14.8.1 Create Block Entry

In the create block, it is necessary to fill in the interval text box with the time in minutes between the arrival of customers to the system. In our model we will use:

$$expo(5)$$

This means that the time separating the arrivals of customers in the system is exponentially distributed with a mean of 5 min.

We can actually default all of the other text boxes at this point. If this is our intention, then the batch size will be one. This means that the customers arrive individually. If we default first create text box to nothing, an entity will arrive at 0.0 time. This is not necessarily a valid assumption. We can say that the time for the first creation follows the same distribution as the interarrival rate. So in the first arrival text box, we can also put:

$$expo(5)$$

14.8.2 Assign Block Entry

We will use the assign block to record what time the entity arrived in the system. This information is necessary to calculate later how much total time the entity spent in the system. To enter this value, it is first necessary to press the add button. In the new variables window, locate the variable or attribute text box and enter the name:

arrtime

In the value text box, enter the variable name:

tnow

Tnow is the system variable that holds the value for the simulation time. When we set the entity attribute *arrtime* to *tnow*, each entity will remember what time he or she arrived in the system. Exit all of the assign windows.

14.8.3 Queue Block Entry

The next step is to identify the queue that the customer will be entering. In the queue block, you can use the queue ID drop-down box to select among the queues previously defined in the queues element. This will prevent any misspelling of the queue name and speed the modeling process. The defaults for capacity and balk label blocks can be accepted at this time.

14.8.4 Seize Block Entry

After the queue block, it is time to specify the name of the resource that is at the head of the queue. This is performed in the seize block. The entity actually seizes the resource. In the seize block window it will be necessary to press the add button. In the new resource window, the resource ID can be specified with the drop-down text box as:

clerk

This will result in any entities seizing the resource clerk whenever the clerk is idle. Exit the resource window and the seize block window.

14.8.5 Delay Block Entry

The delay block is used to represent the service time for the entities' transaction with the clerk. To insert this delay, locate the duration text box in the delay block window and enter the following information:

Norm(5,1)

This will result in a service time delay for the clerk that is normally distributed with a mean of 5 min and a standard deviation of 1 min. All of the other fields can be defaulted. Exit the delay block window.

14.8.6 Release Block Entry

The release block will free the resource from the entity. In the release block window click the add button. In the new resource window, use the drop-down text box to select the clerk. The quantity to release text box should be defaulted to 1. Close the resource and the release block windows.

14.8.7 Tally Block Entry

The tally block is used to collect observational data for calculating summary statistics. In our example we use it to keep a running average of the customers' system time. In the tally block window use the tally ID drop-down text box to select systime. Systime was the tally previously defined in the tallies experiment element. The tally block also needs to know what to tally. Because we are interested in system time, we must calculate the difference in time from when the entity arrived in the system and when it reached the tally block. To calculate this value, insert the following equation in the value text box:

tnow − arrtime

Because the system variable for the simulation time *tnow* will always be a greater value than the entity arrival time recorded in the attribute *arrtime*, the system time difference will always be positive. Close the tally block.

14.8.8 Dispose Block Entry

There are actually no data that need to be entered in the dispose block. All of the text boxes can be left with the default values.

14.9 Running the Model

Once all of the model blocks and experiment elements are assembled and completed, the practitioner can attempt to run the model. This is performed by clicking the VCR-like run button on the upper tool bar. The program can also be run by the run-go menu sequence or by pressing the F5 key. ARENA will display the status of the program in the lower right corner. The program must initialize itself before it can run. During this process, the program is compiled and linked.

As with any new program, it is highly likely that one or more errors exist during the compiling and linking processes. The number of programming errors that are likely to be present increases with the complexity of the program. Even with this small program, do not be surprised if one or more of the following errors is present:

- Unedited module
- Unconnected exit point
- Incorrect number of arguments
- Linker error

14.9.1 This Module Has Not Been Edited

This error is caused by the failure to provide necessary information for a model block or an experiment element. Sometimes it is possible to click the edit button at the bottom of the screen.

14.9.2 Unconnected Exit Point

This error means that the practitioner has neglected to connect properly all of the blocks in the model. This frequently occurs when it is necessary to put additional blocks on another row of the flow view window.

14.9.3 Incorrect Number of Arguments

This error may occur when the practitioner does not complete all of the required fields in a model block or an experiment element. Another cause for this error is if an incorrect queue or resource name is inadvertently entered.

14.9.4 Linker Error

This error occurs when a corresponding model or experiment block does not exist. For example, if the practitioner has a queue block but no queues element, a linker error may be generated.

14.10 Basic Animation of the Model

The mathematical model that we have developed up to this point is a good beginning. However, we must also take advantage of the animation features of ARENA to obtain a taste for its true potential. To begin this process, the practitioner should move down to a clean part of the flow chart view window.

The basic animation of the model requires a few additional steps. As an absolute minimum, we must specify:

- Entity pictures
- Queue positions
- Resource pictures

14.10.1 Entity Pictures

Entity pictures are a graphic representation of the entity as it flows through the model. Typical entity pictures include:

- People
- Work in progress

- Paperwork
- Telephone calls

In order to designate an entity picture, the picture must have a specific name and be assigned to the entity. There must also be a graphic image associated with the picture's name. This process requires the use of:

- A pictures element
- An assign block
- The edit-entity picture menu sequence

14.10.1.1 Pictures Element

The pictures element is used to define the names of different pictures used throughout the simulation model. When a picture is declared in the pictures element, it activates an entity attribute called picture. To declare a name for a picture, the practitioner must drag a pictures element from the project tool bar to the flow chart screen. The pictures element windows are illustrated in Figure 14.19.

To declare a picture, the practitioner must press the add button in the pictures element window. In the new picture window, the practitioner will need to enter the name of the picture graphic in the name text box. Typical names of picture entities are:

- Customer
- Parts
- Box

For our example, it is sufficient to enter the name of the picture as *customer*. Once the entity name is entered, exit from the pictures and pictures element windows.

14.10.1.2 Assign Block

An assign block is used to associate the picture name with the entity. In our example, we can reuse the existing assign block in the model. When the assign block is open, click the add button. In the variable or attribute text box enter:

picture

In the value block enter the picture name of our previously declared:

customer

Close the variables and assign block windows.

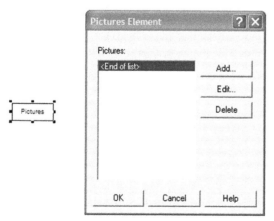
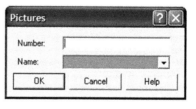

FIGURE 14.19 Pictures element.

14.10.1.3 The Edit-Entity Picture Menu Sequence

The last step in setting up the entity animation is to specify the graphic image to the name of the picture. The graphic window is accessed by following the edit-entity picture menu sequence. When this sequence is followed, the window illustrated in Figure 14.20 appears.

On the left side of the screen are the graphic images available in the model. On the right side of the screen is a library of images. If there is no library, the practitioner can load one of the libraries such as people.plb or workers.plb. The general process consists of copying an image from the right-side library area to the left-side name area.

To associate a graphic image with the customer picture, press the add key near the names text box on the left side of the screen. When a new unused panel appears, use the names drop-down text box to select the name of the previously defined picture:

Customer

When you are performing this process, make sure that the unused panel is selected. This ensures that the name entered in the names text box will be associated with the new panel. Identify a suitable graphic image from the library file on the right side of the screen. When a suitable image is found, click on the image so that its image appears depressed. When both the customer panel and the image panel are depressed, the image can be copied by pressing the double "<<" button on the current library side of the entity picture placement window.

If you are using graphics from a library file, it will sometimes be necessary to change the direction that a graphic is facing. This is performed by flipping the graphic around the vertical axis. To perform this action, it is necessary to access the ARENA picture editor. You can do this by double clicking on the picture of the customer. To flip the graphic, click on the outline of the image and then press the arrange-vertical flip menu sequence. The image should change directions. The picture editor also has a number of other features that also allow you to change the color and fills of existing drawings. These can be investigated at a later time.

When you are done with the entity, close the entity picture placement window by clicking the OK button. The entity part of the animation process is now complete.

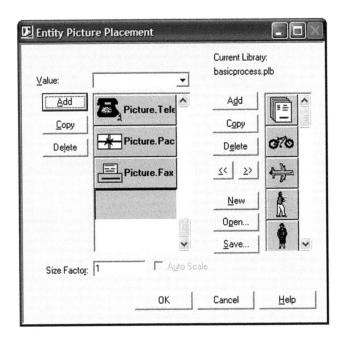

FIGURE 14.20 Entity picture placement.

14.10.2 Queue Positions

To animate the entities waiting in a queue, it is necessary to place a queue graphic on the flow chart window. To place a queue graphic on the window, click the queue button from the tool box. The queue button is illustrated in Figure 14.21.

When the queue button is clicked, the queue window shown in Figure 14.22 will appear on the screen. In the queue window, the practitioner must identify the queue that should be animated. This is performed by using the identifier drop-down text box to select the previously defined queue:

Clerkq

The queue window has a number of other options. The practitioner can specify that the queue be animated by either a line or a point queue. If a line queue is used, as many entities that can be graphically fit along the queue will be animated. In contrast, the point queue is defined by moving and clicking the mouse. Each time the mouse is clicked, a position in the queue for a single entity is created. For our purposes, the normal line queue is sufficient. The practitioner may also check the rotate and flip check boxes. These help properly orient the movement of the entities in the queue. When the queue window is closed by clicking the OK button, the mouse cursor becomes a crosshair. The queue is created by pressing the left mouse button and dragging the mouse. It is possible to specify the length of the queue and the orientation by using the mouse in this manner. When the button is released and reclicked, the queue is positioned. The position, orientation, and size of the queue can also be changed once the queue is actually created. This is accomplished by clicking on one of the sides of the queue and moving the mouse.

FIGURE 14.21   Queue icon button.

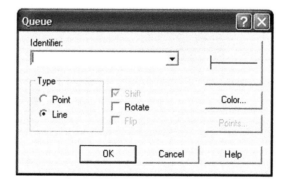

FIGURE 14.22   Queue animation window.

14.10.3 Resource Pictures

The final step to creating our minimal animation is to designate graphic images for the clerk resource. This is accomplished by pressing the resource button. The resource button is located in the toolbar. It is illustrated in Figure 14.23.

When the resource button is clicked, the resource picture placement window appears. This window is illustrated in Figure 14.24.

The resource picture placement window works in a similar manner to the entity picture placement window. However, instead of one graphic, the resource has a total of four different graphics, each one representing a different possible state of the resource. These include:

- Idle
- Busy
- Inactive
- Failed

FIGURE 14.23 Resource icon button.

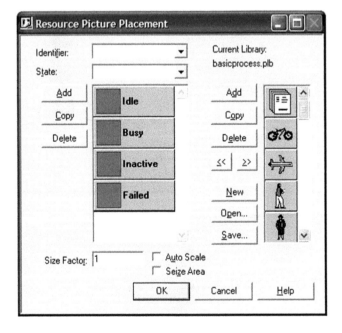

FIGURE 14.24 Resource picture placement window.

The idle graphic is used when no entities are currently seizing the resource. The busy graphic is used when an entity is currently seizing the resource. Inactive is usually used to model a resource that is working but unavailable for use. An example is a worker that is on a break. The failed state is for a machine that is nonfunctional. Although a sophisticated model could possibly include all four states, we will be concerned with only the idle and busy states for our example.

To associate the panels for a resource, you must use the identifier drop-down text box to specify the previously declared resource name. The state must also be selected. This is performed by clicking on one of the four state panels. When the state panel is selected, the actual graphic from the current library side of the resource picture placement window can be transferred to the state panel. The busy state offers the practitioner the option of specifying a seize point on the graphic. This is the position on the busy resource graphic where the entity that is seizing the resource can appear. If the resource is a clerk, the practitioner can position a seize point directly in front of the resource.

Begin designating the graphic name by selecting the resource clerk from the drop-down box. Transfer any picture you wish for the resource idle and busy states. Figure 14.25 illustrates the selection of graphics for these resource states from the people.plb library.

Once the resource idle and busy graphics have been assigned to the resource clerk, close the resource picture placement window. The seize area may now need to be adjusted. The seize area is represented by a small circle with a connecting line. This is illustrated in Figure 14.26.

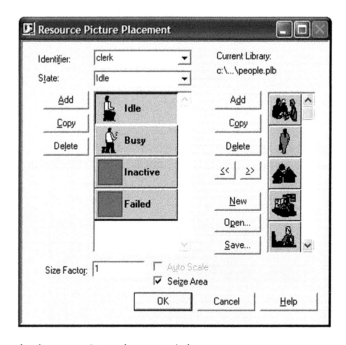

FIGURE 14.25 Completed resource picture placement window.

FIGURE 14.26 Resource sieze point.

The actual seize area is the round circle. If it is not positioned correctly, it can be moved by clicking on the round circle when the resource is not active. If the resource is active, you will not be able to click on the seize area. If the resource is active it can be made inactive by clicking on it. When the seize area has been clicked on, the point can be moved by dragging the mouse. It may be necessary to move the seize area a number of times to obtain the proper position. Your final simple animated model should look something like Figure 14.27.

14.10.4 Running the Animated Model

We are now ready to run our basic animated model. Click the go button on the VCR-like control. If everything has been performed correctly, you should see images of the resource and entities in the queue. The image of the resource should change between idle and busy. The screen should look like Figure 14.28 at the end of the run. Note that at the end of the 480-min run, the clerk is busy, and there are still six customers waiting in the line.

14.11 Additional Queue Techniques

More sophisticated queuing systems will require correspondingly more sophisticated queue-modeling blocks. Three additional queue-related blocks are:

- Select
- Pickq
- Qpick

14.11.1 Select Block

The select block is used when the entity is in a single snake queue and must choose from a number of different resources. The select block icon and the select block window are illustrated in Figure 14.29.

When the entity arrives at the select block, the decision as to which resource to choose is controlled by a resource selection rule. The most commonly used resource selection rules for the select block are:

- POR
- CYC
- RAN

FIGURE 14.27 Simple animated model.

FIGURE 14.28 Simple animated model running.

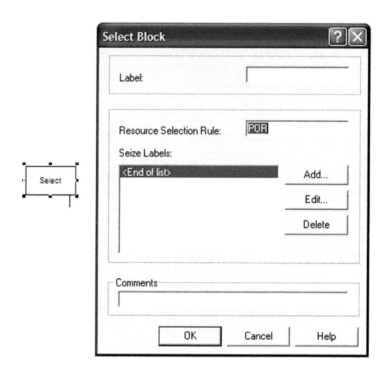

FIGURE 14.29 Select block.

In the select block, the POR (preferred order rule) means that if more than one resource is available to the entity, the entity will pick the resource that is highest ranking on the list. CYC (cyclic) means that the entity will rotate among available resources if more than one is idle. Last, the RAN (random rule) means that the entity will select randomly among the available resources if more than one is idle.

In addition to specifying the resource selection rule, the practitioner must also specify the seize labels. These are the label names for the seize blocks from which the entity can choose. If, for example, there are three resources called clerk1, clerk2, and clerk3, the corresponding seize labels for these resources must be entered in the seize labels section of the select block window. The completed select block is illustrated in Figure 14.30.

Figure 14.31 illustrates the relationship between the preceding queue block and the subsequent seize blocks. Note how the seize blocks are not directly attached to the select block.

14.11.2 Pickq Block

The pickq block is used to select among several parallel queue server sequences. The pickq block icon and the pickq block window are illustrated in Figure 14.32.

When the entity arrives at the pickq block, the entity chooses a queue server sequence based on a specified queue selection rule. The most commonly used queue selection rules for the pickq block are:

- POR
- LNQ
- SNQ
- CYC
- RAN

With the pickq block, the POR will send the entity to the first available queue in the queue labels list. lNQ (largest number in queue rule) will send the entity to the parallel queue with most entities in the

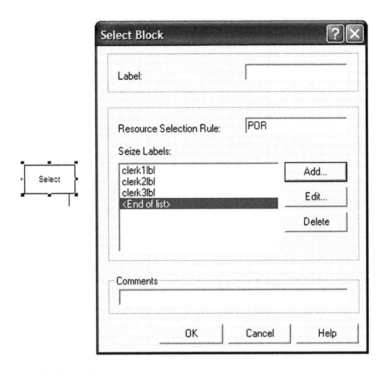

FIGURE 14.30 Select block details.

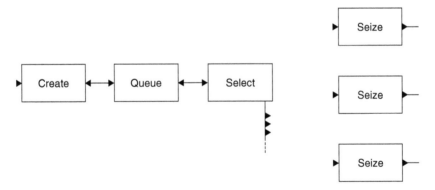

FIGURE 14.31 Select block use.

queue. Conversely, SNQ (smallest number in queue rule) will send the entity to the parallel queue with the fewest entities. CYC will send the entity to the different queues on a rotating basis. Last, RAN will send the entities to the queues in random order.

Of these queue selection rules, in customer entity type systems the most useful rule is the SNQ rule. This is because customer type entities will naturally gravitate toward the parallel queue with the smallest number of other customers.

To identify the parallel queues, the practitioner will also have to specify the queue labels by clicking the add button. The completed pickq block is illustrated in Figure 14.33.

In this model, the entity will select among clerkq1, clerkq2, or clerkq3. The selection is based on the smallest number in each of these queues. Figure 14.34 illustrates the pickq block inserted into the model.

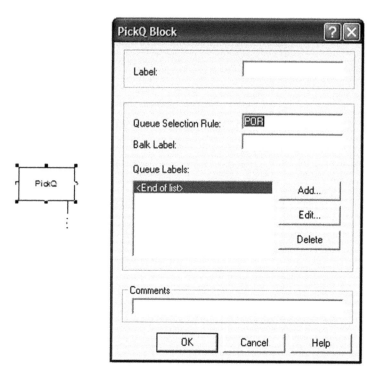

FIGURE 14.32   PickQ block.

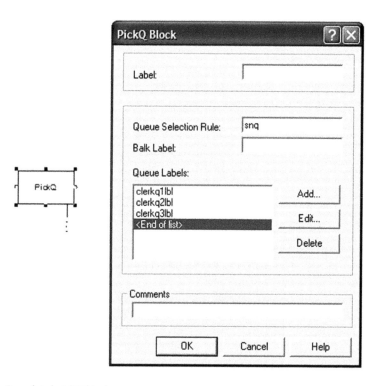

FIGURE 14.33   Completed pickQ block.

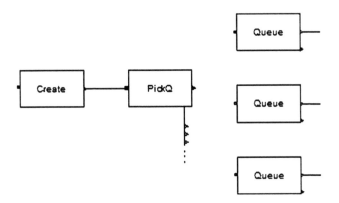

FIGURE 14.34 PickQ block usage.

14.11.3 Qpick Block

The qpick block is used to select an entity from one of several upstream parallel queues to be processed through a single seize block. The qpick block icon and the qpick block window are illustrated in Figure 14.35.

The entity is chosen according to a queue selection rule in a similar manner as the pickq block. Common queue selection rules for the pickq block also include:

- POR
- LNQ
- SNQ
- CYC
- RAN

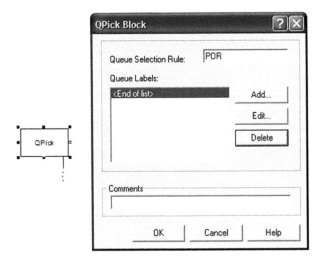

FIGURE 14.35 Qpick block.

In the qpick implementation of the POR, the resource will use the queue that is listed first in the set of queue labels. If there are no entities in that queue, the next queue on the list will be utilized. INQ rule will result in an entity being served from the queue with the largest number of entities. Conversely, the SNQ rule will result in an entity served from the queue with the smallest number of entities. RAN will extract an entity from the available queues at random. Last, CYC will cause an entity to be removed from the available queues in rotation.

The practitioner must also specify the queue labels for each queue in the queue label box. If we have three queues, the completed qpick block window will be as illustrated in Figure 14.36.

In this model, the entity will be taken from among serviceq1, serviceq2, or serviceq3. The queue priority rule is set at POR. With the qpick block this means that the entity if one exists will be taken first from serviceq1, then serviceq2, then serviceq3. this would represent a system where there is only one clerk, but first-, second-, and third-class lines. Naturally, the first-class customers would be served first. Figure 14.37 illustrates the qpick block inserted into the model.

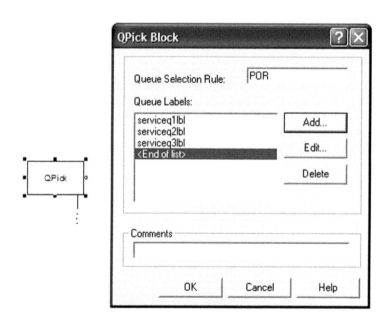

FIGURE 14.36 Completed Qpick block.

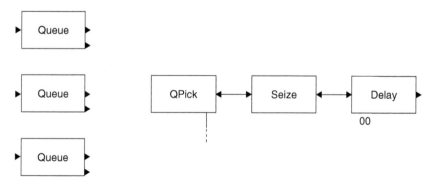

FIGURE 14.37 Qpick block usage.

14.11.4 Match Block

The match block is used to match entities from two different queues. This block is frequently used when an original entity is split or duplicated earlier in the program and the separate entities are processed independently. When the independent processing is complete, the match block allows the two entities to be reassembled for further processing. Examples of match block applications include:

- Passengers going through the metal detector and carry-on baggage being x-rayed
- Court paperwork being processed while defendants wait in the courtroom
- Fast food being prepared while customers wait for their order

The match block icon and the match block window are illustrated in Figure 14.38.

The match block window contains two important text boxes. The first important text box requires that the practitioner input the match attribute. This is the entity attribute that is used to identify uniquely the two entities as one. A typical name for this entity attribute is idnum for identification number.

The second important text box is where the practitioner must identify the label names of the queues where the entities are held. When the labels add button is clicked, the labels window in Figure 14.39 appears.

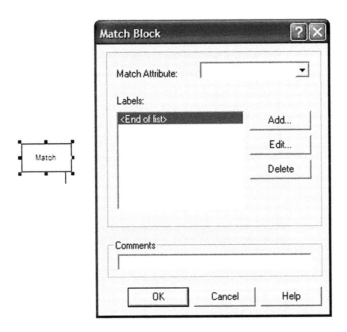

FIGURE 14.38   Match block.

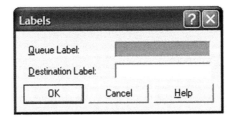

FIGURE 14.39   Match block labels window.

In this labels window, the practitioner must specify both the queue label where the entity is residing and the destination label to send the entity. A queue and destination label set must be added for each of the entities in the match. When all of the entities are available for a match in the designated queues, all of the entities will be sent to their individual destination labels. In the case of a security checkpoint, the passenger entity and his or her carry-on luggage entity are recombined. This means that the customer entity who just finished with the metal detector can continue through the system. Figure 14.40 illustrates a completed label window.

At the same time, the luggage entity can be disposed of by sending it to an appropriately labeled disposal block destination. The completed match block is illustrated in Figure 14.41.

The entire application of the match block with two holding queues, the systemfinishedlbl station destination, and the displbl disposal destination are illustrated in Figure 14.42.

14.12 Modeling Transporters

Transporters are moving resources that can move entities between locations. Examples of transporters include:

- Hand
- Forklifts

FIGURE 14.40 Completed match block labels window.

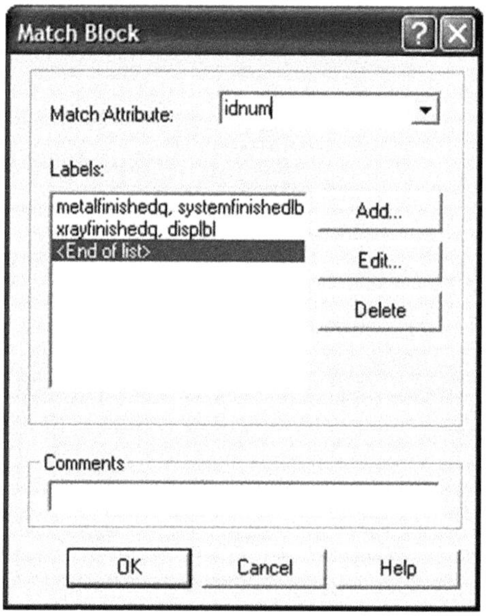

FIGURE 14.41 Completed match block.

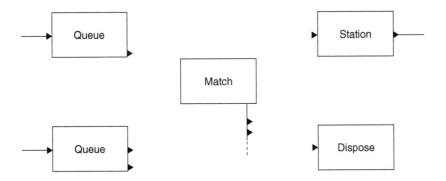

FIGURE 14.42 Match block usage.

To model a transporter in ARENA, you will need to use a number of new model blocks and elements. Model blocks include:

- Request
- Transport
- Free

Experiment elements include:

- Transporters
- Distances
- Storages

14.12.1 Request Model Block

The request block is used by the entity to request a transporter to its current location. It is somewhat analogous to the normal seize block. The request block is normally preceded by a queue. The entities remain in the queue until a transporter is available. When a transporter is available, it is assigned the entity that is first in the queue. The entity is then removed from the queue and put in a storage area until the transporter arrives at the entity's location. The most important text box for the request block is the name of the transporter. If the entity is to be animated while it is in storage waiting for an allocated transporter, the storage text box must also be completed. Defaults can be accepted for the other text boxes. The request block icon and window are illustrated in Figure 14.43.

14.12.2 Transport Model Block

Once the transporter arrives at the entity's location, the transport block specifies where the transporter should take the entity. The transport block is analogous to the delay block. When the entity is moving to another station, there is a time delay before it arrives. The important text boxes in the transport window are the transport unit and the destination text boxes. The default velocity value specified in the corresponding transporters element can be accepted. The transport icon and window are illustrated in Figure 14.44.

14.12.3 Free Model Block

The free block is analogous to the release block. Once the entity has arrived at the destination station, the entity will normally free the transporter. The only significant text box is the transporter unit text box. Defaults can be accepted for all of the other boxes. The free block is illustrated in Figure 14.45.

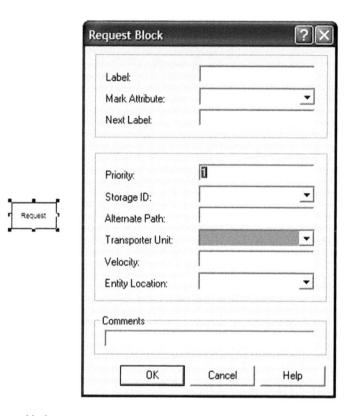

FIGURE 14.43 Request block.

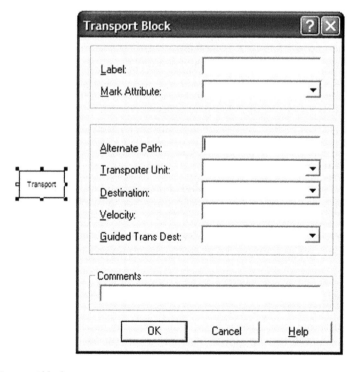

FIGURE 14.44 Transport block.

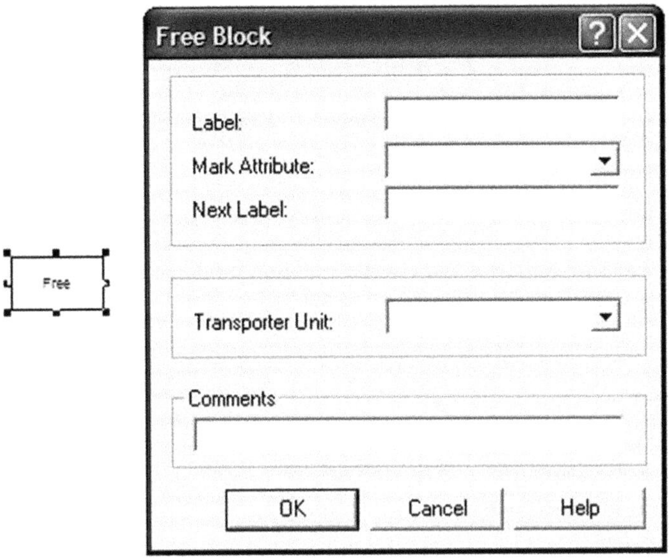

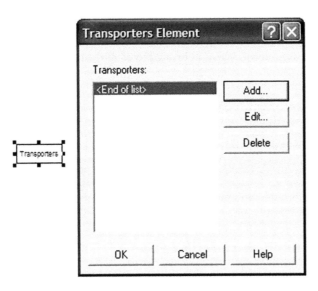

FIGURE 14.45   Free block.

14.12.4 Transporters Experiment Element

The transporters element is used to define the characteristics of the transporters in the model. To create a transporter you will have to click the add button in the transporters element window. In the new transporter window there are several important text boxes. The first important text box is name. The second text box is the number of units. The third text box is a drop-down box for the system map type. There are distance and network options. The default distance option should be kept. The fourth important text box is for the map ID. The entry for the text box will be the name of the distance map that the transporter will operate within. The final important text box is for the velocity of the transporter. The transporters icon and the transporters element window are illustrated in Figure 14.46.

FIGURE 14.46   Transporters element.

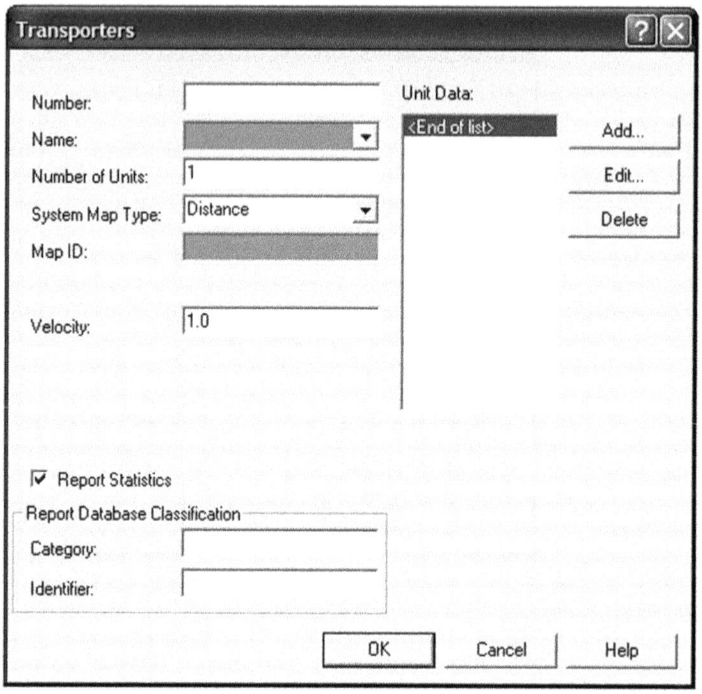

FIGURE 14.47 Transporters element detail.

The transporters window is illustrated in Figure 14.47.

14.12.5 Distances Experiment Element

The distances experiment element specifies the map over which the transporter travels. A distance is specified by clicking the add button in the distances element window. Once the distances window appears, the name for the distance must be entered in the identifier text box. In the distance window, the stations on the distance must also be specified. This means that the stations add button must be clicked to specify the distances between any two stations. When the add button is clicked, the stations window appears. In the stations window, all of the text boxes must be completed. If the appropriate stations have already been defined, then the drop-down boxes can be used to enter the names of the stations. In the stations window, you will also need to specify the distance between the two stations. The process of adding the distances between each pair of stations needs to be repeated until all of the expected paths between two stations have been entered. Figure 14.48 illustrates the distances icon and the distances element window. Figure 14.49 illustrates the distance and stations windows.

14.12.5.1 Storages Experiment Element

The storages experiment element is used to hold the entities that have been allocated a transporter but have not yet been picked up for transport. The storages element is relatively simple. In the storages element window click the add button. In the new storages window enter the name of the storage. The storages icon, storages element window, and storages window are illustrated in Figure 14.50.

14.12.6 Building the Transporter Model

Now that we have been introduced to the various transporter-related blocks and elements, it is time to build a small sample model. To keep things simple, we will use many of the parts from the previous models that we have built. If you so desire, you may delete unwanted parts from the previous model and

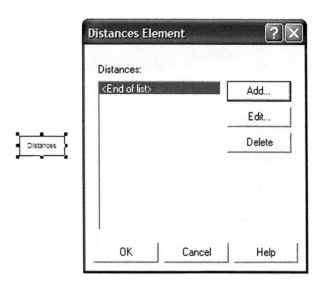

FIGURE 14.48 Distances element.

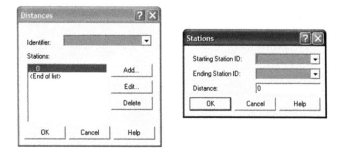

FIGURE 14.49 Distances element detail.

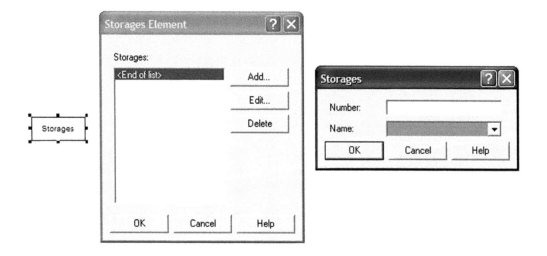

FIGURE 14.50 Storages element.

insert the appropriate transport-related blocks. The transporter model should be built in the same manner as any other conventional model. This means that it is most effective to create the blocks and elements, complete the elements, and then complete the model last.

Our model will need the following blocks. We will essentially replace the seize-delay-release sequence of the previous example model with a request-transport-free sequence. Begin by deleting the seize, delay, and release blocks. Drag, drop, and reconnect the icons for the model so that you have the following model.

- Create
- Assign
- Station
- Route
- Station
- Queue
- Request
- Transport
- Station
- Free
- Route
- Station
- Tally
- Dispose

When you have completed placing and connecting all of the blocks, your model should look something like Figure 14.51.

The corresponding elements for this experiment are listed below. Because we are replacing the resource with the transporter, we can delete the resource element. We will also have to add the transporters and distances elements to our experiment.

- Attributes
- Queues
- Tallies
- Dstats
- Replicate
- Transporters
- Distances

The completed set of experiment elements should appear as in Figure 14.52.

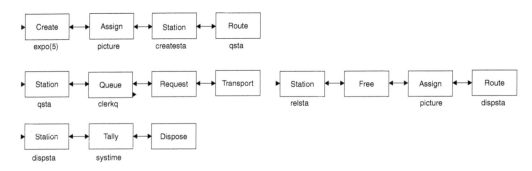

FIGURE 14.51 Transporter model.

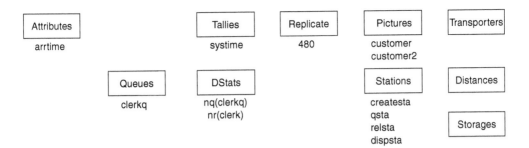

FIGURE 14.52 Transporter elements.

14.12.6.1 Complete the Experiment Elements

We are now ready to complete the experiment elements. The two elements that we must complete are the transporters and distances elements.

14.12.6.2 Transporter Element Entry

In the transporter element, we will need to enter the name of the cart in the name text box. Enter the name:

Trolley

In the system map type, ensure that the text box displays distance. In the map ID text box insert the name:

Trolleymap

Note that if we had edited the distances map first, the drop-down box would have had the name trolleymap listed. The velocity of the transporter can be defaulted at 1.0. When you are complete with these entries, the transporter window should be similar to Figure 14.53.

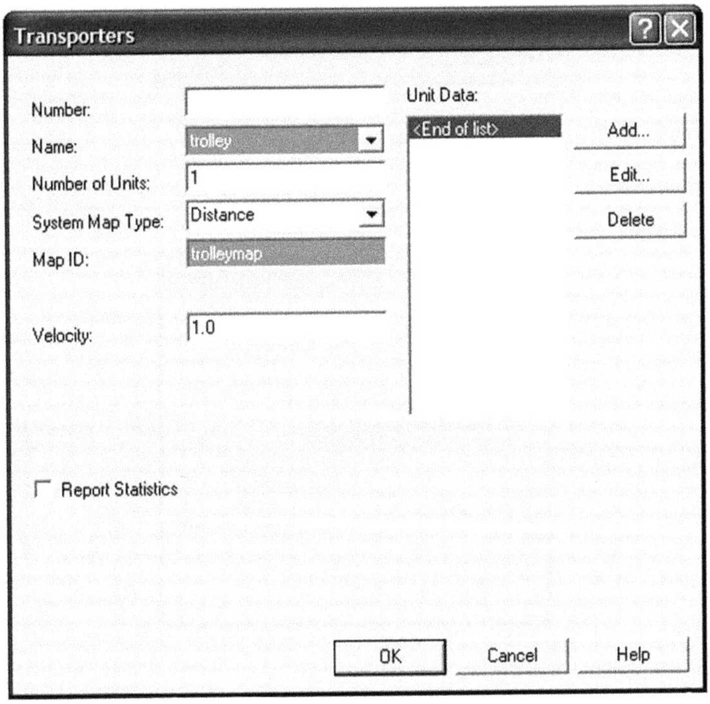

FIGURE 14.53 Completed transporter element.

14.12.6.3 Distances Element Entry

The distances element needs the name of the map that the transporter will be operating on. It also needs to know the distance between stations on the map. Insert the following name in the distances window:

Trolleymap

Add one station set with the qsta and relsta stations and a distance of 5 units. The stations window should look like Figure 14.54.

Close the stations window. When you are complete, your distances window should appear as in Figure 14.55.

14.12.6.4 Storage Element Entry

The storage element entry is the last step to defining the sample example experiment. In the storages element window, click the add button. In the new storages window, enter the following name in the name text box:

Trolleysto

The number box can be left blank. The completed storage window should look like Figure 14.56.

When the storages window is closed, the storages element window should look like Figure 14.57.

Your final set of experiment elements should look similar to Figure 14.58.

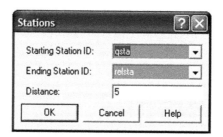

FIGURE 14.54   Completed distance element stations window.

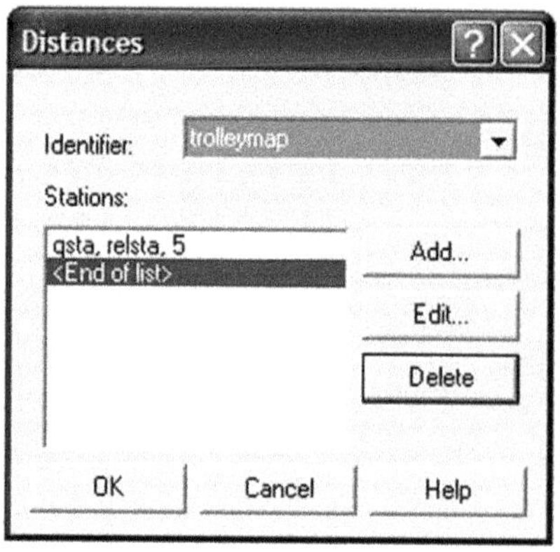

FIGURE 14.55   Completed distances element.

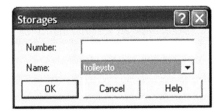

FIGURE 14.56 Completed storages element detail.

FIGURE 14.57 Completed storages element.

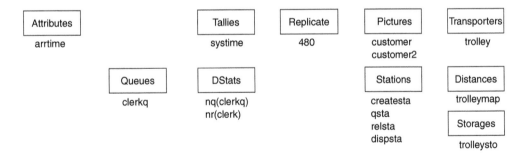

FIGURE 14.58 Completed experiment elements.

14.12.6.5 Completing the Model Blocks

With the experiment elements complete, attention can now be directed toward the model blocks. We will have to provide entries for the request, transport, and free blocks.

14.12.6.6 Request Block

Open the request block in the model. In the transporter ID drop-down text box, select trolley. In the entity location drop-down text box, select qsta. Defaults can be accpeted for all of the other text boxes.

14.12.7 Animating the Transporter Model

To animate the transporter model, we can reuse our previous animation. To reuse the previous animation we will need first need to delete the resource animation icon. We will also need to add the following animation components:

- Transporters
- Distances

14.12.7.1 Transporter Placement

The transporter placement window is selected by clicking the transporter icon on the animate transfer toolbar. Figure 14.59 illustrates the transporter icon.

When the transporter placement icon is clicked, the transporter placement window (Figure 14.60) appears.

This window is similar in function to the resources window. The name of the transporter must also be selected from the identifier drop-down text box. Animation images must be copied from the right to the left for each of the transporter states. In our model, we need to animate only the idle and busy states.

FIGURE 14.59   Transporter icon button.

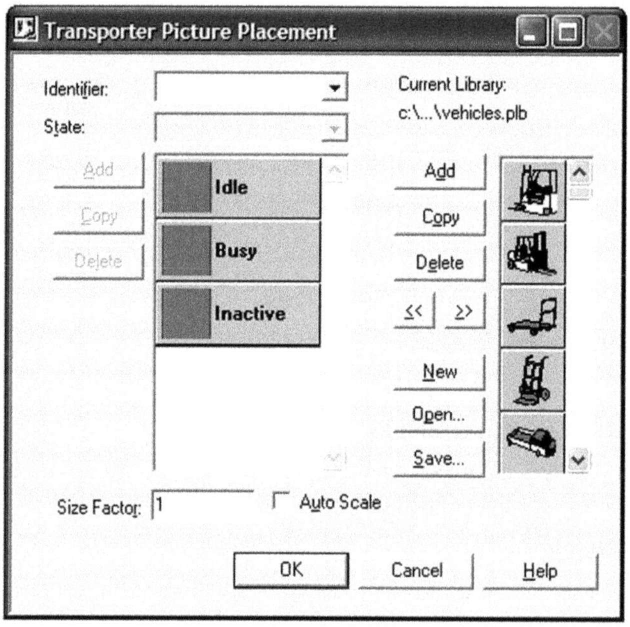

FIGURE 14.60   Transporter picture placement.

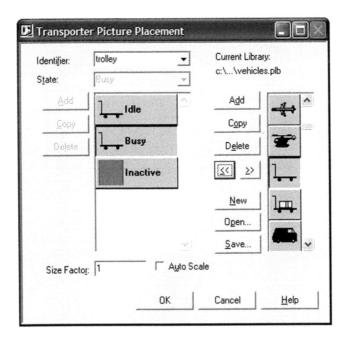

FIGURE 14.61   Completed transporter picture placement.

A number of vehicle pictures can be selected from the vehicle.plb library. To illustrate the actual entity traveling on the transporter, it is necessary to use the same picture for both the idle and busy states. We will later define what is known as the ride point to animate the entity riding on top of the transporter. Remember that the idle and busy panel must be selected in order to execute the transfer.

Proceed by selecting the name trolley in this block and transfer pictures to the trolley animation. Your transporter placement window should look something like Figure 14.61.

The final step in the transporter placement window is to specify the ride point for the entity. This is accomplished by double clicking on the busy panel for the trolley. When the picture editor window comes up, execute the menu sequence object-ride point. The mouse cursor will be come a crosshair. Move the crosshair to the point at which you want the entity to ride on the transporter. Some trial and error is necessary to make the entity ride on the desired point on the transporter. As a result, the ride point may have to be moved a number of times.

When the transporter placement window is closed, the mouse cursor becomes a crosshair. When the mouse button is clicked, the transporter appears on the flow chart window. The transporter can be stretched to the desired size.

14.12.7.2 Distance Placement Element

The distance placement process begins when the distance placement icon is clicked. The distance placement window appears as illustrated in Figure 14.62 and Figure 14.63 illustrates the distance placement icon.

In the distance placement window, the only necessary action is to use the drop-down identifier text box to select the trolleymap option. The distance placement window should now look like Figure 14.64.

When the distance placement window is closed, the mouse cursor becomes a crosshair. The crosshair should be positioned above the station icon for the qsta station and clicked once. The crosshair should then be moved to the station icon for the relsta station and clicked once. This will fix the distance that represents the trolley map. At this point, your model should look like Figure 14.65.

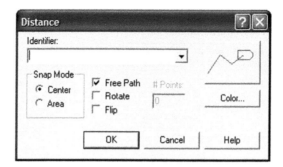

FIGURE 14.62 Distance window.

FIGURE 14.63 Distance icon button.

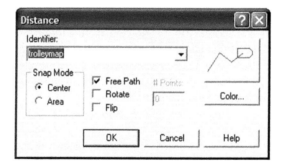

FIGURE 14.64 Completed distance window.

14.12.7.3 Storage Area

The final step in animating the model is to animate the storage point for entities that have been allocated a transporter but that have not yet been transported. The storage animation process begins by pressing the storage icon. This icon is illustrated in Figure 14.66.

When the storage icon is clicked, the storage window shown in Figure 14.67 appears.

The storage window is relatively simple. The only necessary action is to select trolleysto in the identifier drop-down text box. The storage window should now look like Figure 14.68.

When the storage window is closed, the cursor becomes a crosshair. The mouse now acts in a similar fashion as when placing a queue. The first mouse click will fix the head of the storage. The length and orientation of the storage is defined by first moving the mouse and then clicking. The length of the storage line does not need to be very long. In our case with one transporter, only one entity can be allocated a transporter and wait in the storage at any given time. A good place to position the storage is at the head of the queue. This way, when the entity moves from the queue to the storage, it appears as

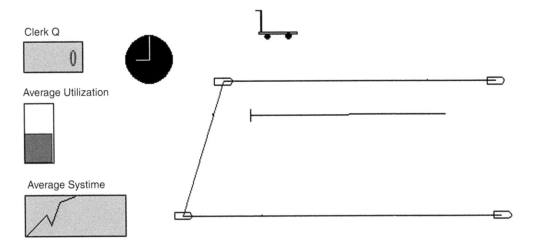

FIGURE 14.65 Transporter model animation.

FIGURE 14.66 Storage icon button.

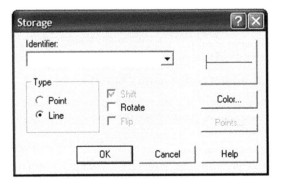

FIGURE 14.67 Storage window.

though he or she is advancing. The final transporter animation window should look something like the illustration in Figure 14.69.

14.12.7.4 Transporter DSTATS

Because we no longer are using the clerk, we will need to replace any clerk-related statistics with statistics related to the transporter. This means that we will need to change the SIMAN expression in the clerk utilization dstat to the following SIMAN expression:

Nt(trolley)

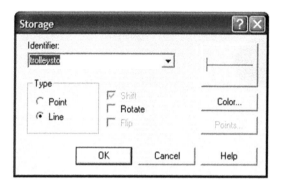

FIGURE 14.68 Completed storage window.

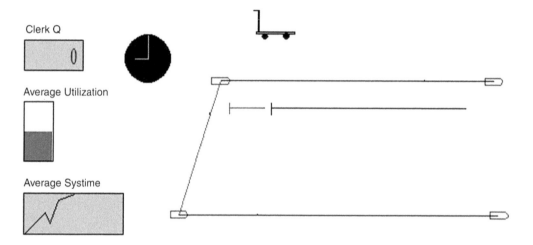

FIGURE 14.69 Transporter model animation with storage.

This will yield the utilization rate of the trolley instead of that of the clerk.

14.12.8   Running the Transporter Animation Model

The transporter model is now ready to run. As usual, it is very likely that one or more programming errors have occurred in the model development stage. Once the model is running, it should appear similar to Figure 14.70.

Note how the entities are moved between the qsta station and the relsta station by the transporter. Also note how entities appear in the storage position once the transporter has dropped off the entity that was riding on the transporter. If the ride point is not to your liking, you can edit the position in the picture editor.

14.13   Modeling Conveyors

ARENA can also be easily used to model conveyors. There are actually two different types of conveyors. The first type is the nonaccumulating conveyor. This means that when objects are placed on the conveyor, the spacing between the objects is maintained. An example of a nonaccumulating conveyor is the moving sidewalk. The second type of conveyor is the accumulating conveyor. The spacing between objects on the accumulating conveyor can be reduced from their original positioning. An example of an accumulating conveyor is the baggage conveyor at an airport. We will use the simpler nonaccumulating conveyor for demonstration purposes.

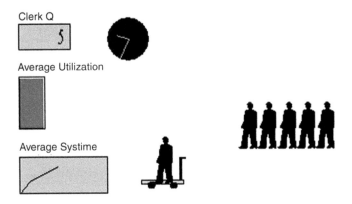

FIGURE 14.70 Running transporter model.

14.13.1 Building the Conveyor Model

The conveyor model should be built in the same manner as any other conventional model. This means that it is most effective to create the blocks and elements, complete the elements, and then complete the model last. To model a conveyor in ARENA, you will need to use a number of new model blocks and elements.

Model blocks include:

- Access
- Convey
- Exit

Experiment elements include:

- Conveyors
- Segments

14.13.2 Access Model Block

The access block is used by the entity to request an entry onto a conveyor. It is somewhat analogous to the normal seize block. The access block is normally preceded by a queue. The entities remain in the queue until a conveyor segment is available. When a conveyor segment is available, it is assigned the entity that is first in the queue. The entity is then removed from the queue and put on the conveyor. The most important text box for the access block is the name of the conveyor. The other text boxes can be defaulted. The access block icon and window are illustrated in Figure 14.71.

14.13.3 Convey Model Block

Once the entities are on the conveyor, they are conveyed to a particular station. The convey block is analogous to the transport block. The important text boxes in the convey block window are the conveyor name and the destination station text boxes. The convey icon and the convey block window are illustrated in Figure 14.72.

14.13.4 Exit Model Block

The exit block is analogous to the free block. Once the entity has arrived at the destination station, the entity will normally exit the conveyor. The only significant text box is the conveyor name text box. Default values may be accepted for all of the other boxes. The exit icon and exit block are illustrated in Figure 14.73.

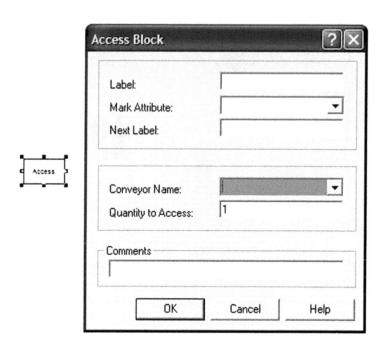

FIGURE 14.71   Access block.

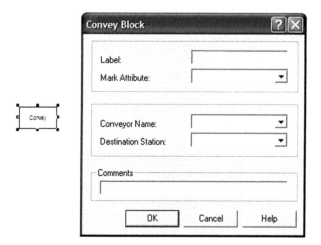

FIGURE 14.72   Convey block.

14.13.5 Conveyors Experiment Element

The conveyors element is used to define the characteristics of the conveyor in the model. To create a conveyor you will have to click the add button in the conveyor element window. In the new conveyor window there are two important text boxes. The first important text box is for the name of the conveyor. The second important text box is for the segment set ID. The conveyors icon, conveyors element window, and conveyors window are illustrated in Figure 14.74.

14.13.6 Segments Experiment Element

The segments experiment element specifies the stations between which the conveyor moves. A segment is specified by pressing the add button in the segment element window. Once the segment window

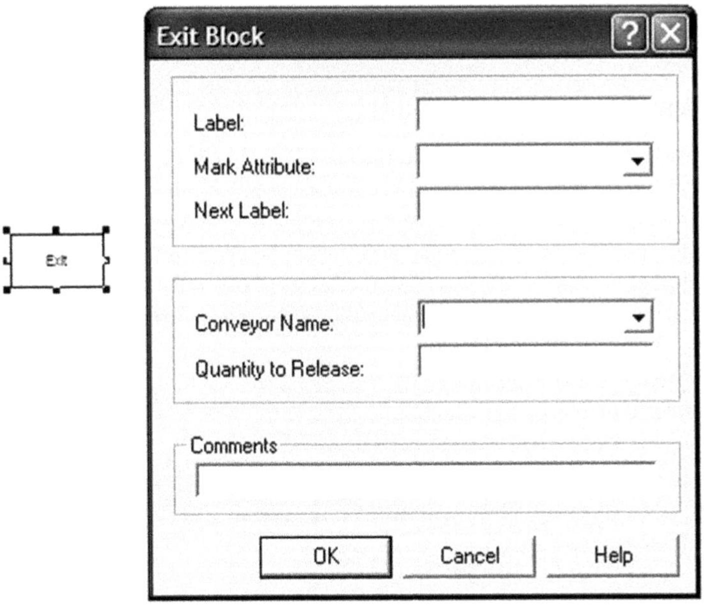

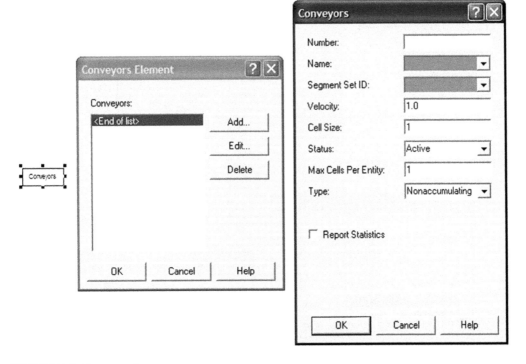

FIGURE 14.73 Exit block.

FIGURE 14.74 Conveyors element.

appears, the name for the segment must be entered in the identifier text box. Next, the beginning station drop-down text box must be completed. Last, the next stations add button must be clicked. When the add button is clicked, the next stations window appears. In this window, the next station drop-down box and the length text box must be completed. If there are additional stations on the segment, the additional stations can be specified in the same manner. Figures 14.75 and 14.76 illustrate the segment icon, segment element window, segments window, and next stations window.

14.13.7 Building the Conveyor Model

Now that we have been introduced to the various conveyor-related block and elements, it is time to build a small sample model. Once again, to keep things simple, we will use many of the parts from the previous models that we have built. If you so desire, you may delete unwanted parts from the either the resource-based model or the transporter-based model and insert the appropriate conveyor-related blocks. The conveyor model should be built in the same manner as any other conventional model. This means that it is most effective to create the blocks and elements, complete the elements, and then complete the model last.

Our model will need the following blocks. We will essentially replace the seize-delay-release or request-transport-free sequence of the earlier example models with an access-convey-exit sequence. Begin by

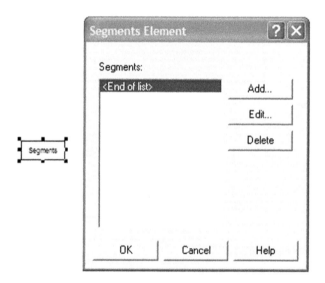

FIGURE 14.75 Segments element.

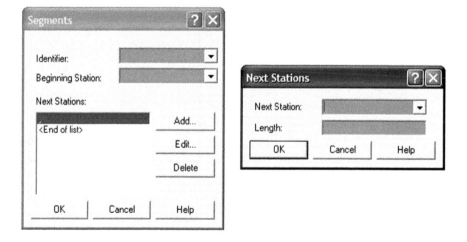

FIGURE 14.76 Segments element detail.

deleting the unnecessary blocks. Drag, drop, and reconnect the icons for the model so that you have the following model.

- Create
- Assign
- Station
- Route
- Station
- Queue
- Access
- Convey
- Station
- Exit
- Route
- Station
- Tally
- Dispose

When you have completed placing and connecting all of the blocks, your model should look something like Figure 14.77.

The corresponding elements for this experiment are listed below. If you are recycling a previous model, make sure that you delete the unneeded elements. This means that if you are reusing the resources model, delete the resources element. If you are recycling the transporters model, delete the transporters, distances, and storages elements. Note that we are adding the conveyors and segments elements to our experiment.

- Attributes
- Queues
- Tallies
- Dstats
- Replicate
- Conveyors
- Segments

The completed set of experiment elements should appear as in Figure 14.78.

14.13.7.1 Complete the Experiment Elements

We are now ready to complete the experiment elements. The two elements we must complete are the conveyors and the segments elements.

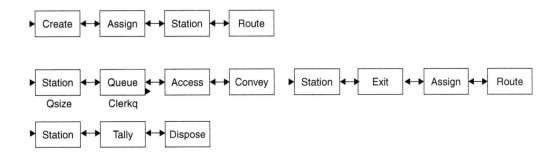

FIGURE 14.77 Completed conveyor model.

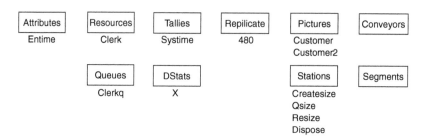

FIGURE 14.78 Completed conveyor experiment.

14.13.7.2 Conveyor Element Entry

In the conveyor element, we will need to enter the name of the conveyor in the name text box. Enter the name:

conv

In the segment set ID text box, insert the name:

convseg

Default values may be accepted for the rest of the text boxes. However, you should ensure that the type text box contains the value nonaccumulating. When you are complete with these entries, the conveyors window should be similar to Figure 14.79.

14.13.7.3 Segment Element Entry

The segments element needs the name of the segment that the conveyor will be operating on. It also needs the beginning station. Insert the following name in the segments window:

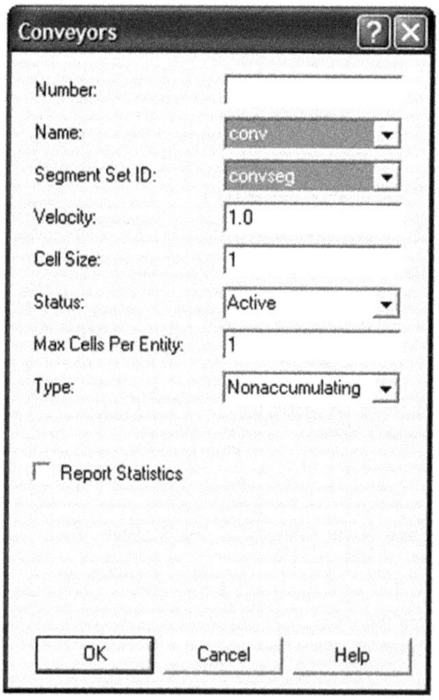

FIGURE 14.79 Completed conveyors element.

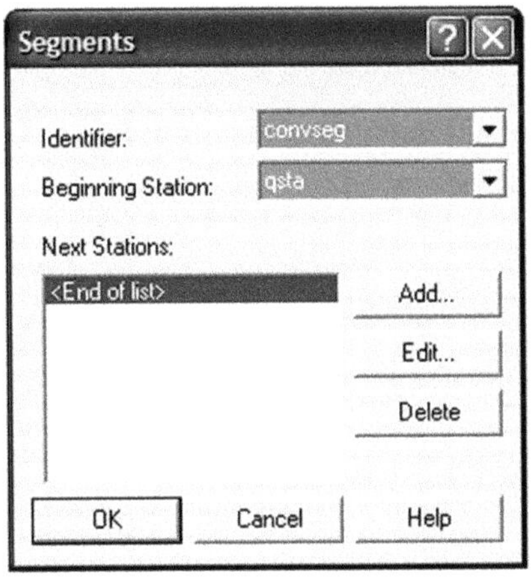

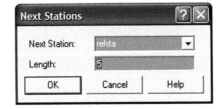

FIGURE 14.80 Completed segments element.

> convseg

Use the text drop-down box for the beginning station to specify:

> qsta

The segments window should now look like Figure 14.80.

Next, we will need to add a station to the segment. Click add in the segments window and use the next station drop-down text box in the next stations window to specify:

> Relsta

Last, in the length text box in the next stations window, enter the value:

> 5

The resulting next stations window should look similar to Figure 14.81.

Close the next stations window. When you are complete, your segments window should appear as in Figure 14.82.

Your final set of experiment elements should look similar to Figure 14.83.

14.13.7.4 Completing the Model Blocks

With the experiment elements complete, attention can now be directed towards the model blocks. We will have to provide entries for the access, convey, and exit blocks.

FIGURE 14.81 Segments next stations window.

FIGURE 14.82   Completed segments window.

FIGURE 14.83   Completed experiment.

14.13.7.5 Access Block Entry

Open the access block in the model. In the conveyor name drop-down text box, select:

Conv

Default values can be accepted for all of the other text boxes. Your completed access block should look like Figure 14.84.

14.13.7.6 Convey Block Entry

Click on the convey block. In the conveyor name drop-down text box, select:

Conv

In the destination drop-down text box, select:

Relsta

The convey block should look like Figure 14.85.

14.13.7.7 Exit Block Entry

The last block is the exit block. Open this block and use the conveyor name drop-down text box to specify:

Conv

All of the other text boxes can be defaulted. Your exit block should look the same as Figure 14.86.

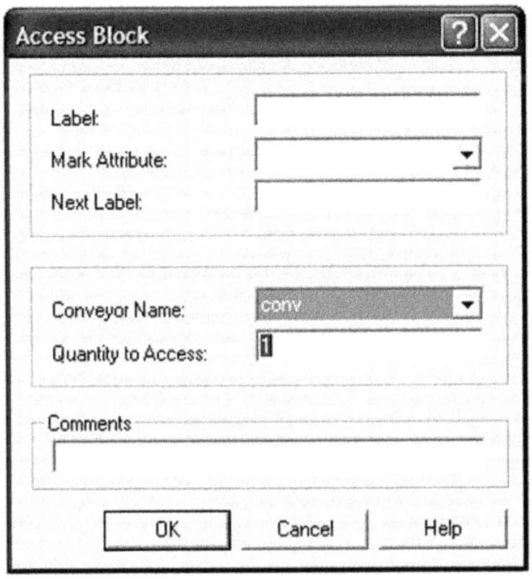

FIGURE 14.84 Completed access block.

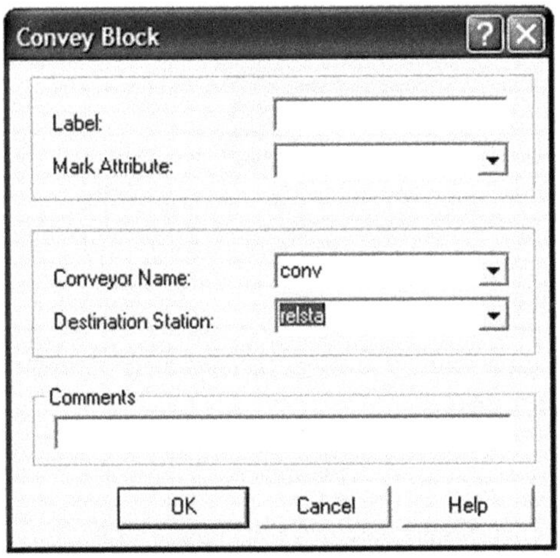

FIGURE 14.85 Completed convey block.

14.13.8 Animating the Conveyor Model

To animate the conveyor model, we can again reuse much of our previous animations. To reuse the previous animation we will first need to delete all of the other unnecessary animation icons. To make the animation realistic, we will also reposition some of our stations so that the conveyor moves horizontally between the qsta and relsta stations. We will also need to add the following animation components.

- Segments
- Static conveyor drawing

14.13.8.1 Segments

The segments placement window is selected by clicking the segments icon on the animate transfer toolbar. Figure 14.87 illustrates the segments icon.

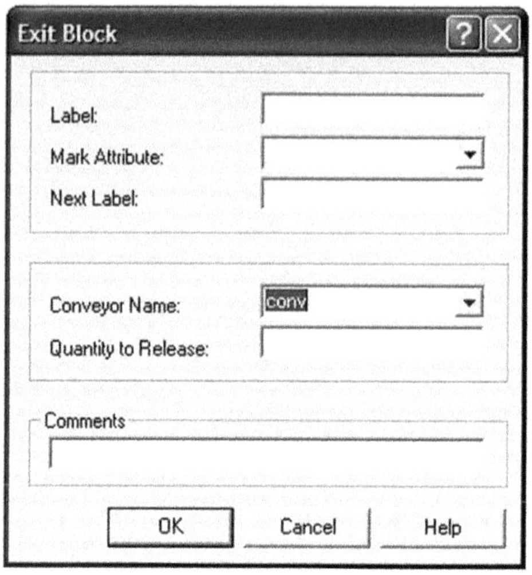

FIGURE 14.86 Completed exit block.

FIGURE 14.87 Segment icon button.

When the segments animation icon is pressed, the segement animation window appears. Use the identifier drop-down text box to select:

Convseg

The segments animation window should now resemble Figure 14.88.

When the segments window is closed, the mouse cursor becomes a crosshair. To define one of the stations on the segment, move the crosshair above the qsta station. Click once to set one end of the

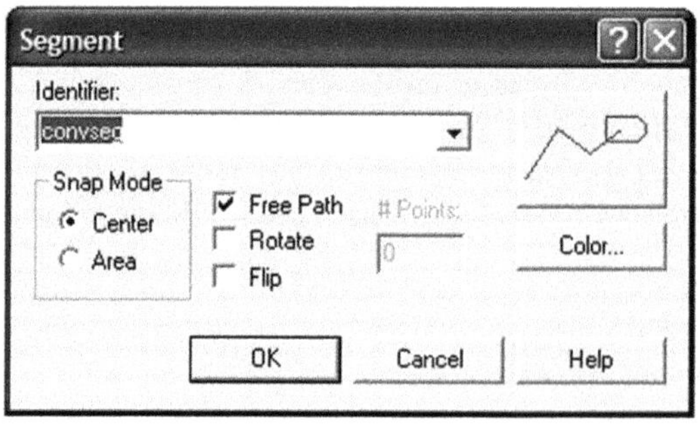

FIGURE 14.88 Completed segment window.

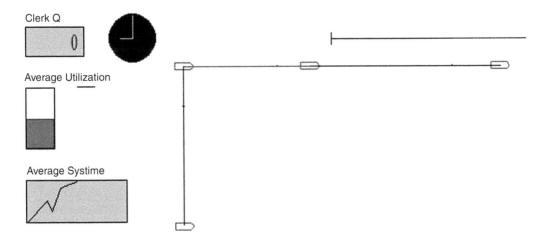

FIGURE 14.89 Conveyor model.

segment. Next, move the crosshair to the second station, the relsta station. Click once more to fix the end of the segment. The animation should now look like Figure 14.89.

14.13.8.2 Static Conveyor Drawing

When conveyors are animated, it appears as though the entities are moving along by themselves. In reality, the entities are actually moving along by the conveyor. To accurately animate the use of a conveyor, it is necessary to provide a static drawing of the conveyor. With this type of drawing, it will appear as though the entities are actually riding on the conveyor even though the conveyor drawing is static. Arena has a number of drawing tools to enhance the animation in this manner. Figure 14.90 has a static drawing of the conveyor positioned underneath the conveyor segment.

14.13.8.3 Conveyor DSTATS

Because we no longer are using the clerk resource or the transporter, we will need to replace any clerk- or transporter-related statistics with statistics related to the conveyor. This means that we will need to change the SIMAN expression in the utilization dstat to the following SIMAN expression:

$$ICS(conv)$$

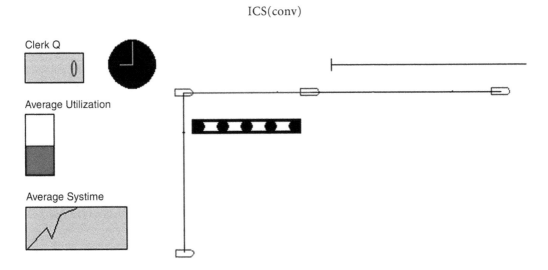

FIGURE 14.90 Conveyor model with static conveyor graphic.

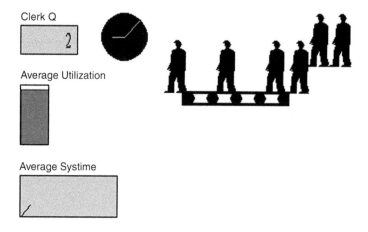

FIGURE 14.91 Running conveyor model animation.

This will yield the utilization rate of the conveyor instead of that for the clerk or transporter.

14.13.9 Running the Conveyor Animation Model

The conveyor model is now ready to run. As usual, it is very likely that one or more programming errors have occurred in the model development stage. Once the model is running, it should appear similar to Figure 14.91.

   Note how the entities who are on the conveyor appear to be riding it. The entities in the queue are waiting for space to access the conveyor.

14.14 Outputting Data to a File for Validation or Comparison Purposes

The practitioner will have to output data to a file for two distinct purposes. These purposes are:

- Model validation
- Model comparison

14.14.1 Model Validation

In a typical model validation process, the practitioner will be interested in comparing system times between the actual system and the simulation model. The actual system data are collected manually. The model data are generated by the simulation program. To write individual system times to a file, it is necessary to use the original already existing tallies element. This will result in individual entity times being written to the specified file each time an entity passes through the corresponding tally block. The use of the tallies element to write these data is illustrated in Figure 14.92. This example uses the file c:\valid.dat as the file name. Note that in the tallies element, it is necessary to enclose the file name with double quotes.

   Although it may not make as much sense, it is also possible to use individual time-dependent statistics for the validation data. Individual time-dependent statistics do not necessarily make sense because you are really interested in the average output measure of performance. This information is already provided to you at the end of the simulation run. However, if you insist on proceeding with individual time-dependent statistics, this means that you would be interested in using either the number in queue or the utilization for validation purposes.

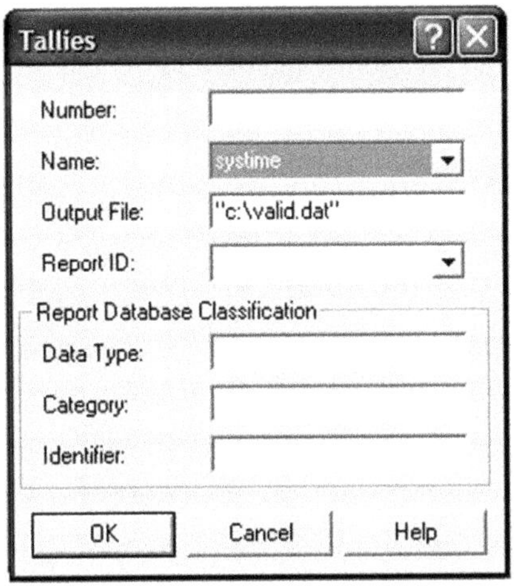

FIGURE 14.92 Tallies element.

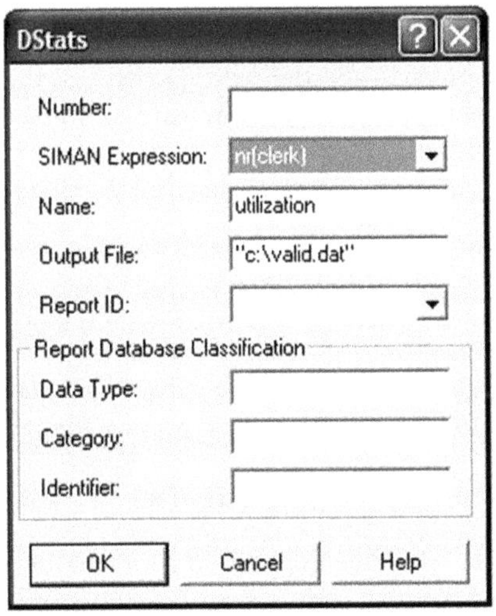

FIGURE 14.93 DStats element.

The technique for writing individual time-dependent data to a file is similar to that of the tallies. Figure 14.93 presents an example of using the average utilization for validation purposes.

The file format for both the tallies and dstats output files is proprietary. However, it can be easily exported into an ASCII file by the ARENA Output Analyzer, which we cover later in this chapter.

14.14.2 Model Comparison

The second purpose for outputting a file is for comparing the output performance of different models. Here we must determine whether our output analysis is based on a nonterminating or a terminating system. If the system is nonterminating, then we would use the same output approach as described in the previous section. This means that we would use the tallies element if we were interested in observational data or the dstats element if were interested in the time average number in queue or average utilization rate.

If we are working with terminating simulations, then we will need to use the outputs element. This element is illustrated in Figure 14.94.

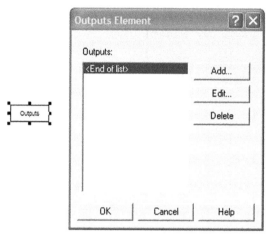
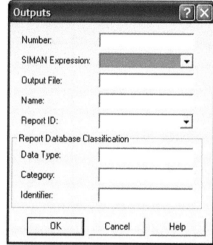

FIGURE 14.94 Outputs element.

Each time a simulation replication is run, the outputs element will write the value defined in the SIMAN expression text box. In order to use the outputs element, there must be a previously defined corresponding tally or dstat. In the SIMAN expression text box you will need to use either:

$$Tavg(tally\ name)$$

or

$$Davg(discrete\ event\ statistic\ name)$$

The function tavg contains the current tally average for the defined tally. For example, if we had a tally name systime, the correct entry to write the replication average for systime would be:

$$Tavg(systime)$$

The use of the outputs element to write tally replication data to a file is illustrated in Figure 14.95.

Similarly, the function davg contains the current time-dependent average for the defined discrete event statistic. For a time-dependent statistic named utilization, the correct entry to write the replication average for the average utilization would be:

$$Davg(utilization)$$

The use of the outputs element to write dstat replication data to a file is illustrated in Figure 14.96.

When the output element is used, the resulting file will need to be exported to ASCII format using the ARENA Output Analyzer.

14.15 Input Analyzer

The Input Analyzer will normally be used by the practitioner to fit observed distributions to theoretical distributions for use in the simulation model. This process requires the following steps.

- Enter observed data into an ASCII file.
- Import the ASCII file in the Input Analyzer.
- Obtain summary statistics on the input data.
- Use the Input Analyzer to generate a best theoretical fit.

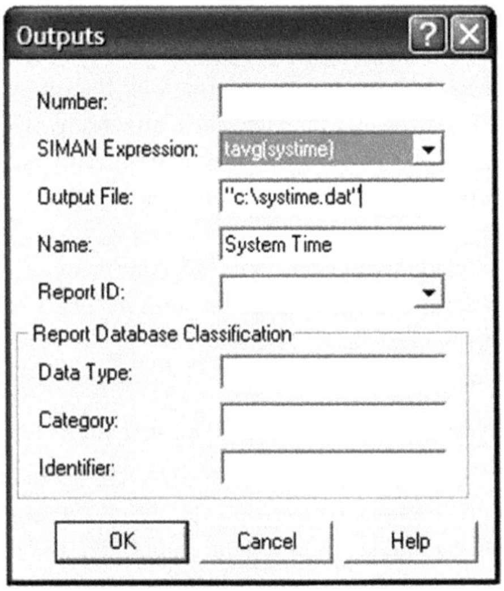

FIGURE 14.95 Completed outputs element for system time.

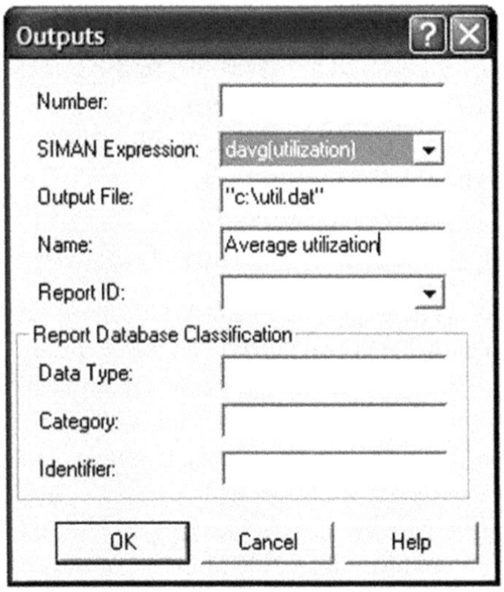

FIGURE 14.96 Completed outputs element for average utilization.

14.15.1 Enter Observed Data into an ASCII File

American Standard Code for Information Interchange (ASCII) files are easily imported and exported between applications. These files are often known as text files. If you are using historical data, it is possible that your data are already in an electronic file in ASCII format. If this is the case, you can skip to the next section. If not, you will have to hand enter your data into what will eventually become an ASCII file. At this point you have two options.

1. Enter directly into a text file.
2. Enter in an electronic worksheet and export as a text file.

14.15.1.1 Enter Directly into a Text File

Entering the data directly into a text file is easily performed with an ASCII text editor such as the notepad accessory that comes with Microsoft Windows. To enter data in the proper format, simply enter a single value on each line of the file. This means that after each number you will press the enter key or the carriage return key. The file will appear as a single column with many data values. The file name should be saved with the extension of ".dat." This is the extension that the Input Analyzer will be expected when you attempt to read in the data.

Although a text editor such as notepad is very easy to use, it lacks the ability to perform any preliminary data analysis. A text editor also cannot keep track of the number of data values that have been entered, nor can it format the data values. These disadvantages clearly make the use of an ASCII text editor less desirable than other data entry means.

14.15.1.2 Enter in an Electronic Worksheet and Export as a Text File

The practitioner may also opt to use an electronic worksheet such as Microsoft Excel for data entry. This is a much wiser choice because it does not suffer from the disadvantages inherent in a text editor. This means that preliminary summary statistics can easily be generated, the data can be formatted, and the number of data values can be tracked. The only disadvantage is that the electronic worksheet format must be exported into an ASCII file. This is actually not that difficult a task to perform.

The practitioner should use a separate workbook file for each different type of input data. The data should be entered in only the first column starting with the first row. The name of the file should clearly reflect the source of the data. Once the data values have been entered, the file should be saved in its native format as a.wks worksheet file. Once this has been performed, the data can be exported. Use the file-save as menu sequence to export the data to an ASCII file. In the save as type drop-down text box, it is essential that text (tab delimited) be chosen. This is what exports the file in ASCII format. In the name text box, ensure that you use the extension ".dat" so that the file can be imported into the Input Analyzer. This process is illustrated in Figure 14.97.

14.15.2 Import the ASCII File in the Input Analyzer

Regardless of which method you used to generate your ASCII input data file, you should have the data in a single column with one data value on each line. The input file should also have the extension of ".dat." If all is well then we are ready to import the file into the Input Analyzer.

The Input Analyzer can be launched from either within the ARENA environment or from the Rockwell Software program group. When the Input Analyzer loads, it appears as in Figure 14.98.

When the Input Analyzer comes up, most of the menu features are not active. In order to use properly the Input Analyzer, you will have to open a new session. This is performed by executing the menu sequence file-new or the new icon on the toolbar. The Input1 window now appears as illustrated in Figure 14.99.

Once the Input1 window comes up, it will be possible to execute the menu sequence: file-data file – use existing…. This will bring up a file open box. You can navigate to the appropriate subdirectory to open up the data file. This is illustrated in Figure 14.100.

14.15.3 Obtain Summary Statistics on the Input Data

When the data file is opened, a new window will appear. On the top half of the window, the Input Analyzer will create a histogram of the input data. On the bottom half of the window, the Input Analyzer will list summary statistics. This window is illustrated in Figure 14.101.

14.15.4   Use the Input Analyzer to Generate a Best Theoretical Fit

The input data can be fit individually to different distributions, or the Input Analyzer can attempt to make a best theoretical fit to the input data. These options are accessed through the fit menu. If the fit-fit all menu sequence is selected, the window shown in Figure 14.102 will appear.

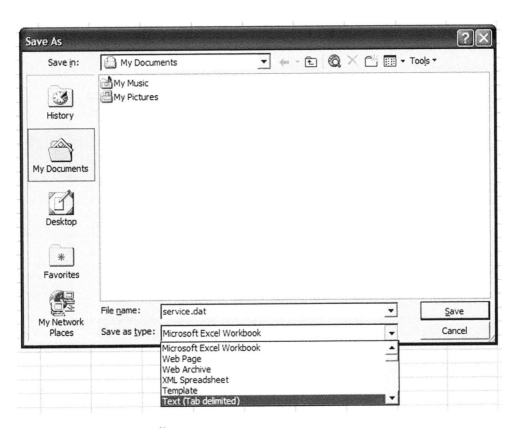

FIGURE 14.97 Importing ASCII files.

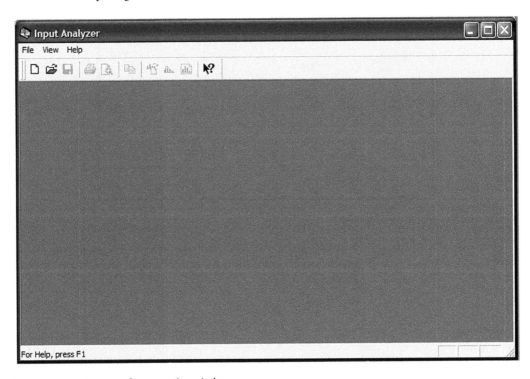

FIGURE 14.98 Input analyzer opening window.

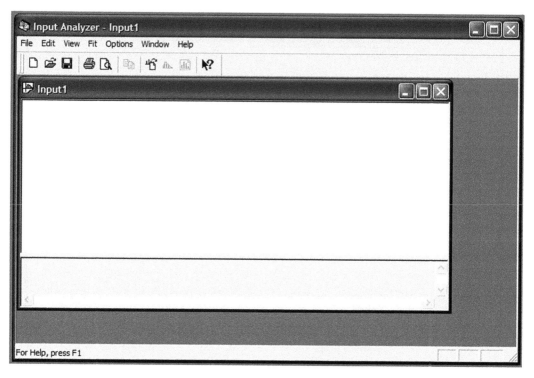

FIGURE 14.99   Input analyzer new file window.

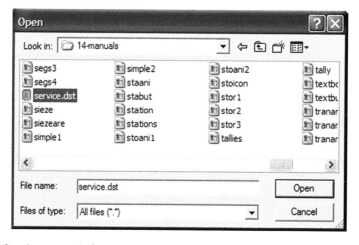

FIGURE 14.100   Opening a new window.

The results of the fit-all process indicates that the Input Analyzer believes that the best theoretical distribution fit is a triangular distribution with a minimum value of 33.5, a mode of 76, and a maximum value of 96.5. This would be represented in ARENA as:

$$TRIA(33.5, 76, 96.5)$$

To see the complete results of all the data distributions, press the window-fit all summary menu sequence. The window in Figure 14.103 now appears.

The summary sorts the theoretical fits in ascending order of square error between the observed data and the theoretical distribution. The lower the error, the better is the fit. The summary indicates that

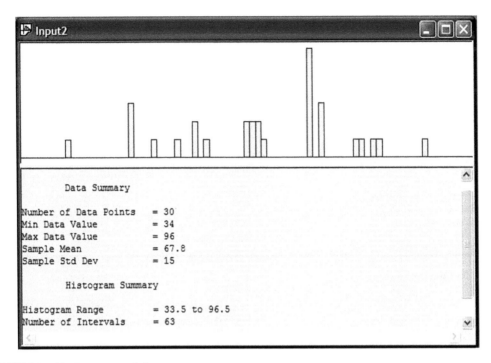

FIGURE 14.101 Summary statistics.

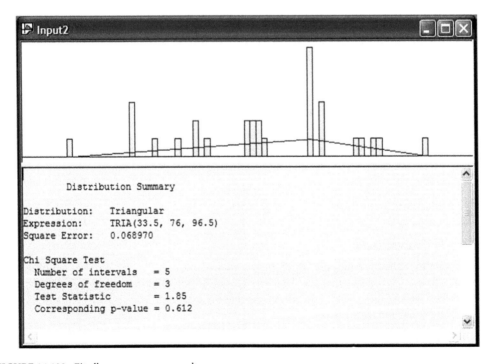

FIGURE 14.102 Fit all menu sequence results.

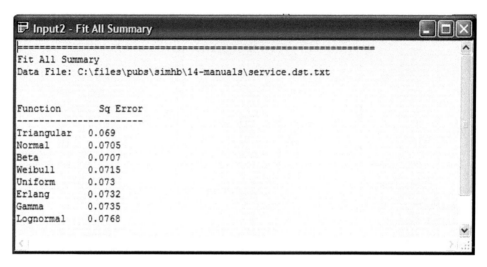

FIGURE 14.103 Fit all summary.

the triangular distribution was the best fit, but the normal distribution was a very close second. As with any data set, the more data values that are available, the greater is the reliability of the fit.

14.16 Output Analyzer

The Output Analyzer is used to obtain statistical summary data and compare output measures of performance. The general process for performing these tasks is:

- Write the output data to a file in arena.
- Launch the output analyzer.
- Select the type of analysis.
- Perform the analysis.

14.16.1 Write the Output Data to a File in ARENA

The first step in the output analysis process is to ensure that the output data have been properly written to a file. For detailed instructions on this process, practitioners are directed to review the output performance section of this chapter. The process is summarized briefly in this section. The output data file can be based on either:

- Individual data
- Replication data

14.16.1.1 Individual Data

If the data are for validation purposes or to analyze nonterminating systems, the data are likely to be individually based. This means that the data were written to the file using either the tallies or dstats element. When writing these data to the file, insure that the extension ".dat" is used so that the Output Analyzer can readily import the data.

14.16.1.2 Replication Data

If the data are for replication analysis of terminating systems or comparing terminating systems, the data will be replication based. This means that only one summary data value will be written to the output file for each replication. In order to write this type of data to the file, the outputs element must have been used. When using the outputs element, insure that the ".dat" extension was used.

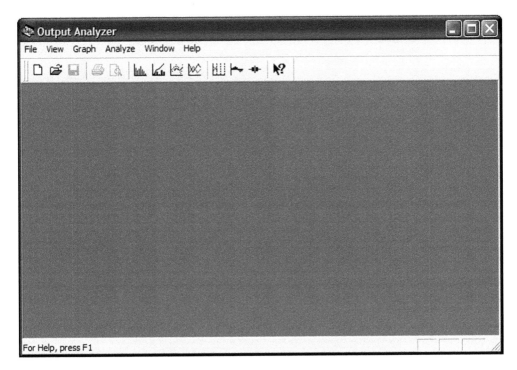

FIGURE 14.104 Output analyzer opening window.

14.16.2 Launch the Output Analyzer

Once the output file has been created, the Output Analyzer can be launched. This is performed either by using the menu sequence tools-Output Analyzer from within ARENA or by using the Rockwell Software/ARENA program group. When the Output Analyzer is launched, the screen shown in Figure 14.104 appears.

14.16.3 Select the Type of Analysis

The output analyzer has a large number of statistical analysis features. The most useful of these are:

- Confidence intervals on the mean
- Compare means
- Compare variances
- One-way ANOVA
- Correllogram
- Batch/truncate means

These analysis options are available through the analyze menu. The options are illustrated in Figure 14.105.

14.16.3.1 Confidence Intervals on the Mean

The confidence interval analysis is used to calculate summary statistics on either individual observation or replication data. When the confidence interval classical menu sequence is executed, the screen shown in Figure 14.106 appears.

To perform the analysis it is necessary to click the data file add button in the classical confidence interval on mean window. When the add button is clicked, the window shown in Figure 14.107 appears.

In the new data file window, you will have to specify the data file name and the type of replications. It may be necessary to click the browse button initially to locate the correct data subdirectory. Once the

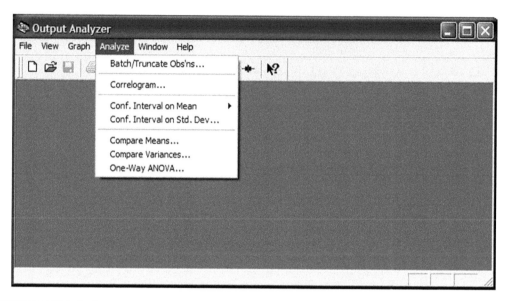

FIGURE 14.105   Analyze menu.

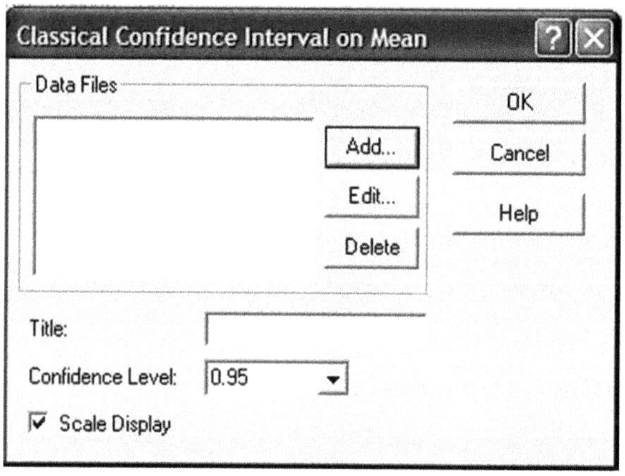

FIGURE 14.106   Classical confidence interval on mean window.

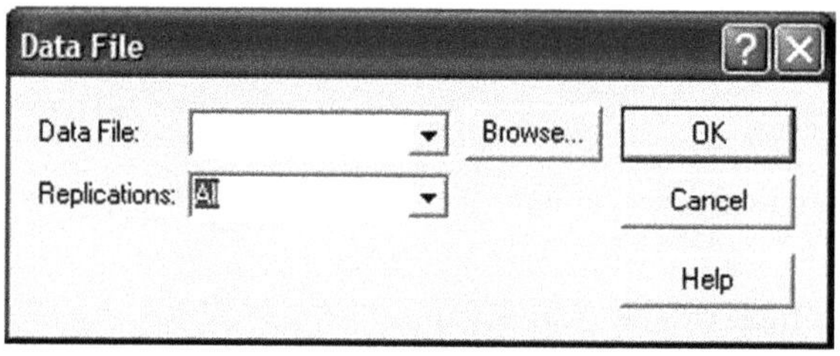

FIGURE 14.107   Data file window.

correct subdirectory is specified, subsequent files can be added by using the drop-down box. If the data are for individual observations, you should select "all" in the type of replication drop-down box. Conversely, if the data are for replications, you should select "lumped" in this box. In Figure 14.108, we will be analyzing the output file for replications.

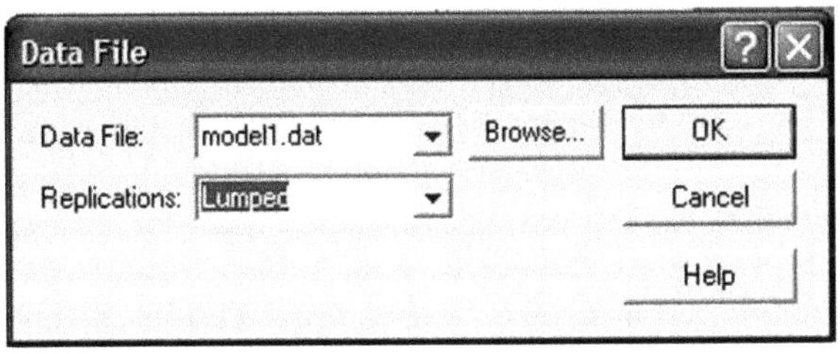

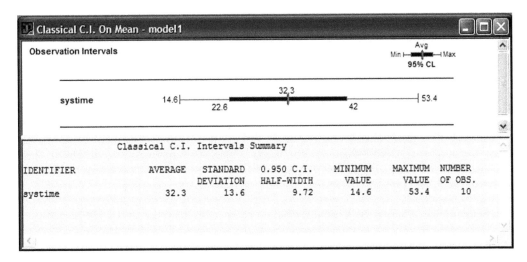

FIGURE 14.108 Completed data file window.

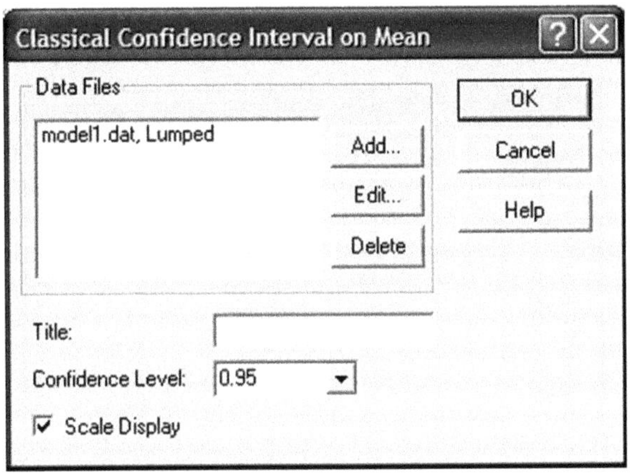

FIGURE 14.109 Completed classical confidence interval on mean window.

FIGURE 14.110 Classical C.I. on mean results.

When we close this window, the original classical confidence interval on mean window appears, as seen in Figure 14.109.

When this window is closed, the confidence interval is displayed as in Figure 14.110.

In addition to illustrating graphically the 95% confidence interval, the Output Analyzer also provided some useful confidence interval summary information. One very useful piece of summary information is the half-width confidence interval. This information is used to calculate the number of replications required for a statistical comparison.

14.16.3.2 Compare Means

The compare means analysis tool is used to generate a confidence interval on the difference in means between two models. If the confidence interval covers the value 0, there is no statistically significant difference between the models. On the other hand, if the confidence interval does not cover 0, then there is a statistically significant difference between the two models. When the compare means menu sequence is executed, the window in Figure 14.111 appears.

To begin the comparison process it is necessary to click the data file add button. This opens the new data file window, illustrated in Figure 14.112. In this window the two files that are to be compared must be specified. If we have replication output data in files model1.dat and model2.dat, the data file window should appear as in Figure 14.113. The replications text box should also be changed to lumped to perform a comparison based on replications.

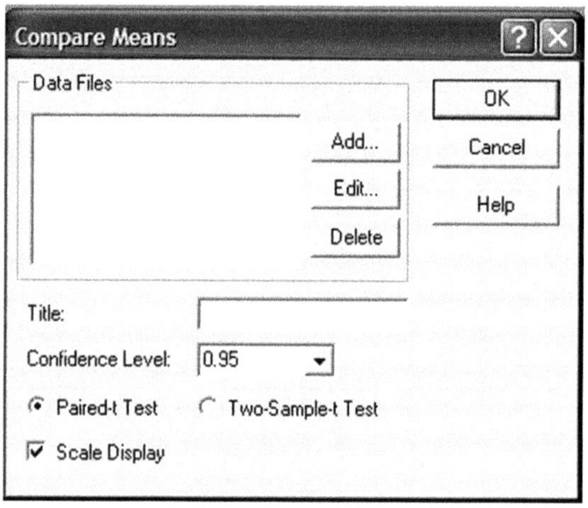

FIGURE 14.111   Compare means window.

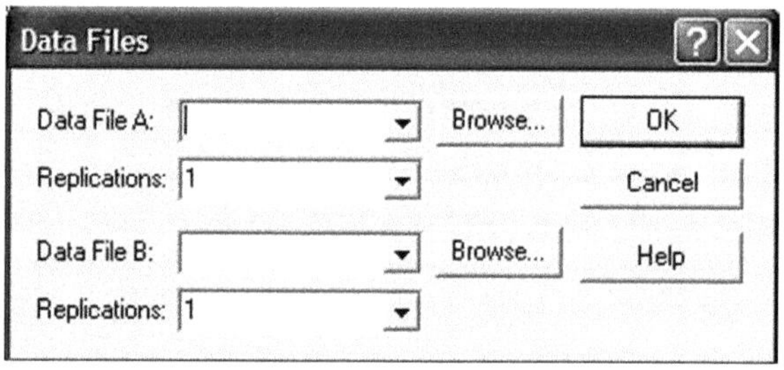

FIGURE 14.112   Compare means data files window.

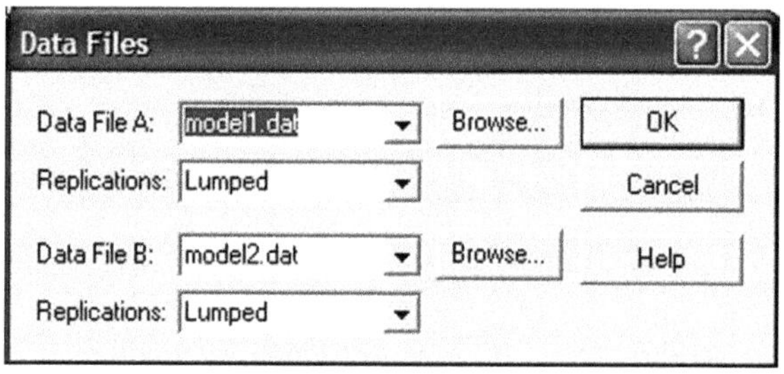

FIGURE 14.113   Completed compare means data files window.

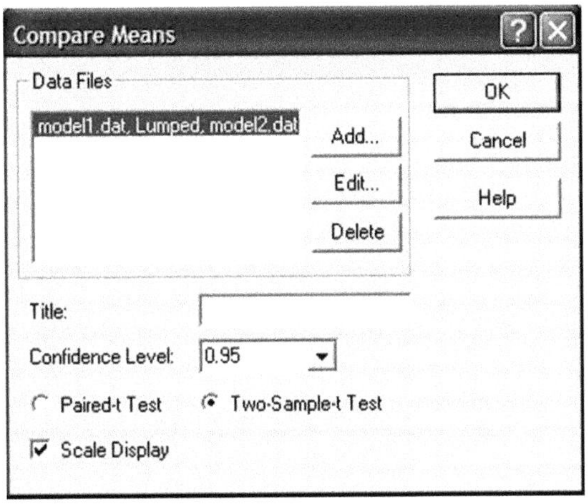

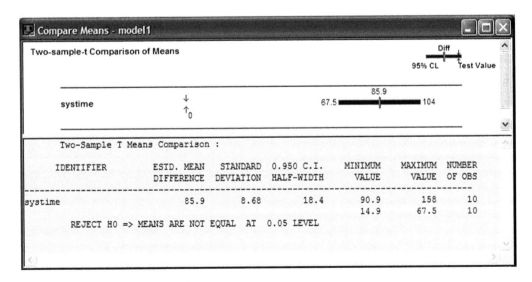

FIGURE 14.114   Completed compare means window.

When the data file window is closed, we return to the compare means window. In this window, you will also need to change the type of test from paired *t* test to two-sample *t* test under normal circumstances. If and only if you deliberately paired the comparisons should you use the paired *t*-test option. The final compare means window is illustrated in Figure 14.114.

When the window is closed, the confidence interval on the difference between means is illustrated (Figure 14.115).

In this particular example, the compare means confidence interval does not cover 0. This means that there is a statistically significant difference between the two models.

14.16.3.3 Compare Variances

The compare variances analysis tool is used to generate a confidence interval on the difference between two models. As with the compare means, if the confidence interval covers the value 0, there is no statistically significant difference between the models. Similarly, if the confidence interval does not cover 0, then there is a statistically significant difference between the two models. When the compare variances menu sequence is executed, the window shown in Figure 14.116 appears.

FIGURE 14.115   Compare means results.

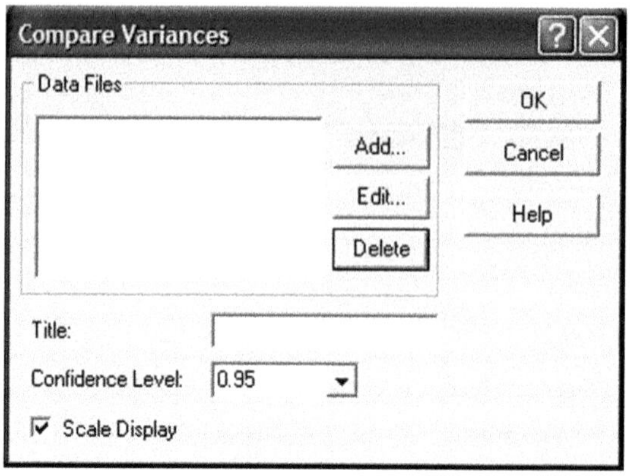

FIGURE 14.116 Compare variances window.

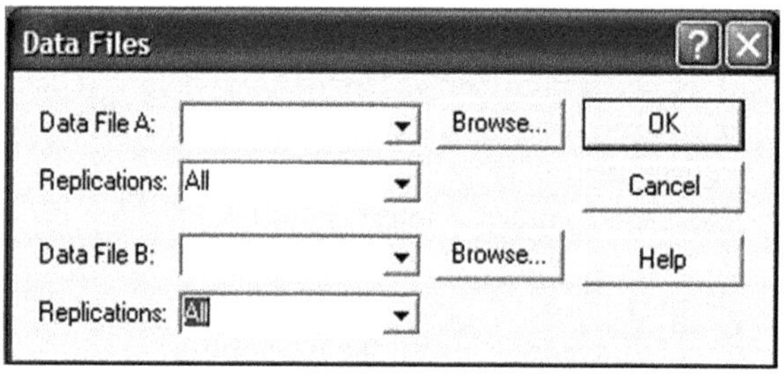

FIGURE 14.117 Compare variances data files window.

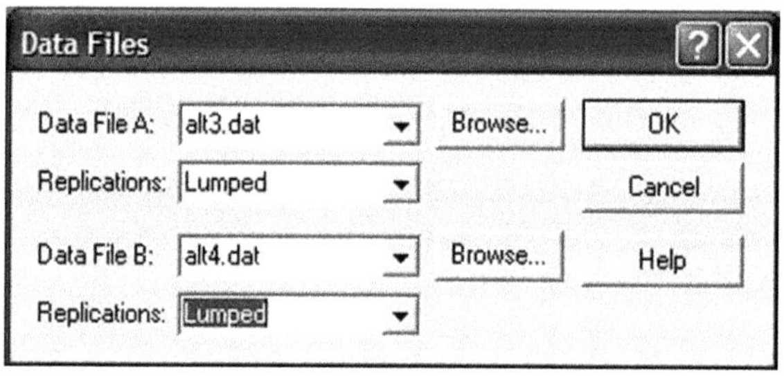

FIGURE 14.118 Completed compare variances data files window.

It is necessary to specify the files to observe. When the add files button is clicked in the compare variances window, the data file window appears (Figure 14.117).

The data file window requires that two data files be specified. We can specify the alt3.dat and the alt4.dat windows as lumped replication treatments in the data file window. The resulting window appears as in Figure 14.118.

When the data files window is closed, the Output Processor returns to the compare variances window. This window appears as in Figure 14.119.

When we execute the compare variances function, the Output Processor returns with Figure 14.120.

Note that the confidence interval does not cover 0 in the window. This means that the variances are statistically significantly different between the two data sets.

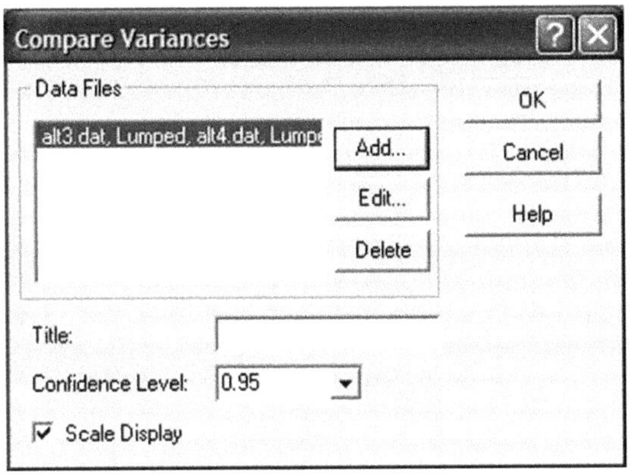

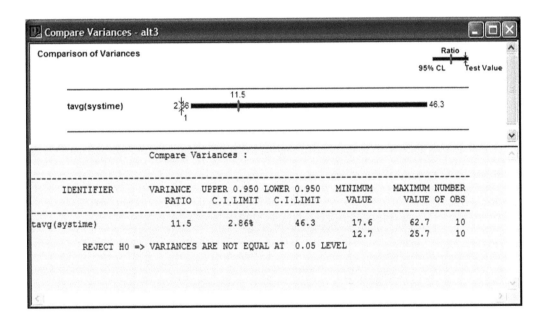

FIGURE 14.119   Completed compare variances window.

FIGURE 14.120   Compare variances results.

14.16.3.4 One-Way ANOVA

The one-way analysis of variance (ANOVA) process is used to determine if there is a statistically significant difference among one or more of the means. The ANOVA procedure does not identify which of the means is statistically significantly different. When the one-way ANOVA option is selected, the window shown in Figure 14.121 appears.

The first step is to specify the data files by clicking the add data file button in the one-way ANOVA window. When the add button is clicked, the data file window appears (Figure 14.122).

The data file window is used to specify each of the data sets for the ANOVA procedure. It is also necessary to change the replications text box to lumped. Figure 14.123 illustrates the file alt3.dat specified as a lumped file.

When all the files have been added, the one-way ANOVA screen should look like Figure 14.124.

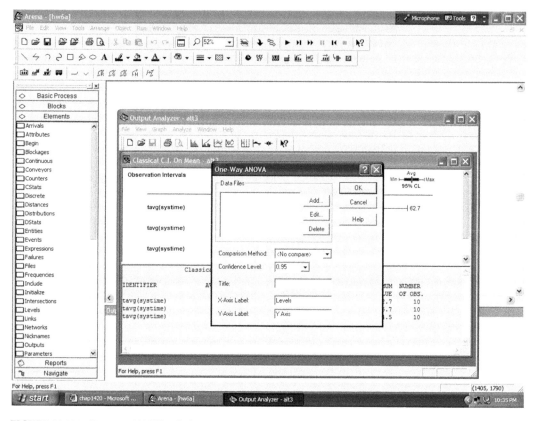

FIGURE 14.121 One-way ANOVA window.

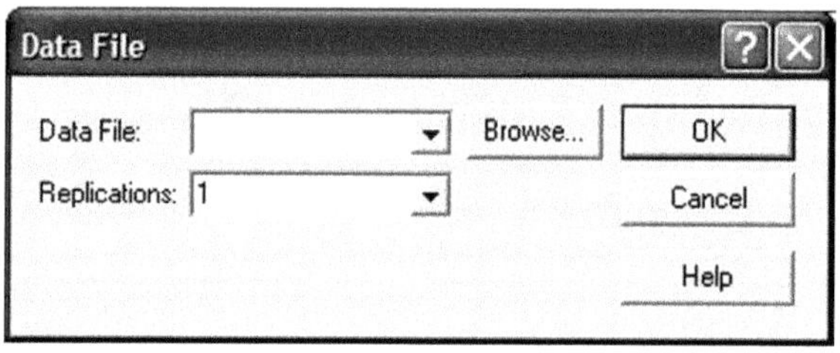

FIGURE 14.122 One-way ANOVA data file window.

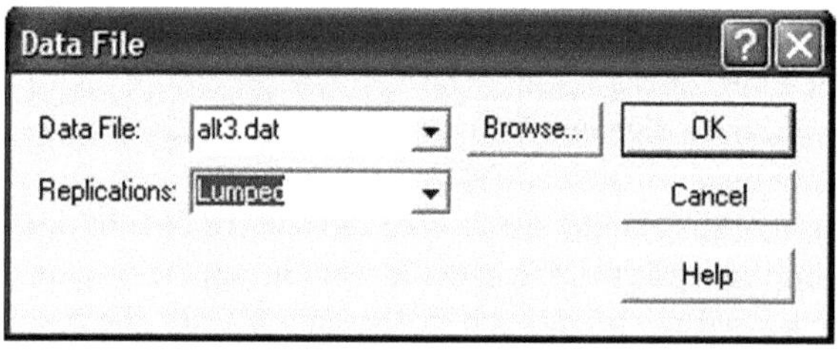

FIGURE 14.123 Completed one-way ANOVA data file window.

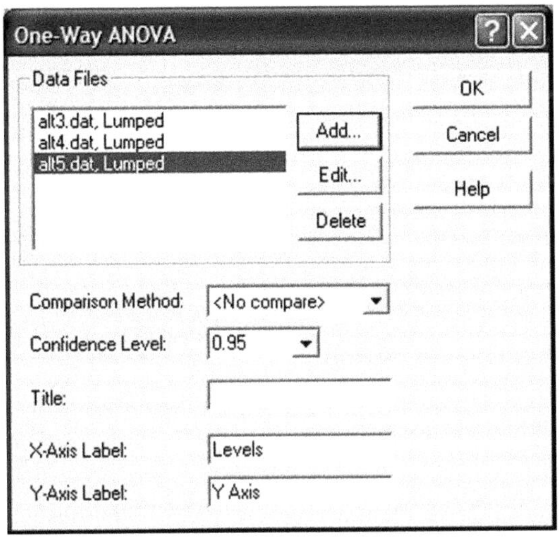

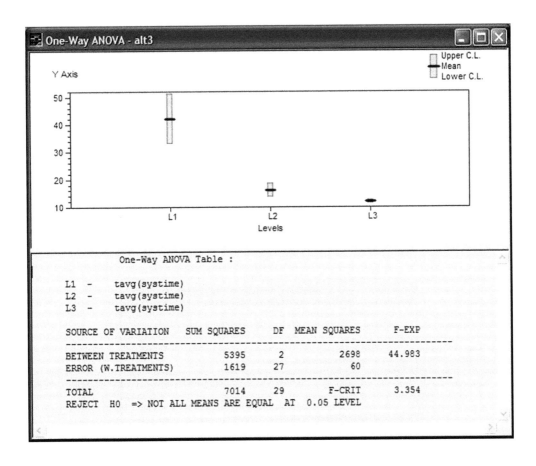

FIGURE 14.124 Completed one-way ANOVA window.

FIGURE 14.125 One-way ANOVA results.

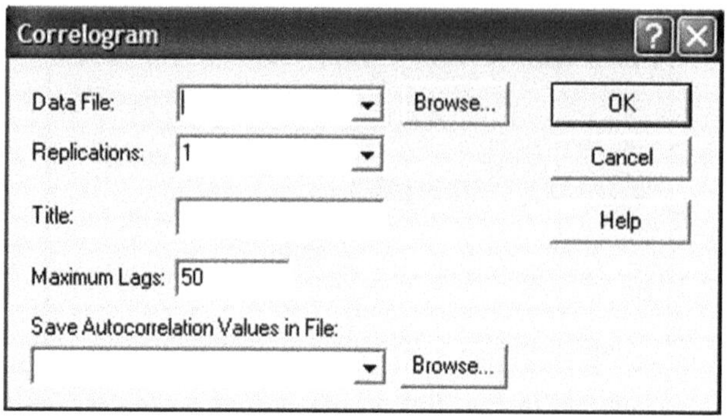

FIGURE 14.126 Correlogram window.

When the ANOVA is executed, the Output Analyzer displays the 95% confidence intervals and the results of the ANOVA test (Figure 14.125). In this example, the F statistic is 44.98 while the critical value is 3.35. The Output Analyzer indicates that the null hypotheses of all means being equal should be rejected.

14.16.3.5 Correlogram

The correlogram function is used to determine the lag at which there is no correlation. This is used to reduced the effect of autocorrelation. When the correllogram function is selected, the window seen in Figure 14.126 appears.

To use the correlogram function, the data file name and maximum lags text boxes must be completed. All of the other boxes can be left with the default values. We will use the file nonterm.dat with a maximum lag of 200. Some trial and error may be necessary in determining the maximum lag. If the maximum lag size chosen is not large enough to cover the lag required for a zero correlation, the maximum lag size should be increased. Figure 14.127 illustrates the completed correlogram window.

When the OK button is clicked, the correlogram is displayed in the upper window (Figure 14.128). In the lower window the Output Processor displays correlogram-related statisitics.

In order to identify the lag at which the correlation becomes insignificant, we can observe the correlogram. A more precise value can be obtained by scrolling down in the bottom window. This is illustrated in Figure 14.129.

At a lag of approximately 175 observations, the lag becomes insignificant.

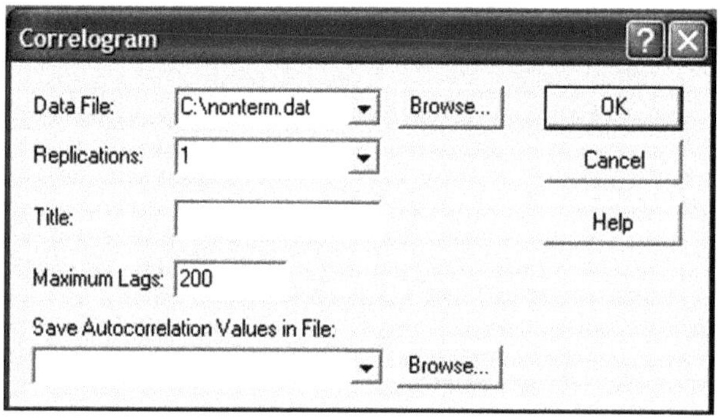

FIGURE 14.127 Completed correlogram window.

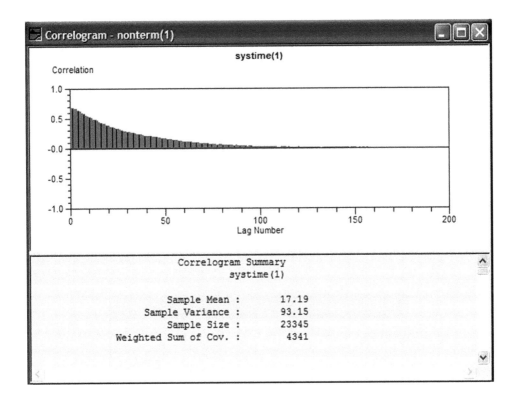

FIGURE 14.128 Correlogram summary results.

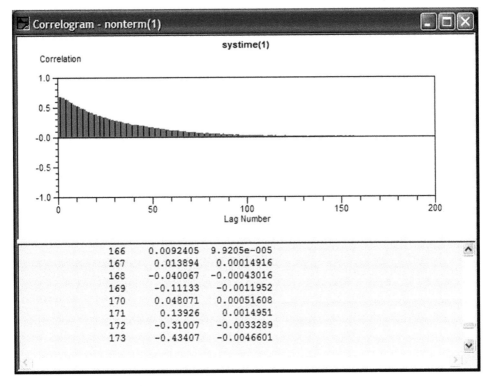

FIGURE 14.129 Nonsignificant lag.

14.16.3.6 Batch/Truncate Means

The batch/truncate means function is used in conjunction with the correlogram function. The batch/truncate means function is used to remove the initial transient phase from nonterminating data. This function is also used to split the data into equal batches of specific size for subsequent statistical analysis. When this function is selected, the window shown in Figure 14.130 appears.

We can use the batch/truncate means function to convert our nonterminating data file nonterm.dat into 10 independent batches. This is accomplished by making each batch 10 times the size of the nonsignificant lag. Because we have seen that this point occurs at 175 observations, we can make each batch 1,750 observations. This also means that our file nonterm.dat must be large enough to have 17,500 observations after the initial transient period. In our example below, we will delete the first 100 observations. The means of these batches will be written to the file nonterm2.dat. The completed batch/truncate window is illustrated in Figure 14.131.

When the batch/truncate function is executed, the screen shown in Figure 14.132 appears.

This screen indicates that a total of 13 batches were created from the original data set nonterm.dat. The means of the 13 batches have been written to the file nonterm2.dat. Nonterm2.dat can now be examined like any other replication mean file using the other statistical functions in the Output processor.

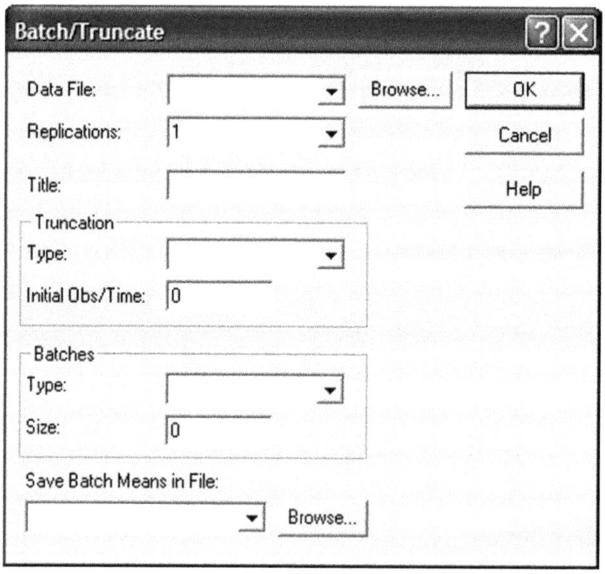

FIGURE 14.130 Batch/truncate window.

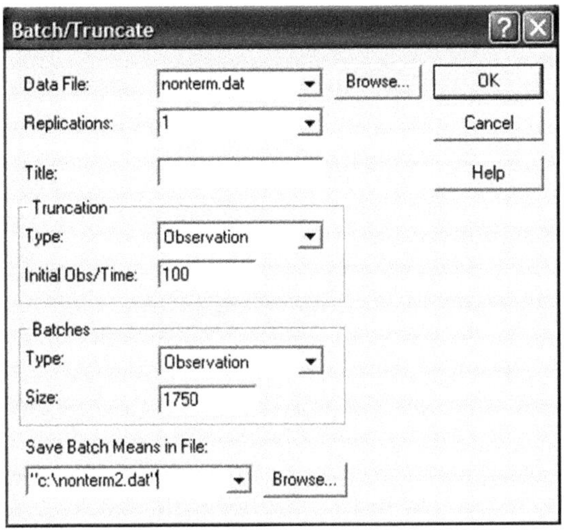

FIGURE 14.131 Completed batch/truncate window.

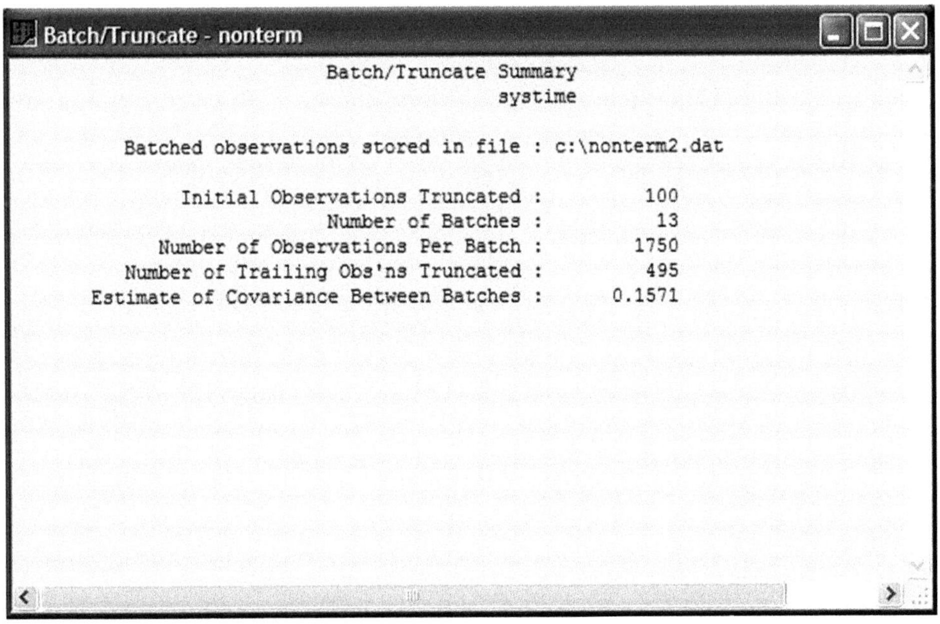

FIGURE 14.132 Batch/truncate results.

15

Simulation Using AutoMod and AutoStat

Matt Rohrer
Brooks-PRI Automation

0-8493-1241-8/04/$0.00+$1.50
© 2004 by CRC Press LLC

15.1 Introduction

This chapter provides an introduction to simulation using both the AutoMod and AutoStat software. AutoMod combines three-dimensional graphics with the most comprehensive set of templates and objects for modeling many different applications. You can use AutoMod to construct simulations for planning and design or day-to-day operations analysis and controls development testing. After designing and building a model within the AutoMod software, you can use the AutoStat software to conduct extensive output analysis on the model.

This chapter is designed to give you a basic understanding of the key elements of AutoMod and AutoStat. For more information, please refer to the *AutoMod User's Guide* and the *Getting Started with AutoMod* textbook, both of which are installed with the educational license of the AutoMod software. (To access online manuals, select Documentation > Online Manuals > Welcome from within the Auto-Mod program group in the Start menu.)

The chapter begins with a tutorial designed to teach you how to create, edit, and run a simple simulation model within the AutoMod software. The chapter is split into several sections so you can complete the tutorial in small increments. After the tutorial, the chapter provides an introduction to statistical analysis using the AutoStat software.

15.2 Introduction to the AutoMod Tutorial

This tutorial provides step-by-step procedures for building a model. You will build the model in verifiable stages, adding more processes and detail in each lesson. The model is a small system with eight processing steps. To move products (represented by loads) from process to process, you will create a conveyor system and an automated guided vehicle (AGV) system (Figure 15.1).

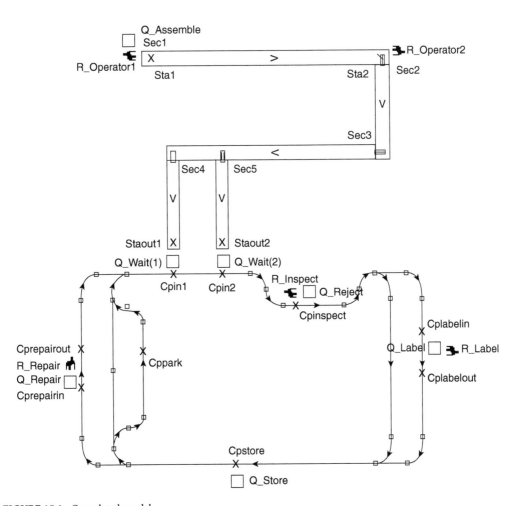

FIGURE 15.1 Completed model.

15.3 Using the Mouse[1]

This guide often refers to using the main mouse button. Unless you have changed your system defaults, the main mouse button is the left mouse button.

The main mouse button has four functions: selecting items for editing, clicking buttons, using the menus, and placing graphics. Unless otherwise specified, use the main mouse button to perform the steps in the tutorial.

15.3.1 Centering and Zooming

If you are using a three-button mouse, the middle mouse button is used for centering and zooming. If you are using a two-button mouse, the right mouse button is used for centering and zooming.

To center the display at the cursor's position on the grid, place the cursor at the desired location and click the appropriate mouse button. To zoom in on an area, place the mouse cursor in the Work Area window, hold down the button indicated above for centering, drag the mouse to "fence" the area, and release the mouse button.

Note: You can undo the last zoom (or other view change) with the keyboard command Control+Shift+U. Also, you can return to the top view by pressing v.

15.4 Section 1: Getting Started

15.4.1 What You Will Learn

In this section, you will learn how to:

- Create a new model.
- Name and create systems within a model.
- Save your work.

15.4.2 Model Description

In the next few sections, you will build a simple model that simulates product assembly in a conveyor system. Later, you will expand the model by adding vehicles to transport the assembled products to inspection, labeling, and storage areas. The model's conveyor system is outlined in Figure 15.2.

In the conveyor system, products are assembled in two steps. The first step of the assembly occurs in queue Q_Assemble. After completing this step, the assembly operator places the products onto the conveyor at station sta1. Products then travel to station sta2, where a second operator completes the product's assembly on the conveyor.

The finished products then continue on the conveyor to either staout1 or staout2, where they transfer to automated guided vehicles (AGVs).

15.4.3 Opening the AutoMod Software

There are two different environments within the AutoMod software:

- Edit environment
- Simulation environment

---

[1]This tutorial was designed for use with the standard graphics version of the AutoMod software. If you are using the VR graphics version of the software, the mouse and view controls differ from those documented here.

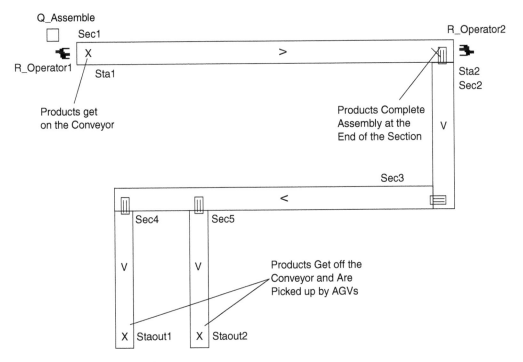

FIGURE 15.2 Conveyor system layout.

The edit environment is used to create and define a model. The simulation environment is used to view and analyze the actions of a compiled model.

Now you will enter the edit environment and begin creating a model.

From the Start menu, select AutoMod from the AutoMod program group. The Work Area window (Figure 15.3) opens. You are now in the edit environment, where you create and define your model.

15.4.4 Creating a New Model

To create a new model

1. From the File menu, select New.
2. Create a new directory for the model.
3. Name the model Demo1; then press Enter. The name appears in the title bar of the Work Area window. The Process system palette appears and the Selection window opens.

15.4.5 Creating New Systems within the Model

There can be many different systems within each model. Systems fall into three categories: process, movement, or static. The process system contains the model logic that controls how products (loads) are processed in a model. When you create a new model, a process system with the same name is automatically created. In addition to a process system, you can add an unlimited number of movement systems. The following types of movement systems are used in this tutorial:

- Conveyor
- AGV (path mover)

These movement systems are used to move loads from one area of your model to another. A model can contain either one movement system or a combination of several movement systems.

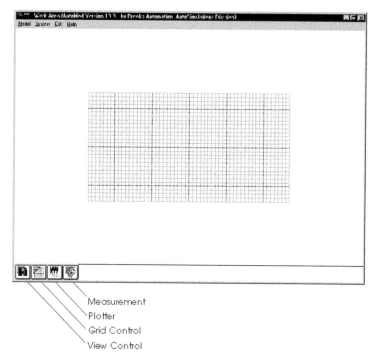

FIGURE 15.3 Work area window.

Finally, static systems can be used to show walls or other nonmoving elements of a structure. A static system enhances model graphics. This tutorial does not cover static systems.

You are now going to add a conveyor system to your model.

15.4.5.1 Creating a Conveyor System

To create a conveyor system:

1. From the System menu, select New. The Create A New System window opens.
2. From the System Type drop-down list, select Conveyor.
3. Name the system Conv; then press Enter. *Note: System and entity names are case sensitive.*
4. Click Create. The Conveyor System palette opens. This palette contains the drawing tools needed to create conveyor sections and stations.

15.4.6 Exporting Your Model

At this point, you will export the model to save your changes. Exporting does the following:

- Saves the model.
- Creates a backup (archive) model to be used in case the power goes out.
- Provides upward compatibility with newer versions of AutoMod.

Tip: Exporting is equivalent to saving in other applications. Therefore, export your model frequently.

1. From the File menu, select Export.
2. Click Yes to confirm that you want to export the model.

15.4.7 Review

You have defined a model named Demo1. This model consists of two systems: a process system called Demo1 and a conveyor system called Conv. You have exported the model.

15.5 Section 2: Building a Model

15.5.1 What You Will Learn

In this section, you will learn how to:

- Draw straight conveyor sections.
- Define products (loads).
- Define processes.
- Move products down a conveyor.
- Build and run the model.

15.5.2 Model Description

In the first phase of the model, you will draw a conveyor section and send products from one end to the other. In the AutoMod software, loads represent a quantity of product or material.

You want loads to enter the model at the first conveyor station, sta1, and travel to the second conveyor station, sta2, at the other end of the conveyor (see Figure 15.4). After arriving at sta2, loads leave the system (die). You are going to create a new load at station sta1 every 5 s.

15.5.3 Drawing a Conveyor System

Before you draw the conveyor section, zoom in on the upper left half of the grid. *Note: To undo a zoom, you can either press Control+Shift+U to undo the last change or press "v" to return to the top view.*

1. Using the middle mouse button (right mouse button on a two-button mouse), click and drag a box around the upper half of the grid.[2]
2. On the Conveyor palette, click Single Line. The Single Line window opens.

Note: The default names of sections start with the base name "sec," followed by a number that increments every time you place another section. If you wish to change the name of the section you are about to place, type in the name and press Enter before placing it.

3. Select the Orthogonal check box so the section you are adding will be horizontal or vertical in the Work Area window.

Note: You will use the default attributes for conveyor speed, width, and so on.

15.5.4 Drawing to Scale

It is important that you draw the conveyor section using the correct scale; the section must be 80 feet long. Use both the grid and the Measurement window to help you. By default, the grid lines are five feet apart, with thicker lines every 50 feet.

FIGURE 15.4 The first section of conveyor.

---

[2]To undo a zoom, you can either press Control+Shift+U to undo the last change or press "v" to return to the top view.

1. In the Work Area window, click Measurement to open the Measurement window (you might need to move the Single Line window to see the icon). You will see the screen in Figure 15.5.
2. Move the Measurement window below the Work Area window.
3. Select Track Mouse.

Tip: Selecting the Snap option in the Measurement window causes the graphics you are placing to move in increments determined by the grid spacing. Throughout this tutorial, you should clear Snap before drawing or placing objects that require fine positioning. Do not clear the Snap option now.

4. In the Work Area window, click once in the upper left corner of the grid where the two thicker lines intersect (refer to Figure15.6).

Tip: If you want to delete the section you are drawing and have placed only one point, press Esc.

5. Move the mouse to the right. Watch the Measurement window's Length field to see how long the conveyor section is. When the Length field is around 80, use the grid to locate the 80-ft mark and click the mouse again.

Note: The Measurement window tracks the mouse, not the section length, so the value in the window may not be exactly 80. That is why you should also use the grid. If you cannot tell from the grid whether the section is the right length, refer to "Editing sections" below to change the section's length.

Notice that the direction of travel for the conveyor is from the left to right, or first mouse click to second mouse click.

Tip: If you need to change the direction of a conveyor section, use the Select tool on the Conveyor palette to select the desired section; then select Change Direction from the Edit menu.

6. Close the Measurement window.

FIGURE 15.5 Measurement icon button.

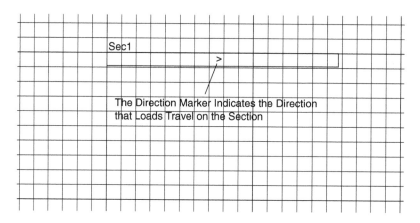

FIGURE 15.6 Measurement screen.

15.5.4.1 Editing Sections

If you want to delete a conveyor section and have placed only one of the two ends, press Esc to delete it. If you have already placed both of the ends of the path, you can move, delete, or edit it:

1. On the Conveyor palette, click Select.
2. Click on the conveyor section you want to move, delete, or edit.
3. Select Move, Delete, or Edit from the Edit menu.
4. Drag or redraw the section, or edit the section's ending value to adjust its length.

15.5.5   Placing Stations

Now that you have drawn the first conveyor section, you are ready to place the first two conveyor stations. Stations are locations at which loads can enter and leave a conveyor section, and stations are also where work is performed. In this model, loads get on the conveyor at station sta1 and are unloaded at station sta2 (refer to the illustration below).

1. On the Conveyor palette, click Station. The Station window opens (Figure 15.7).

Note: The default names of stations start with the base name "sta," followed by a number that increments every time you place another station. If you want to change the name or attributes of the station you are about to place, do so before placing it. For now, use the default names and attributes for the stations.

2. In the Work Area window, drag the first station into position at the beginning of the conveyor section (the graphic for the station appears when you click the mouse button and is placed when you release the mouse button). You have now placed sta1.
3. Drag the second station into position at the end of the conveyor section. You have now placed sta2.

The Work Area window contains a single section, as shown in Figure 15.8.

15.5.5.1 Moving Stations

If you need to move a station you have already placed, use the following procedure:

1. On the Conveyor palette, click Select.
2. Click the station you want to move.

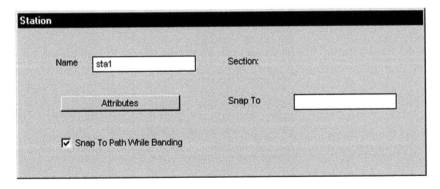

FIGURE 15.7 Defining a station.

FIGURE 15.8 Conveyor section with stations.

3. From the Edit menu, select Move.
4. Click the station and drag it with the mouse to the desired location.
5. Click OK.

Tip: Always a good idea to export your work frequently.

15.5.6 Opening the Process System

Now that you have created a simple conveyor system in your model, you need to write the logic that moves loads on the conveyor. To do this, open the process system.

1. From the System menu, select Open. The Open a System window opens.
2. Select Demo1 in the select list; then click Open. You have moved from the conveyor system (Conv) to the process system (Demo1). The Process System palette (Figure 15.9) opens.

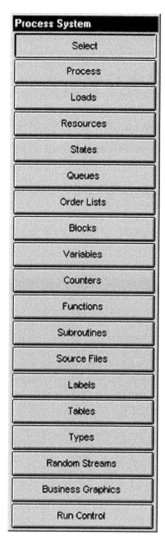

FIGURE 15.9 Process system palette.

15.5.6.1 Defining Processes

The process system allows you to define your manufacturing procedures, called processes. Processes are used to direct and control load movement in a model.

In this section you will define two processes: P_EnterAssembly, which places loads on the beginning of the conveyor, and P_CompleteAssembly, which removes loads from the conveyor.

1. On the Process System palette, click Process. The Process window (Figure 15.10) opens.
2. Click New. The Define A Process window (Figure 15.11) opens.
3. Name the process P_EnterAssembly; then press Enter. [Hyphens and spaces are not allowed in names, so use the underscore (_) instead.]

Note: You can name a process with any set of alphanumeric characters and underscores.

4. Click OK/New. The window is now ready for you to define the second process.
5. Name the process P_CompleteAssembly; then press Enter.
6. Click OK. The two processes you have defined appear in the select list (Figure 15.12).

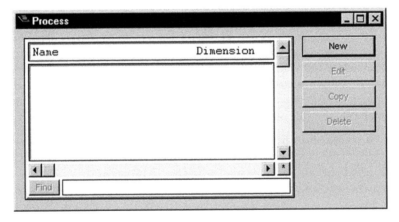

FIGURE 15.10   Process window.

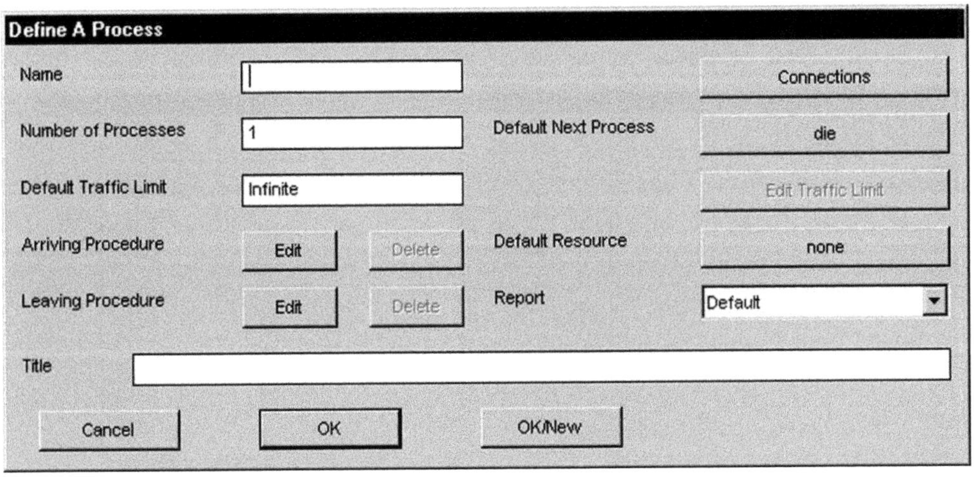

FIGURE 15.11   Define a process window.

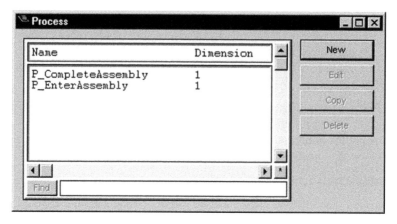

FIGURE 15.12 Defining processes.

15.5.6.2 Defining Loads

Now you are ready to create the loads that are transported in the model. In the AutoMod software, you can define as many load types as necessary to distinguish between different types of products in a system. In this model, we want to process only one type of load.

When you define a load type, you can also define a load creation specification that defines the number of loads of that type that are created during a simulation and the time between load arrivals. The load creation specification also specifies which process the loads execute first. Once created, the loads are sent from process to process, carrying out process procedures.

1. On the Process System palette, click Loads. The Loads window (Figure 15.13) opens.
2. Because you are defining a new type of product (load), click New to the right of the Load Types select list. The Define A Load Type window (Figure 15.14) opens.
3. Name the load type L_Carton; then press Enter. The "L" represents load and "Carton" represents the load type.

Note: Remember that names of entities (which include processes and loads) cannot have spaces or hyphens in them. For this reason, use the underscore (_).

Now you need to define the creation specification for this load type.

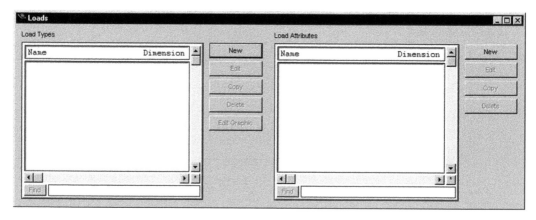

FIGURE 15.13 Loads window.

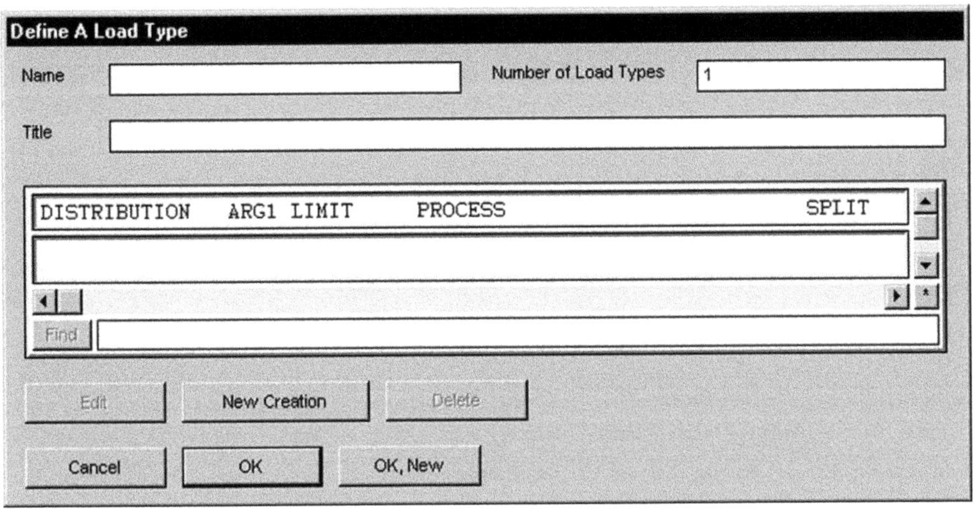

FIGURE 15.14 Define a load type window.

4. Click the New Creation button. The Define A Creation Spec window (Figure 15.15) opens.
5. Click First Process. The Pick A First Process window opens.
6. Because you want your loads to start at process P_EnterAssembly, select P_EnterAssembly in the Pick A First Process window; then click OK. The process P_EnterAssembly appears to the right of the First Process button in the Define A Creation Spec window.
7. A generation limit is the total number of load creation cycles that occur before this creation specification stops producing loads. Leave the generation limit at "infinite," so this load type is created continuously throughout the simulation.
8. The distribution specifies the interval of time in the creation cycle. Leave the constant distribution with a mean of 5 s. This distribution creates a load every 5 s and sends it to process P_EnterAssembly.
9. Click OK. The creation information now appears in the Distribution select list.
10. Click OK. L_Carton now appears in the Load Types select list.

15.5.7 Defining Load Graphics

Now you are ready to define the size of the loads in the model. The default load size is $1 \times 1 \times 1$ ft. You will define a slightly larger load.

1. Select L_Carton; then click the Edit Graphic button. The Edit Load Type Graphics window opens.

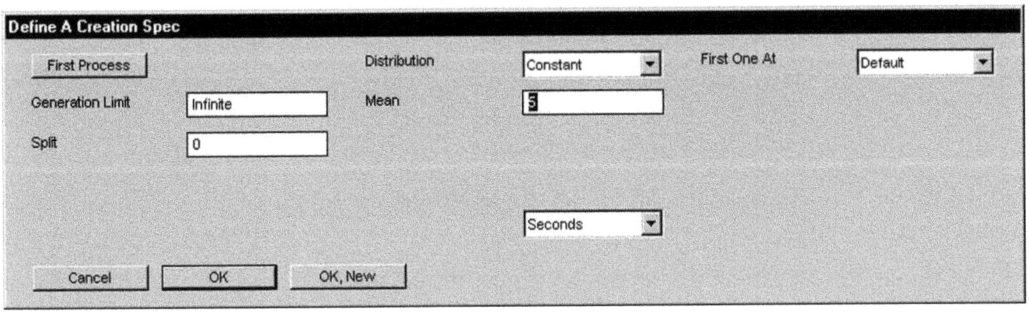

FIGURE 15.15 Defining a creation specification.

Define A Load Type

| Name | L_Carton | | Number of Load Types | 1 |

Title

DISTRIBUTION	ARG1 LIMIT	PROCESS	SPLIT
Constant	5sec Infinite	P_EnterAssembly	0

Find

| Edit | New Creation | Delete |

| Cancel | OK | OK, New |

FIGURE 15.16 Defining a new load type.

2. Click Place.
3. In the Work Area window, click above station sta1 to place the load's graphic; a small green box appears, representing a load of type L_Carton.
4. In the Edit Load Types Graphics window, select the Scale All check box (see Figure 15.17). This allows you to change all three dimensions of the load at once.
5. Change the Scale value to "2," then press Enter. Notice that the X, Y, and Z scale boxes all change to "2," making the load scale 2 × 2 × 2 and doubling the size of the load graphic in the Work Area window. Now you will change the dimensions of the load to 2 × 3 × 2.
6. Clear the Scale All check box. This allows you to change one load dimension at a time.
7. To increase the Y dimension of the load, click the Y Scale button (the second button under "Scale") to scale the Y dimension.
8. Change the Scale value to "3"; then press Enter. The Y dimension of the load increases to three.

You will now hide the load's picture until the model is running.

9. Click Hide. Click Yes to confirm. The load graphic no longer appears in the Work Area window.
10. Click Done.

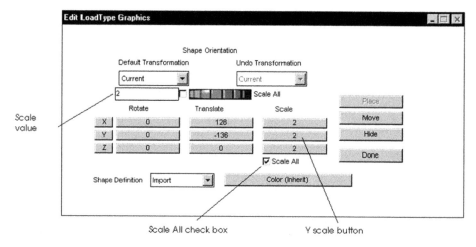

FIGURE 15.17 Scaling load graphic.

15.5.8 Arriving Procedures

You have defined loads for the model and sent them to their first process, P_EnterAssembly. Now you need to write the instructions for P_EnterAssembly, telling the loads what to do when they get there. You write instructions for processes in arriving procedures. An arriving procedure is a load's instructions for a process, including which resources to use and how long each action takes. Loads execute arriving procedures when they enter a process. Arriving procedures are written in a text file called a source file. A source file can contain one or more arriving procedures. In this model, you will write all procedures in the same source file, which makes editing the procedures easy.

15.5.9 Defining Source Files

To define the source file for your model:

1. On the Process System palette, click Source Files. The Source Files window opens.
2. Click New. Name the file mycode.m.

Note: Source file names must end with a .m extension.

3. Click OK.
4. Select mycode.m in the select list; then click Edit. A text editor opens a blank file.

Note: Understanding simple procedural language is a necessary part of modeling. This tutorial explains some of the most commonly used syntax in the AutoMod software. For more information about AutoMod syntax, refer to the AutoMod syntax help. You can open the help file from the Start menu or by selecting Syntax Help from the Help menu in the software.

15.5.10 Writing Arriving Procedures

You have defined two processes in the model: P_EnterAssembly and P_CompleteAssembly. When a load is sent to a process, it immediately executes the process' arriving procedure. Because new loads are sent to process P_EnterAssembly, this process' arriving procedure provides the first instructions for load activity during a simulation.

You will write the arriving procedure for P_EnterAssembly to cause loads to get on the conveyor at station sta1. The arriving procedure will then send loads to the P_CompleteAssembly process.

The arriving procedure for P_CompleteAssembly will cause loads to travel from station sta1 to sta2 on the conveyor. The procedure will then cause loads to leave the simulation (die).

Each arriving procedure must begin with the line

begin <processname> arriving procedure

and must end with the line

end

Between these lines, you can type the instructions for the loads that enter the process. The procedures for this section use three actions: move, travel, and send.

move The move action places the load in a specific place, such as in a queue or at a station on a conveyor system. You can use the move action to get loads into a movement system and to change loads from one movement system to another (you will use this extensively when you add the AGV system to the model later in the tutorial).

travel Once a load is in a movement system, you can use the travel action to cause the load to use the system to travel from one location to another. The load automatically finds the shortest path between locations in a movement system.

send   The send action sends a load to another process. It is usually the last action in an arriving procedure.

1. Type the following procedures in the source file. AutoMod syntax is case sensitive and must be typed exactly as shown.

Tip: Any remarks placed between the symbols /* and */ in the code are comments and do not affect the simulation. Comments are notes that help you understand what the code is doing.

```
beginP_EnterAssembly arriving procedure
    move into Conv.sta1/* Get on the conveyor at station sta1 */
    send to P_CompleteAssembly/* Leave the current process */
end

begin P_CompleteAssembly arriving procedure
    travel to Conv.sta2/* Travel from current location to station sta2 */
    send to die/* Leave the simulation */
end
```

The first action in the P_EnterAssembly procedure causes each load to move into station sta1 in the conveyor system. The second action sends the load to the next process, P_CompleteAssembly. The load then starts performing the arriving procedure of P_CompleteAssembly.

P_CompleteAssembly's arriving procedure causes the load to move on the conveyor from the first station to the second (sta2). Once there, the load leaves the model.

Note: Throughout this tutorial, you will expand these arriving procedures to simulate more complex processes.

2. From the File menu, select Save, then Exit. You have now defined a source file containing the arriving procedures.

Note: If you get any error messages when quitting a source file, they are caused by typing errors. To fix the typing errors, click the Return to Edit button in the Error Correction window to return to fix the mistakes on the lines specified. When you are finished, save your changes and exit the file editor.

15.5.11 Running the Model

At this point, you have completed everything necessary to build and run the model.

1. Export the model.
2. From the Run menu, select Run Model. Confirm by selecting Yes. It takes a few moments for the model to compile and link; then the simulation environment opens.

The simulation environment has three windows:

Status window	Shows whether the model is paused or running and what the current simulation time is.
Message window	Shows messages and errors as the model runs.
Simulation window	Shows the animation of the model.

3. Zoom in on the conveyor section by using the middle mouse button (right mouse button on a two-button mouse) to band the desired area.
4. To start the simulation, select Continue from the Control menu. Watch the model for a few minutes to see loads traveling on the conveyor from sta1 to sta2.

Tip: You can toggle between pausing or continuing the simulation by pressing "p" (lowercase). By default, entities are displayed in wireframe (only the outlines of entities appear in the Simulation window). You can toggle between wireframe and solid views by pressing "w" (lowercase).

15.5.12 Editing the Model

To return to the edit environment:

1. Press "p" to pause the model.
2. From the File menu, select Run Model Editor. You are now back in the edit environment.

15.5.13 Closing the Model

As you work through the tutorial, you may want to close your model at the end of a section and resume building your model later. Therefore, exit AutoMod now for practice:

From the File menu, select Exit.

You will open the model at the beginning of the next section.

15.5.14 Review

You have drawn your conveyor system and placed two stations.

You have created all of the logic necessary to move loads from the process P_EnterAssembly to the process P_CompleteAssembly, that is, from the beginning of the conveyor to the end of the conveyor:

- You have defined your loads.
- You have built and run the model.

15.6 Section 3: Changing Views

15.6.1 What You Will Learn

In this section, you will learn how to:

- Open the model.
- Rotate the picture.
- Expand the drawing grid.
- Change the spacing of the drawing grid.
- Turn text and graphics on and off.
- Use keyboard commands and the Help menu.
- Save the configuration of windows and views.

The following material shows how to control the view of the model.

15.6.2 Opening Your Model

To open your model, do the following:

1. Start the AutoMod software.
2. From the File menu, select Open. A navigation window opens.
3. Navigate to the model's.dir directory.
4. Double-click Demo1.mod.

Note: When you open a model, the grid is automatically resized to fit the entities currently placed in the model.

In this section, you will learn how to expand the drawing grid.

15.6.3 Rotating the Picture

The View Control window allows you to change the view of your model. You can view the model from a variety of positions. This window also lets you adjust the size of the grid. To open the View Control window:

1. In the Work Area window (Figure 15.18), click View Control.
2. The View Control window (Figure 15.19) opens.

The View Control window contains the following options:

Rotate Rotating moves the model around the different axes.

Translate Translating moves the model along the different axes.

Scale Scaling the model makes it bigger or smaller on the screen.

Child Windows on Top This prevents the palette and dialog windows from getting lost behind other windows.

Perspective Perspective is a method of drawing in which all lines lead to a vanishing point; this is how we naturally view the world.

Orthogonal Orthogonal is a method of drawing in which all lines are at right angles to each other; this is a projection of a two-dimensional drawing into a three-dimensional view.

Solid Solid displays model entities as solid objects. When solid is off, only the wireframe outlines of shapes are displayed.

Friction Friction controls continuous movement, including translation, rotation, and scaling. When friction is on, the graphics move only when you indicate. When friction is off, any movement command causes the model to move continuously until you explicitly stop it, either by toggling friction on again or by pressing the space bar.

Axis Display Axis display causes a triad (X,Y,Z) to be displayed at the model's origin.

Screen Screen controls the model movement. The Screen check box controls the movement of the model in relation to screen coordinates or world coordinates. The world coordinates are in relation to the model's origin (Axis Display). The screen coordinates are in relation to the current screen view.

The axes of the following views originate from the world coordinates:

Top View the entity from the positive Z axis.

Front View the entity from the negative Y axis.

Bottom View the entity from the negative Z axis.

Back View the entity from the positive Y axis.

Right Side View the entity from the positive X axis.

Left Side View the entity from the negative X axis.

Create Views It is possible to define views of your model and name them so that you can display that view later. To do this, adjust your model to the desired view and click Create Views. Name the view. When you want to return to this view later, use the Set Views menu.

Set Views This drop-down list lists the views you have defined using Create Views.

Set Limits This button opens the Set Limits window, which allows you to adjust the size of the drawing grid (discussed below).

FIGURE 15.18 View control icon button.

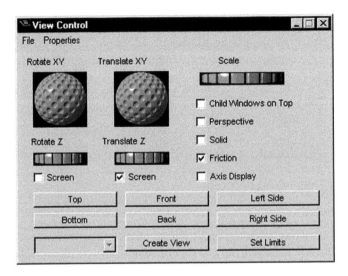

FIGURE 15.19 View control window.

Take a few minutes to familiarize yourself with the view control.

Tip: If at any time you lose sight of your model graphics, press "v" to restore the top view.

3. Click Top to return to the default view.
4. Select Child Windows on Top.

15.6.4 Expanding the Drawing Grid

When you open a model, the drawing grid is automatically sized to fit the entities in the model. If you are adding entities, you can expand the drawing grid to help you align and position the new entities. For this tutorial, you need to expand the grid so you can draw the rest of the conveyor system and the AGV system.

Note: If you close and reopen the model at any time during this tutorial, refer to this section to reset the size of the drawing grid.

To expand the drawing grid:

1. In the View Control window, click Set Limits. The Set Limits window appears.
2. Change the drawing limits for each dimension to the values shown in Figure 15.20.
3. Click OK to close the Set Limits window.
4. Close the View Control window.

You are now ready to customize further the drawing grid using the grid control.

15.6.5 Changing the Spacing of the Drawing Grid

The Grid Control window allows you to change the position of the drawing grid and the distance between grid lines. All entities are drawn or placed on the plane of the current grid. Use the following steps to open the Grid Control window:

1. In the Work Area window, click Grid Control (Figure 15.21).

The Grid Control window opens (Figure 15.22).

Origin/Orientation Origin and Orientation allow you to adjust the position of the grid. Origin should be thought of as translational movement of the grid, and Orientation as rotational grid movement.

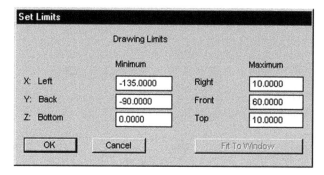

FIGURE 15.20   Defining drawing grid limits.

Align Grid to... This allows you to move the grid to predefined positions: top, bottom, front, back, left side or right side. Changing the grid alignment allows you to place graphics on different axes.

Display Use the Display check box to graphically turn the grid on or off.

Measure Opens the Measurement window.

Grid Spacing Specifies the number of units between successive grid lines.

Minor Line Specifies the distance between minor lines (fine gray lines). This distance is calculated by multiplying the minor line value by the Grid Spacing value (in model units).

Major Line Specifies the spacing of major lines (heavy grid lines) in terms of the number of minor lines.

Take a few minutes to familiarize yourself with the grid control.

2. When you are finished, type zeros in all Origin and Orientation values of the Grid Control window; then align the grid to "Top."
3. Set the Grid spacing and line values to the defaults (shown in the illustration above).
4. Close the Grid Control window.

15.6.6 Turning Text and Graphics On and Off

The Display window allows you to hide or display each system or system entity. It also allows you to change colors of system entities, such as conveyor sections, queues, resources, and so on.

1. From the Model menu, select Display. The Display window (Figure 15.23) opens.
2. To change colors of system entities, select the color from the drop-down list.
3. To see the entity change color in the Work Area window, click Apply.
4. You can also make entities visible or invisible by changing the Visible check box. For example, if you do not want the text in your process system displayed, clear the Visible check box next to Text.

Take a few minutes to familiarize yourself with the Display features.

5. When you are finished, reset all system entities to Visible.
6. Click Ok.

FIGURE 15.21   Grid control icon button.

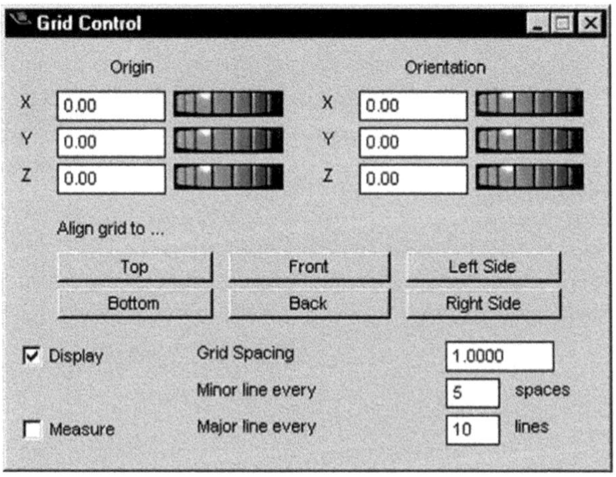

FIGURE 15.22 Grid control window.

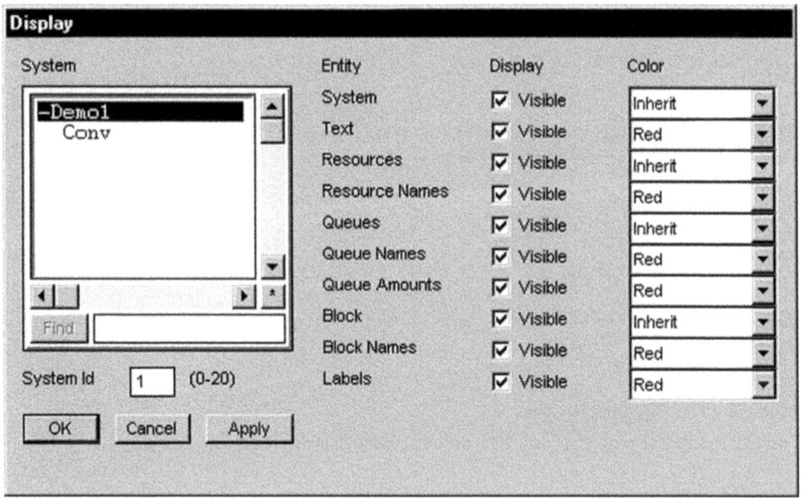

FIGURE 15.23 Display window.

15.6.7 Using Keyboard Commands

Many of the viewing changes you have just made can also be accomplished using the keyboard. There is an online help file that you can use as a quick reference for keyboard shortcuts.

1. In the Work Area window, type h. The AutoMod Help displays a list of keyboard commands. Take a few minutes to get familiar with some of the basic commands, such as the following:

   h = Help menu display
   X = rotate X axis clockwise
   x = rotate X axis counterclockwise
   Y = rotate Y axis clockwise
   y = rotate Y axis counterclockwise
   Z = rotate Z axis clockwise
   z = rotate Z axis counterclockwise
   P = perspective
   G = grid toggle
   f = friction toggle

2. Close the Help file.

15.6.8 Saving the Configuration of Windows and Views

To save the current size and position of your windows, save the startup configuration. Then every time you open this model, the view will appear as you saved it. For example, you can organize the placement of the windows on your screen in any manner you desire: resize the Work Area window, reposition the palette, change system display preferences, and set options from the View Control.

To save the Startup Configuration:

Once your preferences are set, select Save Startup Config from the File menu.
A new file called ".am2rc" is created in your directory; it contains the instructions for setting up the model. When you open any model within the directory containing the ".am2rc" file, the screen appears as you previously set it.

15.6.9 Review

- You have now learned how to change the orientation of your picture.
- You have expanded the drawing grid and learned how to change its spacing.
- You have learned how to adjust the appearance and size of windows, text, and graphics. You have also learned how to save the configuration of these items in the startup configuration.
- You have learned keyboard commands.

15.7 Section 4: Adding Queues and Resources

15.7.1 What You Will Learn

In this section, you will learn how to:

- Change the load creation rate.
- Define resources.
- Define queues.
- Edit process arriving procedures.
- Change the animation step to make the simulation run faster or slower.

15.7.2 Model Description

In this section, you will add operators and time delays to your model to simulate the assembly of loads in the conveyor system. You will make the following changes to the model:

1. Change the load creation rate from one load every 5 s to one load every 2 min.
2. Add two resources to assemble the loads in the system.
3. Add a queue Q_Assemble to hold loads at the beginning of the conveyor.

15.7.3 Changing the Load Creation Rate

Because you will be adding assembly delays to the model, you need to slow the load creation rate from one every 5 s to one every 2 min.

1. On the Process System palette, click Loads. The Loads window (Figure 15.24) opens.
2. From the Load Types select list, select L_Carton.
3. Click Edit to the right of the Load Types select list. The Edit A Load Type window opens.
4. Select the creation specification that reads:
   Constant 5 sec Infinite P_EnterAssembly 0 stream0
5. Click Edit in the bottom of the window. The Define A Creation Spec window opens.
6. Change the Mean to 2 min, as shown in Figure 15.25.

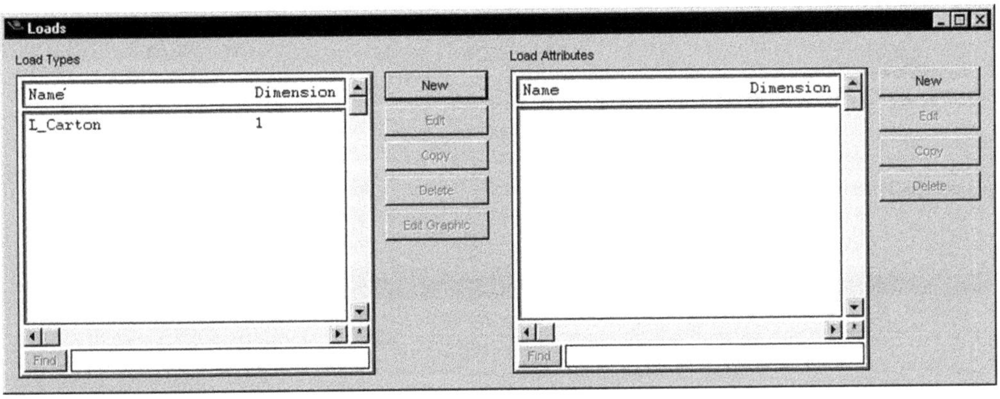

FIGURE 15.24 Loads window.

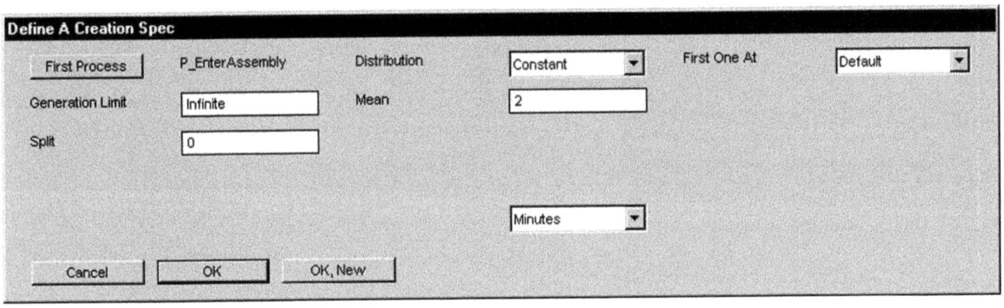

FIGURE 15.25 Changing the load creation rate.

7. Click OK. The Edit A Load Type window opens.
8. Click OK. The Loads window opens.

15.7.4 Defining Resources

Resources are used to represent machines, operators, tools, fixtures, and other entities that process loads. In your model, you will use two resources to represent operators. The first operator assembles loads and places them on the conveyor at station sta1. The second operator completes the assembly on the conveyor at station sta2.

1. On the Process System palette, click Resources. The Resources window opens.
2. Click New to the right of the Resources select list. The Define A Resource window opens.
3. Name the resource R_Operator1; then press Enter. The default capacity represents how many loads a resource can work on at one time. In this model, the Resource R_operator1 can assemble only one load at a time, so leave the Default Capacity set to "1."
4. Click OK, New. You are now ready to define the second resource.
5. Name the resource R_Operator2; then press Enter.
6. Click OK.

You have just defined two resources: R_Operator1 and R_Operator2. You will instruct loads to claim and use these resources when you edit the process-arriving procedures later in this section.

15.7.5 Placing Resource Graphics

You are going to import a human graphic to make the representation of the operators more realistic. You will place the operators at either end of the conveyor (see Figure 15.26).

FIGURE 15.26   The conveyor and resources.

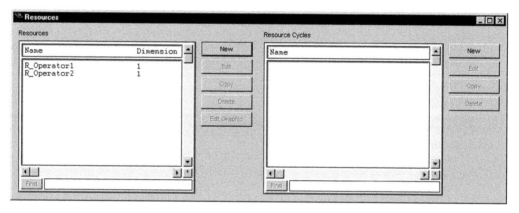

FIGURE 15.27   Resources window.

1. From the Resources select list (Figure 15.27), select R_Operator1.
2. Click Edit Graphic. The Edit Resource Graphics window opens.
3. Zoom in on the conveyor.
4. From the Shape Definition drop-down list in the Edit Resource Graphics window, select Import.
5. Navigate to the "demos/graphics/cell" directory in the software installation directory ("AutoMod" by default).
6. Double-click the file "man.cel."

Tip: To position the operator without snapping to nearby grid lines, open the Measurement window, and then clear the Snap option.

7. Click Place; then drag the operator into position at the beginning of the conveyor (see Figure 15.26).
8. Click Done.

15.7.5.1 Placing R_Operator2

Tip: If the right end of the conveyor is out of view, click the middle mouse button (use Control + the left button on a two-button mouse) in the right side of the window. This centers the screen where you clicked the mouse, moving the right side of the picture into view. This is an alternative to pressing "v" and rezooming the view.

1. Repeat steps 1–8 above to place the second operator (only complete steps 1–8; do not close the Edit Resource Graphics window). Once the operator is placed, you need to rotate him so he is facing the conveyor.
2. Click the Rotate Z button (the button's current value is 0).
3. Type "180;" then press Enter. This rotates the operator so he is facing the conveyor.
4. Click Done.

15.7.6 Defining Queues

Queues represent a temporary holding area, such as a loading dock or a workbench. In order for a load to get off a conveyor section or vehicle, it must be moved into a queue or a movement system (or sent to die).

You will now define a queue Q_Assemble at the beginning of the conveyor where the first step of each load's assembly takes place.

1. On the Process System palette, click Queues. The Queues window opens.
2. Click New. The Define A Queue window (Figure 15.28) opens.
3. Name the queue QAssemble; then press Enter.
4. Change the Default Capacity to "i;" then press Enter. The word Infinite appears, meaning that there is no limit to the number of loads that can be in the queue at one time.
5. Click OK. The Queues window (Figure 15.29) opens, with the newly defined queue listed.

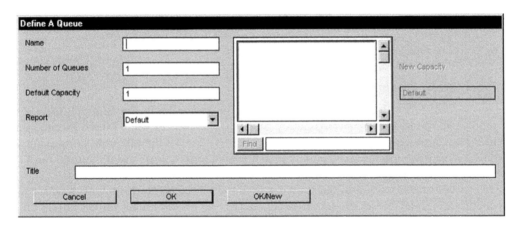

FIGURE 15.28 Define a queue window.

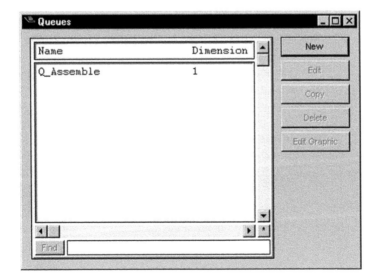

FIGURE 15.29 Queues window.

FIGURE 15.30   The conveyors, resources, and queues.

15.7.7 Placing Queue Graphics

You need to place the assembly queue graphically in order for it to be visible during a simulation.

1. Select Q_Assemble in the Queues select list.
2. Click Edit Graphic. The Edit Queue Graphics window opens.
3. Click Place.
4. Drag the queue into position above the resource (see fig. 15-30).
5. Click Done.

After you have finished placing Q_Assemble, your model appears as shown in Figure 15.30.

6. Export the model.

15.7.8 Editing Process Arriving Procedures

To instruct loads to use the resources and queue that you have defined, you must edit the arriving procedures for processes P_EnterAssembly and P_CompleteAssembly. Loads executing the P_EnterAssembly process must first move into the queue Q_Assemble and wait while being assembled. The first assembly step requires an amount of time that is uniformly distributed between 85 and 115 s. After the step is complete, R_operator1 takes 10 s to remove the load from the queue and place it on the conveyor. Because the capacity of R_Operator1 is one, the operator can assemble only one load at a time. Any additional loads that are awaiting assembly accumulate in Q_Assemble until the operator is available (loads are served on a first come, first served basis). Once placed on the conveyor, the loads are sent to the process P_CompleteAssembly.

Loads executing the P_CompleteAssembly process travel down the conveyor to station sta2. Each load remains on the conveyor at sta2 while using R_Operator2 for a time that is uniformly distributed between 50 and 70 s, which is the amount of time required to complete the second step of the assembly. After using the operator, the assembled loads are sent to die. Writing the procedures to simulate these activities requires using the following AutoMod actions:

get The get action claims one unit of a resource's capacity. If a resource has no available capacity when a load tries to claim it, the load is delayed until a unit of the resource's capacity becomes available.

free The free action frees one unit of a resource's capacity. The same load that gets a resource must also free it.

wait The wait action delays a load for a specified amount of time. You can create random delays by using a distribution with the wait action. For example, in this model, loads delay in the first assembly step for a time that is uniformly distributed between 85 and 115 s. The AutoMod syntax for defining a uniform distribution requires you to calculate the mean and offset for the range of possible values.

To find the offset, subtract the minimum value from the maximum value and divide by two:

$$(115 - 85)/2 = 15$$

To find the mean, subtract the offset from the maximum value:

$$115 - 15 = 100$$

The syntax for delaying the load is:

wait for uniform 100,15 sec

Note: The AutoMod software supports several distributions for generating random values in a simulation, including constant, uniform, normal, and so on. For more information about distributions and the syntax required to use them in a model, see the AutoMod syntax help.

use As an alternative to using the get, wait, and free actions, you can perform the same claim, delay, and release of a resource using a single use action.

To edit the P_EnterAssembly and P_CompleteAssembly arriving procedures:

1. On the Process System palette, click Source Files. The Source Files window opens.
2. Double-click mycode.m.
3. Modify the procedures to appear as follows:

```
begin P_EnterAssembly arriving procedure
      move into Q_Assemble/* Moves the load into the queue */
      get R_Operator1/* Claim the operator */
      wait for uniform 100,15 sec/* Delay for first step of assembly */
      wait for 10 sec/* Delay to place the load on the conveyor */
      move into Conv.sta1/* Get on the conveyor at station sta1 */
      free R_Operator1/* Release the load's claim on the operator */
      send to P_CompleteAssembly/* Leave the current process */
end

begin P_CompleteAssembly arriving procedure
      travel to Conv.sta2/* Travel from current location to station sta2 */
      use R_Operator2 for uniform 60,10 sec/* Delay to complete assembly */
      /* use = get, wait, and free combined */
      send to die/* Leave the simulation */
end
```

4. From the File menu, select Save, then Exit.
5. Export the model.
6. From the Run menu, select Run Model.
7. Click Yes to build the model.

The model takes a few minutes to compile. When the Message window reads "Ready to simulate," you are ready to run the model and watch the animation.

15.7.9 Changing the Animation Step

The animation step is the period of simulated time between animation updates. The longer the animation step, the faster the simulation. Conversely, setting a shorter animation step slows the simulation because graphics need to be redrawn more frequently. The animation step at the beginning of a simulation is set to 1 s.

To change the animation step:

1. From the Control menu, select Animation Step. The Change Animation Step window opens.
2. Change the animation step to "0.5," as shown in Figure 15.31; then press Enter.

Note: The Change Animation Step window also allows you to synchronize the simulation to a multiple of real time. This option is not discussed in this tutorial.

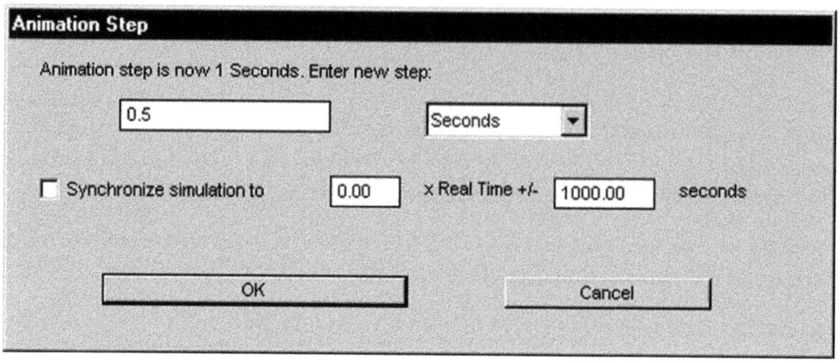

FIGURE 15.31   Changing the animation step.

3. Click OK to close the Change Animation Step window. The animation step has now been changed to redraw graphics every half-second (this results in a slower simulation than the default).
4. Zoom in on the conveyor; then press "p" to continue the simulation.

Tip: You can double the animation step during a simulation by pressing "D" (uppercase). You can halve the animation step during a simulation by pressing "d" (lowercase).

During the simulation, resources change color to represent their state:

- Green is busy.
- Blue is idle.

Notice that the operators change states from idle to busy, or blue to green, when they are claimed by loads in the simulation.

When you have watched the simulation to your satisfaction, return to the editing environment:

5. Press "p" to pause the model.
6. From the File menu, select Run Model Editor.

15.7.10 Review

- You have defined two operators. One operator assembles and places loads on the beginning of the conveyor; the other operator completes the loads' assembly at the end of the conveyor.
- You have defined a queue where loads wait for assembly at the beginning of the conveyor.
- You have written arriving procedures for P_EnterAssembly and P_CompleteAssembly that tell loads which resources to use, where to travel, and so on.
- You learned how to change the animation step to make a simulation run faster or slower.

15.8 Section 5: Completing the Conveyor System

15.8.1 What You Will Learn

In this section, you will learn how to:

- Add conveyor sections and stations.
- Define and use load attributes.
- Define and use variables.
- Use Run Control snaps.
- Examine statistics from a standard report.

15.8.2 Model Description

In this section, you will make the following changes:

1. Add conveyor sections to complete the conveyor (refer to the conveyor illustration on the next page).
2. Edit R_Operator2's arriving procedure to send assembled loads to either station staout1 or station staout2, with equal probability of going to either station.
3. Define a load attribute to track how long each load is in the system. Print the value to the Message window.
4. Define a variable to track the total number of loads processed.
5. Define a run control to automatically stop the simulation after 8 h.

15.8.3 Adding Conveyor Section

You are now ready to add the remaining sections in the conveyor system.

1. Open the conveyor system "Conv."
2. On the Conveyor palette, click Single Line. The Single Line window (Figure 15.32) opens.
3. If the Orthogonal check box is not selected, select it now.
4. Click the Snap to Section button.
5. Open the Measurement window.
6. Select Track Mouse.

Use Figure 15.33 for steps 7–9.

7. Click below the right end of sec1. This is the starting point for sec2.
8. Move the cursor down 30 feet to the end of sec2 (refer to the illustration); then click the mouse button again. The conveyor sections are perpendicular. A transfer (a rectangle) is automatically drawn between the two sections.

Tip: If you want to delete a section you are drawing and have only placed one end, press Esc. If you have placed both ends, delete and redraw the section.

9. Draw sec3 and sec4.
10. To draw section 5, click Select on the Conveyor palette. Select section sec4; then select Copy from the Edit menu.
11. Change the X To value to "16" (this moves the section 16 feet on the x axis); then click OK. as shown in Figure 15.34. The copied section is automatically named sec4_1.
12. To rename the section, select sec4_1, select Edit from the Edit menu, then change the section name to sec5. Click OK, Quit Edit Each.
13. Close the Measurement window.

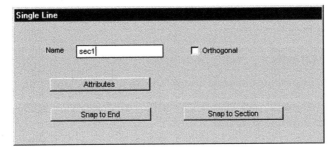

FIGURE 15.32 Single line window.

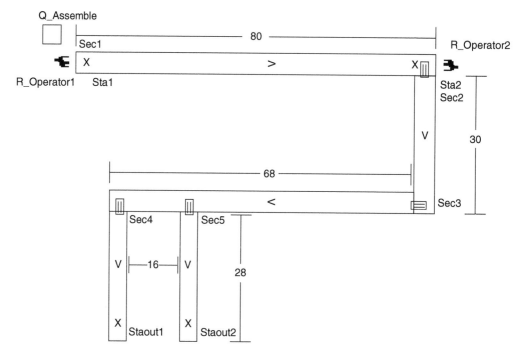

FIGURE 15.33 Conveyor illustration.

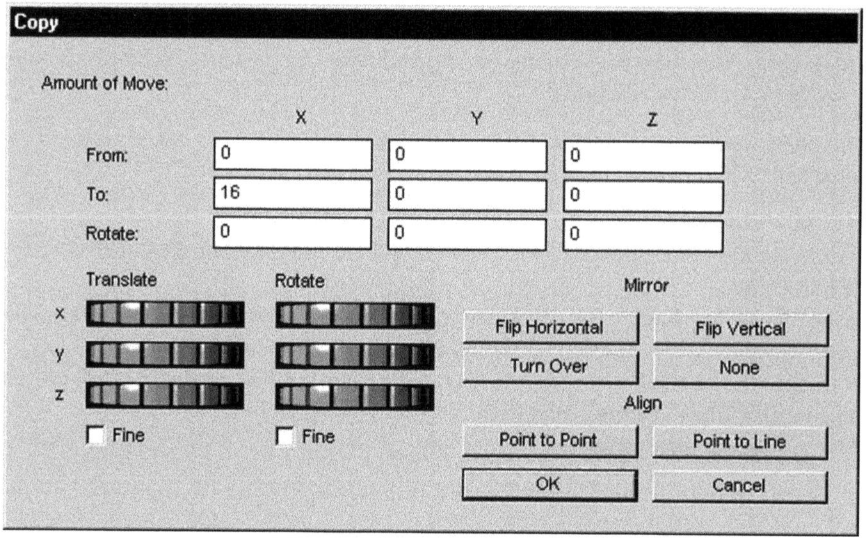

FIGURE 15.34 Moving a copied section.

15.8.4 Adding Stations

Now you need to add exit stations where assembled loads can transfer out of the conveyor system.

1. On the Conveyor palette, click Station. The Station window (Figure 15.35) opens.
2. Change the station name to staout1.
3. Drag the station into the correct position on section sec4 (refer to Figure 15.36).
4. Drag station staout2 into the correct position on section sec5.

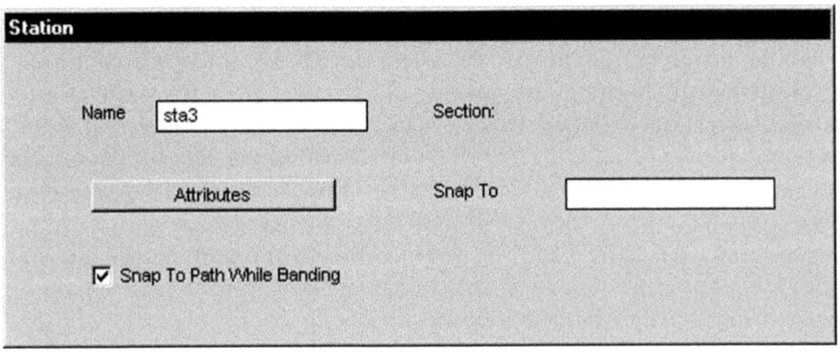

FIGURE 15.35 Station window.

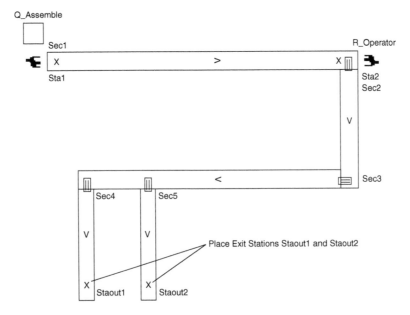

FIGURE 15.36 Define a process window.

Tip: If you are not happy with the placement of any station, you can move it:

- *On the Conveyor palette, click Select.*
- *Select the station you want to move.*
- *Select Move from the Edit menu.*
- *Use the mouse to click and drag the station to the desired location.*
- *Click OK.*

Your conveyor path and stations are now drawn.

5. Export the model.

15.8.5 Editing the Process System

You are now ready to edit the process system and cause loads to travel down the rest of the conveyor before exiting the simulation. You will also learn how to track the throughput and cycle time of loads in the simulation.

15.8.6 Adding an Arrayed Process (P_Out)

Instead of allowing assembled loads to leave the system immediately, you now need to send the loads to one of the exit stations at the end of the conveyor. To direct the loads, you could create two separate procedures (one to send loads to station staout1 and the other to send loads to station staout2). However, because the only difference between the two procedures would be the name of the destination station, you can save time by creating two arrayed processes that share the same arriving procedure.

Arrayed processes are copies of the same process. Other entities (such as queues and resources) can also be arrayed. Arrays are useful when you need to create a group of entities that share the same characteristics. Defining the processes as an array makes modeling easier because you need to write and edit only one arriving procedure for both processes.

Another benefit of using arrays is that they allow you to align multiple entities of different types to form assembly lines or work cells. In this model, you will associate each member of the process array with a corresponding exit station on the conveyor. That is, loads that are sent to the first process will also travel to the first exit station (staout1). Similarly, loads that travel to the second process will travel to the second exit station (staout2).

1. Open the Process System.
2. On the Process System palette, click Process. The Process window opens.
3. Click New to define a new process. The Define A Process window opens.
4. Name the process P_Out; then press Enter.

Now you are ready to create the process array.

5. Change the Number of Processes to "2"; then press Enter. The Define A Process window appears as shown in Figure 15.37.
6. Click OK to close the Define A Process window.

You have now defined two arrayed processes. In the AutoMod syntax, you can refer to an individual process by appending the process number (in parentheses) to the name. For example, P_Out(1) and P_Out(2).

15.8.6.1 Sending Loads to a Member of a Process Array

When loads have completed assembly, they must be sent from the P_CompleteAssembly process to one of the two arrayed P_Out processes. In this model, each load has a 50/50 chance of going to either the first arrayed process, P_Out(1) or the second arrayed process, P_Out(2).

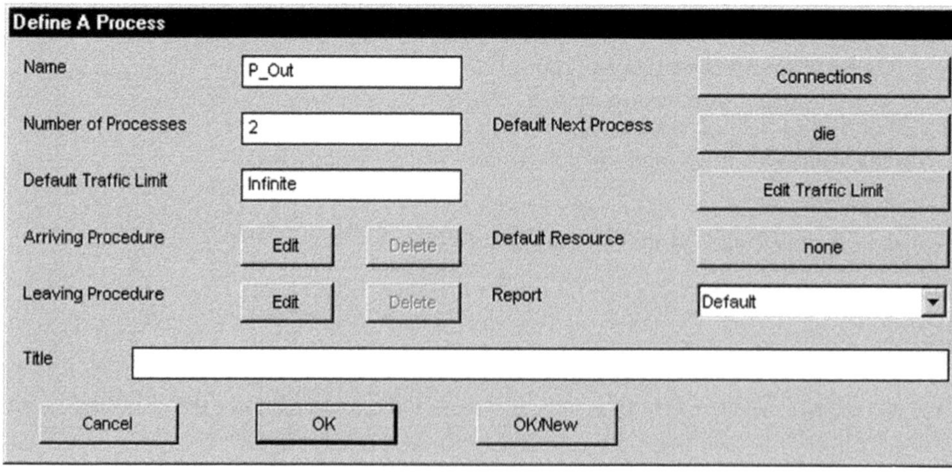

FIGURE 15.37 Placing the exit stations.

To distribute loads evenly between the processes, you can use a oneof distribution in the P_CompleteAssembly arriving procedure. The oneof distribution allows you to select randomly from a series of values or entities based on the frequency of each selection. The syntax for selecting P_Out1 and P_Out2 with equal probability is shown in the procedure below.

Note: For more information about the oneof distribution, see the AutoMod syntax help online.

To edit P_CompleteAssembly's arriving procedure:

1. Edit the source file mycode.m.
2. Edit the last line of the procedure to appear as shown below:

```
begin P_CompleteAssembly arriving procedure
    travel to Conv.sta2/* Travel from current location to station sta2 */
    use R_Operator2 for uniform 60,10 sec/* Delay to complete assembly */
    /* use = get, wait, and free combined */
    send to oneof(50:P_Out(1), 50:P_Out(2))/* Each load has a 50% chance of going to either
        P_Out(1) or P_Out(2) */
end
```

Do not close the editor, you will define the P_out processes' arriving procedure next.

15.8.6.2 Sending Loads to the Correct Exit Station

The arrayed processes P_Out(1) and P_Out(2) share the same arriving procedure. In the procedure you need to align each process with a corresponding exit station on the conveyor. If loads are sent to P_Out(1), they need to travel to station staout1. If loads are sent to P_Out(2), they need to travel to station staout2 (see Figure 15.38).

To align a member of the process array with a conveyor station, you can use the system keyword procindex.

15.8.7 Using Procindex to Align Arrayed Entities

A system keyword is a name that represents a numeric value in the software. You can use the system keyword procindex in the arriving procedure of an arrayed process to represent the current process' index number. For example, if a load is sent to process P_Out(1), the value of procindex is 1. If a load is sent to process P_Out(2), the value of procindex is 2.

You can use the keyword procindex to align arrayed entities of different types. For example, assume loads are sent to an arrayed process, P_Burnish. The arriving procedure for P_Burnish is written as follows:

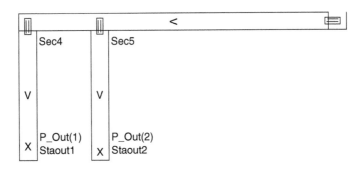

FIGURE 15.38 Conveyor illustration.

```
begin P_Burnish arriving procedure
    move into Q_Wait(procindex)
    use R_Machine(procindex) for 10 sec
    send to die
end
```

If a load is sent to the first member of the process array, P_Burnish(1), the value of procindex is 1. Consequently, the load moves into the first member of the queue array, Q_Wait(1), and uses the first member of the resource array, R_Machine(1).

If a load is sent to the second member of the process array, P_Burnish(2), the value of procindex is 2. Consequently, the load moves into the second member of the queue array, Q_Wait(2), and uses the second member of the resource array, R_Machine(2).

In this tutorial, you will use the keyword procindex to align loads in the P_Out procedures with the correct conveyor exit stations. Later, you will add arrayed queues, which will also be aligned using procindex.

15.8.8 Defining P_Out's Arriving Procedure

To define the arriving procedure that aligns each arrayed process with a corresponding conveyor exit station, do the following:

1. At the end of the source file, type P_Out 's arriving procedure, as shown below:

```
begin P_Out arriving procedure
    travel to Conv.staout(procindex)
    send to die
end
```

Note: If you are using procindex, the keyword must be enclosed in parentheses to distinguish it from the location name. This syntax is valid, because the AutoMod software allows you to include parentheses around the numeric portions of location names. For example, when writing a procedure, you could move loads into Conv.staout1 or Conv.staout(1).

2. From the File menu, select Save, then Exit.
3. Export the model.

15.8.9 Tracking Cycle Time

Now you will track cycle time for each load using a load attribute. A load attribute is a user-defined entity that stores data. All loads have the same attributes; however, each load's copy of an attribute may contain data that is unique to that load (such as color, part type, and cycle time).

To track cycle time, you will put a time stamp on each load as it enters the system. As each load leaves the system, you will compare the load's time stamp to the current simulation time to determine how long the load was in the system.

15.8.9.1 Defining a Load Attribute

You can create a time stamp for each load by creating a load attribute, A_Time, and setting it to the current time when each load enters the system in P_EnterAssembly. When the load finishes P_Out, you will subtract the value in A_Time from the current time to calculate cycle time.

To define the load attribute:

1. On the Process System palette, click Loads. The Loads window opens.
2. Click New to the right of the Load Attributes select list. The Load Attributes window opens.
3. Name the attribute A_Time; then press Enter.

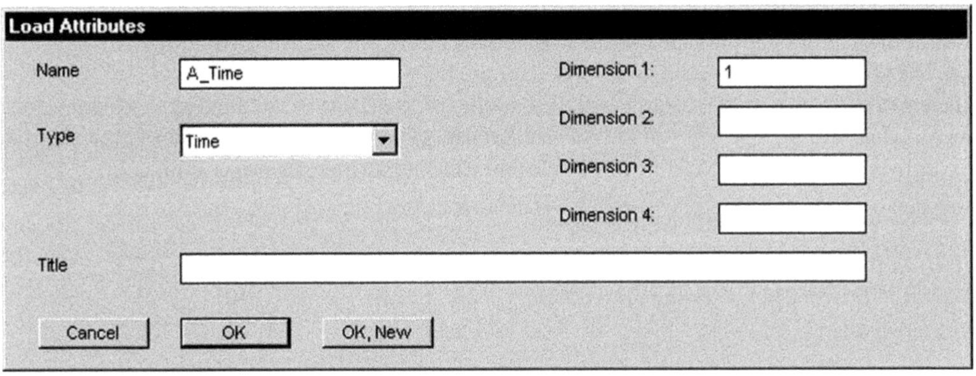

FIGURE 15.39 Load attributes window.

A load attribute can store many types of data (real, integer, time, and so on). In this model, the attribute A_Time represents the time a load spends in the system, and is therefore of type Time.

4. From the Type drop-down list, select Time. The Load Attributes window appears as in Figure 15.39.
5. Click OK. The attribute A_Time now appears in the Load Attributes select list.

Before editing the model logic to track each load's cycle time, you will create a variable to track the number of loads that complete processing in the simulation (throughput).

15.8.10 Tracking Throughput

Tracking throughput involves counting the number of loads that complete processing in the simulation. To count each load, you will create a variable, V_Numdone and increment the variable each time a load leaves the system. A variable is a user-defined entity that stores data. All loads have access to the same variables, and (unlike load attributes) the value of a variable is the same for every load in the model.

Tip: When determining how to store data in a model, use variables to track information that applies to the entire model. Use load attributes to track information that is specific to each load.

15.8.11 Defining a Variable

To define the variable V_Numdone:

1. On the Process System palette, click Variables. The Variables window opens.
2. Click New. The Define A Variable window opens.
3. Name the variable V_Numdone; then press Enter.

Like attributes, variables can also be of different types. This variable is tracking integer data, so you do not need to change the type.

4. Click OK. The newly defined variable appears in the Variables select list.

Note: Refer to Chapter 3, "Process System," in volume 1 of the AutoMod User's Guide *for more information about load attributes and variables.*

15.8.12 Editing the Arriving Procedures to Track Statistics

You are now ready to edit the model logic to use the load attribute and variable that you have created. To track cycle time, you will set the value of the load attribute A_Time to the current simulation time whenever a load enters the simulation. When a load leaves the simulation, you will compare the current simulation time with the attribute value to determine the cycle time.

To track throughput, you will increment the value of the V_Numdone variable, each time a load leaves the simulation. Editing the arriving procedures requires using the following AutoMod syntax:

set The set action sets the value of a variable or attribute to a defined value.

increment The increment action increases the value of a variable or attribute by a defined value.

ac The clock attribute ac refers to the current simulation time of the (absolute) clock. You can use the attribute ac in model logic to get the current time at any point during a simulation.

1. Edit the source file mycode.m.
2. Insert the third line, shown below, into the P_EnterAssembly arriving procedure:

```
begin P_EnterAssembly arriving procedure
     move into Q_Assemble/* Moves the load into the queue */
     set A_Time to ac/* Stamp the load with its creation time */
     get R_Operator1/* Claim the operator */
     wait for uniform 100,15 sec/* Delay for first step of assembly */
     wait for 10 sec/* Delay to place the load on the conveyor */
     move into Conv.sta1/* Get on the conveyor at station sta1 */
     free R_Operator1/* Release the load's claim on the operator */
     send to P_CompleteAssembly/* Leave the current process */
end
```

The new line places a time stamp on each load as it enters the system. For example, a load that enters this process at simulation time 60 s has its A_Time attribute set to 60.0.

3. Modify the P_Out arriving procedure to appear as shown below:

```
begin P_Out arriving procedure
     travel to Conv.staout(procindex)/* Travel to exit station */
     set A_Time to ac - A_Time/* Calculate cycle time */
     print this load, "Time in system =" A_Time to message
     inc V_Numdone by 1/* Count throughput */
     send to die
end
```

The first two new lines finish tracking the cycle time for A_Time by subtracting the initial value of A_Time from the current clock (ac) and printing it to the Message window. The last bold line increments the variable V_Numdone to track throughput.

4. From the File menu, select Save; then Exit.

15.8.13 Defining Run Control

Currently, the model is defined to simulate indefinitely, until you pause or stop the simulation. You can define a run control to determine how long the model simulates and how often statistics are reported. A snap is a period of time after which statistics are written to the AutoMod report and are possibly reset. The AutoMod report is named "Demo1.report" and is saved in the model directory.

1. From the Model menu, select Run Control. The Run Control window (Figure 15.40) opens.
2. Click New. The Define Snap Control window (Figure 15.41) opens.
3. Change the Number of Snaps to "8"; then press Enter. By default, each snap is 1 h long. Therefore, the model writes statistics to the report file every hour for 8 h.
4. Click OK. The Run Control window opens, with the snap control description in its select list. After eight snaps, the simulation will stop automatically.

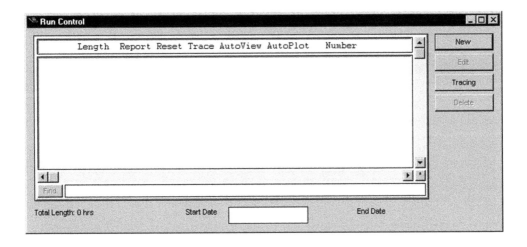

FIGURE 15.40 Run control window.

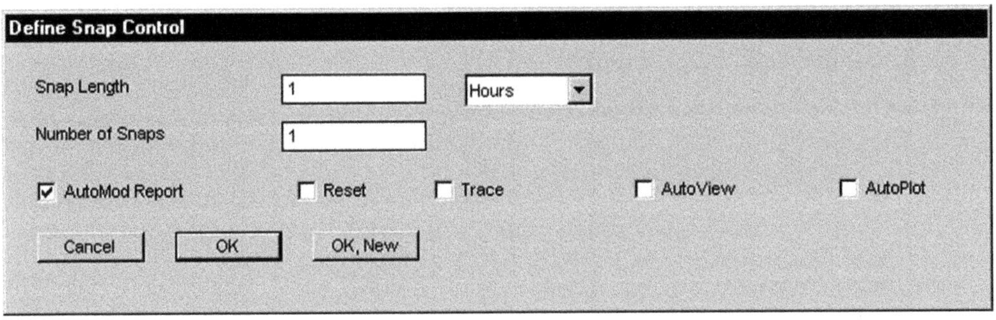

FIGURE 15.41 Define snap control window.

15.8.14 Building and Running the Model

You are now ready to run the model and see the effect of your changes in the simulation.
To run the model:

1. Export the model.
2. From the Run menu, select Run Model; then build the model. The Simulation environment opens.
3. Zoom in on the conveyor.
4. Press "p" to continue the simulation.

Watch to make sure loads travel down sections 4 and 5. Notice that a message is printed to the Message window every time a load leaves the system. Each load's time in system varies slightly because of the random assembly times.

5. To run the simulation to completion, press "g" to turn off the animation. The animation stops, and the simulation continues at an accelerated speed.

Once the simulation has run for the length of time defined in the Run Control (eight snaps of 1 h each), an "End of run" message appears in the Message window, and the simulation automatically stops.

15.8.15 Checking Statistics

Once the model has finished, you can check the simulation statistics by displaying them to the screen or by viewing the saved AutoMod report.

Note: Depending on your placement of the conveyor stations and the version and build of the software you are using, your statistics may vary slightly from those listed in this section.

15.8.16 Displaying Statistics to the Screen

When displaying statistics to the screen, you can view the system-generated statistics for entities in the simulation, as well as the custom throughput statistic you defined using a variable in the model. In this section, you will display statistics for:

- Resources
- Queues
- Processes

You will also display the value of the variables used in the simulation.

Note: You can display statistics and variable values both during the run and at the end of simulation. For more information about statistics, refer to Chapter 7, "Running a Model," in volume 1 of the AutoMod *User's Guide.*

15.8.17 Displaying Resource Statistics

To view the resource statistics for the two operators:

1. From the View menu, select Report. The Reports window opens.
2. In the entity type list, select "Resources"; then click Display. The Resources Report window opens. Some of the resource statistics are shown below:

Name	Total	Util	Av_Time	Av_Wait
R_Operator1	240	0.91	109.58	0.39
R_Operator2	238	0.49	59.46	0.00

R_Operator1 assembled 240 loads (Total). In the model, R_Operator1's assembly time is uniformly distributed between 85 and 115, with a mean of 100 s. Because the operator takes an additional 10 s to place each load on the conveyor, his average time of 109 s is reasonable. The average wait of loads requiring assembly is very small (less than 1 s). Loads are required to wait for assembly because the operator is highly utilized; the operator is utilized about 91% of the time (Util).

R_Operator2 has assembled fewer loads than R_operator2 (238 compared to 240). This is because he must wait for loads to travel down the conveyor before he can begin work. R_Operator2's assembly time is uniformly distributed between 50 and 70 s, so his average time is approximately the mean (60 s). R_Operator2 is utilized only about 50% of the time, and, consequently, loads have no waiting time to complete assembly.

Now take a look at the queue statistics.

15.8.18 Displaying Queue Statistics

To view the queue statistics:

1. From the View menu in the Simulation window, select Report. The Reports window opens.
2. In the entity type list, select "Queues"; then click Display. The Queues Report window opens. Some of the queue statistics are shown below:

Name	Total	Cur	Average	Max	Av_Time
Q_Assemble	240	1	0.92	2	109.98
Space	240	0	0	1	0.00

The queue Space is a default queue where all loads start the simulation. You can see that a total of 240 loads were created in the simulation. Newly created loads immediately execute the first action in the arriving procedure of their first process. In this model, their first action is to move into queue Q_Assemble. As a result, loads spend no time in Space.

All created loads (240) entered Q_Assemble. When the simulation stopped, there was one load still in the queue (the Cur statistic indicates the queue's current quantity). The Max statistic indicates that a maximum of two loads were in the queue simultaneously during simulation (one load being assembled, and another load awaiting assembly). The average time that loads spent in the queue is equal to the average time that loads spent being assembled by R_Operator1 (109.58) plus the average time they spent waiting for the operator (0.39); displayed values are rounded to the nearest hundredth.

Now take a look at the process statistics.

15.8.19 Displaying Process Statistics

To view the process statistics:

1. From the View menu in the Simulation window, select Report. The Reports window opens.
2. In the entity type list, select "Processes"; then click Display. The Processes Report window opens. Some of the process statistics are shown below:

Name	Total	Cur	Av_Time
P_CompleteAssembly	239	2	136.89
P_EnterAssembly	240	1	109.98
P_Out(1)	118	0	125.50
P_Out(2)	119	1	109.43

The statistics indicate that all created loads (240) entered the P_EnterAssembly process, and one load is currently processing.

Notice that only 239 loads entered the P_CompleteAssembly process, and two loads are currently processing. To understand the difference between the process total (239 loads) and the resource total for R_Operator2 (238 loads), remember that the P_CompleteAssembly process includes a travel action. The currently processing load is traveling to station sta2 but has not yet reached the operator. The travel time also explains why loads spent more time on average in P_CompleteAssembly (136.89 s) than in P_EnterAssembly (109.98 s).

From the total statistics for P_Out(1) and P_Out(2), you can see that a fairly equal number of loads are being sent to both processes (119 vs. 118). The totals are often not exactly equal because the oneof distribution defines a unique probability for each load (the frequencies are not normalized).

Now take a look at the custom statistic you defined in the model.

15.8.20 Displaying Variable Values

To view the value of the variable defined in the model:

1. From the View menu in the Work Area window, select Variables. The Variables Report window opens, as shown in Figure 15.42.

The value of the V_Numdone variable indicates that 236 loads completed the simulation.

2. Edit The Model.

15.8.21 Viewing the AutoMod Report

AutoMod automatically summarizes statistics during the run and writes them to a report file at the end of each snap. The report for your model is named "Demo1.report." To view the report, close the simulation environment and open the report file (Demo1.report) with any text editor.

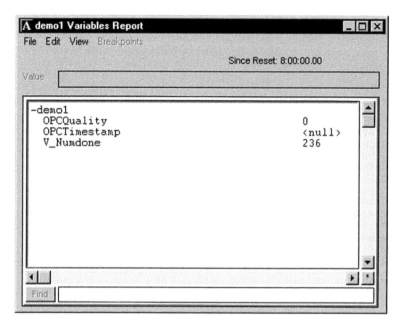

FIGURE 15.42   Variables report window.

15.8.22 Review

You have now completed the first phase of the model, which defined three processing operations: P_EnterAssembly, P_CompleteAssembly, and P_Out.

The next phase adds an AGV system that carries loads through four other processes:

- Inspection
- Labeling
- Repair
- Storage

15.9 Section 6: Creating an AGV System

15.9.1 What You Will Learn

In the next few sections, you will add an automated guided vehicle (AGV) system to your conveyor model and add some more processing steps for your products. You will learn how to:

- Create an AGV system.
- Draw paths.
- Assign vehicle travel options.
- Link the AGV system to the conveyor system.

15.9.2 Introduction

In AutoMod, automated guided vehicles are modeled using a path mover system. Path movers can represent vehicles, manually-operated lift trucks, or people. The path mover drawing tools can be used to create any type of vehicle system in which vehicles follow a specific path or route. When modeling a vehicle system, the concerns are movement, collision avoidance, and vehicle routing and scheduling.

15.9.3 Terms

There are several terms that you will need to understand before reading this part of the tutorial.

> **control point** A location along a path at which vehicles can stop and receive instructions.
> **work location** A location at which a vehicle can find work.
> **park location** A location at which a vehicle can park when idle.
> **block** A barrier to limit how many vehicles can be in a physical area at one time.
> **vehicle scheduling lists** Lists that schedule a vehicle's actions, such as where it can work and where it can park.
> **transfer** A connection between two sections of path.

15.9.4 Vehicle Motion

Vehicles are controlled in three ways:

- Vehicle scheduling lists
- Vehicle procedures and functions
- Job scheduling

This tutorial discusses how to create and use vehicle scheduling lists; however, it does not teach vehicle procedures and functions or job scheduling. Vehicles traveling on the same path avoid collisions by automatically decelerating and accumulating behind preceding vehicles. Every time a vehicle encounters congestion or reaches a destination, it adjusts its velocity to meet the circumstances. Vehicles decelerate to stop at destination control points, work locations, and park locations but do not slow down to claim intermediate control points or blocks. Vehicles accelerate to their defined velocity whenever possible.

For a vehicle to avoid colliding with other vehicles at an intersection, it must claim strategically placed blocks; blocks limit the number of vehicles that can travel in the bounded (blocked) area at the same time. Once paths are defined and the appropriate blocks are placed, AutoMod automatically calculates the shortest route between control points.

15.9.5 System Description

You will expand the model by adding an AGV system, in which vehicles sequentially transport loads from the conveyor system to an inspection area, a repair area (if they fail inspection), a labeling area, and a storage area. Stored loads delay temporarily and then leave the system.

The completed model is shown in Figure 15.43.

15.9.6 Adding an AGV System to the Model

To define an AGV system in the model:

1. From the System menu, select New. The Create A New System window opens.
2. From the System Type drop-down list, select Path Mover.
3. Name the path mover system "AGV"; then press Enter.
4. Click Create. The Path Mover palette (Figure 15.44) opens.

The Path Mover palette allows you to accomplish several tasks: draw paths, place control points, define vehicles, and create scheduling lists. The palette also contains a Select tool.

15.9.7 Adjusting the Grid

Before you draw a path, change the grid spacing so that the minor lines are 10 ft apart.

1. In the Work Area window, click Grid Control (Figure 15.45).

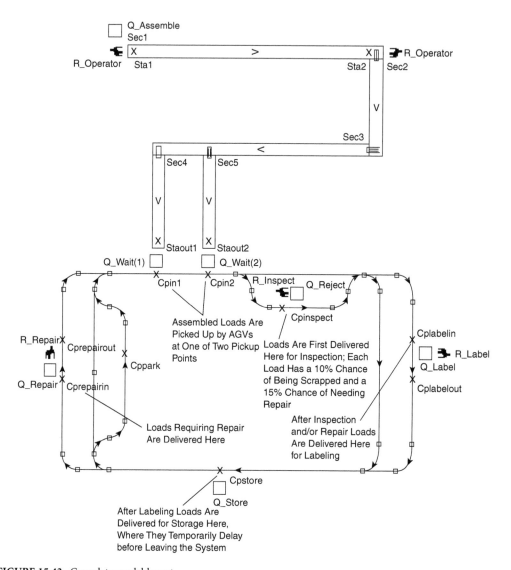

FIGURE 15.43 Complete model layout.

2. Edit the grid spacing values to:

Grid Spacing	10 (press Enter)
Minor line every	1 (press Enter)
Major line every	10 (press Enter)

3. Close the Grid Control window.

15.9.8 Drawing Paths

The process of drawing a path is similar to that of drawing a conveyor.

1. On the Path Mover palette, click Single Line. The Single Line window (Figure 15.46) opens.

Note: Paths have attributes. Each path segment can be assigned a direction of travel as well as a vehicle travel option. You will use the default path attributes in this tutorial. Straight paths can be placed at any angle and can be of any length.

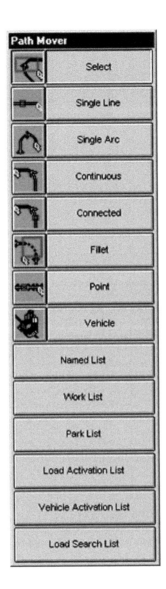

FIGURE 15.44 Path mover palette.

FIGURE 15.45 The grid controller opens.

2. If necessary, select the Orthogonal check box to draw lines that are perpendicular to each other and to the grid.
3. Click the Measurement icon; then select Track Mouse. Clear the Snap check box. Use this window to draw the path to scale.
4. Zoom in on the drawing.
5. Draw four lines to represent the skeleton of your path (paths are drawn using the same procedure as conveyor sections). Use the grid to place them the correct distance from each other (see Figure 15.47).

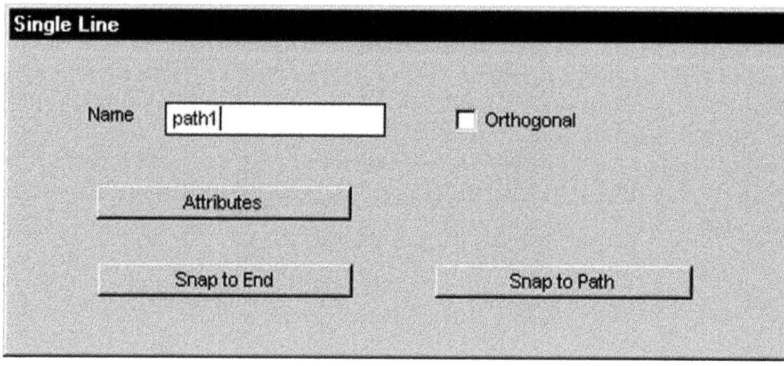

Single Line

Name [path1]   □ Orthogonal

[Attributes]

[Snap to End]   [Snap to Path]

FIGURE 15.46 Single line window.

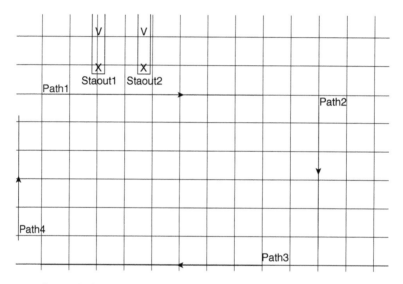

FIGURE 15.47 First four paths for the AGV system.

Note: Make sure the direction of the paths is correct. If you need to change the direction of a path, use the Select tool on the Path Mover palette to select the desired path; then select Change Direction from the Edit menu. If you need to make the grid larger, refer to Section 15.6.4 for information on how to resize the grid.

15.9.8.1 Editing and Deleting Paths

If you need to edit or delete a path, use the Select tool to select the path. Then, using the right mouse button, use the Edit menu to change the path's direction, edit its length, or delete the path.

15.9.8.2 Filleting Paths

The fillet tool connects two straight lines by drawing a curved section of path. Use this tool to connect the individual paths.

1. On the Path Mover palette, click Fillet.
2. Drag across the top and right paths with the mouse or select them each individually. The Fillet window opens.

The Trim option extends or shortens the endpoint of the section to match the endpoint of the arc.

3. Click OK. The two paths are now connected.

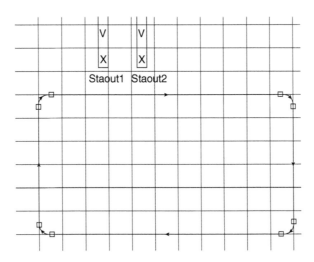

FIGURE 15.48 Outer loop of path.

Note: The box that appears between two sections is called a transfer. Transfers indicate that the paths are joined and that vehicles can move from one path to the other.

4. Fillet the remaining paths until the system looks like the one in Figure 15.48.

Note: It is not important what the paths are named or what order they are drawn in for this system.

5. Draw the straight lines of the inner loops as shown in Figure 15.49 (do not draw the inner loop for the parking area yet).

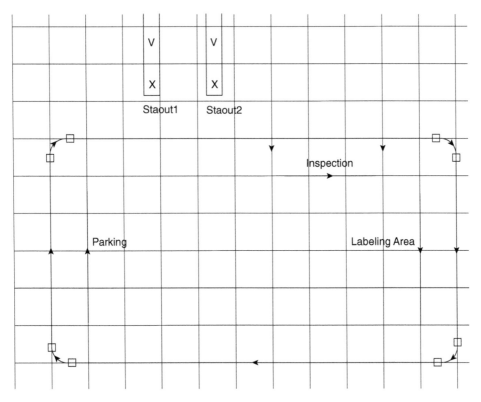

FIGURE 15.49 Inner loop of path.

6. Fillet the new paths (do the inspection paths last).

Tip: When using the fillet tool on the last inspection path, you may need to decrease the fillet path's radius; the inspection path should connect to the outer path (not the inner labeling area path).

7. Draw the parking loop's straight lines, as shown in Figure 15.50.
8. Fillet those paths to complete the path (Figure 15.51).

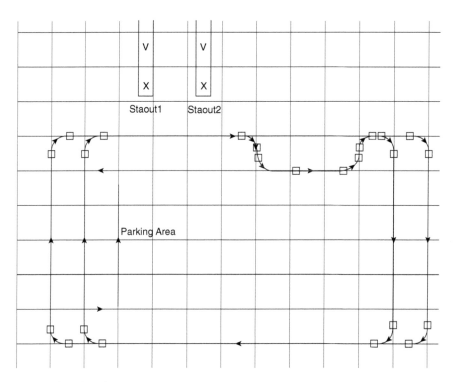

FIGURE 15.50 Parking loop lines.

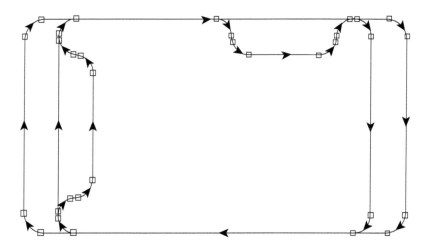

FIGURE 15.51 Completed path.

15.9.9 Defining Blocks

Now that you have drawn the path system, you need to define and place blocks to prevent vehicle collisions. A block is a boundary that you place on a path to limit the number of vehicles that can travel in the bounded (blocked) section of path at the same time. Blocks are necessary at intersections where vehicles merge from one path to another. In the system that you have drawn, there are four intersections where paths merge that require blocks.

To define the blocks:

1. Open the Process system. The Process System palette opens.
2. On the Process System palette, click Block. The Blocks window opens.
3. Click New to define a new block. The Define A Block window opens.
4. Name the block B_Merge; then press Enter.
5. Change the Number of Blocks to "4"; then press Enter.
6. The capacity of the blocks is already set to the default value of one (to limit the number of vehicles that can travel in each block to one vehicle), so click OK to close the Define A Block Window.

You are now ready to place the block graphics.

15.9.9.1 Placing Block Graphics

The size of the block graphic determines the area of the path that is blocked. The default block graphic size is $1 \times 1 \times 1$ ft. You will define larger blocks and place them on intersections where paths merge in the system.

1. In the Blocks window, click Edit Graphic. The Edit Block Graphics window opens.
2. Select the first block, B_Merge(1).
3. Click Place; then place the graphic anywhere in the Work Area window (after changing the size of the block, you will move it to the correct location on the path).
4. In the Edit Block Graphics window (Figure 15.52), select the Scale All check box.
5. Change the Scale value to "2"; then press Enter.

Notice that the X, Y, and Z scale boxes all change to "2," making the block scale $2 \times 2 \times 2$ and doubling the size of the block graphic in the Work Area window.

Now that you have scaled and placed the block graphic, you can move it into the correct position on the path.

6. Click Move in The Edit Block Graphics window.

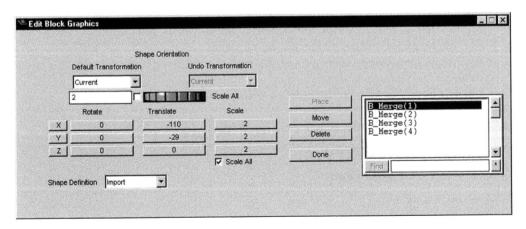

FIGURE 15.52 Edit block graphics window.

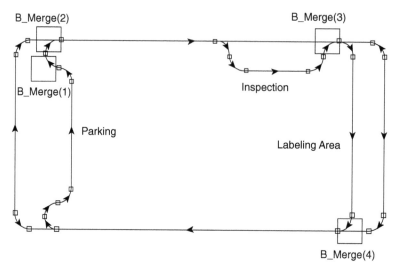

FIGURE 15.53   Placing blocks at intersections where paths merge.

7. In the Work Area window, drag the block's graphic so that its top edge is aligned with the transfer where the outer parking loop merges with the inner loop, as shown in Figure 15.53.
8. Place, scale, and move the remaining blocks so that they are aligned with the transfers where the parking, inspection, and labeling loops merge with the outer path, as shown in Figure 15.53.
9. Click Done to close the Edit Block Graphics window.

15.9.9.2 Hiding Block Graphics

Because blocks are logical entities, block graphics are often hidden during animation (blocks are used to prevent collisions in the model and are not visible in a real-world system).

To hide block graphics:

1. From the Model menu, select Display. The Display window opens.
2. Select "Demo1" in the system select list.
3. Clear the Visible check boxes for Block and Block Names.
4. Click OK to close the Display window. The block graphics disappear in the Work Area window.
5. Export your model.

15.9.10 Review

- You have created an AGV system. In this system, you have drawn paths on which vehicles travel.
- You have connected multiple paths using the fillet tool.
- You have learned what transfers are and that they get created automatically when you connect paths.
- You have placed blocks to prevent vehicle collisions when merging.

15.10 Defining and Placing Control Points

15.10.1 What You Will Learn

In this section, you will learn:

- What a control point is
- How to name and place control points

15.10.2 Control Points

A control point is a location on the guidepath at which a vehicle can stop and receive instructions about where to work and park; vehicles can search lists and execute procedures only when stopped at a control point.

In addition to allowing vehicles to receive instructions, control points can also be used to limit the number of vehicles that can travel on a path at the same time. Each control point has a user-defined capacity that specifies the number of vehicles that can simultaneously claim the control point as a destination. The default capacity for control points is infinite, which allows an unlimited number of vehicles to travel to a control point at the same time. To limit the number of vehicles that can travel on the path leading to a control point, change the control point capacity to an integer value (for example, a capacity of 2 limits the number of vehicles that can travel on the path leading to the control point to two vehicles).

15.10.3 How Vehicles Claim Control Points

Each stopped vehicle claims at least one control point. A moving vehicle may have multiple control points claimed at once: the control point to which it is currently traveling, and one or more control points ahead of the vehicle (on the route to its destination). For example, for a vehicle to move from control point A to control point B, it must claim B before leaving A. If B has a capacity of one, this ensures that two vehicles are not traveling on the same segment of path at the same time. To avoid stopping at control point B, the vehicle attempts to claim control point C before reaching the place where it must begin decelerating to stop at control point B.

15.10.4 Defining Control Points

Now add control points for each process area. These points allow vehicles to stop at the processing areas and park when idle.

1. Open the AGV system.
2. On the Path Mover palette, click Point. The Control Point window (Figure 15.54) opens.
3. Name the control point cpin1; then press Enter. You do not need to edit the control point attributes. By default, the control point capacity is set to infinite, which allows an unlimited number of vehicles to claim and travel to a control point at the same time.
4. Click the mouse on the path below staout1 of the conveyor system (refer to Figure 15.55) to place cpin1.

You have now placed control point cpin1.

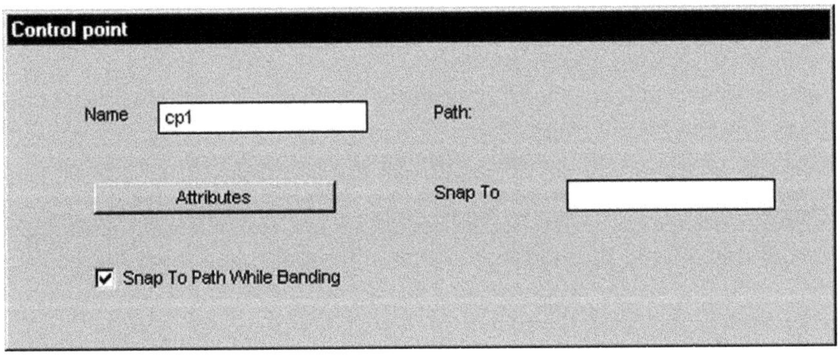

FIGURE 15.54 Control point window.

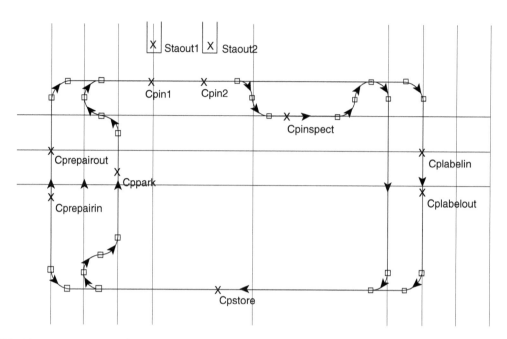

FIGURE 15.55 Main control points.

5. Place control point cpin2 under conveyor station staout2.
6. Name the next point cppark; then place it on the parking loop (refer to the previous illustration).
7. Place the points cpinspect, cplabelin, cplabelout, cpstore, cprepairin, and cprepairout, as shown in the previous illustration.
8. Export your model.

15.10.5 Adding Queues to the Model

The assembled loads need a way to get from the conveyor exit stations into the AGV system. You will define an array of queues and place a queue at the end of each conveyor section to hold loads while they wait for a vehicle to pick them up.

15.10.6 Defining an Array of Queues

1. Open the Process System.
2. On the Process System palette, click Queues. The Queues window opens.
3. Click New. The Define A Queue window opens.
4. Name the queue Q_Wait; then press Enter.
5. Change the Number of Queues to "2"; then press Enter. The queue Q_Wait is now an array of two.
6. Click OK. The Queues window opens.
7. Select Q_Wait; then click Edit Graphic. The Edit Queue Graphics window opens.
8. Select Q_Wait(1) in the right select list.
9. Click Place. Place the queue between the end of the conveyor and the path as shown in Figure 15.56 (you may need to clear the Snap option in the Measurement window before placing the queue).
10. Select Q_Wait(2); then place the graphic.
11. Click Done.
12. Export your model.

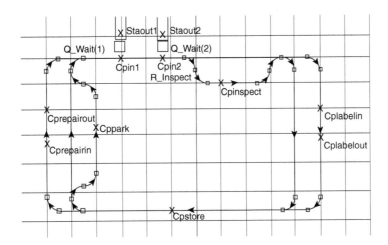

FIGURE 15.56 Path with control points and queues.

15.10.7 Review

You have placed control points on the path where vehicles can stop and pick up or set down loads. The points all have an infinite capacity, allowing multiple vehicles to claim and travel to a control point at the same time. Q_Wait(1) and Q_Wait(2) provide an area for loads to wait while transferring from the conveyor system to the AGV system.

15.11 Defining Vehicles

15.11.1 What You Will Learn

In this section, you will learn how to:

- Define a vehicle.
- Define vehicle starting locations.

15.11.2 Defining a Vehicle

In this section you will edit the default vehicle type, "DefVehicle" and set the number of vehicles in the system to three. Each vehicle requires 5 s to pick up and drop off loads.

1. Open the AGV system.
2. On the Path Mover palette, click Vehicle. The Vehicles window opens.
3. Click Edit to edit the default vehicle type's characteristics. The Edit A Vehicle Definition window (Figure 15.57) opens.

Vehicles have the following attributes:

Vehicle Type The name of the vehicle you are currently editing.
Vehicle Capacity The number of loads a vehicle can transport at one time. The default is one load. If the capacity is greater than one, the vehicles are multiload vehicles. Multiload vehicle scheduling is based on the concept of closest task for both picking up and dropping off loads.
Load Pick Up Time The amount of time a vehicle takes to pick up a load.
Load Set Down Time The amount of time a vehicle takes to set down or drop off a load.
Number of Vehicles The number of vehicles of this type in the system.

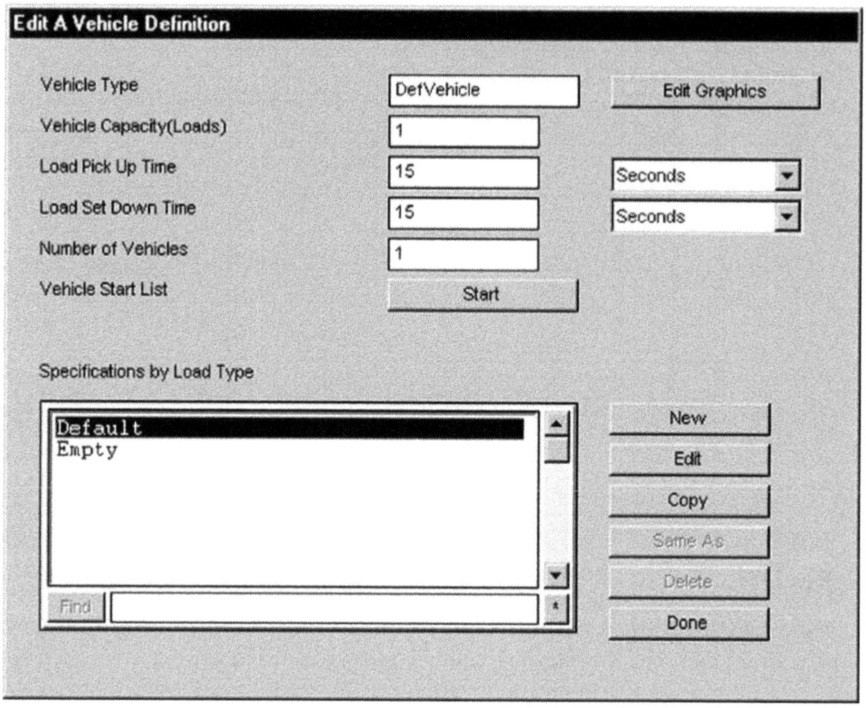

FIGURE 15.57 Edit a vehicle definition window.

Vehicle Start List By default, vehicles can start at any location in the system. Those locations are chosen randomly at the beginning of the run. It is also possible to force your vehicle to originate from only one or from several control points.

Edit Graphics Allows you to edit the graphics of the vehicle.

Specifications by Load Type window In addition to defining the vehicles themselves, you may describe the vehicles when they are carrying different load types. For example, vehicles might be required to carry very heavy or very wide loads, which can change characteristics such as speed. The standard load types are Default and Empty.

4. Change the Load Pick Up Time to "5"; then press Enter.
5. Change the Load Set Down Time to "5"; then press Enter.
6. Change the Number Of Vehicles to "3"; then press Enter to create three vehicles.
7. Click Done. You have now edited all of the necessary vehicle attributes.

15.11.3 Defining Starting Locations for Vehicles

By default, vehicles can start at any location in the system. Those locations are chosen randomly at the beginning of the run. In this simulation, however, you want the vehicles to start at the parking location cppark. To do this, you will create a list that includes only that point and assign it as the vehicle start list. You create the list using a named list.

A named list is a list of points that can be used in another scheduling list, such as a work list or park list (you will learn about these other lists in the next section).

1. On the Path Mover palette, click Named List. The Named List window opens.
2. Click New. The New Named List window opens.
3. Type "Start"; then press Enter. This is the name of the list. Click create. The edit named list window opens. There are two areas in the Edit Named List window: List Items (on the left) and Location Selection List (on the right). The Location Selection List displays all available locations in the

system. you select names from the list on the right and add them to the list on the left, which becomes your named list.

4. In the Location Selection List, select cppark.
5. Click Add After. The location cppark appears in the list on the left.
6. Click done. The named lists window opens.
7. On the Path Mover palette, click Vehicle. The Vehicles window opens.
8. Click Edit. The Edit A Vehicle Definition window opens.
9. Click the Vehicle Start List button (it currently says "Random"). The Vehicle Start List window opens.
10. Select "Start," then click OK.

The vehicle type "DefVehicle" now has a start list that starts all vehicles at cppark.

15.11.4 Defining a Vehicle Graphic

You are now ready to import a graphic to represent the vehicles in the simulation.

1. Click Edit Graphics. The Edit Vehicle Graphics window opens.
2. From the Shape Definition drop-down list, select Import.
3. Navigate to the "demos/graphics/cell" directory in the software installation directory.
4. Double-click on "AGVK.CEL." The picture is imported into the AGV system.
5. Click Place.
6. Click anywhere in the Work Area window to place the graphic.
7. Click Hide. Click Yes to confirm. The vehicle graphic you just placed is hidden from view until you run the model.
8. Click Done to close the Edit Vehicle Graphics window.
9. Click Done to close the Edit A Vehicle Definition window
10. Export the model.

15.11.5 Review

In this section, you edited the default settings for the vehicles in the system and defined cppark as their starting location. You also imported a graphic for the vehicles.

15.12 Scheduling Vehicles

15.12.1 What You Will Learn

In this section, you will learn:

- How to control vehicles through scheduling lists
- The definition of a work list
- The definition of a park list
- How to sort lists
- How to move loads from the conveyor to the AGV system

15.12.2 Scheduling Vehicles

In the AutoMod software, you can schedule vehicles by creating scheduling lists at locations in the vehicles' movement system. Scheduling lists are lists of locations where vehicles can search for their next task. Vehicle tasks include:

- Delivering loads
- Picking up (retrieving) loads
- Moving to park
- Parking (idle)

The time that vehicles spend in each of these states is reported in the vehicle statistics.

Note: When moving, vehicles automatically search for the shortest route to their destination.

15.12.3 Scheduling Lists

There are five types of scheduling lists that you can define for a location in a movement system:

- Work lists
- Park lists
- Load activation lists
- Vehicle activation lists
- Load search lists

Named lists, which you learned about in the previous section, can be used in any of these other lists.

Note: In this tutorial, you will define only work lists and park lists.

Definitions of work and park lists, and the order in which vehicles search the lists, are discussed in the next section.

15.12.3.1 Work and Park Lists

work list A list of locations at which a vehicle looks for work (a load to pick up). When a vehicle becomes idle or is awakened, the first thing it does is look for work at the current location (this does not require a work list). If you want the vehicle to look for work at other locations, list those locations in a work list. For example, if a vehicle is idle at cp1, it checks to see if any loads are waiting to be picked up at its current location (cp1). If none are found, it searches all control points on cp1's work list until it finds a task. If there isn't any work, then the vehicle searches cp1's park list.

park list A park list is a list of locations where an empty vehicle can park. The vehicle searches the list if it fails to find work. When a parking location with available capacity is found, the vehicle claims the point (reducing its available capacity) and moves to park at the location. If no parking location is found, the vehicle parks at its current control point.

Scheduling lists are always searched from the top down.

15.12.4 Building the System in Phases

You have drawn all paths and placed all control points in the system. In this section, you will begin routing loads and scheduling vehicles in the system. To get started, you will schedule vehicles to pick up loads at the conveyor and take them to the storage area. Once that is working, you will implement the other areas: inspection, rejection, labeling, and repair.

15.12.5 Creating Work Lists

All locations where vehicles can park or set down loads in the system require a work list. Because you are not yet implementing all of the processing areas, you can ignore locations other than the model's pickup, storage, and parking areas. Within these areas, there are two control points where you need to create work lists: the parking location (cppark) and the storage location (cpstore).

Vehicles begin the simulation at cppark and will return to this location when idle (that is, when they can't find work on a work list). Therefore, cppark needs a list of locations where parked vehicles can look for work. After picking up loads, vehicles deliver them to cpstore and then must look for more work, so cpstore also needs a work list.

Tip: Vehicles will be more efficient if they search for work at pickup locations in the order of the locations' distance to the vehicles' current control point. Consequently, you should add control points to work lists in order of distance, with the closest control point listed first.

The two locations in the system where loads wait to get on a vehicle are cpin1 and cpin2 (the ends of the conveyors); these are the locations where work is available. Therefore, the work lists at cppark and cpstore must include these locations. Without a work list, the vehicles would remain parked during the entire simulation.

To create the work lists, you will define a work list for point cppark and then copy the same list for point cpstore. Because both points are closest to cpin1, vehicles should search this location for work first, then cpin2. To define a work list at point cppark:

1. On the Path Mover palette, click Work List. The Work Lists window opens.
2. Click New. The New Work List window opens.
3. Select cppark, then click New. The Edit Work Lists window opens.
4. Click Add After. The Add Work List Location window opens.

Note: Each location specified within a list must have a sorting option associated with it; sorting options are useful when adding groups of locations to a list. In this tutorial, you will add only single locations and will use the default sorting option "at."

5. Select cpin1. This is the first location where vehicles look for work.
6. Click Add. Location cpin1 is added to the work list.
7. Click Add After. The Add Work List location window opens.
8. Select cpin2; then click Add.
9. Click Done. You have now define the required work list for cppark. If vehicles do not find work at cpin1, they will look at cpin2.

To copy the work list for point cpstore:

10. In the Work Lists window, select cppark; then click Copy. The Copy Work Lists window opens.
11. Select cpstore; then click Copy. The work list for cppark is copied for cpstore.

Now that you have created the required work lists in the system, you need to create a park list to instruct vehicles to travel to cppark when idle.

15.12.6 Creating a Park List

All locations where vehicles can set down a load in the system require a park list. If, after dropping off a load, a vehicle does not find any work, the empty vehicle can determine where to park so that it is not blocking the path. In this section, vehicles will drop off loads only at cpstore. From there, send the vehicles to cppark.

1. On the Path Mover palette, click Park List. The Park Lists window opens.
2. Click New. The New Park List window opens.
3. Select cpstore; then click New. The Edit Park Lists window opens.
4. Click Add After. The Add Park List Location window opens.
5. Select cppark in the Add Park List Location window.
6. Click Add. The location cppark is added to the list.

7. Click Done.
8. Export your model.

15.12.7 Modifying the Model

Now that you have created the required scheduling lists in the path mover system, you need to edit the process system and route loads through the path mover system. You must make the following changes to the model:

1. Add a queue for the storage area.
2. Move loads from process P_Out to a new process, P_Store, which represents the storage area. The loads will move from the conveyor system to the AGV system. You must then adjust the cycle time statistic to include the time spent moving the loads to the storage area.

15.12.8 Adding the Storage Queue

To add the queue for the storage area:

1. Open the process system.
2. On the Process System palette, click Queues. The Queues window opens.
3. Define a new queue named Q_Store using the default queue attributes.
4. Place the default queue picture below cpstore on the bottom section of the path (Figure 15.58; you may need to turn Snap off in the Measurement window).

15.12.9 Adding a New Storage Process

To route loads from the conveyor system to the storage area, you need to define a new process: define a new process called P_Store that uses the default process settings. You are now ready to define the process arriving procedure.

15.12.10 Defining the Storage Process Arriving Procedure

Currently, the last process that loads execute in the model is the P_Out process. You will need to edit this process to send loads to the process P_Store. You will also need to move the cycle time and throughput calculations to the P_Store arriving procedure so the calculations will include the time that loads spend in the AGV system.

Edit the source file mycode.m.

1. Define a new arriving procedure for P_Store at the end of the file:

   ```
   begin P_Store arriving procedure
   end
   ```

2. Edit the P_Out arriving procedure and cut the cycle time and throughput calculations:

   ```
   set A_Time to ac - A_Time/* Calculate cycle time */
     print this load, "Time in system =" A_Time to message
     inc V_Numdone by 1/* Count throughput */
   send to die
   ```

3. Paste the cycle time information into the P_Store procedure. To move loads into the AGV system, you will edit the P_Out arriving procedure to move loads first into a queue (where they can wait for a vehicle) and then into one of the two pickup control points (either cpin1 or cpin2). You will use procindex in the procedure to align a conveyor exit station with an arrayed queue and control point.
4. Modify the P_Out arriving procedure to appear as shown below:

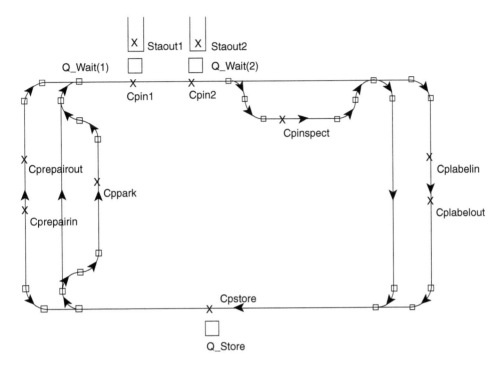

FIGURE 15.58 Storage queue illustration.

```
begin P_Out arriving procedure
  travel to Conv.staout(procindex)
  move into Q_Wait(procindex)/* Location to wait for a vehicle */
  move into Agv.cpin(procindex)/* Get on vehicle */
  send to P_Store
end
```

The first two lines cause the loads to move into the appropriate queue and AGV location based on their arrayed process. The last line sends loads to the process P_Store.

5. Modify the P_Store arriving procedure to appear as shown below. The inserted lines move loads through the AGV system and into the storage queue:

```
begin P_Store arriving procedure
  travel to Agv.cpstore/* Travel onboard vehicle to cpstore */
  move into Q_Store/* Get off vehicle */
  wait for 20 sec
  set A_Time to ac - A_Time
  print this load, "Time in system = " A_Time to message
  inc V_Numdone by 1
  send to die
end
```

6. From the File menu, select Save; then Exit.
7. Export your model.

15.12.11 Verifying the Model

Now run the model to verify that loads are moving through the AGV system correctly. Vehicles pick up loads at Q_Wait and then take them to Q_Store. When idle, the vehicles park at cppark.

1. Run the model with the animation on.
2. Turn off the animation and let the simulation run to completion.
3. From the View menu, select Variables. V_Numdone should show approximately 236 loads completed (your model may vary slightly depending on path distances and control point locations).
4. Edit the model.

15.12.12 Review

You have changed the processing flow so that loads no longer leave the simulation in P_Out; instead they are delivered to the storage area by a vehicle. The storage area consists of a process (P_Store) and a queue (Q_Store).

Vehicles start at cppark. The vehicles search cpin1 and cpin2 for work, and, when loads are waiting to be picked up, the vehicles transport them to cpstore. After dropping off loads at cpstore, the vehicles look for more work at cpin1 and cpin2. If no work is found, the vehicles return to park at cppark.

You have also moved the code that tracks cycle time and throughput from P_Out to P_Store, so that the statistics include the time that loads spend in the AGV system.

In the next section, you will expand the model to include the inspection, labeling, and repair processes.

15.13 Adding the Inspection, Labeling, Repair, and Rejection Processes

15.13.1 What You Will Learn

In this section, you will review how to:

- Create processes, queues, and resources.
- Place graphics.
- Edit work and park lists.

15.13.2 Model Description

In this section you will add three new processes: inspection, labeling, and repair (Figure 15.59). In addition, some loads will be rejected and scrapped from the model in a fourth process: reject.

These new processes use resources and queues. The new processes also require some changes to the work and park lists in the model.

15.13.3 Adding Queues

Before adding the new processes, add three new queues to your model: one for the labeling area, one for the repair area, and one for the rejection process.

1. Define the following three new queues with a capacity of infinite:
   - Q_Repair
   - Q_Label
   - QReject
2. Place the queue graphics (refer to Figure 15.60). You may need to turn Snap off in the Measurement window.
3. Export the model.

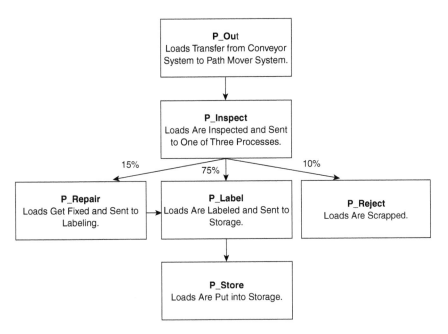

FIGURE 15.59 Process flow.

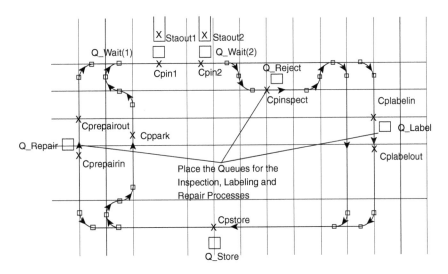

FIGURE 15.60 Layout for queues.

15.13.4 Adding Resources

Now you will add three new resources to your model: one for inspecting the loads, one for labeling the loads, and one for repairing the loads.

1. Define three new resources. Use the default resource attributes for each resource.
   - R_Inspect
   - R_Label
   - R_Repair
2. Edit the graphic for the R_Inspect.
3. Select Import from the Shape Definition drop-down list in the lower left corner of the window.

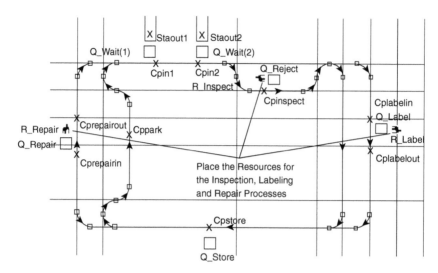

FIGURE 15.61 Layout for resources.

4. Navigate to the demos/graphics/cell directory in the software installation directory ("AutoMod" by default).
5. Double-click the file man.cel.
6. Place the operator next to Q_Reject, as shown in Figure 15.61.
7. Repeat this process to place graphics for the other resources. You will need to rotate the graphics around the Z axis, as shown in the table below:

Resource	Z rotation amount
R_Label	180 degrees
R_Repair	270 degrees

15.13.5 Creating the New Processes

Add four new processes to accommodate the labeling, repairing, storage, and inspection/rejection areas in the model.
Define four new processes:

- P_Label
- P_Repair
- P_Reject
- P_Inspect

15.13.6 Defining the Process Arriving Procedures

Instead of sending loads from the conveyor system directly to storage, loads must now first travel to the inspection process. You will edit the model logic to reroute loads and define the procedures for each of the new processing areas.

1. Edit the source file mycode.m.
2. Edit the P_Out arriving procedure and edit the last line to send loads to P_Inspect instead of P_Store:

```
begin P_Out arriving procedure
travel to V_Convloc(procindex)
move into Q_Wait(procindex)/* Location to wait for a vehicle */
move into Agv.cpin(procindex)/* Get on vehicle */
```

```
    send to P_Inspect
  end
```

3. Define a new arriving procedure for the inspection process P_Inspect:

```
    begin P_Inspect arriving procedure
      travel to Agv.cpinspect/* Travel to the inspection area */
      use R_Inspect for 30 sec/* Inspect while onboard the vehicle */
      send to oneof(10:P_Reject, 15:P_Repair, 75:P_Label)
    end
```

The inspection procedure sends loads to the control point cpinspect. There the loads are inspected while still on the vehicle. The inspection process sends loads to one of three processes using a oneof distribution (each load has a 10% chance of being scrapped, a 15% chance of needing repair, and a 75% chance of passing inspection and going directly to the labeling area).

4. Define the arriving procedure for the rejection process P_Reject.

```
    begin P_Reject arriving procedure
      move into Q_Reject/* Get off vehicle */
      wait for 30 sec
      set A_Time to ac - A_Time/* Calculate cycle time for rejected loads */
      print this load, "was rejected. Total time in system was"
      A_Time to message
      send to die
    end
```

Rejected loads get off the vehicle in the inspection area and delay temporarily in Q_Reject before leaving the simulation. Because rejected loads leave the simulation, their cycle time must be calculated and printed in the P_Reject arriving procedure.

5. Define the arriving procedure for the repair process P_Repair:

```
    begin P_Repair arriving procedure
      travel to Agv.cprepairin/* Travel to repair area */
      move into Q_Repair/* Get off vehicle */
      use R_Repair for normal 6,1 min/* Repair delay */
      move into Agv.cprepairout/* Get on vehicle */
      send to P_Label
    end
```

Loads that require repair travel by vehicle to cprepairin. Loads get off the vehicle and are repaired in queue Q_Repair for a time that is normally distributed with a mean of 6 min and a standard deviation of 1 min.

After repair, loads wait to be picked up by a vehicle at cprepairout and then travel to the labeling process.

6. Define the arriving procedure for the labeling process P_Label:

```
    begin P_Label arriving procedure
      travel to Agv.cplabelin/* Travel to labeling area */
      move into Q_Label/* Get off vehicle */
      use R_Label for 2 min/* Labeling delay */
      move into Agv.cplabelout/* Get on vehicle */
      send to P_Store
    end
```

Loads get off the vehicle at cplabelin and delay for 2 min while being labeled. The loads then wait to be picked up by a vehicle at cplabelout and travel to the storage area, where they leave the simulation.

7. From the File menu, select Save, then Exit.
8. Export the model.

15.13.7 Modifying the Work and Park Lists

Now you must create work and park lists to cause vehicles to service the new areas.

Inspection After picking up a load from the conveyor system, each vehicle travels directly to the inspection area. Inspection takes place on board the vehicle. Vehicles that drop off rejected loads in the inspection area must look for work or a parking location at the other points in the system.

Repair Vehicles carrying loads that need repair drop off loads at cprepairin. The vehicles must then look for additional work or a parking location.

Label Vehicles carrying loads that need labeling drop off loads at cplabelin. The vehicles must then look for additional work or a parking location.

Because there are new locations where work becomes available, you must edit the model's existing work lists to cause loads to look for work at these locations. You will also create new work and park lists at the new setdown locations.

The locations where vehicles set down loads or park are:

- cppark (where vehicles park when idle)
- cpinspect (where rejected loads get off the vehicle to leave the system)
- cplabelin (where loads get off the vehicle to be labeled)
- cpstore (where loads get off the vehicle to be stored)
- cprepairin (where loads get off the vehicle to be repaired)

The locations where work becomes available are:

- cpin1 (where loads transfer from the conveyor system)
- cpin2 (where loads transfer from the conveyor system)
- cplabelout (where labeled loads get back on a vehicle)
- cprepairout (where repaired loads get back on a vehicle)

15.13.8 Editing Work Lists

To edit the existing work lists so vehicles look for work in the labeling and repair areas:

1. Open the AGV system.
2. Edit the work list for cppark and append the labeling and repair pickup points, as shown below:

Work list for cppark
At cpin1
At cpin2
At cplabelout
At cprepairout

Note: For optimal efficiency, locations are added to the vehicle list in order of their distance to the current control point, with closest locations listed first. Vehicles search locations on the list from the top down.

3. Edit the work list for cpstore and insert the repair and labeling pickup points, in the order shown:

Work list for cpstore
At cprepairout
At cpin1
At cpin2
At cplabelout

15.13.9 Defining Work Lists

To cause vehicles that set down loads in the inspection, labeling, and repair processes to look for work, you must define new work lists.

1. Define a work list for cpinspect, as shown below:

Work list for cpinspect
At cplabelout
At cprepairout
At cpin1
At cpin2

2. Copy the cpinspect work list for the control point cplabelin. Vehicles look for work at cplabelin in the same order of locations as vehicles at cpinspect.
3. Define a work list for cprepairin, as shown below:

Work list for cprepairin
At cprepairout
At cpin1
At cpin2
At cplabelout

15.13.10 Defining Park Lists

The only location at which idle vehicles are allowed to park in the system is cppark. You have already created a park list for cpstore that causes vehicles to look for parking at this location.

To create the park lists for the remaining locations where vehicles set down loads:

1. Copy the park list of cpstore for cpinspect, cplabelin, and cprepairin.
2. Export the model.

15.13.11 Verifying the Model

1. Run the model. After picking up loads from the conveyor system, vehicles travel through the inspection area and to the other three processes.
2. Watch the Message window. Reject messages appear, stating how long loads were in the system before being rejected.

15.14 AutoMod Tutorial Summary

You have now created a simple model that uses conveyors and vehicles to move loads through several processing steps. You moved the loads through the system using process-arriving procedures.

You also learned how to read standard statistics as well as create your own variables and attributes to track custom statistics. You are now ready to learn how to use the AutoStat software to analyze a simulation.

15.15 Basic Statistical Analysis Using AutoStat

When random samples are used as input for a model, a single simulation run may not be representative of the true behavior of the real-world system. Therefore, you could make erroneous inferences about the system if you make only one run. The AutoStat software helps you apply the proper statistical sampling techniques to your model in order to estimate accurately performance under random conditions.

The following sections discuss how to perform basic statistical analysis on a model using the AutoStat software, teaching you how to calculate confidence intervals for several performance metrics.

15.16 Why Use AutoStat?

The key to a successful simulation study is proper analysis. However, in too many simulation studies, analysts spend most of their time developing the model and not enough time analyzing the simulation results. Sometimes decisions are incorrectly based on just one run of the model.

The AutoStat software helps you to determine which results and alternatives are statistically significant, which means that with high probability the observed results are not caused by random fluctuations but by the change you are experimenting with.

15.17 Calculating Confidence Intervals

The example model "ASExample" illustrates why confidence intervals are important when analyzing a random model. In this system, products have an interarrival time that is uniformly distributed between zero and 10 min. The products go through two processes: checking and processing.

The checking process is modeled using a resource with a capacity of two. Checking each load takes a time that is uniformly distributed between 6 and 10 min.

The processing step uses a resource with a capacity of eight. The resource can process each load in a time that is uniformly distributed between 25 and 35 min.

The source file for the example model is shown below:

```
begin P_init arriving
    while 1 = 1 do
    begin
        clone 1 load to P_checkers nlt L_job
        wait for u 5,5 min
    end
end

begin P_checkers arriving
    move into Q_checkers/* capacity is infinite */
    use R_checkers for u 8,2 min/* capacity of 2*/
    send to P_processors
end

begin P_processors arriving
    move into Q_processors/* capacity infinite */
    use R_processors for u 30,5 min/*capacity of 8 */
    send to die
end
```

Suppose you simulated this system for 12 h and then looked at the average time spent in the checking process. Because there is so much randomness in the model (including the arrival times and the service time of each process), the average time reported for the checking process for that run may or may not be an accurate estimate of how the system behaves over a longer period of time.

Now suppose you ran the model 10 times, each time using different random numbers (such runs are called replications). The following table shows the results of 10 such replications.

Replication Number (I)	Average Time in the Checking Process
1	752.23
2	785.49
3	645.13
4	639.96
5	610.13
6	661.42
7	645.28
8	606.32
9	677.74
10	584.53

Each of these times is a valid average time for the checking process. But the range between the lowest and highest value is 200.96 s. So what is the "correct" value, and how do you make decisions based on these numbers?

Using the average value from a single replication can be misleading. It's called a point estimate, and it can be very different from the true mean of the process. The appropriate action in this case is to report a confidence interval, not just a single number.

15.17.1 How AutoStat Calculates Confidence Intervals

The AutoStat software uses the replication/deletion technique for computing confidence intervals. The replication/deletion method strives to use steady-state data in the formation of point estimates and confidence intervals for the various responses, which is accomplished by obtaining the average level of the response for each replication after a warmup period. The averages obtained after each warmup period are independent and are approximately normally distributed random numbers. Thus, they can be used to construct a confidence interval for the steady-state mean value of the response.

To generate the average times shown in the table above, the model was run using an 8-h warmup followed by a 24-h snap length. Thus, the simulation was run for 8 h, at which time statistics were reset (all statistics, except for Current values, were set to zero), to remove the warmup bias. The simulation was run for 24 h, during which statistics were gathered. This procedure was followed 10 times, each using different random numbers.

15.18 Performing Statistical Analysis with AutoStat

The following steps outline the process for conducting an analysis with AutoStat:

1. Open a model in AutoStat.
2. Define an analysis (for example, a single scenario analysis to determine confidence intervals).
3. Make the runs.
4. Define the responses (statistics that you want to measure).
5. Display the results.

The rest of this chapter walks you through these steps to determine the confidence intervals for example model "ASExample."

15.19 Opening a Model in AutoStat

To use the AutoStat software, you must open a model that has been compiled in AutoMod.

To open example model "ASExample" for use in the AutoStat software:

1. In the AutoMod software, import "ASExample."
2. From the Run menu, select Run AutoStat.
3. In the confirmation window, click Yes to build the model. After the build is complete, AutoStat opens, and the AutoStat Setup wizard opens.

The first time you open a model in AutoStat, use the AutoStat Setup wizard to set the parameters for your analysis.

15.19.1 Using the AutoStat Setup Wizard

Whenever you open a model for the first time in the AutoStat software, the AutoStat Setup wizard asks you to:

- Indicate whether the model contains randomness.
- Determine a time limit to stop models that are potentially in an infinite loop.
- Estimate the model's warmup length.
- Define how long to run the model to collect statistics.
- Indicate whether to use common random numbers for the analyses.

The information in the wizard is used to define several model properties. The following sections explain the questions that the wizard is asking.

Note: While using the AutoStat software, click Help if you want more information about a particular screen.

In the AutoStat Setup Wizard window, click Next to advance to the first question in the wizard.

15.19.1.1 Is the Model Random or Deterministic?

As discussed earlier, when a simulation model contains random input (such as the random arrival and service times in the example model), the statistics from one run may not be representative of long-term average system output. Therefore, you need to perform multiple replications (runs using different random numbers) to get an accurate understanding of the system's behavior, including meaningful confidence intervals.

Note: A deterministic model contains no random input and therefore no variability in output from run to run, so only a single run of each scenario is necessary. Very few simulation models are deterministic.

1. Select Model is random.
2. Click Next.

15.19.1.2 Do You Want To Stop Runs That May Be in an Infinite Loop?

An infinite loop causes a model to repeat the same section of logic so that the simulation never ends. If you are using AutoStat to make runs, the software can automatically stop a run that seems to be taking longer than you expected.

The example model runs very quickly (in about 1 or 2 s) if it is working correctly. If a run takes substantially longer than that (for example, 30 s), it might indicate that there is an infinite loop in the model and that the run should be canceled.

To set an infinite loop time of 30 s:

1. Select Yes to indicate that you want to set up an infinite loop time limit.
2. Click Next.
3. Type "30" in the Maximum run time text box and select seconds from the drop-down list.
4. Click Next.

15.19.1.3 Does the Model Require Time To Warm Up?

If a model does not reflect the state of the system being modeled at the beginning of the simulation, it needs to warm up, or "reach steady state," before you gather statistics.

Some systems start "empty," such as service centers. In these systems, there are no customers left over from the day before, and new customers cannot enter the system until the center opens in the morning.

Systems such as the factory in the example model, however, are usually full of products that are spread throughout the factory and that are in various stages of being manufactured. In the example model, loads do not get created until the simulation starts, so it will take some time before the system is primed with jobs. Therefore, you want to set up a warmup time and discard statistics gathered during that time.

1. Select Yes to indicate that the model requires a warmup time.
2. Click Next.

15.19.1.4 What Is the Estimated Warmup Time?

AutoStat can perform an analysis to determine the amount of time the model requires to warm up; however, conducting warmup analyses is not discussed in this chapter. For this example, make a rough estimate without the help of AutoStat and say that the simulation takes 8 h to warm up.

To define the warmup time:

1. Type "8" in the warmup estimate text box and select hours from the drop-down list.
2. Click Next.

15.19.1.5 Do You Want to Create the Warmup Analysis?

For this example, assume that you have estimated the warmup correctly and do not need to conduct a warmup analysis.

1. Select Do Not Create Analysis.
2. Click Next.

15.19.1.6 What Is the Snap Length for Collecting Statistics?

After the warmup time, the statistics are going to be reset and then collected for some amount of time. The statistics-gathering time is called the snap length. The length of the snap varies depending on whether the system modeled is a terminating or a nonterminating system. In a terminating system, the snap length is equal to the length of the system's operation. For example, if you were simulating a bank that is open from 9:00 A.M. to 5:00 P.M., the snap length would be 8 h.

In a nonterminating system, the snap needs to be long enough to collect meaningful, representative data from the system. The time required varies from system to system and is best determined by someone familiar with the system's operation.

The example model currently completes about 300 loads in 24 h, which is meaningful enough to generate confidence intervals for this small model.

To define the snap length:

1. Type "24" in the Snap length text box and select hours from the drop-down list. *Note: The run control in AutoStat always overrides the run control defined in AutoMod.*
2. Click Next.

15.19.1.7 Do You Want to Use Common Random Numbers?

If you are analyzing a simulation to compare multiple system configurations (scenarios), you might want to design your model to use common random numbers. This method of analysis is also known as correlated sampling. Using common random numbers allows you to analyze configuration changes while duplicating the experimental conditions for each configuration (that is, random events do not change from one configuration to the next). For the purposes of this example, do not use common random numbers.

1. Select No.
2. Click Next. The last screen of the wizard displays.
3. Click Finish.

The information in the wizard is used to set up the model properties. You can change any of these settings, such as the snap length, warmup length, and so on, at any time by editing the model properties (discussed next).

4. From the File menu, select Save to save the properties.

15.19.2 Editing Model Properties

If you need to change a model property after you have run the Model Setup wizard, edit the model properties.

To edit the model properties:

1. From the Properties menu, select Edit Model Properties. The Model Properties window opens.
2. When conducting your own analyses, you may want to change some of these values. For this example model, however, do not change any of the properties.
3. Click Cancel to close the Model Properties window.

15.19.3 The AutoStat File System

When you open a model and set it up in AutoStat, several directories and files are created (see Figure 15.62).

The main AutoStat directory is the .sta directory. The .sta directory contains all the AutoStat information for a model. The .sta directory contains a file called astat.sta.xml, which contains all the information about the model properties you set using the wizard, as well as information for analyses (which you will learn how to define in this chapter).

Once you have made runs for your analyses, numbered run directories are created, which contain message files, reports, and information about each run.

15.20 Defining a Single Scenario Analysis

As mentioned earlier, AutoStat can conduct several types of analyses. In this chapter, you will learn how to conduct a single scenario analysis, in which you can run your model "as is" (without any changes) in order to compute confidence intervals and other statistics.

To define a single scenario analysis:

1. From the Create New Analysis of Type drop-down list, select Single Scenario.
2. Click New. The Single Scenario Analysis window (Figure 15.63) opens.
3. Name the analysis "Example Single Scenario."
4. Type "10" in the Number of Replications text box and press Tab. AutoStat will make 10 runs, with each run using different random numbers for each random event in the model.

For confidence intervals, you should use at least three to five replications. The greater the number of replications or the longer the snaps, the narrower the confidence intervals.

For the example model, you want to use the default run control, which you set up using the wizard, so do not make any other changes in this window.

You have defined the single scenario analysis. Now you are ready to make the runs.

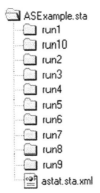

FIGURE 15.62 AutoStat file system.

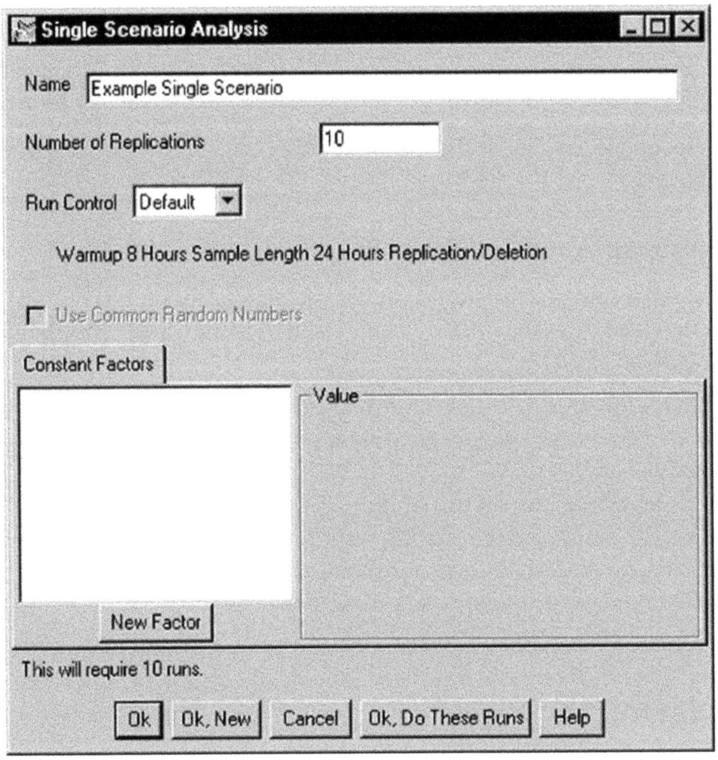

FIGURE 15.63 Single scenario analysis.

15.21 Making Runs

There are several ways to make runs for an analysis in AutoStat. In this example, you can make the runs directly from the Single Scenario Analysis window.

Click OK, Do These Runs. AutoStat makes 10 runs, recording the statistics.
You can also make runs using either the Execution menu or the Runs tab using the following options:

Do All Runs Makes all runs that have been defined for all analyses.
Do Some Runs You can select the analysis for which you want to make runs.

Do Runs Until Defines a time to stop making runs. For example, make runs until tomorrow at 8:00 A.M.

Parallel Execution Makes runs on more than one computer (not discussed in this chapter).

While the runs are being made, or after they are finished, you can define responses, which are the statistics you are interested in analyzing.

15.22 Defining Responses

A response is a statistic in your model that you want to study in an analysis. For example, you might be interested in the utilization of a resource, the average time spent waiting for a tool, the time in system, and so on.

There are three types of responses:

AutoMod A standard statistic from the report file.

User A statistic from a user-defined report (not discussed in this chapter).

Combination The combination of two or more responses, such as the time spent in several processes, or the average utilization of a group of equipment. An example is shown later in this chapter.

You can define responses at any time (either before or after making runs). You can also define responses from most output windows.

15.22.1 Defining an AutoMod Response

In the example model, you want to determine confidence intervals for the average time that loads spend in the checker process. Therefore, you need to define a response to track loads' average time in that process.

To define an AutoMod response:

1. Click the Responses tab.
2. Click New to create a new response of type AutoMod Response. The AutoMod Response window (Figure 15.64) opens.
3. Name the response "Checker Average Time."
4. Select the system, entity, and statistic you want to define as a response. By default, the process P_checkers and the statistic Av_Time are already selected.
5. Click OK. The response name appears in the Defined Responses list.

15.23 Displaying the Results

Each analysis has several ways of displaying output. The output is calculated for all defined responses. For single scenarios, the types of output are:

Summary Statistics A table of statistics for each response, including the average, standard deviation, minimum and maximum values, and other information.

Bar Graph A graph of a response's values.

Confidence Intervals The confidence interval for each response by confidence level (90%, 95%, and so on).

Runs Used A list of the runs that are used for this analysis (useful if you have more than one analysis defined).

Run Results A table of each response by run.

For this example, you will look at confidence intervals and summary statistics.

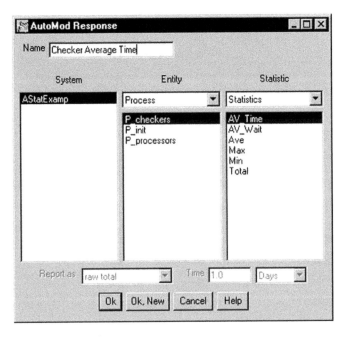

FIGURE 15.64 AutoMod response window.

15.23.1 Viewing Confidence Intervals

To view confidence intervals:

1. From the Analysis tab, click the plus (+) sign to expand the list of output options.
2. Double-click Confidence Intervals. The Confidence Intervals window opens.

By default, the confidence intervals (Figure 15.65) displayed are for a 90% confidence level. The window displays the response name, the low and high values in the interval, and how many runs are used to generate the interval.

You can adjust the confidence level using the drop-down list. The following table shows the 95% and 99% intervals for the Checker Average Time response:

Measure	95%	99%
CI Low (seconds)	589.072	578.647
CI High (seconds)	636.828	647.253
# of Runs	10	10

Note: The values displayed on your machine might vary slightly from those shown in this chapter, depending on which version and build of the software you are using.

3. Close the Confidence Intervals window.

15.23.2 Narrowing the Confidence Interval

The narrower the range of a confidence interval, the more accurate the estimate. To decrease the width of a confidence interval by approximately half, you must increase the sample size by four. You can increase the sample size one of two ways:

- Make more runs (four times the number of runs)
- Make longer runs (increasing the run length by a factor of 4)

Either of these methods provides more information to AutoStat, allowing it to narrow the range of the interval.

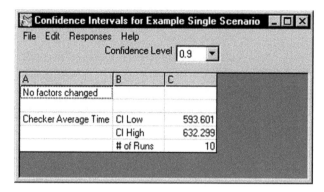

FIGURE 15.65 Ninety-percent confidence level for checker average time.

15.23.3 Making More Runs

It is possible to make additional runs for an analysis at any time.
    To make additional runs for an analysis:

1. Edit the analysis.
2. Edit the number of replications.
3. Make the additional runs.

15.23.4 Making Longer Runs

You can increase the length of runs used for an analysis in two ways:

- Edit the model properties and change the default sample time.
- Edit the analysis and create a custom run control.

Both approaches require you to redo existing runs so that your analysis is based on the new information.

15.23.5 Changing the Default Sample Time

Editing a model's properties changes the sample time for all analyses that use the default run control.
    For this example, you are going to define a custom run control, not change the default sample time, so do not edit the model's properties.

15.23.6 Defining a Custom Run Control

For this example, you want to make the runs for the analysis four times as long by defining a custom run control.
    To define a custom run control for an analysis:

1. From the Analyses tab, select Example Single Scenario Analysis and click Edit. The Single Scenario Analysis window opens.
2. From the Run Control drop-down list, select Custom. The Custom Run Control window (Figure 15.66) opens.
3. You want to lengthen the snap length by a factor of 4, so change the Snap Length from 24 to 96 h.
4. Click OK to close the Custom Run Control window.
5. Click OK, Do These Runs to make the new runs. Ten new runs are made using the new run control (the 10 runs using a 24-h snap length are still saved, as well).
6. Click OK when the runs have finished.

FIGURE 15.66 Custom run control.

7. Double-click Confidence Intervals to display the confidence intervals.
8. Change the confidence level to 99%.

The following table shows the 99% intervals for the Checker Average Time response for 24-h and 96-h snap lengths:

Measure	99%	99%
CI Low (s)	578.647	594.817
CI High (s)	647.253	637.243
Snap Length	24 h	96 h

The original interval has a range of 68.606 s, whereas the new snap length of 96 h has an interval of only 42.426 s.

15.23.7 Viewing Summary Statistics

To view summary statistics for the single scenario analysis:

From the Analysis tab, expand the Example Single Scenario analysis and double-click Summary Statistics (Figure 15.67).

The resulting table shows you the average, standard deviation, minimum, maximum, and median values of the average time response for the 10 runs.

15.23.8 Defining a Combination Response

You can combine two or more responses into a single response using a combination response. Combination responses are useful when you want to combine statistics, such as averaging several machines' utilizations into one statistic for an equipment group, or summing several different WIP levels.

In this example, you have already defined a response for the average time that loads spend in the checking process. If you create a response for the average time in the processor process and then add the two together in a combination response, the combination response is the average time that loads spend in the whole system.

To define a combination response for the time in system:

1. From the Responses tab, define an AutoMod response, named "Processor Average Time," for the average time loads spend in P_processors. Click OK.
2. From the New Response of Type drop-down list, select Combination Response and click New. The Combination Response window opens.
3. Name the combined response "Time in System."
4. Click Append Term to add a duplicate row below the Checker Average Time response.
5. Double-click the Name in the new row.
6. Select Processor Average Time.

The combination response is now a sum of the two AutoMod factors, as shown by the formula in the window (Figure 15.68).

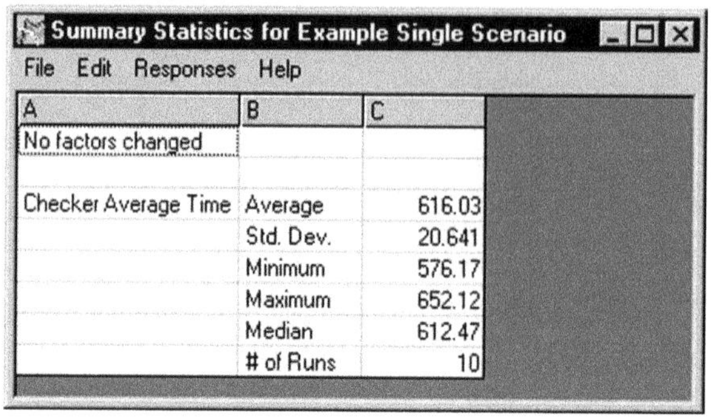

FIGURE 15.67 Summary statistics.

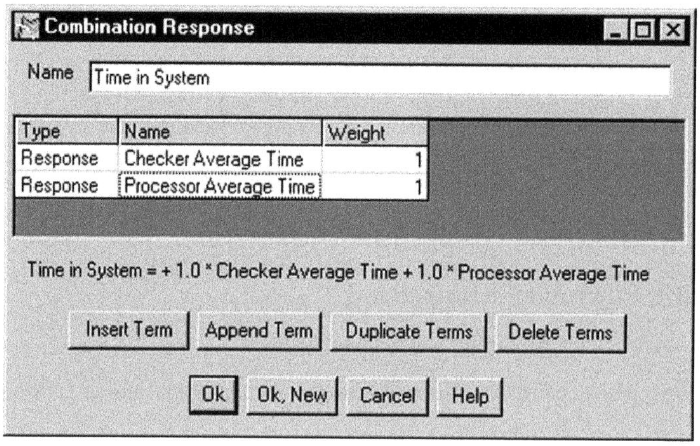

FIGURE 15.68 Viewing the relationship between terms in a combination response.

7. Click OK to close the Combination Response window.
8. Display the confidence intervals and summary statistics for the analysis. Notice that the two new responses, Processor Average Time and Time in System, are now shown in addition to the original response, Checker Average Time.

15.23.9 Weighting Terms in a Combination Response

When using a combination response, you can define a weight (or relative importance) for each of the terms used to calculate the combined response (Figure 15.69). This is often useful when analyzing cost versus payback situations, such as whether the cost of a new piece of equipment is paid for by an increase in throughput.

You can also use the weight value to convert the statistics measured into different units. For example, you can convert time units from seconds to minutes, or convert one monetary unit to another.

In this example, the terms in the Time in System response are currently calculated in seconds (all standard AutoMod statistics are reported in seconds). By default, the terms are each given a weight of one, which means the terms have equal value and are being used in their default units.

Suppose you wanted to report the Time in System in minutes instead of seconds.

To report the Time in System in minutes:

1. From the Responses tab, select Time in System in the Defined Responses list and click Edit. The Combination Response window opens.
2. Double-click the Weight for Checker Average Time.
3. Enter a Weight of 0.0167 (1 min/60 s) as a multiplier to convert seconds to minutes.

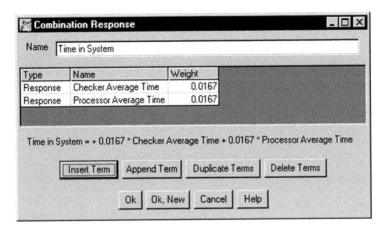

FIGURE 15.69 Weighting terms in a combination response.

    4. Enter 0.0167 as the weight for Processor Average Time.

Converting seconds to minutes using the Weight column:

    5. Click OK to close the Combination Response window.
    6. View the confidence intervals and summary statistics for the analysis. Notice that the values for the Time in System are now in minutes instead of seconds.

Tip: You could rename the analysis to show the units displayed. For example, you could call this analysis "Time in System (minutes)." Then you would know when viewing output that this response's units are different from the other responses being displayed.

15.24 Statistical Analysis Using AutoStat Summary

You have now learned how to conduct a simple analysis to vary the random numbers used in a simulation and calculate confidence intervals for reporting simulation results. AutoStat is a powerful statistical analysis tool and should be used on every project to ensure you draw sound conclusions from analyzing your simulation projects.

16

Simpak User's Manual

Charles E. Donaghey
University of Houston

Randal W. Sitton
University of Houston

16.1 Introduction

SIMPAK is a collection of BASIC subprograms that are used to develop simulation models. Using BASIC to develop and run simulation models allows a user to utilize a personal computer with a QBASIC interpreter or compiler. This chapter describes SIMPAK, and the disk that is furnished with this manual will have the SIMPAK code. It is suggested that the user load SIMPAK in a directory where it can be easily accessed by the BASIC system that is utilized. However, the system can also be used from a floppy disk drive.

When you want to utilize SIMPAK, you enter the BASIC environment that will be used and load the SIMPAK program. You create the desired model as the main program in SIMPAK and call on the appropriate subprograms that are part of the package to aid you in this creation. SIMPAK has a random variate generator, a scheduler, an event remover, and subprograms for list processing. The main program for a SIMPAK model should be placed after the random variate functions that are at the beginning of the SIMPAK package.

It can be questioned if SIMPAK has any advantage over existing simulation languages. One obvious advantage is the cost. Existing simulation languages can be expensive, and most have protection devices that allow them to be used only on machines where this device has been installed. SIMPAK can be used on any machine that has some version of BASIC. Another advantage is that SIMPAK gives the user full control over the development of the model and the output that results from it. Existing simulation languages become complex when the user attempts to develop output that is not in the standard output format. Another academic advantage is that students who are introduced to simulation get a better perspective of the elements that are required in constructing a model. This perspective is not obtained by starting with an existing simulation language.

16.2 Random Variate Routines

The routines for generating random variates from specified distributions begin at the front of the SIMPAK program. There are eight functions that are used to generate random values from specified probability distributions. The names of the functions, the probability distributions to which they apply, and the required parameters are shown in Table 16.1.

For example, if it is desired to set variable Y to a random value from a normal distribution with a mean of 54.8 and a standard deviation of 3.9, the user would use the statement:

$$Y = FNNORM(54.8, 3.9)$$

Figure 16.1 contains a short program and its output that demonstrate the use of the random variable generator in SIMPAK.

The first statement in the program (RANDOMIZE TIMER) seeds the random number generator using the internal clock. The user wants to generate 1000 values from a normal distribution, so the variable N is set to 1000. In the loop, 1000 random values from a normal distribution with a mean of 54.8 and a standard deviation of 3.9 are created. The sum of these values and the sum of the squares of the values are stored in SUM and SUMSQ. The program then calculates and prints the mean and the standard deviation of the 1000 values that have been created. From the output it can be seen that there is close agreement between the input parameters and the sample mean and standard deviation.

```
RANDOMIZE TIMER: N=100

FOR I=1 TO N

Y= FNNORM(54.8, 3.9) SUM=SUM+ Y: SUMSQ=SUMSQ+ Y1\2 NEXT I

LPRINT "MEAN="; SUM / N

LPRINT "STD.DEV.= ";SQR ( ( N* SUMSSQ – SUM 1\2) / (N-1) STOP

MEAN=54.803 STD. DEV.=3.856
```

FIGURE 16.1 SIMPAK program and its output that demonstrates use of the random variate functions.

TABLE 16.1 Descriptions of Random Variable Functions in SIMPAK

Distribution	Function Name	Arguments
Normal	FNNORM	Mean, Std. Dev.
Exponential	FNEXPO	Mean
Weibull	FNWEIB	m, Theta
Log-normal	FNLOGN	Mean, Variance
Binomial	FNBINOM	N, P
Poisson	FNPOIS	Lambda
Uniform	FNUNIF	Min, Max
Erlang	FNERLANG	Mean, K

16.3 Schedule Initializer Subprogram (INITIAL)

During a simulation the process of scheduling events on a schedule and removing these events from the schedule goes on constantly. Before a simulation is started, the schedule should be initialized. This initialization process removes any events that are in the schedule from the last simulation. The initialization process is done quite simply with the statement:

CALL INITIAL

There are no input arguments that have to be specified, and the subprogram returns no values.

16.4 SIMPAK Scheduler (SCHED)

The schedule that is maintained by SIMPAK keeps a chronological list of event codes and the times of occurrence of each event. The subprogram is called SCHED, and the two arguments that must be supplied are the event code and the time of occurrence of the event. When developing a simulation model, the model builder should determine the types of events that can take place and assign an arbitrary event code for each of these event types. For example, in a simple manufacturing situation, the event types might be:

1. New job arrival
2. Job begins processing on A machine
3. Job leaves A machine
4. Job begins processing at B machine
5. Job leaves system
6. Intermediate report (every 8 h)
7. End of simulation (240 h)

The event codes for each of these events will be 1 through 7. The modeler could schedule the first intermediate report at $t = 8.0$ with the statement:

CALL SCHED (6,8.0)

The end of the simulation could be scheduled at $t = 240.0$ with the statement:

CALL SCHED (7, 240.0)

The schedule that is maintained by SIMPAK is capable of storing 100 events. If an attempt to add another event is made when there are currently 100 in the schedule, the system will display the message "SCHEDULE FILLED," and the simulation will stop. There is usually a logic error in the model when this occurs. Most simulation models will have far fewer than 100 future events scheduled at any time during their execution.

16.5 Event Remover (REMOVER)

The companion subprogram to the scheduler is the subroutine REMOVER. The scheduler routine SCHED puts events on the schedule, and the subprogram that removes the earliest event in chronological order from the schedule is REMOVER. There are two arguments returned when this subprogram is called. The two arguments that are set when this subprogram is referenced are:

Event code, Time of occurrence

The first variable is set by the subprogram to the event code of the event in the schedule that has the earliest occurrence time. The second variable is set by the subprogram to the occurrence time of this earliest event. The subprogram also makes available, on the schedule, the space that this earliest event required. This space can then be used by the Scheduler to schedule future events. To demonstrate the use of the SCHED and the REMOVER routines, consider the following statements:

```
CALL SCHED (4, 3.9)
CALL SCHED (2, 6.2)
CALL SCHED (10,1.4)
CALL REMOVER (K, T) : PRINT K,T
CALL REMOVER (K, T) : PRINT K,T
CALL REMOVER (K, T) : PRINT K,T
```

The three lines that would be printed would be:

10	1.4
4	3.9
2	6.2

The event codes and their associated times are printed with the times in chronological order

16.6 Example Problem 1

There are many simulation models that can be developed using only the Initializer, Scheduler, and Event Remover subprograms that have been discussed so far. This example will demonstrate such a model.

Customers arrive at a service center in a random manner with the mean number of arrivals per hour = 7.2. There is a single server, and the required service time for a customer is normally distributed with a mean of 6.2 min and a standard deviation of 1.9 min. The system is to be simulated for 500 h, and statistics are to be gathered on the utilization of the server and the customer queue that forms in front of the server.

Solution: Because the arrivals are occurring in a random manner, the time between arrivals will have an exponential distribution with the mean time between arrivals = 1/(mean number of arrivals per hour). Therefore, the mean time between arrivals, in hours, will be 1/7.2. In any simulation model there has to be consistency in the time units that are used, and in this model the units will be hours. Therefore, the mean service time, in hours, will be 6.2/60, and the standard deviation of the service time, in hours, will be 1.9/60. There are only three event types that are needed in this model, and they are listed below:

Event Code	Event
1	Customer Arrival
2	Service Completion
3	End of Simulation

The variables that will be used in the model and their definitions are:

TBARV	Mean time between arrivals (hours).
SVCM	Mean service time (hours).
SVCSD	Service time standard deviation (hours).
CUMWT	Total customer queue time for all customers.
CUMWK	Total server work time.
BFLAG	Set to 0 when server is idle, 1 when server is busy.
NSERVED	Total number of customers served during simulation.
NWAIT	Number of customers currently in queue.
MAX	Maximum number of customers in queue during simulation.
CLK	Current value of simulation clock.
LSIM	Length of simulation (hours).

The SIMPAK listing for this model is shown in Figure 16.2, and the resulting output is shown in Figure 16.3. There are comments attached to almost all of the lines of code in the model that show the purpose and the logic flow of the model.

```
'****** INITIALIZE ******
RANDOMIZE TIMER: CALL INITIAL            'INITIALIZE SCHEDULE
TBARV=117.2: SVCM=6.2/60: SVCSD = 1.9/60 'SET PARAMETERS TO HOURS
LSIM = 500: CALL SCHED( 3,LSIM )         'SCHED. END OF SIM. @ 500 HRS
T = FNEXPO(TBARV): CALL SCHED(l, T)      'SCHED. 1ST ARRIVAL


'****** REMOVE EARLIEST EVENT *****
ABC: CALL REMOVER(CODE, TCLK)            'REMOVE EARLIEST EVENT
DELTA = TCLK - CLK: CLK = TCLK           'FIND ELAPSED TIME, ADV CLOCK
CUMWT = CUMWT + NW AIT * DELTA           'UPDATE CUMULATIVE WAIT TIME
CUMWK = CUMWK + BFLAG * DELTA            'UPDATE CUMULATIVE WORK TIME
ON CODE GO TO ARVL, DEPT, EOS            'BRANCH ON EVENT CODE


'******** JOB ARRIVAL ********
ARVL: T = FNEXPO(TBARV) + CLK            'FIND TIME OF NEXT ARRIV AL
CALL SCHED(l, T)                         'SCHED NEXT ARRIV AL
IF BFLAG = 0 THEN                        'CHECK FOR SERVER BUSY 'FIND
A = FNNORM(SVCM, SVCSD)                   SERVICE TIME
CALL SCHED(2, CLK+A)                     'SCHED. END OF SERVICE
     BFLAG=l                             'SET SERVER FLAG BUSY
   ELSE
       NWAIT = NWAIT + 1                 'INCREASE NUMBER IN LINE
IF NW AIT > MAX THEN MAX = NW AIT        'CHECK FOR MAX LINE LENGTH
END IF GOTO ABC                          'BRANCH TO NEXT EVENT


'****** JOB DEPARTURE ******
DEPT: NSERVED = NSERVED + 1              'INCREASE NUMBER SERVED
IF NW AIT = 0 THEN
     BFLAG = 0                           'SET SERVER IDLE IF NONE WAITING
ELSE
  NW AIT = NW AIT -1: BFLAG = 1          'DECREASE NUMBER IN LINE 'FIND
      A = FNNORM(SVCM, SVCSD)             SERVICE TIME 'SCHED. END OF
      CALL SCHED(2, A + CLK)             SERVICE
  END IF
  GOTO ABC                               'BRANCH TO NEXT EVENT
```

—*continued*

FIGURE 16.2 SIMPAK program for example problem.

```
'****** END OF SIMULATION ******

EOS: LPRINT "LENGTH OF SIMULATION
(HOURS) "; LSIM

LPRINT "NUMBER SERVED "; NSERVED

LPRINT "AVERAGE LINE LENGTH "; CUMWT 1
LSIM

LPRINT "AVERAGE WAIT TIME (HOURS) ";
CUMWT 1 NSERVED LPRINT "FRACTION TIME
BUSY ";CUMWK 1 LSIM

LPRINT "MAXIMUM LINE LENGTH "; MAX

STOP
```

FIGURE 16.2   continued.

```
LENGTH OF SIMULATION (HOURS) 500

NUMBER SERVED 3580

AVERAGE LINE LENGTH 1.219

AVERAGW WAIT TIME (HOURS) .170

FRACTION BUSY TIME .746

MAXIMUM LINE LENGTH 10


LENGTH OF SIMULATION (HOURS) 500

NUMBER SERVED 3635

AVERAGE LINE LENGTH 1.172

AVERAGE WAIT TIME (HOURS) .161

FRACTION BUSY TIME .751

MAXIMUM LINE LENGTH 10
```

FIGURE 16.3   Output of two runs of example problem 1.

The output from two runs of this model is shown in Figure 16.3. The length of the simulation (LSIM) in both runs was 500 h. It should be remembered that the output from a stochastic simulation model has variability, and the results will differ because of this variability. This model is stored on your SIMPAK disk as EXMP1.BAS.

16.7 List Processing Initializer (LISTINIT)

When a model is going to be developed that will require list processing, the data structures for list processing have to be initialized, and this is accomplished with the statement:

CALL LISTINIT

There are no input or output arguments that are required or generated with this subprogram. Once it is called, the SIMPAK system is ready to process lists.

16.8 List Creator (GRAB)

When a list is to be created in a SIMPAK model, it is done with subprogram GRAB. There are two arguments that are used with GRAB. The first argument is the variable name that the user wants associated with the list, and the second argument is the size of the list that will be created. For example, if a list is to be created that is to have space for 12 words, and the name of the list is to be LSTNAME, the statement would be:

CALL GRAB (LSTNAME, 12)

A list capable of storing 12 pieces of data will be created, and the address of the list will be stored in a variable called LSTNAME. Each time a new list is created, the address of this new list will be stored in the first argument. For example, consider the following segment of code in Figure 16.4.

There are now three separate lists current in memory: A, B, and C. List A has 12 words, list B has 15, and list C has four. All values in the list will initially be set to zero.

16.9 List Inserter (LISTPUT)

The List Inserter subprogram puts values into selected positions in a created list. The three input arguments that must be specified are:

List name, List position for value to be stored, Value

Consider the segment of code shown in Figure 16.4. There are three lists that are current in memory after that code has been executed. Suppose we would like to place the value 2.5 into the third position of list A, the value 0.0076 in the fifth position of list B, and −129 into the third position of list C. Figure 16.5 shows the code that will perform these tasks.

An error message, "ATTEMPT TO INSERT BEYOND LIST LENGTH" will appear if a user attempts to place a value in a position that is beyond the size of the list that has been created. For example, list A was created with a list size of 12. The error message would appear if an attempt were made to place a value in position 13 or higher. The simulation will stop after the message is printed.

CALL GRAB (A, 12)

CALL GRAB (B, 15)

CALL GRAB (C, 4)

FIGURE 16.4   Segment of SIMPAK code to create three lists.

CALL LISTPUT (A, 3, 2.5) CALL LISTPUT (B, 5, .0076)

CALL LISTPUT (C, 3, -129)

FIGURE 16.5   Segment of code to place values into lists created in Figure 16.3.

16.10 List Retriever (LISTR V)

The List Retriever subprogram is used to retrieve a value that have been previously stored in a list. The three arguments that must be specified are:

List name, Position in list, Variable name that will take on value

For example, consider the lists that were created in Figure 16.4 and then had some values inserted in them in Figure 16.5. Figure 16.6 shows the code that would retrieve the values from the lists. The variables V1, V2, and V3 would be set to 2.5, 0.0076, and −129, respectively.

It should be remembered that all the positions in a list contain zero when they are created, and they will remain zero unless they are replaced by a value that is not zero in the List Insertion subprogram. If an attempt is made to retrieve a value beyond the list length, the error message "ATTEMPT TO RETRIEVE BEYOND LIST LENGTH" will be displayed, and the program will stop.

16.11 List Extender (LISTEXTN)

A list can have its length expanded by use of the List Extender subprogram (LISTEXTN). For example, in Figure 16.3, list A was created with a length of 12 words. At some point in the execution of a model, suppose the length of list A has to be expanded to a length of 21 words. There are two input arguments that have to be set before the List Extender subprogram is called. These arguments are:

Address of list to be expanded, Number of words to be added

The code that would have to be executed to accomplish the task in the example is:

CALL LISTEXTN (A, 9)

16.12 List Remover (LISTFREE)

During a simulation there might be several thousand lists created of various sizes. Each list might contain information on an entity that is processed through the model. Whenever that entity has moved completely through the model, the list that pertained to it is no longer necessary. There is only a finite amount of space available for storing all the list data in SIMPAK, and once this space has become full, no other lists can be created. The error message, "OUT OF LIST SPACE" will be displayed when an attempt is made to create a new list, and execution of the model will stop. It is necessary that the modeler remove a list from the model when it is no longer needed, and this is done with the List Remover subprogram. It returns the space that the list was using to the available list space and makes it available for new lists. The only argument that has to be specified is the address (name) of the list that is being removed. For example, suppose the user wanted to remove the three lists that were created in Fig. 16-4. This could be accomplished by the code in Figure 16.7.

```
CALL LISTRTV (A, 3, VI)
CALL LISTRTV (B, 5, V2)
CALL LISTRTV (C, 3, V3)
```

FIGURE 16.6 Segment of code to retrieve values stored in lists.

```
CAL LISTFREE (A) CALL LISTFREE (B)
CALL LISTFREE (C)
```

FIGURE 16.7 List removal of three lists created in Figure 16.3.

16.13 Example Problem 2

Jobs arrive in a random manner at a single server processing system in a random (Poisson) manner. The mean number of arrivals per hour is 15.5. The processing time (in minutes) has been found to have the following equation:

$$T = X + 0.02ta - 0.03N$$

where X is a random value from a normal distribution with a mean of 1.5 min and a standard deviation of.25 min, ta is the time of arrival (in hours) of the job in the system, and N is the number in the waiting line when service begins on the job. The arrival time, ta, can vary from 0.0 to 8.0. It is desired to keep statistics on the average processing time, the proportion of time the server is busy, and the maximum length of the line for an 8-h simulation.

Solution: The arrival time of the job has to be carried along with the job. When processing begins, it has to be used to calculate the processing time. Therefore, a list will be used in this problem. The event codes that will be used will be:

Event Code	Event
1	Customer Arrival
2	Customer Departure
−1	End of Simulation

Each job will have a list that is two words in length. The items that will be stored in the list will be:

Word	Contents
1	Event code (1 or 2)
2	Time of arrival

Figure 16.8 shows the SIMPAK listing for this example (EXMP2.BAS on Disk). A dimension statement is provided for customers waiting for service (CUSTWT). Because the service time for a customer depends on his or her arrival time, it is necessary to keep the customer identity for each one in line. In the "INITIALIZE" section of the program, the first job is scheduled. It should be noted that it is a list address for jobs that are placed in the schedule, not the event code itself. A "1" is placed in the first position in the list, and the job arrival time in the second. List addresses will always be positive numbers. The end of the simulation is scheduled for t = 8 h. The event code for this event is −1, so any time an event with a negative event code is removed from the schedule, it will be the end of the simulation.

The earliest event in the simulation is removed from the schedule in the "REMOVE EARLIEST EVENT FROM SCHEDULE" segment. The elapsed time since the last event is calculated (DELTA), and the simulation clock (CLK) is moved to the time of this next event. The busy time for the server is accumulated (BUSYTM). BUSYF is zero when the server is idle and 1 when he is working. A check is made for the end of the simulation by looking for a negative event code. If the code is not negative it must be a list address, and CODE contains the list address. The event code is in the first position of the list, and a branch to the arrival segment is made if the code is "1" or to the departure segment if it is "2."

The first task in the "JOB ARRIVES AT SERVER" segment is to create the job that will follow the one that has just arrived. A new list is created, the time of the job arrival is found, and the job is placed on the schedule. After this has been done, attention is focused on the job that has just arrived. If the server is currently busy (BUSYF = 1), the list address of the job is placed at the end of the waiting line, and control is sent back to the next event segment. If the server is not busy (BUSYF = 0), the newly arrived job can have its processing begin immediately. Its processing time is calculated, event code "2" is placed in the first position of the list, and the list address is placed on the schedule to occur at the end of the processing.

Control is sent to the "JOB DEPARTURE" segment when a job has completed processing. The number of completed jobs (NCOMP) is increased by one. The list that contained the information concerning

```
'****** INITIALIZE *******
DIM CUSTWT(100)                          'DIMENSION FOR CUST. IN LINE
RANDOMIZE TIMER: LSIM = 8                'RESEED RAN NUM GEN, SET SIM TIME
CALL INITIAL: CALL LISTINIT              'INITIALIZE SCHED & LIST PROc.
T = FNEXPO(1 / 15.5)                     'FIND TIME 1ST ARRIVAL
CALL GRAB( LISTAD, 2)                    'GET LIST FOR 1ST JOB
CALL LISTPUT (LISTAD, 2, T)             'PUT ARRVL. TIME IN 2ND POSITION
CALL LISTPUT (LIST AD, 1, 1)            'PUT "1" IN 1ST POS IN LIST
CALL SCHED(LISTAD, T)                    'SCHEDULE 1ST ARRIVAL
CALL SCHED( -1, LSIM)                    'SCHED END OF SIMULATION
'*** REMOVE EARLIEST EVENT FROM SCH ***
AA: CALL REMOVER( CODE, TCLK)            'REMOVE EARLIEST EVENT
DELTA = TCLK - CLK: CLK = TCLK          'FIND DELTA & UPDATE CLOCK.
BUSYTM = BUSYTM + DELTA * BUSYF         'UPDATE SERVER WORK TIME
IF CODE < 0 THEN GOTO XX                 'CHECK FOR END OF SIMULATION
CALLLISTRV (CODE, 1, VV)                'FIND EVENT CODE VV
IF VV= 1 THEN GOTO BB                    'BRANCH FOR ARRIVAL
IF VV= 2 THEN GOTO CC                    'BRANCH FOR DEPARTURE
'***** JOB ARRIVES AT SERVER *********
BB: CALL GRAB(LISTAD, 2)                 'GET LIST FOR NEXT JOB
T = FNEXPO(1 / 15.5) + CLK               'GET TIME TILL NEXT JOB
CALL LISTPUT (LISTAD, 2,T)              'PUT ARRVL TIME IN LIST
CALL LISTPUT (LISTAD, 1, 1)             'PUT "1" IN ISTPOS IN LIST
CALL SCHED( LISTAD, T)                   'SCHEDULE NEXT ARRIVAL
IF BUSYF = 1 THEN                        'CHECK IF SERVER BUSY
N = N + 1: CUSTWT(N) = CODE              'PUT LIST # IN LINE
IF N > MAX THEN MAX =N                   'CHECK FOR MAX LINE
ELSE                                     'JOB CAN BE PROCESSED
TA=CLK                                   'ARRIV AL TIME = CLOCK
X = FNNORM( 1.5, .25)                    'CREATE X VALUE
ST = (X + .02 * T A - .03 * N) / 60     'FIND SERVICE TIME
CALLLISTPUT( CODE, 1,2)                 'PUT "2" IN 1 ST POS
CALL SCHED(CODE, ST +CLK)                'SCHED. DEPARTURE
        BUSYF = 1                        'SERVER BUSY
END IF
GOTO AA                                  'GET NEXT EVENT
'***** JOB DEPARTURE
*****************************
```

FIGURE 16.8   SIMPAK program for example problem.

- continued

Figure 16.8 -- continued

CC: NCOMP = NCOMP + 1: BUSYF=O	'ACC. NUM FIN, SERVER IDLE
CALL LISTFREE(CODE)	'GET RID OF LIST
IF N = 0 THEN	'CHECK FOR JOBS WAITING
GOTO AA	'GO FOR NEXT EVENT
ELSE	
N = N - 1: LISTAD = CUSTWT(1)	'GET CUST. LIST IN FRONT OF LINE
FOR I= 1 TON:	'MOVE UP LINE
CUSTWT(I) = CUSTWT(I + I)	
NEXT I	
CALL LISTRTV(LISTAD, 2, TA)	'FIND ARRIV AL TIME
X = FNNORM(1.5, .25)	'CREATE X FROM NOR. DIST.
ST = (X + .02 * T A - .03 * N) / 60	'FIND SERVICE TIME
CALL LISTPUT(LISTAD, 1,2)	'PUT "2" IN ISTPOS
CALL SCHED(LISTAD, ST +CLK)	'SCHED. DEPARTURE
BUSYF= 1: GOTO AA	'SERVER BUSY, NEXT EVENT
END IF	

```
'***** END OF SIMULATION
*************************
```

XX: LPRINT "NUMBER OF JOBS COMPLETED ";
NCOMP

LPRINT "MAXIMUM NUMBER WAITING "; MAX

LPRINT "SERVER UTILIZATION "; BUSYTM /
LSIM

STOP

FIGURE 16.8 continued.

the job is no longer needed, and it is removed. The server is tentatively set idle (BUSYF = 0), and if there are no jobs waiting in line (N = 0), control is sent back to the "EARLIEST EVENT" segment. If there are jobs waiting (N > 0), the job in the first position in the line is removed from the line, all jobs are moved up one position, the server is put back in the busy status, and the job is processed.

At the "END OF SIMULATION" segment, the results of the model are printed. Figure 16.9 shows the results from two runs of the model.

NUMBER OF JOBS COMLETED 121

MAXIMUM NUMBER OF WAITING 4

SERVER UUTILIZATION .403

NUMBER OF JOBS COMPLETED 132

MAXIMUM NUMBER WAITING 4

SERVER UTILIZATION

FIGURE 16.9 Output from two runs of Example 2.

16.14 Discussion

SIMPAK can be a useful tool in constructing simulation models. It allows models to be quickly developed on almost any personal computer that has a BASIC module. The output can be tailored to suit a particular model. Many users have designed some simple animation to go with their models using the GRAPHICS commands in BASIC. The animation capabilities of many simulation languages are pushed by their developers extensively. Many people feel that these animation capabilities are overstressed. They cause the development of the model to be much more complex (usually requiring a consultant) and slow the execution time.

It is very helpful to use comment statements in your SIMPAK models, so that the logic can be followed. A single quotation mark (') appearing on a line of BASIC code indicates that the remainder of the line is a comment. The RANDOMIZE TIMER statement that was used in the example problems in this manual seeds the random number generator. An internal clock value will be used for the seed. This seed will fix the sequence of generated random numbers. There are some situations in which a modeler would like to keep the same sequence of random numbers. The statement RANDOMIZE is used in these situations. The user will be prompted to furnish the random number seed in response to the command. All random numbers that are generated are uniformly distributed in the interval [0, 1]. The random variate generator uses these uniformly distributed numbers to create values from various distributions.

When a model is first developed, it can be useful to insert a statement after the point where the earliest event in the schedule is removed. This statement will cause the event code and the time of the event to be displayed on the screen. Consider Example 1 in this chapter. The earliest event in the schedule is removed from the schedule on the line labeled ABC:

ABC: CALL REMOVER(CODE, TCLK)

Immediately after this statement, the following line could be entered:

PRINT TCLK, CODE: INPUT A$

The time of the event and the event code will be displayed on the screen, and they will remain there until the user hits the enter/return key on the keyboard. The simulation will then continue. This information can be helpful in developing the model and can quickly point out logic errors. The line should be removed from the model when the development phase is completed.

The remainder of this chapter is a reference sheet that shows the function of each of the subprograms, its required input arguments, the call statement, and the return arguments. This sheet can be helpful while a model is constructed.

16.15 SIMPAK Reference Sheet

16.15.1 Random Variable Routines

Distribution	Function Name	Arguments
Normal	FNNORM	Mean, Std. Dev.
Exponential	FNEXPO	Mean
Weibull	FNWEIB	m, Theta
Log-Normal	FNLOGN	Mean, Variance
Binomial	FNBINOM	N, p
Poisson	FNPOIS	Lambda
Uniform	FNUNIF	Min, Max
Erlang	FNERLANG	Mean, K

16.15.2 Initializers

For Schedule   CALL INITIAL
For List Processing   CALL LISTINIT

16.15.3 Scheduling

Place Event in Schedule CALL SCHED (Event Code, Time of Occurrence)
Remove Earliest Event   CALL REMOVER (Event Code, Time)

16.15.4 List Processing

Create a List   CALL GRAB(Name for List, Number of Words)
Put Value in List   CALL LISTPUT(List Name, Position, Value)
Retrieve Value   CALL LISTRTV(List Name, Position, Value)
Extend List   CALL LISTEXTN(List Name, Additional Words)
Get Rid of List   CALL FREE (List Name)

Appendix 1
Statistical Tables

The statistical tables in the Appendix were developed using a Microsoft Excel worksheet. The following functions were utilized to generate the values for the chi-square, normal, t, and F distribution tables.

Chi-square: Chiinv(probability, deg_freedom)
Where:
  Probability = Probability of an observation to the right of the critical value
  Deg_freedom = Number of degrees of freedom

Normal Normdist(x, mean, standard deviation, cumulative)
Where:
  X = critical value
  Mean = average for the normal distribution
  Standard Deviation = Standard deviation for the normal distribution
  Cumulative = Yes or No
  Note: For the standard normal Z table, the mean is 0 and the standard deviation is 1.

t TINV(probability, deg_freedom)
Where:
  Probability = Probability of an observation to the right of the critical value
  Deg_freedom = Number of degrees of freedom

F FINV(probability, deg_freedom1, deg_freedom2)
Where:
  Probability = Probability of an observation to the right of the critical value
  Deg_freedom1 = Number of degrees of freedom in the numerator
  Deg_freedom2 = Number of degrees of freedom in the numerator

Chi-Square Goodness of Fit Table for Selected α Values

d.f.	α = 0.01	α = 0.05
1	3.841	6.635
2	5.991	9.210
3	7.815	11.345
4	9.488	13.277
5	11.070	15.086
6	12.592	16.812
7	14.067	18.475
8	15.507	20.090
9	16.919	21.666
10	18.307	23.209
11	19.675	24.725
12	21.026	26.217
13	22.362	27.688
14	23.685	29.141
15	24.996	30.578
16	26.296	32.000
17	27.587	33.409
18	28.869	34.805
19	30.144	36.191
20	31.410	37.566
21	32.671	38.932
22	33.924	40.289
23	35.172	41.638
24	36.415	42.980
25	37.652	44.314
26	38.885	45.642
27	40.113	46.963
28	41.337	48.278
29	42.557	49.588
30	43.773	50.892

These values are for the right-hand side critical value with α values of 0.01 and 0.05. Practitioners will ordinarily use the 0.05 α value.

t-Distribution Critical Values for Selected α Values

d.f.	$\alpha = 0.025$
1	12.706
2	4.303
3	3.182
4	2.776
5	2.571
6	2.447
7	2.365
8	2.306
9	2.262
10	2.228
11	2.201
12	2.179
13	2.160
14	2.145
15	2.131
16	2.120
17	2.110
18	2.101
19	2.093
20	2.086
21	2.080
22	2.074
23	2.069
24	2.064
25	2.060
26	2.056
27	2.052
28	2.048
29	2.045
Inf.	1.960

These are critical values for the right-hand side of the *t* distribution. The $\alpha = 0.025$ value is used for all two-sided tests run at $\alpha = 0.05$. This includes validation hypothesis tests and confidence interval comparisons.

Table of Normal Distribution Z Values

Z	0.00	0.01	0.02	0.03	0.04	0.05	0.06	0.07	0.08	0.09
0.0	0.5000	0.5040	0.5080	0.5120	0.5160	0.5199	0.5239	0.5279	0.5319	0.5359
0.1	0.5398	0.5438	0.5478	0.5517	0.5557	0.5596	0.5636	0.5675	0.5714	0.5753
0.2	0.5793	0.5832	0.5871	0.5910	0.5948	0.5987	0.6026	0.6064	0.6103	0.6141
0.3	0.6179	0.6217	0.6255	0.6293	0.6331	0.6368	0.6406	0.6443	0.6480	0.6517
0.4	0.6554	0.6591	0.6628	0.6664	0.6700	0.6736	0.6772	0.6808	0.6844	0.6879
0.5	0.6915	0.6950	0.6985	0.7019	0.7054	0.7088	0.7123	0.7157	0.7190	0.7224
0.6	0.7257	0.7291	0.7324	0.7357	0.7389	0.7422	0.7454	0.7486	0.7517	0.7549
0.7	0.7580	0.7611	0.7642	0.7673	0.7704	0.7734	0.7764	0.7794	0.7823	0.7852
0.8	0.7881	0.7910	0.7939	0.7967	0.7995	0.8023	0.8051	0.8078	0.8106	0.8133
0.9	0.8159	0.8186	0.8212	0.8238	0.8264	0.8289	0.8315	0.8340	0.8365	0.8389
1.0	0.8413	0.8438	0.8461	0.8485	0.8508	0.8531	0.8554	0.8577	0.8599	0.8621
1.1	0.8643	0.8665	0.8686	0.8708	0.8729	0.8749	0.8770	0.8790	0.8810	0.8830
1.2	0.8849	0.8869	0.8888	0.8907	0.8925	0.8944	0.8962	0.8980	0.8997	0.9015
1.3	0.9032	0.9049	0.9066	0.9082	0.9099	0.9115	0.9131	0.9147	0.9162	0.9177
1.4	0.9192	0.9207	0.9222	0.9236	0.9251	0.9265	0.9279	0.9292	0.9306	0.9319
1.5	0.9332	0.9345	0.9357	0.9370	0.9382	0.9394	0.9406	0.9418	0.9429	0.9441
1.6	0.9452	0.9463	0.9474	0.9484	0.9495	0.9505	0.9515	0.9525	0.9535	0.9545
1.7	0.9554	0.9564	0.9573	0.9582	0.9591	0.9599	0.9608	0.9616	0.9625	0.9633
1.8	0.9641	0.9649	0.9656	0.9664	0.9671	0.9678	0.9686	0.9693	0.9699	0.9706
1.9	0.9713	0.9719	0.9726	0.9732	0.9738	0.9744	0.9750	0.9756	0.9761	0.9767
2.0	0.9772	0.9778	0.9783	0.9788	0.9793	0.9798	0.9803	0.9808	0.9812	0.9817
2.1	0.9821	0.9826	0.9830	0.9834	0.9838	0.9842	0.9846	0.9850	0.9854	0.9857
2.2	0.9861	0.9864	0.9868	0.9871	0.9875	0.9878	0.9881	0.9884	0.9887	0.9890
2.3	0.9893	0.9896	0.9898	0.9901	0.9904	0.9906	0.9909	0.9911	0.9913	0.9916
2.4	0.9918	0.9920	0.9922	0.9925	0.9927	0.9929	0.9931	0.9932	0.9934	0.9936
2.5	0.9938	0.9940	0.9941	0.9943	0.9945	0.9946	0.9948	0.9949	0.9951	0.9952
2.6	0.9953	0.9955	0.9956	0.9957	0.9959	0.9960	0.9961	0.9962	0.9963	0.9964
2.7	0.9965	0.9966	0.9967	0.9968	0.9969	0.9970	0.9971	0.9972	0.9973	0.9974
2.8	0.9974	0.9975	0.9976	0.9977	0.9977	0.9978	0.9979	0.9979	0.9980	0.9981
2.9	0.9981	0.9982	0.9982	0.9983	0.9984	0.9984	0.9985	0.9985	0.9986	0.9986
3.0	0.9987	0.9987	0.9987	0.9988	0.9988	0.9989	0.9989	0.9989	0.9990	0.9990
3.1	0.9990	0.9991	0.9991	0.9991	0.9992	0.9992	0.9992	0.9992	0.9993	0.9993
3.2	0.9993	0.9993	0.9994	0.9994	0.9994	0.9994	0.9994	0.9995	0.9995	0.9995
3.3	0.9995	0.9995	0.9995	0.9996	0.9996	0.9996	0.9996	0.9996	0.9996	0.9997
3.4	0.9997	0.9997	0.9997	0.9997	0.9997	0.9997	0.9997	0.9997	0.9997	0.9998
3.5	0.9998	0.9998	0.9998	0.9998	0.9998	0.9998	0.9998	0.9998	0.9998	0.9998

F Distribution Critical Values for 0.05 α Value

Den.\ Num.	1	2	3	4	5	6	7	8	9	10
1	161	199	215	224	230	233	236	238	240	241
2	18.51	19.00	19.16	19.25	19.30	19.33	19.35	19.37	19.38	19.40
3	10.13	9.55	9.28	9.12	9.01	8.94	8.89	8.85	8.81	8.79
4	7.71	6.94	6.59	6.39	6.26	6.16	6.09	6.04	6.00	5.96
5	6.61	5.79	5.41	5.19	5.05	4.95	4.88	4.82	4.77	4.74
6	5.99	5.14	4.76	4.53	4.39	4.28	4.21	4.15	4.10	4.06
7	5.59	4.74	4.35	4.12	3.97	3.87	3.79	3.73	3.68	3.64
8	5.32	4.46	4.07	3.84	3.69	3.58	3.50	3.44	3.39	3.35
9	5.12	4.26	3.86	3.63	3.48	3.37	3.29	3.23	3.18	3.14
10	4.96	4.10	3.71	3.48	3.33	3.22	3.14	3.07	3.02	2.98
12	4.75	3.89	3.49	3.26	3.11	3.00	2.91	2.85	2.80	2.75
15	4.54	3.68	3.29	3.06	2.90	2.79	2.71	2.64	2.59	2.54
20	4.35	3.49	3.10	2.87	2.71	2.60	2.51	2.45	2.39	2.35
24	4.26	3.40	3.01	2.78	2.62	2.51	2.42	2.36	2.30	2.25
30	4.17	3.32	2.92	2.69	2.53	2.42	2.33	2.27	2.21	2.16
40	4.08	3.23	2.84	2.61	2.45	2.34	2.25	2.18	2.12	2.08
60	4.00	3.15	2.76	2.53	2.37	2.25	2.17	2.10	2.04	1.99
120	3.92	3.07	2.68	2.45	2.29	2.18	2.09	2.02	1.96	1.91
1000	3.85	3.00	2.61	2.38	2.22	2.11	2.02	1.95	1.89	1.84

Den. \Num.	12	15	20	24	30	40	60	120	1000
1	243	245	248	249	250	251	252	253	254
2	19.41	19.43	19.45	19.45	19.46	19.47	19.48	19.49	19.49
3	8.74	8.70	8.66	8.64	8.62	8.59	8.57	8.55	8.53
4	5.91	5.86	5.80	5.77	5.75	5.72	5.69	5.66	5.63
5	4.68	4.62	4.56	4.53	4.50	4.46	4.43	4.40	4.37
6	4.00	3.94	3.87	3.84	3.81	3.77	3.74	3.70	3.67
7	3.57	3.51	3.44	3.41	3.38	3.34	3.30	3.27	3.23
8	3.28	3.22	3.15	3.12	3.08	3.04	3.01	2.97	2.93
9	3.07	3.01	2.94	2.90	2.86	2.83	2.79	2.75	2.71
10	2.91	2.85	2.77	2.74	2.70	2.66	2.62	2.58	2.54
12	2.69	2.62	2.54	2.51	2.47	2.43	2.38	2.34	2.30
15	2.48	2.40	2.33	2.29	2.25	2.20	2.16	2.11	2.07
20	2.28	2.20	2.12	2.08	2.04	1.99	1.95	1.90	1.85
24	2.18	2.11	2.03	1.98	1.94	1.89	1.84	1.79	1.74
30	2.09	2.01	1.93	1.89	1.84	1.79	1.74	1.68	1.63
40	2.00	1.92	1.84	1.79	1.74	1.69	1.64	1.58	1.52
60	1.92	1.84	1.75	1.70	1.65	1.59	1.53	1.47	1.40
120	1.83	1.75	1.66	1.61	1.55	1.50	1.43	1.35	1.27
1000	1.76	1.68	1.58	1.53	1.47	1.41	1.33	1.24	1.11

Den. = Degrees of freedom for the denominator
Num. = Degrees of freedom for the numerator
The F distribution critical values are for right-hand-side values at $\alpha = 0.05$. When the two-sided F test statistic is calculated with the larger variance divided by the smaller variance, a 0.10 α test uses only the right-hand-side $\alpha = 0.05$ values.

Duncan Multiple-Range Test r_p Values at $\alpha = 0.05$

d.f.\ p	2	3	4	5	6	7	8	9	10
24	2.92	3.07	3.16	3.23	3.28	3.31	3.35	3.37	3.39
30	2.89	3.03	3.13	3.20	3.25	3.29	3.32	3.35	3.37
40	2.86	3.01	3.10	3.17	3.22	3.27	3.30	3.33	3.35
60	2.83	2.98	3.07	3.14	3.20	3.24	3.28	3.31	3.33
120	2.80	2.95	3.04	3.12	3.17	3.22	3.25	3.29	3.31
Inf.	2.77	2.92	3.02	3.09	3.15	3.19	3.23	3.27	3.29

The degrees of freedom is the same value as the number of degrees of freedom for the mean square error (MSE) from the ANOVA output. Practitioners should round to the lower listed number of degrees of freedom.

Appendix 2
Course Outline

Lecture 1: Introduction

1.1 Agenda

- An introduction to simulation modeling and analysis
- Other types of simulation
- Purposes of simulation
- A discussion of the advantages and disadvantages of simulation
- Famous simulation quotes

1.2 Examples of Systems That Can Be Simulated

- Manufacturing systems
- Service systems
- Transportation systems

1.3 Examples of Manufacturing Systems

- Machining operations
- Assembly operations
- Materials handling equipment
- Warehousing

1.4 Examples of Service Systems

- Hospitals and medical clinics
- Retail stores
- Food or entertainment facilities
- Information technology
- Customer order systems

1.5 Examples of Transportation Systems

- Airport operations
- Port shipping operations
- Train and bus transportation
- Distribution and logistics

1.6 Other Types of Simulation Models

- Management training simulators
- Equipment operation simulators

1.7 Purposes of Simulation

- Gaining insight into the operation of a system
- Developing operating or resource policies to improve system performance
- Testing new concepts and/or systems prior to implementation
- Gaining information without disturbing the actual system

1.8 Advantages to Simulation

- Experimentation in compressed time
- Reduced analytical requirements
- Models easily demonstrated

1.9 Disadvantages to Simulation

- Simulation cannot give accurate results when the input data are inaccurate
- Simulation cannot provide easy answers to complex problems
- Simulation cannot solve problems by itself

1.10 Other Considerations

- Simulation model building can require specialized training
- Simulation modeling and analysis can be costly
- Simulation results involve a lot of statistics

1.11 Famous Simulation Quotes

- "You cannot study a system by stopping it."
- "Run it again."
- "Is that some kind of game you're playing?"

Lecture 2: Basic Simulation Process

2.1 Agenda

- Problem definition
- Project planning
- System definition/model formulation
- Input data collection and analysis
- Model translation
- Verification
- Validation
- Experimentation
- Analysis
- Conclusions and recommendations

2.2 Problem Definition

- Basic understanding of what we are trying to find out and/or solve
- Orientation
- Objectives

2.3 Project Planning

- Work breakdown structure
- Linear responsibility chart
- Gantt chart

2.4 Work Breakdown Structure

- Division of work into individual work packages for which responsibility can be assigned

2.5 Linear Responsibility Chart

- Who is responsible for what and to what degree
- Appended to the right of the WBS
- Uses codes to assign responsibility and accountability

2.6 Gantt Chart

- Duration
- Sequencing

2.7 System Definition/Model Formulation

- Flow chart of the system
- Identification of system components to be modeled
- Input data requirements
- Output measures of performance
- How much detail
- How much flexibility
- Variables
- Interactions

2.8 Input Data Collection and Analysis

- Collection
- Existing data
- New data
- Analysis
- Fitting theoretical distributions to the raw data

2.9 Model Translation

- Programming the model into a computer programming language
- Programming language choices
- General purpose language
- Simulation package

2.10 Verification

- Ensuring that the model behaves in the way it was intended
- Debugging

2.11 Validation

- Ensuring that the model behaves the same as the real system
- Face validity
- Statistical validity

2.12 Experimentation

- Provide answers to project objectives
- Design alternative system configurations

2.13 Analysis

- Determine how many simulation runs to make
- Execute simulation runs
- Statistical comparison of alternative system performance

2.14 Conclusions and Recommendations

- Experimental results
- Identification/recommendation of the best course of action
- Justification as to why is it the best
- Report and/or presentation

Lecture 3: Introduction to Models

3.1 Agenda

- Basic simulation concepts
- A comprehensive example of a manual simulation

3.2 Basic Simulation Concepts

- Basic simulation model components
- Simulation event lists
- Measures of performance statistics

3.3 Examples of Simple Systems

- A customer service center with one representative
- A barber shop with one barber
- A mortgage loan officer in a bank
- A piece of computer numerically controlled machine in a factory
- An ATM machine

3.4 Each Simple System Contains

- Entities
- Queues
- Resources

3.5 Entities

- Batches
- Interarrival times
- Attributes

3.6 Queues

- Lines for entities
- Queue priorities
- Queue behavior

3.7 Resources

- Resource examples
- Resource states
- Service delays

3.8 Resource Examples

- Customer service representatives
- Barbers
- Loan officers
- Factory machines
- ATMs

3.9 Resource States

- Idle
- Busy
- Inactive
- Failed

3.10 Inactive State

- Scheduled work breaks
- Meals
- Vacations
- Preventive maintenance periods

3.11 Failed State

- Broken machines
- Inoperative equipment

3.12 Resource Service Delays

- Time to total an order
- Receive payment
- Process a loan
- Machine a part

3.13 Simulation Event List

- Keeps track of all system events
- System events change the status of the system
- Discrete systems hop between events
- Arrivals
- Service starts
- Service ends

3.14 Measures of Performance Statistics

- Observational or time dependent
- System time
- Queue time
- Time average number in queue
- Utilization

3.15 System Time

- Observational
- Begins when entity arrives in the system
- Ends when the entity leaves the system
- Average obtained by dividing by number of entities

$$\text{Average System Time} = \frac{\sum_{i=1}^{n} T_i}{n}$$

where

T_i = The system time for an individual entity, arrival time – departure time

n = The number of entities that are processed through the system

3.16 Queue Time

- Observational
- Begins when entity arrives in the queue
- Ends when the entity leaves the queue and begins service
- Average obtained by dividing by number of entities

$$\text{Average Queue Time} = \frac{\sum_{i=1}^{n} D_i}{n}$$

where

D_i = The queue time for an individual entity, queue arrival time – service begin time

n = The number of entities that are processed through the queue

3.17 Time Average Number in Queue

- Time-dependent statistic
- Number of entities observable in the queue at any given time
- Function of how long a given number of entities are in the queue
- Can have fractional values or values less than one
- Average obtained by dividing by total time

$$\text{time average number in } Q = \frac{\int_0^T Q\,dt}{T}$$

where

Q = The number in the queue for a given length of time

dt = The length of time that Q is observed

T = The total length of time for the simulation

3.18 Utilization

- Time-dependent statistic
- Resource is either idle or busy
- Idle = 0, busy = 1
- Average obtained by dividing by total time

$$\text{Average resource utilization} = \frac{\int_0^T B\,dt}{T}$$

where

B = Either 0 for idle or 1 for busy

dt = The length of time that B is observed

T = The total length of time for the simulation

3.19 Manual Simulation Example

- Single server, single queue
- Interarrival times (min): 1, 4, 2, 1, 8, 2, 4, 3
- Service times (min): 2, 5, 4, 1, 3, 2, 1, 3
- Calculate:
  - Average time in system
  - Time average number in queue
  - Average utilization

3.20 Additional Basic Simulation Issues

- Simulation can be difficult to understand by competent programmers
- Programmers think from an external viewpoint
- View simulation from the perspective of the entity

Lecture 4 : Problem Formulation

4.1 Agenda

- A formal problem statement
- An orientation of the system
- The establishment of specific project objectives

4.2 Formal Problem Statement

- Increasing customer satisfaction
- Increasing throughput
- Reducing waste
- Reducing work in progress

4.3 Tools for Developing the Problem Statement

- The Fishbone chart
- The Pareto chart

4.4 The Fishbone Chart

- Cause and effect diagram, man-machine-material chart, and Isikawa chart
- Used to identify the cause of the problem or effect of interest
- Looks like a fish skeleton
- Bones are possible causes or source of problem
- For a manufacturing process there are major bones for:
  - Man
  - Machine
  - Material
  - Methods
- Major bones can have sub-bones
  - Man — shift 1, shift 2, shift 3, etc
  - Raw materials — component 1, component 2, etc

4.5 The Pareto Chart

- Useful for several sources or causes of the problem or problems of interest
- A few factors are the cause of many problems
- 80–20 rule
- Need to know:
  - The number of each source of defect
  - The cost associated with the defect
- Multiplied together, the true cost of each defect can be determined

4.6 Pareto Example

- Large number of inexpensive easily repairable defects
- Large number of expensive defects or nonrepairable product
- Small number of inexpensive easily repairable defects
- Small number of expensive defects or nonrepairable product

4.7 Orientation

- Familiarization with the system
- View from a simulation standpoint

4.8 Orientation Process

- Initial orientation visit
- Detailed flow orientation visit
- Review orientation visit

4.9 Initial Orientation Visit

- High level understanding of the system
- Basic inputs and outputs
- Guided tour by system representative
- Avoiding information overload
- Noting processes requiring in depth orientation

4.10 Detailed Flow Orientation Visit

- Gather detailed information on the operation of the system
- The types of entities that are processed by the system
- The number and type of queues
- The number and type of system resources
- The sequence of processes as experienced from the eyes of the entities
- How the system performance can be measured
- Do not record input data
- May require multiple visits
- Does not necessarily have to have system representative present

4.11 Review Orientation Visit

- Ensure that the understanding of the system operation is consistent with the actual system
- May observe new aspects of the system
- Should be performed with system representative

4.12 Aircraft Loading Orientation Example

- The number of passengers on the plane
- The type of passenger, regular, family, handicapped
- The rate at which passengers travel down the aisle
- How far back the passenger must travel down the aisle
- Whether the passenger has an aisle, center, or window seat
- The percentage of passengers with carry on bags
- The time it takes passengers to load each piece of luggage in the overhead bins

4.13 Often Overlooked Details

- The number of times a passenger must get out of his or her seat to let other passengers by
- How long it takes a passenger from an aisle or center seat to let the other passenger by
- Whether the passenger has to search the luggage compartments for space
- How long it takes the passenger to search for the space and place his or her luggage

4.14 Tools for the Practitioner's Orientation

- Pen and pad
- Digital camera
- Camcorder

4.15 Project Objectives

- Developed in close cooperation with the individuals commissioning the simulation study
- Dynamic

4.16 Example Project Objectives

- Performance-related operating policies
- Performance-related resources policies
- Cost-related resource policies
- Equipment capabilities evaluation

4.17 Performance-Related Operating Policies

- Interested in making the best use of the existing resources
- Priority scheduling
- Scheduling quoting
- Preventive maintenance

4.18 Performance-Related Resources Policies

- Determining the best performance among different resource level alternatives
- Smallest average system times
- The least number of jobs that are tardy with respect to their due dates

4.19 Cost-Related Resource Policies

- Reducing costs while still maintaining a given level of performance
- How many or how few resources are needed at any given point in time

4.20 Equipment Capabilities Evaluation

- Test or determine the capabilities of a proposed type of new equipment
- Some organizations requiring equipment vendors to demonstrate simulation models

4.21 Decision-Making Tools for Determining Project Objectives

- Brainstorming
- Nominal group technique
- Delphi process

4.22 Brainstorming

- Creating a storm of ideas
- Ideas generated by being exposed to others' ideas
- No participant criticizing the validity of another participant's ideas

4.23 Brainstorming Preparation

- Assemble in one physical location
- 5 to 12 participants needed for developing synergy
- Ideas are recorded on a large-format medium

4.24 Brainstorming Process

- Orientation on the brainstorming rules
- Leader puts one idea for the project objectives on the board
- Each group member puts another idea on the board in turn
- Members who do not have an idea pass
- The process repeats until all group members pass
- Result is a list of project objective ideas

4.25 Electronic Brainstorming

- E-mail discussion list
- Electronic bulletin board

4.26 Nominal Group Technique

- Identify the most important project objectives
- Each group member is given an equal number of equal-weight votes
- Group members vote on the project objective ideas
- All votes may be cast for a single idea or spread among a number of ideas
- The ideas with the most votes are used as the project objectives

4.27 Delphi Process

- Undue political influence is present that interferes with the proper selection of the project objectives
- Also used to identify the most important project objectives
- All voting is conducted anonymously

4.28 Delphi Process Procedure

- Administrator distributes a list of all of the brainstorming ideas to the participants
- Participants cast one vote for a particular idea
- Administrator tabulates votes for each idea
- Most popular ideas are sent forward to the next round
- When the ideas converge to a limited number the process is complete

Lecture 5: Project Planning I

5.1 Agenda

- Project management concepts
- Project manager functions
- Project stakeholders
- Work breakdown structures
- Linear responsibility charts
- Gantt charts

5.2 Project Management Concepts

- Project parameters
- Project lifecycles
- Project stakeholders

5.3 Project Parameters

- Time
- Cost
- Technical performance

5.4 Project Life Cycles

- Conceptual
- Planning
- Execution
- Completion

5.5 Project Stakeholders

- Individuals with some sort of vested interest in the project
- Action or lack of action on part of the stakeholder can affect the project
- Internal
- External

5.6 Internal Project Stakeholders

- Individuals who are directly associated with the simulation project team
- Individuals over whom the project manager exercises some degree of control
- Examples

5.7 Examples of Internal Project Stakeholders

- Practitioner/project manager
- Analysts
- Statisticians
- Data collectors
- Functional managers who supervise the project team members

5.8 External Project Stakeholders

- Individuals who are not directly associated with the simulation project team
- Individuals over whom the project manager does not exercise some degree of control
- Examples

5.9 Examples of External Project Stakeholders

- Practitioner/project manager's manager
- Managers of the process being modeled
- Workers in the process being modeled
- Simulation software vendor/developer
- Equipment vendors and manufacturers
- Regulatory agencies
- Competitors
- Public or community in general

5.10 Project Manager Functions

- Planning
- Organizing
- Motivating
- Directing
- Controlling

5.11 Planning

- Work breakdown structure
- Gantt chart

5.12 Organizing

- Identifying, acquiring, and aligning personnel and resources for the project team
- Traditional organizational chart
- Linear responsibility chart

5.13 Motivating

- Motivating the project team
- A highly motivated team has a greater probability of project success
- Motivation theories

5.14 Motivating Theories

- Maslow's hierarchy of needs theory
- Alderfer's ERG theory
- Herzberg's two-factory theory

5.15 Maslow's Hierarchy Theory

- Five-level hierarchy
- Must fulfill lower levels before upper level needs become important
  - Self-actualization
  - Self-esteem and respect
  - Belongingness
  - Safety and security
  - Physiologic

5.16 Alderfer's ERG Theory

- Similar to Maslow, but only three levels
- Must also fulfill lower levels before upper level needs become important
  - Growth
  - Relatedness
  - Existence
- Frustration-regression

5.17 Herzberg's Two-Factor Theory

- Hygiene factors and motivating factors
- Hygiene factors
  - Absence is highly dissatisfying
  - Presence is not highly motivating
- Motivating factors
  - Absence is not dissatisfying
  - Presence is highly motivating

5.18 Hygiene Factors

- Good working conditions
- Pay
- Fringe benefits
- Job security

5.19 Motivators

- Feelings of achievement
- Recognition
- Responsibility
- Opportunity for advancement
- Opportunity for personal growth

5.20 Directing

- Leading the project activities
- Ability to influence others to accomplish a specific goal
- Authority
  - *De jure*
  - *De facto*
- Cannot assign responsibility without authority

5.21 Leadership-Motivation Theories

- Expectancy theory
- Equity theory
- Contingency leadership theory
- Hershey–Blanchard situational leadership theory

5.22 Hershey–Blanchard Situational Leadership Theory

- Leadership technique is based on technical capability and willingness of team members

5.23 Combinations of Team Members

- Able and willing
- Able but unwilling
- Unable but willing
- Unable and unwilling

5.24 Leadership Style

- Able and willing—delegating
- Able but unwilling—supporting
- Unable but willing—coaching
- Unable and unwilling—directing

5.25 *De Jure* Authority

- Awarded by official written organizational document
- Legal power to acquire and use resources
- Project manager, functional manager

5.26 *De Facto* Authority

- Based on an individual's personal knowledge or skills related to a particular task
- Other individuals comply out of respect for the individual's unique knowledge or skills
- Engineer, analyst, technician, mechanic

5.27 Controlling

- Setting project standards
- Observing performance
- Comparing the observed performance with the project standards
- Taking corrective action

Lecture 6: Project Planning II

6.1 Developing the Simulation Project Plan

- A work breakdown structure (WBS)
- A linear responsibility chart (LRC)
- A Gantt chart

6.2 Work Breakdown Structure (WBS)

- Separation of project tasks into successively lower divisions of subtasks
- Tasks are assigned numbers
- Subtask levels within a task are separated by decimals
- Need at least two levels to be meaningful
- Can subdivide until the subtask is a work package
- An element of work for which authority and responsibility can be assigned

6.3 Linear Responsibility Chart (LRC)

- Alternative to traditional organizational charts
- Illustrates who participates and to what degree for each WBS task
- Codes represent degrees of participation
  - P = primary responsibility
  - S = secondary responsibility
  - W = worker
  - A = approval
  - R = review

6.4 Gantt Chart

- Horizontal bar chart
- Each task is an individual bar
- Duration and sequence of project tasks
- Project task relationships

6.5 Project Task Relationships

- Finish to start
- Start to start
- Finish to finish

6.6 Advanced Project Management Concepts

- PERT
- CPM
- Networks

6.7 Network Terminology

- Activities
- Events
- Nodes
- Arcs
- Paths
- Critical path

Lecture 7: System Definition

7.1 Agenda
- Determining the system classification
- How much of the system to model
- What components and events to model
- What input data to collect
- What output data to generate with the model

7.2 System Classifications
- Discrete vs. continuous vs. combined
- Terminating vs. nonterminating

7.3 Discrete vs. Continuous vs. Combined
- Discrete
  - System events occurred according to discrete jumps in the time clock
  - Arrivals, service starts, service ends
- Continuous
  - The status of components are continuously changing with respect to time
  - Use differential equations to track weights and concentrations
- Combined
  - Contains both continuous and discrete components
  - Fluids change to discrete units such as cans or boxes

7.4 Terminating vs. Nonterminating
- Terminating systems
  - Have a natural terminating event
  - Do not keep entities in the system from one time period to the next
- Nonterminating systems
  - Have a terminating event, but keep entities in the system between time periods
  - Do not have a terminating event and run continuously
- Statistical analysis approach differs

7.5 Examples of Terminating-Type Systems
- Stores
- Restaurants
- Banks
- Airline ticket counters

7.6 Examples of Nonterminating-Type Systems
- Most manufacturing facilities
- Repair facilities
- Hospitals

7.7 High-Level Flow Charts

- Essential tool for defining the system
- Helps obtain a fundamental understanding of the system logic
- Graphically depicts how the major components and events interact
- Based on previous observations from orientation activities
- Require a certain level of discipline to create before jumping into programming

7.8 Standard Flow Chart Symbols

- Oval
- Rectangle
- Tilted parallelogram
- Diamond

7.9 Oval

- Used to designate both the start and stop processes
- First and last symbol on the chart
- Start has only one way out from the bottom or side
- Stop has only one way in from the top or side

7.10 Rectangle

- Used to represent general purpose processes
- Entered from the top or the left side
- Exited from either the bottom or right side
- A service time delay

7.11 Tilted Parallelogram

- Used for processes which involve some sort of input or output
- Entered from the top or the left side
- Exited from either the bottom or right side
- Creation or arrival of entities into a system

7.12 Diamond

- Used to represent a decision in the flowchart logic
- One input connector
- Two output connectors
- Input connector should come into the top vertex
- Output can leave through either of the side vertices or the bottom vertex
- Output connectors are labeled as either true or false or yes or no
- Only one output path may be taken at a given time

7.13 Sample High-Level Flow Chart

- A single-queue, single-server system

7.14 System Components To Model

- The model must contain enough content to be properly understood
- A limited amount of time is available for most projects
- Begin with a coarse model and add refinements later

7.15 Specific System Components

- Personnel
- Machines
- Transporters
- Conveyors

7.16 Personnel

- Sales clerks
- Customer service representatives
- Machine operators
- Material handlers

7.17 Machines in Service Systems

- Computer or network systems
- Automatic teller
- Ticket machines
- Scanners
- X-ray machines

7.18 Machines in Manufacturing Systems

- Computer numerically controlled mills
- Machining centers
- Lathes
- Turning centers
- Robots

7.19 Transporters in Service Systems

- Airplanes
- Buses
- Trains

7.20 Transporters in Manufacturing Systems

- Forklifts
- Hand trucks
- Dollies
- Automatically guided vehicles

7.21 Conveyors in Service Systems

- Moving sidewalks
- Escalators
- Chair lifts

7.22 Conveyors in Manufacturing Systems

- Overhead crane systems
- Fixed production assembly lines

7.23 Processes and Events in Service Systems

- Arrival of customers at a processing area
- Customer queue behavior
- Service processing
- Payment for the goods

7.24 Arrival of Customers at a Processing Area

- Interarrival times of batch
- Batch sizes

7.25 Queue Behavior

- Types of queues
- Queue priorities
- Queue entity behavior

7.26 Types of Queues

- Parallel queues
- Single snake queues

7.27 Queue Priorities

- First-in–first-out (FIFO)
- Last-in–first-out (LIFO)
- Shortest processing time (SPT)
- Longest processing time (LPT)
- Lowest value first (LVF)
- Highest value first (HVF)
- User-defined rules

7.28 Queue Entity Behaviors

- Balking
- Reneging
- Jockeying

7.29 Service Processes

- Retail service checkout processes
- Banking service processes
- Restaurant service processes
- Airline ticket counters service processes

7.30 Retail Service Checkout Processes

- Calculating the cost of goods
- Payment

7.31 Calculating the Cost of Goods

- Number of goods being purchased
- Type of goods being purchased

7.32 Payment Process

- Cash
- Check
- Credit
- Debit
- Account

7.33 Banking Service Processes

- Deposits
- Withdrawals
- Money orders
- Cashier's checks

7.34 Other Types of Banking Service Processes

- Opening new accounts
- Closing current accounts
- Safety deposit box transactions
- Mortgages

7.35 Restaurant Service Processes

- Placing an order
- Waiting for the order
- Consuming the order
- Augmenting the order
- Paying for the order

7.36 Airline Ticket Counters Service Processes

- Purchasing tickets
- Checking in
- Changing tickets
- Changing seats

7.37 Purchasing Tickets

- Determining a suitable flight itinerary
- Payment
- Issuing the tickets
- Checking in luggage

7.38 Checking In

- Issuing tickets
- Checking in luggage

7.39 Changing Tickets

- Determining the new flight itinerary
- Canceling the old flight
- Issuing the tickets
- Rerouting baggage

7.40 Changing Seats

- Successful seat change
- Unsuccessful seat change

7.41 Manufacturing System Processes and Events

- Types of work orders
- Machine queue behavior
- Machine processing
- Machine buffers
- Material transportation
- Machine failures
- Preventive maintenance
- Product inspection failures

7.42 Types of Work Orders

- Raw materials
- Components
- Manufacturing processes

7.43 Machine Queue Behavior

- Most regular queue priority schemes can also be used for manufacturing
- SPT
- HVF

7.44 Machine Processing

- Parallel
- Serial

7.45 Machine Buffers

- Limited amount of space between the individual processing machines
- Blocking
- Starving

7.46 Material Transportation by Vehicle

- Waiting until the transporter arrives at the work-in-progess's location
- Loading the work-in-progress onto the transporter
- Actual transportation process
- Unloading from the transporter

7.47 Material Transportation by Conveyor

- Waiting to access the conveyor
- Loading the conveyor
- Actual transportation process
- Unloading from the conveyor

7.48 Machine Failures

- Broken components
- Jammed machines

7.49 Preventive Maintenance

- Calibration
- Tooling replacement
- Lubrication
- Cleaning

7.50 Product Inspection Failures

- When the product becomes defective
- Reworked
- Scrapped

7.51 Events Not To Model

- Events with very limited impact on the system outputs
- As a result of the small importance or infrequent occurrence of the event
- Discovery of an explosive device in a security checkpoint system
- Power outage in a manufacturing facility
- Bus involved in an accident
- Workers going on strike

7.52 Types of Input Data to Collect

- Entity related
- Resource related

7.53 Entity-Related Input Data

- Interarrival times
- Batch sizes
- Classifications
- Balking, reneging, and jockeying
- Movement times

7.54 Resource-Related Input Data

- Service times
- Break times
- Failure rates
- Scheduled maintenance
- Movement times

7.55 Output Data

- Measures of performance
- Average time in the system
- Average time in a queue
- Time average number in queue
- Average utilization rates
- Counters

Lecture 8: Input Data Collection and Analysis

8.1 Agenda

- The use of input data in simulation
- Sources for input data
- Collecting input data
- Deterministic versus probabilistic input data
- The identification of types of input data
- Common input data distributions
- Analyzing input data

8.2 The Use of Input Data in Simulation

- Observe input data
- Fit to theoretical distribution
- Generate data from theoretical distribution

8.3 Sources for Input Data

- Historical records
- Manufacturer specifications
- Vendor claims
- Operator estimates
- Management estimates
- Automatic data capture
- Direct observation

8.4 Collecting Input Data

- Manually or electronic devices
- Data collection devices
- Time collection mode and units
- Other data collection considerations

8.5 Manual Data Collection Devices

- Stop watch
- Clipboard
- Form

8.6 Electronic Data Collection Devices

- Time studies board
- Notebook computer
- Video camcorder

8.7 Time Collection Mode and Units

- Mode
  - Absolute
  - Relative

- Units
  - Hours
  - Minutes
  - Seconds
  - Decimal format

8.8 Other Data Collection Considerations

- You want unbiased data
- You do not want to disrupt the process

8.9 Deterministic vs. Probabilistic Input Data

- Deterministic
- Probabilistic

8.10 Deterministic

- Data occurs in the same manner or in a predictable manner each time
- Data only need to be collected once, since it never varies in value
- Examples
  - Computer numerically controlled machining program processing times
  - Preventive maintenance intervals
  - Conveyor velocities

8.11 Probabilistic

- Process does not yield the same value
- Follows some sort of probability distribution
- Examples
  - Interarrival times
  - Customer service processes
  - Repair times

8.12 Discrete vs. Continuous Input Data

- Discrete
- Continuous

8.13 Discrete

- Can take only certain values
- Usually this means a whole number
- Examples
  - The number of people who arrive in a system as a group or batch
  - The number of jobs processed before a machine experiences a breakdown

8.14 Continuous

- Can take any value in the observed range
- Fractional numbers are a definite possibility
- Examples
  - Time between arrivals
  - Service times
  - Route times

8.15 Input Data Distributions

- Common distributions
- Less common distributions
- Offset combination distributions

8.16 Common Input Data Distributions

- Bernoulli
- Uniform
- Poisson
- Exponential
- Normal
- Triangular

8.17 Bernoulli

- Random occurrence with one of two possible outcomes
- Success or failure
- Examples
  - Pass/fail inspection processes
  - First class vs. coach passengers
  - Rush vs. regular priority orders

$$mean = p$$

$$\text{variable} = p\,(1 - p)$$

where

$$p = \text{The fraction of successes}$$

$$(1\text{-}p) = \text{The fraction of failures}$$

8.18 Uniform

- Specific range of possible values
- Each individual value is equally likely to be observed
- Single six-sided dice
- First cut for modeling the input data with little knowledge of the process

$$mean = \frac{(a + b)}{2}$$

$$\text{variance} = \frac{(b - a)^2}{12}$$

where

$$a = \text{Minimum value}$$

$$b = \text{Maximum value}$$

8.19 Poisson

- Used to model a random number of events that will occur in an interval of time

- One parameter lambda
- Mean and variance are both equal to lambda
- Example
  - How many failures will occur in a given period of time
  - Number of entities in a batch

$$p(x) = \frac{e^{-\lambda}\lambda^x}{x!}$$

where

λ = Mean and variance

x   = Value of the random variable

8.20 Exponential

- Associated with proven or assumed random or Poisson processes
- Interarrival times of observations in Poisson processes are exponentially distributed
- Examples
  - Interarrival of customer interarrival of orders
  - Interarrival of machine breakdowns or failures

$$mean = B$$

$$\text{variance} = B^2$$

The probability is represented by the following formula:

$$f(x) = \frac{1}{B}e^{-x/B}$$

where

B = The average of the data sample

x = The data value

Also,

$$x = -B * \ln(1 - F(x))$$

where

$F(x)$ = Percentage of the data will exist with a value less than or equal to x

8.21 Triangular

- Complete knowledge of the system is not known
- Suspicion that the data is not uniformly distributed; may be normal
- Minimum possible value, the most common value, and the maximum possible value
- Examples
  - Manufacturing processing times
  - Customer service times
  - Travel times

$$mean = \frac{a+m+b}{3}$$

$$variance = \frac{(a^2 + m^2 + b^2 - ma - ab - mb)}{18}$$

where

a = Minimum value

m = Most common value, the mode

b = Maximum value

8.22 Normal

- Many service processes follow the normal distribution
- Many processes actually consist of a number of subprocesses
- Mean and standard deviation
- Symmetric
- Examples
    - Manufacturing processing times
    - Customer service times
    - Travel times

$$f(x) = \frac{1}{\sigma\sqrt{2\pi}} e^{-(x-\mu)^2/2\sigma^2}$$

where

μ = Mean

σ = Standard Deviation

8.23 Less Common Input Data Distributions

- Weibull
- Gamma
- Beta
- Geometric

8.24 Weibull

- Used to represent distributions which cannot have values less than zero
- Occurs with normal distribution that represent service or process times close to 0
- α shape parameter and a β scale parameter

The mean and variance for the Weibull distribution are represented mathematically by:

$$mean = \frac{\beta}{\alpha}\Gamma(\frac{1}{\alpha})$$

$$variance = \frac{\beta^2}{\alpha}\left\{2\Gamma\left(\frac{2}{\alpha}\right) - \frac{1}{\alpha}\left[\Gamma\left(\frac{1}{\alpha}\right)\right]^2\right\}$$

where

α = Shape parameter

β = Scale parameter

$$\Gamma = \int_0^\infty x^{\alpha-1}e^{-x}dx$$

$$f(x) = \alpha\beta^{-\alpha}x^{\alpha-1}e^{-(x/\beta)^\alpha}, \text{ for } x > 0, 0 \text{ otherwise}$$

where

α = Shape parameter

β = Scale parameter

8.25 Gamma

- Used to represent distributions which cannot have values less than zero
- Occurs with normal distribution that represent service or process times close to 0
- α shape parameter and a β scale parameter
- When $\alpha = 1$, distribution becomes exponential

$$mean = \alpha\beta$$

$$variance = \alpha\beta^2$$

where

α = Shape parameter

β = Scale parameter

$$f(x) = \frac{1}{\beta^\alpha\Gamma(\alpha)}x^{\alpha-1}e^{-x/\beta}, \text{ for } x > 0, 0 \text{ otherwise}$$

where

α = Shape parameter

β = Scale parameter

$$\Gamma = \int_0^\infty x^{\alpha-1}e^{-x}dx$$

8.26 Beta

- Able to cover only the range between 0 and 1
- Possible to offset it and or scale it with a multiplier value
- May fit data that is skewed to the right rather than the left
- This represents some maximum time that can be taken for a particular process
- Shape parameter α and the shape parameter β

$$mean = \frac{\alpha}{\alpha+\beta}$$

$$variance = \frac{\alpha\beta}{(\alpha+\beta)^2(\alpha+\beta+1)}$$

$$f(x) = \frac{\Gamma(\alpha+\beta)}{\Gamma(a)\Gamma(\beta)}x^{\alpha-1}(1-x)^{\beta-1}, \ for \ 0 < x < 1, 0 \ elsewhere$$

where

α = Shape parameter 1
β = Shape parameter 2

$\Gamma = \int_0^\infty x^{\alpha-1}e^{-x}dx$

8.27 Geometric

- Discrete
- One parameter p
- P is considered as the probability of success on any given attempt
- 1–p is the probability of failure on any given attempt
- Examples
  - Arrival batch sizes
  - Number of items inspected before a failure is encountered

$$mean = \frac{1-p}{p}$$

$$p(x) = p(1-p)^{x-1}, x = 1, 2...$$

where

p = probability of a success.

8.28 Offset Combination Distributions

- Some types of input data are actually a combination of a deterministic component and a probabilistic component
- Minimum time which constitutes the deterministic component
- Remaining component of the time follows some sort of distribution
- Examples
  - Technical telephone support
  - Oil changes for cars
  - Flexible manufacturing system cycles

8.29 Analyzing Input Data

- Determining underlying theoretical distribution
- Based on some sort of comparison between
  - The observed data distribution
  - A corresponding theoretical distribution
- If the difference is small, the data could have come from the theoretical distribution

8.30 Methods

- Graphic approach
- Chi-square test
- Kolmogorov–Smirnov test
- Square error

8.31 Graphic Approach

- Create a histogram of observed data
- Create a histogram for the theoretical distribution
- Visually compare the two histograms for similarity
- Make a qualitative decision as to the similarity of the two data sets
- Questionable

8.32 How To Decide How Many Cells to Use

- Equal interval approach
- Equal probability approach
    Use a maximum number of cells not to exceed 100
    The expected number of observations in each cell must be at least 5

8.33 Chi-Square Test

- Establish null and alternative hypotheses
- Determine a level of test significance
- Calculate the critical value form the chi-square distribution
- Calculate the chi-square test statistic from the data
- Compare the test statistic with the critical vale
- Accept or reject the null hypotheses

8.34 Kolmogorov–Smirnov Test

- Establish null and alternative hypotheses
- Determine a level of test significance
- Determine the critical K–S value from the D table
- Determine the greatest absolute difference between the two cumulative distributions
- Compare the difference with the critical K–S value
- Accept or reject the null hypotheses

8.35 Square Error

- For comparing observed data to a number of theoretical distributions
- Uses a summed total of the square of the error between the observed and the theoretical distributions
- Error is defined as the difference between the two distributions for each individual data cell

8.36 How Much Data Need To Be Collected

- We want to observe the right data
- Want to have observed the different values that are likely to occur
- Need to have enough data to perform a goodness of fit test

8.37 What Happens If I Cannot Fit the Input Data?

- Why this occurs
- What to do about it

8.38 Why This Occurs

- Not enough data was collected
- Data is a combination of a number of different distributions

8.39 What To Do about It

- Collect more data
- Use the observed data to generate an empirical distribution

Lecture 9: Model Translation

9.1 Agenda

- Deciding what type of computer software to use
- Programming the model into the software

9.2 Deciding What Type of Computer Software To Use

- Advances in computer hardware
- Software cost
- Practitioner preference

9.3 Advances in Computer Hardware

- Early simulation models presented data in numerical format
- Advances in microcomputers allow high speed calculations and photorealistic graphics
- Need for animation may require simulation specific software packages
- Need for calculation intensive models may require general purpose programming

9.4 Software Cost

- General purpose programming languages
  - Usually lower in cost
  - May need to purchase or otherwise acquire subroutines
- Simulation specific software packages
  - Can be very expensive
  - Requires frequent additional cost updates to stay current
  - Reduced cost academic or training versions sometimes available

9.5 Practitioner Preference

- Practitioner's programming ability
- Level of calculations
- Need for graphics

9.6 Simulation-Specific Software Packages

- ARENA
- AUTOMOD
- Simscript

9.7 General Purpose Programming Languages

- Visual Basic
- FORTRAN
- C++
- Pascal

9.8 Programming the Model into the Software

- Getting started
- Version management
- Programming commenting
- Program organization
- Mnemonic naming conventions
- Advanced program planning
- Multiple practitioner program development

9.9 Getting Started

- The practitioner must have a fundamental understanding of the model before programming
- Insure that a function high level flow chart has been developed
- Keep the flow chart up to date as the programming progresses

9.10 Version Management

- Project subdirectories
- Saving simulation programs
- File version management techniques
- Backing up simulation project files

9.11 Project Subdirectories

- Advantages of the simulation program subdirectory are:
  - Project files are immediately accessible after starting the simulation program
  - Project files are in the same place regardless of which specific computer is used
  - Fewer navigation issues exist for accessing supporting simulation libraries
- Advantages to using a dedicated project subdirectory are:
  - Keeping the simulation program subdirectory uncluttered
  - Easier to backup separate simulation project file subdirectories
  - Not affected by program reinstallations or deinstallations

9.12 Saving Simulation Programs

- Start with a relatively simple simulation program
- Add additional components and sophistication
- Incrementally different models
- Save the model on a continuous basis, even during a single development session

9.13 File Version Management Techniques

- Keep overwriting the original file with the same name
  - Easy to do
  - A corrupt file loses all previous work
  - Programming error results in no usable model
- Multiple file versions
  - Requires some effort, disk space
  - Always have the last workable model available
  - Incremental versions—model01, model02, etc.
  - Experimental alternatives—cnfg21, cnfg22, cnfg31, etc.

9.14 Backing Up Simulation Project Files

- Not the same as version management
- Risky to keep all versions on one physical device
- Fire or theft
- Keep on main drive and some other type of media in a different location

9.15 Programming Commenting

- Helps while you are developing the program
- Helps understand the program years from now
- Many simulation models will have code that is almost but not quite identical
- Results in constantly hunting for specific segments of code in large complex models
- Commenting at the structure level
- Commenting at the model level

9.16 Program Organization

- Why simulation programs can easily become disorganized
- How to use subroutines and subroutine views
- Programming hot keys
- Mnemonic naming conventions
- Thinking in advance

9.17 Why Simulation Programs Can Easily Become Disorganized

- Graphic interface seems to promote the disorganized program development
- Result is graphical spaghetti code
- Cannot tell where or why the program goes from one module to another

9.18 How To Use Subroutines and Subroutine Views

- Different levels of views can be used to display the program
- A high-level view will not allow any detail to be seen
- Subroutines can be isolated in individual views
- Related subroutines should be positioned together

9.19 Programming Hot Keys

- Enable the practitioner to move between sections of the model or subroutine views
- Particular keys are programmed to go to specific views

9.20 Mnemonic Naming Conventions

- Use consistent mnemonic naming conventions
- Describe the exact nature of the component
- Resource is called clerk
- Queue for the clerk is called clerkq

9.21 Advanced Program Planning

- Most simulation models are modified for representing different experimental alternatives
- Build in model flexibility with a "right the first time" approach
- Use mach01 instead of mach, etc.

9.22 Multiple Practitioner Program Development

- Can reduce length of overall project development time
- Division of work must be carefully assigned to take advantage of special skills
- Requires additional coordination to prevent overwriting models

9.23 Division of Work with Multiple Practitioners

- User-generated code
- Sequential changes in model modes
- Basic animation layout
- Machine or process failures
- Nontypical events for demonstration purposes

Lecture 10: Verification

10.1 Agenda

- Verification

10.2 Verification

- The process of insuring that the model operates as intended
- Include all of the components specified under the system definition phase
- Actually be able to run without any errors or warnings

10.3 To Include All Components Use

- A divide-and-conquer approach
- Subroutine view approach

10.4 To Run without Any Errors or Warnings

- Use animation
- Manually advance the simulation clock
- Write to an output file

10.5 A Divide and Conquer Approach

- Break the larger more detailed model into a smaller simpler higher-level model

- Make a series of small enhancements to the simple model
- Enhancements to the detail of the existing components
- Expand the model to include other components

10.6 Use of Animation for Verification

- Use different entity pictures for different types of entities
- Following the entities through the system
- Changing entity pictures
- Displaying global variables or entity attribute values
- Displaying plots of global variables or entity attributes
- Displaying levels of system statistics

10.7 Manually Advance the Simulation Clock

- Simulation models quickly run in compressed time
- May be necessary to slow the model to observe problems
- Step through by individual events on the simulation clock
- VCR-like controls

10.8 Write to an Output File

- Potentially useful, but difficult to use in comparison to manually advancing the clock
- Provide a permanent record of
  - Event list
  - Entity attributes
  - Global variables

Lecture 11: Validation

11.1 Agenda

- What is validation
- Need for validation
- Face validation
- Statistical validation

11.2 What Is Validation?

- Process of insuring that the model represents reality
- Building the correct model

11.3 Need for Validation

- Assumptions
- Simplifications
- Oversights
- Limitations
- None of these can be so gross as to prevent the validation of the model

11.4 Face Validation

- The model at least on the surface represents reality
- Normally achieved with the assistance of domain experts
- Not a one shot operation, but a continuous process
- Can also help:
    - Instill a sense of ownership in the model
    - Prevent last-minute "why didn't you…" questions
    - Reduce the number of project progress inquiries
- Face validity is not sufficient by itself

11.5 Statistical Validation

- Objective and quantitative comparison between the actual system and the model
- If there is no statistically significant difference, the model is considered valid
- If there is a statistically significant difference, model is not valid
- Validation data collection
- Validation data analysis process

11.6 Validation Data Collection

- Collect data that reflect the overall performance of the system
- Common method is to collect system or flow time
- Data collection approach
    - Single observation
    - Summary statistics
- Collect data from both
    - Actual system
    - Simulation model

11.7 Single Observation Data Collection Approach

- Uses data from individual entity measures of performance
- Easier to collect a sufficient quantity of data
- Possibility of autocorrelation issues

11.8 Summary Statistics Data Collection Approach

- Uses mean data from multiple sets of observations of individual entity measures of performance
- May require an extensive data collection effort

11.9 Actual System Validation Data Collection

- Record system validation data collection conditions
- Need to know the state of the system when collecting data
    - Number of entities in each queue
    - Number of entities in other parts of the system
    - Number and state of resources
- Unknown how long these entities have been in the system
- Cannot use data from these entities
- Record the system times for new entities arriving into the system

11.10 Simulation Model Validation Data Collection

- Record the same type of output data as was collected for the actual system
- Load the simulation model in the same manner as the actual system
- Discard the model data for same number of entities that were initially in the system

11.11 Validation Data Analysis Process

- Determine the appropriate statistical comparison of means test
- Check both the system and model data sets for normality
  - If both are normal, check data sets for having equal variances
    - If variances are equal perform normal t-test
    - If variances are not equal perform Smith–Satterthwaite test
  - If either data set or both are not normal perform nonparametric rank sum test
- If there is no difference the model is valid, proceed with the study
- If there is a difference, the model or input data are suspect, cannot proceed with the study

11.12 Examining the Validation Data for Normality

- Perform an equiprobable chi-square test for normality on both data sets
- H_o: data are normal with the same mean and std as the sample
- H_a: data are not normally distributed with the same mean and std as the sample
- $\alpha = 0.05$
- Calculate critical value for chi-square distribution
- Calculate test statistic
- If test statistic is less than critical value, cannot reject H_o, data are normal
- If test statistic is greater than critical value, reject H_o, data are not normal

11.13 For Normal Data Examine for Equal Variances

- Perform an *F* test
- H_o: data are normal with the same mean and std as the sample
- H_a: data are not normally distributed with the same mean and std as the sample
- $\alpha = 0.05$
- Calculate the critical *F* distribution value
- Calculate the test statistic with the larger variance in the numerator
- If test statistic is less than critical value, cannot reject H_o, variances are equal
- If test statistic is greater than critical value, reject H_o, variances are not equal

11.13.1 F-Test Equation

$$F = \frac{S_M^2}{S_m^2}$$

where

S_M^2 is the variance of the data set with the larger variance

S_m^2 is the variance of the data set with the smaller variance

11.14 Independent T-Test

- Both data sets are normal
- Variances are equal

11.15 T-Test Procedure

- H_o: mean of the system and model data sets are equal
- H_a: mean of the system and model data sets are not equal
- $\alpha = 0.05$
- Calculate the critical t distribution value for $n_1 + n_2 - 2$ degrees of freedom
- Calculate the test statistic
- If test statistic is within ± critical value, cannot reject H_o, means are equal
- If test statistic is exceeds ± critical value, reject H_o, means are not equal

11.15.1 T-Test Equation

$$t = \frac{(\overline{x}_1 - \overline{x}_2)}{\sqrt{(n_1 - 1)s_1^2 + (n_2 - 1)s_2^2}} \sqrt{\frac{n_1 n_2 (n_1 + n_2 - 2)}{n_1 + n_2}}$$

where

$\quad$ t = Calculated test statistic

$\quad$ $\overline{x}_1$ = The mean of the first alternative

$\quad$ $\overline{x}_2$ = The mean of the second alternative

$\quad$ s_1^2 = The variance of the first alternative

$\quad$ s_{21}^2 = The variance of the second alternative

$\quad$ n_1 = The number of data points in the first alternative

$\quad$ n_2 = The number of data points in the second alternative

11.16 Smith–Satterthwaite Test

- Both data sets are normal
- Variances are not equal

11.17 Smith–Satterthwaite Test Procedure

- H_o: mean of the system and model data sets are equal
- H_a: mean of the system and model data sets are not equal
- $\alpha = 0.05$
- Calculate the critical t distribution value for the adjusted degrees of freedom
- Calculate the test statistic
- If test statistic is within ± critical value, cannot reject H_o, means are equal
- If test statistic exceeds ± critical value, reject H_o, means are not equal

11.17.1 Smith–Satterthwaite Degrees of Freedom

$$d.f. = \frac{[s_1^2 / n_1 + s_2^2 / n_2]^2}{[s_1^2 / n_1]^2 / (n_1 - 1) + [s_2^2 / n_2]^2 / (n_2 - 1)}$$

where

$\quad$ $d.f.$ = degrees of freedom

$\quad$ s_1^2 = Sample variance of the first alternative

$\quad$ s_2^2 = Sample variance of the second alternative

n_1 = Sample size of the first alternative

n_2 = Sample size of the second alternative

11.17.2 Smith–Satterthwaite Equation

$$t = \frac{\bar{x}_1 - \bar{x}_2}{\sqrt{\dfrac{s_1^2}{n_1} + \dfrac{s_2^2}{n_2}}}$$

where

t = The t test statistic for the Smith–Satterthwaite

$\bar{x}_1$ = The mean of the first alternative replications.

$\bar{x}_2$ = The mean of the second alternative replications.

s_1^2 = Sample variance of the first alternative

s_2^2 = Sample variance of the second alternative

n_1 = Sample size of the first alternative

n_2 = Sample size of the second alternative

11.18 Rank Sum Test

- When one or the other or both data sets are nonnormal
- Do not need run an F test on the variances
- Compare the ranks of the two data sets

11.19 Rank Sum Test Procedure

- H_o: mean of the system and model data sets are equal
- H_a: mean of the system and model data sets are not equal
- $\alpha = 0.05$
- Calculate the critical Z distribution value
- Calculate the rank sum test statistic
- If test statistic is within ± critical value, cannot reject H_o, means are equal
- If test statistic exceeds ± critical value, reject H_o, means are not equal

11.19.1 Rank Sum Test Preliminary Steps

- Identify the system data set as set 1, the model data set as set 2
- Sort both data sets in ascending order
- Merge the two data sets into one data set
- Sum the values of the ranks for each data set as w1 and w2
- Perform the rank sum test calculations

11.19.2 Rank Sum Test Calculations

- $U_1 = W1 - \dfrac{n1(n1+1)}{2}, \quad U_2 = W2 - \dfrac{n2(n2+1)}{2}$
- U = min(U1, U2)
- mean = n1*n2/2
- var = n1*n2(n1+n2+1)/12
- z = (U-mean)/std

11.20 Why a Model May Not Be Statistically Valid

- System is nonstationary
- Poor input data
- Invalid assumptions
- Poor modeling

Lecture 12: Experimental Design

12.1 Agenda

- Factors and levels
- Two alternative experimental designs
- One-factor experimental designs
- Two-factor experimental designs
- Multifactor experimental designs
- 2^k experimental designs
- Interactions
- Refining the experimental alternatives

12.2 Factors and Levels

- Factors are different variables thought to have an effect on the performance of the system
- Levels are the values that the different factors may take

12.3 Examples of Factors

- Workers performing specific functions
- Machines which perform specific operations
- Machine capacities
- Priority sequencing policies
- Worker schedules
- Stocking levels

12.4 Examples of Levels

- Four vs. five vs. six workers performing a specific function
- An old vs. a new machine performing specific operations
- A 5-ton vs. a 2.5-ton capacity truck
- First-in–first-out vs. last-in–first-out priority sequence policies
- Five 8-h shifts vs. four 10-h shifts
- Restocking order levels between 10 and 25%

12.5 Two Alternative Experimental Designs

- Simplest experimental design
- A base system exists
- No base system exists

12.6 When a Base System Exists

- First alternative is the base model

- Second alternative is
  - An alternate operating policy
  - An alternate resource policy

12.7 When a Base System Does Not Exist

- Both alternatives are for proposed models
- Examples
  - Equipment for a new process from two different manufacturers
  - Facility layout for a service facility of two different designs

12.8 One-Factor Experimental Designs

- Next level of sophistication
- One specific factor that we are going to examine at three or more levels
- Same level of resources but different operating policies

12.9 One-Factor Experimental Design Resource Example

- Three clerks
- Four clerks
- Five clerks

12.10 One-Factor Experimental Design Operating Policy Example

- Three individual parallel clerk queues
- One single snake queue feeding two clerks and one queue feeding one clerk
- One single snake queue feeding into all three clerks

12.11 Two-Factor Experimental Designs

- Two factors
- Examine each of these factors at a number of different levels
- Number of alternatives is equal to:

Number of levels in factor A × number of levels in factor B

12.12 Two-Factor Experimental Design Resource Example

- Three regular clerks with parallel queues
- Four regular clerks with parallel queues
- Three novice clerks with parallel queues
- Four novice clerks with parallel queues

12.13 Two-Factor Experimental Design Operating Policy Example

- Three clerks with regular parallel queues
- Three clerks with a single queue
- Two clerks with regular parallel queues and one clerk with a single express queue
- Two clerks with a single regular queue and one clerk with a single express queue

12.14 Multifactor Experimental Designs

- Much more complicated type of experiment

- Number of alternatives can quickly explode into an unmanageable level
- Number of alternatives is equal to:

$$Total\ number\ of\ alternative = number\ of\ levels\ ^{Number\ of\ factors}$$

12.15 2^k Experimental Designs

- Reduces the number of levels in each factor to two
- Low level and high level
- High and low levels can be different between factors

12.16 Interactions

- When two particular factors have some sort of synergistic effect
- The effect of both of the factors may be larger than the sum of the effects of each of the individual factors
- Interactions can be examined through sophisticated statistical analysis techniques
- Practitioner will probably want to assume there are no special interactions

12.17 Refining the Experimental Alternatives

- An initial experiment is conducted to test the factors
- Follow-up experiments are conducted that focus on the significant factors
- May need to bracket the point at which the levels in the factors become significant

Lecture 13: Analysis I

13.1 Agenda

- Analysis for terminating models
- Analysis for nonterminating models

13.2 Analysis for Terminating Models

- Replication analysis
- Production simulation runs
- Statistical analysis of the simulation run results
- Economic analysis of statistical analysis results

13.3 Analysis for Nonterminating Models

- Starting conditions
- Determining steady state
- Addressing autocorrelation
- Length of replication
- Batching method

13.4 Replication Analysis of Terminating Models

- Used to establish confidence in the precision of the simulation results
- Select an initial number of replications

- Calculate summary statistics from this initial set of replications
- Calculate the level of precision
- If the precision is less than the desired precision calculate new number of replications
- Run the new required number of replications
- Repeat the process until the level of precision meets the desired level of precision

13.5 Calculating Summary Statistics

- Mean
- Standard deviation
- Half-width confidence interval or standard error

$$\text{Standard Error} = t_{1-\alpha/2,\,n-1} * s / \sqrt{n}$$

where

$t = t$ distribution for $1 - \alpha/2$ and $n - 1$ degrees of freedom

s = standard deviation of the replication means

n = number of observations in the sample

13.6 Calculating Precision

- Absolute precision method
- Relative precision method

13.7 Absolute Precision Method

- Select a tolerable level for the precision
- Same units as the sample data
- Selection of an absolute precision level may appear to be somewhat arbitrary
- The formula for calculating the absolute precision is:

$$\text{Absolute Precision} = t_{1-\alpha/2,n-1} * s / \sqrt{n}$$

where

$t = t$ distribution for $1 - \alpha/2$ and $n - 1$ degrees of freedom

s = standard deviation of the replication means

n = number of observations in the sample

13.8 Number of Replications Is Insufficient for Absolute Precison

- When absolute precision exceeds the desired absolute precision
- Manipulate equation to calculate new required number of replications

$$i = \left[\frac{t_{1-\alpha/2,n-1} * s}{\text{Absolute Precision}} \right]^{1/2}$$

where

$t = t$ distribution for $1 - \alpha/2$ and $n - 1$ degrees of freedom

s = standard deviation of the replication means

i = number of replications needed to achieve the absolute precision

13.9 Relative Precision Method

- More rational approach
- Not necessary to select an arbitrary absolute precision level
- Precision is based on ratio of standard error to mean of replications
- Normal level of precision is 0.10
- The formula for calculating the relative precision is:

$$\text{Relative Precision} = \frac{t_{1-\alpha/2, n-1} * s / \sqrt{n}}{\bar{\bar{x}}}$$

where

$t = t$ distribution for $1 - \alpha/2$ and $n - 1$ degrees of freedom

$s = $ standard deviation of the replication means

$n = $ number of replications used to calculate the summary statistics

x bar bar $= $ mean of the replication means

13.10 Number of Replications Is Insufficient for Relative Precison

- When relative precision exceeds the desired relative precision
- Manipulate equation to calculate new required number of replications

$$i = \left[\frac{t_{1-\alpha/2, n-1} * s}{\text{Relative Precision} * \bar{\bar{x}}} \right]^{1/2}$$

where

$t = t$ distribution for $1 - \alpha/2$ and $n - 1$ degrees of freedom

$s = $ standard deviation of the replication means

$n = $ number of replications used to calculate the summary statistics

$i = $ number of replications needed to achieve the relative precision

13.11 Production Simulation Runs of Terminating Models

- The replication analysis must be performed for each individual alternative
- Some alternative will require fewer replications than others
- All of the alternatives must be run at the highest number of replications required by any single alternative
- Must run at least 10 replications for any alternatives

Lecture 14: Analysis II

14.1 Agenda

- Statistical analysis of the run results of terminating models

14.2 Statistical Analysis of the Run Results of Terminating Models

- Simple two-model comparisons
- Three or more model comparisons

14.3 Simple Two-Model Comparisons

- Can utilize either hypothesis test or a confidence interval approach
- Hypothesis tests limited to either accepting or rejecting the null hypothesis
- Confidence interval approach
  - Modification of the corresponding hypothesis test
  - Provide more information than hypothesis tests
  - Graphically show the statistical results
  - Easier to use and explain than hypothesis tests

14.4 Types of Confidence Interval Approaches

- Welch confidence interval approach
- Paired *t*-test confidence interval approach

14.5 Welch Confidence Interval Approach

- Based on Smith–Satterthwaite test
- Automatically accounts for possible differences in variations
- Must first calculate the degrees of freedom estimator
- Calculate confidence interval
- If the confidence interval covers 0, there is no difference between the models
- If the confidence interval does not cover 0, there is a difference between the models

14.5.1   Welch Confidence Interval Degrees of Freedom

$$d.f. = \frac{[s_1^2/n_1 + s_2^2/n_2]^2}{[s_1^2/n_1]^2/(n_1-1) + [s_2^2/n_2]^2/(n_2-1)}$$

where

$d.f.$ = degrees of freedom

s_1^2 = Sample variance of the first alternative

s_2^2 = Sample variance of the second alternative

n_1 = Sample size of the first alternative

n_2 = Sample size of the second alternative

14.5.2   Welch Confidence Interval Calculations

$$\overline{x}_1 - \overline{x}_2 \pm t_{d.f.,1-\alpha/2} \sqrt{\frac{s_1^2}{n_1} + \frac{s_2^2}{n_2}}$$

where

$\overline{x}_1$ = The mean of the first alternative replications

$\overline{x}_2$ = The mean of the second alternative replications

t = The t value for the degrees of freedom previously estimated and $1 - \alpha/2$

14.6 Paired *T*-Test Confidence Interval Approach

- Used when models have some sort of natural pairing

- Calculate a new variable based on the pairs of replication means
- Calculate new variable confidence interval
- If the confidence interval covers 0, there is no difference between the models
- If the confidence interval does not cover 0, there is a difference between the models

14.6.1   Paired *t*-Test Variable Calculations

$$\overline{X}_{1i} - \overline{X}_{2i} = Z_i$$

where

$\overline{X}_{1i}$ = The *i*th replication mean for the first alternative

$\overline{X}_{2i}$ = The *i*th replication mean for the second alternative

Z_i = The difference in means for the *i*th replication

14.6.2   Paired *t*-Test Confidence Interval Calculations

$$\overline{Z} \pm t_{\alpha/2,\,n-1} \frac{s}{\sqrt{n}}$$

where

$\overline{Z}$ = The mean of the Z values

$t_{\alpha/2,\,n-1}$ = The value of the *t* distribution for $\alpha/2$ and $n - 1$ degrees of freedom

s = The standard deviation of the *Z* values

n = The number of pairs of replication means

Lecture 15: Analysis III

15.1 Agenda

- Three or more model comparisons
- Analysis of variance

15.2 Analysis of Variance

- Determines if one or more alternatives is different than the others
- Based on a ratio of the variance between and within the different alternatives
- If the variation between is large and the variance within is small, the ratio is large
- If the variation between is small and the variance within is large, the ratio is small
- When the ratio is large then it is more likely there is a difference among alternatives

15.2.1   ANOVA Procedure

- H_o: no difference among means
- H_a: difference among means
- Select level of significance
- Determine critical value
- Calculate *F* statistic
- Compare *F* statistic with critical value
- If the *F* test statistic is less than the critical *F* value, cannot reject H_o
- If the *F* test statistic is greater than the critical *F* value, reject H_o

15.2.2   ANOVA Calculations

- Calculate the sum of squares total
- Calculate the sum of squares between
- Calculate the sum of squares within
- Calculate the mean squares between
- Calculate the mean squares within
- Calculate the F statistic
- Compare the F statistic to a critical F value

15.2.3   Calculate the Sum of Squares Total

$$SST = \sum_{i=1}^{k} \sum_{j=1}^{n} (x_{ij} - \bar{x})^2$$

where

SST = Sum of squares total

k = Number of different alternatives

n = Number of replications for each alternative

x_{ij} = A single replication mean for a single alternative

$\bar{x}$ = The grand mean of all replication means

15.2.4   Calculate the Sum of Squares Between

$$SSB = \sum_{i=1}^{k} n * (\bar{x}_i - \bar{x})^2$$

where

SSB = Sum of squares between

k = Number of different alternatives

n = Number of replications for each alternative

$\bar{x}_i$ = The mean of the replication means for a single alternative

$\bar{x}$ = The grand mean of all replication means

15.2.5   Calculate the Sum of Squares Within

$$SST = SSB + SSW$$

$$SSW = SST - SSB$$

15.2.6   Calculate the Mean Squares Between

$$MSB = \frac{SSB}{k-1}$$

where

MSB = Mean squares between

SSB = Sums of squares between

k = Number of alternatives

15.2.7   Calculate the Mean Squares Within

$$MSW = \frac{SSW}{k*(n-1)}$$

where

MSW = Mean squares within

SSW = Sum squares within

k = Number of alternatives

n = Number of replications for each alternative

15.2.8   Calculate the *F* Statistic

$$F = \frac{MSB}{MSW}$$

where

F = F statistic

MSB = Mean square between

MSW = Mean square within

15.2.9   ANOVA Results

- If the H_o is rejected, one or more means is statistically significantly different
- To determine which means are different, use the Duncan multiple-range test

Lecture 16: Analysis IV

16.1 Agenda

- Duncan multiple-range test

16.2 Duncan Multiple-Range Test

- Run after ANOVA H_o has been rejected
- Know one or more means are statistically different
- Duncan test indicates which means are statistically different from the others
- Uses a least significant range value to determine differences for a set of means
- If the range of means is larger than the least significant range value there is a difference

16.2.1   Duncan Multiple-Range Test Procedure

- Sort the replication means for each alternative in ascending order left to right
- Calculate the least significant range value for all of the possible sets of adjacent means
- Compare each set of possible adjacent means with the corresponding least significant range value in descending order with respect to the set size
- Mark the nonsignificant ranges

16.2.2   Least Significant Range Calculations

$$R_p = s_{\bar{x}} * r_p$$

where

$s_{\bar{x}}$ = The Duncan standard deviation of the replication means

r_p = The Duncan multiple-range multiplier for a given level of significance, set size, and degrees of freedom

p = The size of the set of adjacent means

16.2.3   Standard Deviation of X-Bar Calculations

$$S_{\bar{x}} = \sqrt{\frac{MSE}{n}}$$

where

MSE = The mean square error of the replication means from the ANOVA results

n = The number of replications in a single alternative

16.2.4   Comparison of Adjacent Means

- Begin with the largest number of adjacent means
- If the range is less than the least significant range value there is no difference
- The range is underlined to represent all the data being the same
- If the range is greater than the least significant range value there is a difference
- The range must be split into smaller ranges and examined individually

16.3 Duncan Multiple-Range Example

Alternative	1	2	3	4
Time (min)	23.5	26.2	27.5	28.1

The following statistically significant conclusions may be stated from this table:

- There is a difference between alternative 1 and all of the other alternatives
- There is a difference between alternative 2 and alternative 4
- There is no difference between alternatives 2 and 3
- There is no difference between alternatives 3 and 4

Lecture 17: Analysis V

17.1 Agenda

- NonTerminating System Analysis

17.2 Nonterminating System Analysis

- Starting conditions
- Determining steady state
- Autocorrelation

- Length of replication
- The batch method

17.2.1   Starting Conditions

- Begin with the system empty
- Begin with the system loaded

17.2.2   Determining Steady State

- Must eliminate the initial transient
- Graphic approach
- Linear regression

17.2.3   Graphic Approach

- Visually determine when the slope of the initial transient approaches 0
- Highly subjective and influenced by individual interpretation
- Not recommended

17.2.4   Linear Regression

- Uses the least-squares method to determine where the initial transient ends
- If the observations' slope is not zero, advance the range to a later set of observations
- Eventually the range of data will have an insignificant slope coefficient
- Steady-state behavior has been reached

17.2.5   Autocorrelation

- Correlation between performance measure observations in the system
- Possible issue with non-terminating systems
- Practitioner may underestimate variance
- Results in the possibility of concluding that there is a difference between systems when there actually is not
- Can be accounted for by complex calculations
- Can be avoided by special techniques

17.2.6   The Batch Method

- Identify the nonsignificant correlation lag size
- Make a batch 10 times the size of the lag
- Make the steady-state replication run length 10 batches long

Lecture 18: Reports and Presentations

18.1 Agenda

- Written report guidelines
- Executive summaries
- Equations
- Screen captures and other graphics
- Presentation guidelines
- Presentation media
- Electronic presentation guidelines
- Electronic software presentation issues
- Actual presentation

18.2 Written Report Guidelines

- Follows the same format as the simulation study process
- Contents
  - Problem statement
  - Project planning
  - System definition
  - Input data collection and analysis
  - Model formulation
  - Model translation
  - Verification
  - Validation
  - Experimentation and analysis
  - Recommendations and conclusions

18.3 Executive Summaries

- Readers will not necessarily have a technical background
- 2 to 3 pages
- Condensed information from:
  - Project objectives
  - Results
  - Recommendations and conclusions

18.4 Equations

- All equations should be included in the report
- Equations should be generated electronically
- Electronically generated equations can be used in both the report and presentation

18.5 Screen Captures and Other Graphics

- Professional quality reports will include:
  - System photographs
  - Simulation software screen captures
  - Flow charts
  - Data plots
  - Statistical confidence intervals
- These graphics can also be used in the presentation
- Different methods for capturing the graphics

18.6 Capturing Graphics

- File import method for digital photographs
- Buffer method for screen captures
- File import method for screen captures

18.7 Presentation Guidelines

- Same basic format as the report
- Must determine level of detail based on:
  - Objective of the presentation
  - Time for the presentation
  - Technical level of the audience

18.8 Presentation Media

- LCD projectors
- Transparencies

18.8.1 LCD Projectors

- May not be readily available
- Allocate additional preparation time for LCD presentations
- LCD projector and notebook compatibility issues
- Other LCD projector presentation issues

18.8.2 Transparencies

- Transparency projectors are more commonly available
- Technology proof
- Corrections are more difficult to make

18.9 Electronic Software Presentation Issues

- Presentations masters
- Use of colors
- Use of multimedia effects
- Speaker's notes
- Use of presentation handouts

18.10 Actual Presentation

- Rehearsal
- Dress
- Positioning
- Posture
- Presentation insurance

Index